Campbell

Chemiluminescence

Ellis Horwood Series in Biomedicine

Series Editor: Dr. Alan Wiseman, Department of Biochemistry,
University of Surrey

Published Titles

Forthcoming Titles

The Blood-Brain Brain Barrier: Pharmacological and Therapeutic Aspects
J. P. Bates and N. G. Burnet

Sterol Biosynthesis Inhibitors: Pharmaceutical and Agrochemical Aspects
D. Berg and M. Plempel (Eds.)

Sulphur Drugs and Related Organic Compounds: Chemistry, Biochemistry and
Toxicology   L. A. Damani (Ed.)

Handbook of Quality Control in Pharmaceutical Products   S. Denyer and R. Baird

Methods for Studying the Neuronal Basis of Behaviour: A Comparative Treatise   M. Fillenz

Neurochemical Pathology of the Human Brain   C. J. Fowler and J. A. Hardy

Clinical Pharmacy: Pathology and Therapeutics   R. J. Greene and N. D. Harris

Affinity Labelling in Steroid and Thyroid Hormone Research: Techniques, Biological Application
and Cloning   H. Gronemeyer (Ed.)

Electrochemical Detection: Techniques in the Fundamental and Applied Biosciences
G. A. Junter (Ed.)

Blood Substitutes: Physiology and Medical Applications   K. C. Lowe (Ed.)

Handbook of Cardiac Glycosides: Chemistry, Pharmacology and Toxicology of Nonrenal Digitalis
Substances   N. Rietbrock and B. G. Woodcock

The Modification of Primary Translation Products by Glycosylation and Acylation   M. F. G. Schmidt
and R. T. Schwarz

Brain Imaging: Techniques and Applications   N. F. Sharif and R. Quirion (Eds.)

Leukaemia Cytochemistry   C. S. Scott (Ed.)

Extending Drug Uses: Postmarketing Development of Medicines   E. Snell (Ed.)

Critical Factors in Haemostasis: Evaluation and Development   C. E. Williams, P. E. Short,
A. J. George and M. P. B. Entwistle

Distribution:

Great Britain and Ireland: VCH Publishers (UK) Ltd., 8 Wellington Court, Wellington Street,
  Cambridge CB1 1HW (Great Britain)

USA and Canada: VCH Publishers, Suite 909, 220 East 23rd Street, New York,
  NY 10010-4606 (USA)

Switzerland: VCH Verlags-AG, P.O. Box, CH-4020 Basel (Switzerland)

All other countries: VCH Verlagsgesellschaft, P.O. Box 1260/1280, D-6940 Weinheim
  (Federal Republic of Germany)

ISBN 3-527-26342-X (VCH Verlagsgesellschaft)   ISBN 0-89573-501-6 (VCH Publishers)

A. K. Campbell

Chemiluminescence
Principles and Applications in Biology and Medicine

VCH

ELLIS HORWOOD
international publishers in science and technology

Dr. A.K. Campbell
Reader in Medical Biochemistry
University of Wales
College of Medicine
Department of Medical Biochemistry
Heath Park
Cardiff, CF4 4XN
UK

Deutsche Bibliothek Cataloguing-in-Publication Data

Campbell, Anthony K.: Chemiluminescence: principles and applications in biology and
medicine/A. K. Campbell. – Weinheim; Deerfield Beach, Fl.: VCH; Chichester: Horwood, 1988.
 (Ellis Horwood health science series)
 ISBN 3-527-26342-X (VCH, Weinheim)
 ISBN 0-89573-501-6 (VCH, Publishers)

ISSN 0930-3367

British Library Cataloguing in Publication Data
Campbell, A. K. (Andrew Keith), *1945–*
Chemiluminescence.
1. Chemiluminescence
I. Title
541.3'5

Library of Congress Card No. 88–8828

Published jointly in 1988 by
Ellis Horwood Ltd., Chichester, England
and VCH Verlagsgesellschaft mbH, Weinheim, Federal Republic of Germany

Printed in Great Britain by The Camelot Press, Southampton

Table of contents

Preface

These *fluorescent* animals you collect on the beach at night are rather rare aren't they? Is this something you do just for fun? How on earth are they connected to your investigations into rheumatoid arthritis and multiple sclerosis?

This book is about chemiluminescence, chemical reactions that produce light. Those of us who enthuse over such reactions have, I am sure, faced many times questions such as these from family and friends. They highlight several areas of confusion and ignorance in both the layman and non-expert scientist. Many people are confused about the difference between fluorescence and chemiluminescence. These are related phenomena, but quite distinct. Switch the illumination of a home aquarium off and the beautiful blue neon fish cease to glow. Yet there are other fish in the oceans which can generate their own light. To find them, you have to dive around a coral reef at night or send a net several hundreds of metres below the ocean surface. Luminous animals may be relatively uncommon on land but in the sea they are abundant. You will find a few, such as dinoflagellates, bacteria on decaying flesh, worms and hydroids, on most beaches in the world. Over half of the earth's surface is covered by ocean more than 1000 metres deep. Here the only light is living light. Why otherwise would the animals here have retained their eyes fully intact through evolution? Bioluminescence, as it is known, thus plays a major role in the ecology of the oceans.

Recently, chemi- and bio-luminescence have had an impact on biochemical analysis, on cell biology, and on medicine. The light-emitting proteins from luminous jelly fish enable us to measure free calcium in living cells. This has led not only to fundamental discoveries about the role of intracellular calcium in controlling the behaviour of cells in healthy tissues but also about its role in cell pathology. It has lead us to propose a role for intracellular calcium in cell injury induced by cells and components of the immune system, relevant to demyelination in multiple sclerosis. The firefly system is now a standard ultrasensitive method for measuring ATP, an indicator of cell viability and biomass, since all living cells must have high ATP, but dead cells do not. Chemi- and bio-luminescent compounds have also had a recent impact on biotechnology, as replacements for radio-active labels such as ^{125}I and ^{32}P.

The markets involve thousands of millions of dollars per year on a world-wide basis! In spite of this fundamental importance and exciting application in biomedicine, most standard chemistry, biology, physiology, and biochemistry textbooks either do not mention chemiluminescence at all or deal with it scantily.

Possibly one of the earliest descriptions of bioluminescence is to be found in *Shih Ching* or *Book of odes*, one of the Thirteen Classics of China dating from around 1500–1000 BC. Luminous animals and fungi were described by Greek and Roman authors, such as Aristotle and Pliny. However, serious scientific investigation did not begin until the 16th century. The first book devoted entirely to luminescence was written by Conrad Gesner in 1555. Translated, the title reads: *On the rare and marvellous plants that are called lunar either because they shine at night or for other reasons; and also on other things that shine in darkness.*

Since then there have been just a few monographs by a single author devoted entirely to chemiluminescence, the most notable being by Horace Raphael Dubois in 1914 (*La vie et la lumière*, 338 pp, Paris), and the magnificent collection by E. Newton Harvey, including his scholarly account of some 800 plus pages on the history of luminescence. More recently, there have been several multi-author volumes, and conference proceedings, dealing with both fundamental and applied aspects of chemi- and bio-luminescence. So, even if there has not been a single-author text on this subject for more than 30 years, how can I justify this one? Surely one person, however enthusiastic, cannot readily do justice, and cover in depth, a field which stretches from the biology of the deep oceans, through the biochemistry, chemistry, and physics of reactions producing an excited state and their application to human good, to the nature of light itself?

It is my belief that science, both its practice and presentation, is not only about in-depth experimentation and consensus, but also, like the arts, about individual expression through 'thought, word, and deed'. The ever-increasing specialisation in science and fragmentation of disciplines may help one to find a single focus, but, unless one is careful, it may give us only a rather narrow perspective on Nature. The first true scientists were natural philosophers, not only capable of tussling with a single problem, but also having the wisdom to step back and view its wider perspective. They recognised the importance of getting the philosophy right if one was to appreciate Nature in all its aspects.

The study of chemiluminescence provides us with an ideal opportunity to return to, and yet develop further, natural science as a whole. It also provides illuminating insights into the historical development of science. Everyone at school learns Boyle's Law. Yet how many know that Boyle was the first to demonstrate the need for oxygen, some 100 years before its discovery, in bioluminescence? Who has heard of Horace Raphael Dubois? Not only did his experiments form the foundation for the investigation of the biochemistry, and chemistry, of bioluminescence over the past 100 years, but also he was one of the pioneers of modern biophysics.

The aim, therefore, of this book is to provide a wide, but balanced, perspective on chemical reactions which produce light: their discovery, where they can be found, and what is so special about a chemiluminescent reaction (Chapter 1), how to study and measure it (Chapter 2), its biology (Chapter 3), and its use as a laboratory tool in research and biotechnology (Chapters 4–9). Any technique in science is only as powerful as the questions asked by the user. I have, therefore, attempted to highlight the unique aspects of chemiluminescence in analysis, and to pinpoint where real discoveries have arisen from its use in biology and medicine. The last chapter looks to the future, and summarises particularly my own enthusiasm for studying chemiluminescence. I have deliberately repeated a few points in more than one chapter, so that each may be read independently.

The book is aimed at researchers and undergraduates in the biological disciplines including medicine, and in chemistry. In view of the exciting prospects for commercial applications of chemiluminescence I have tried to highlight these in such a way that the text will also be of interest to industrialists. The style I have adopted is first to try to ask simple questions fundamental to an understanding of the phenomenon, and then to delve more deeply into its chemistry, physics, and biology. I have tried to explain how things work, including the apparatus. Walk into any laboratory and you are likely to find at least one piece of equipment whose analytical basis involves the measurement of light. Yet how many students know how a photomultiplier, or other types of light detector, work? Scientists must understand their tools. This is essential if we are to realise the strengths and weaknesses of our current experimental approach to a problem, and highlights the need for inventing new devices and new analytical procedures.

My first introduction to chemiluminescence analysis was some twenty years ago whilst studying with Dr (now Professor) Nick Hales, for my PhD, in the Department of Biochemistry, University of Cambridge. I needed to measure tiny amounts of ATP, and had not a great deal of success using a fluorescence assay. Another member of the group, Paul Luzio, now a close friend and collaborator, drew my attention to a paper which showed how the firefly luciferin–luciferase reaction control be used to measure ATP. I was surprised, but delighted, to find that a firefly extract was available commercially. My first assay was a revelation. It worked first time! Using a Heath Robinson scintillation counter I immediately succeeded in establishing an ATP assay some two to three orders of magnitude more sensitive than I had achieved previously. I owe a great debt to many other people for helping me with my research over the past 20 years. First I should like to thank all who have been members of my research group or who have interacted closely with it. Some are still with me, I am pleased to say: Trevor Baines, Peter Evans, Bob Dormer, Maurice Hallett, Steve Simpson, Richard Daw, Chris Davies, Ashok Patel, Ian Weeks, Alun Davies, Stephanie Matthews, Mary Holt, Steve Edwards, Paul Morgan, Ann Roberts, Jan Knight, Shirley Barrow, Ashvin Patel, David Jenner, Graciela Sala-Newby, and Alan Houston, Eryl Davies, Thomas Müller, and Miguel Lucas. I thank Malcolm E. T. Ryall for all his hard work constructing our chemiluminometers, and more recently, Peter Mason. I have been lucky enough to collaborate with many interesting people. I thank the Director, Professor Sir Eric J. Denton, FRS, and all the staff and my friends at the Marine Biological Laboratory, Plymouth, where I have spent so many happy and stimulating hours. Particular thanks go to Skipper Chris Knott and the crew of the *Gammarus* for their help in collecting in my first luminous animals, *Obelia geniculata*, and Commander Bax of Plymouth Ocean Projects and the UWIST sub-aqua club for providing divers to collect them. I also thank particularly Peter J. Herring, whose invitation to join cruises on RRS *Discovery* have enabled me to discover the wonders of living light. I also thank Professor Frank McCapra and his group for their much appreciated collaboration;

Chris Ashley for many enjoyable hours in his laboratory, Nick Hales and those originally in his group, particularly Ken Siddle, Paul Luzio, and Gerry Brenchley, and David Yates and Professor H. Gutfreund, FRS in Bristol, Jean-Marie Bassot for a happy trip to Station Biologique Roscoff, and Alan Walton and George Reynolds and Geoff Newman. I thank Nick Hales for much support, particularly in my early years, the late Leslie Cooper, FRS, for his wisdom, encouragement, and introduction to the Marine Biological Association, Plymouth, and also to his wife Daloni for their wonderful hospitality over the years. I am also very grateful for the long-standing support of the members of the Department of Medical Biochemistry, University of Wales College of Medicine (formerly Welsh National School of Medicine), particularly Professor George Elder for all his support, J. Stuart Woodhead, Bob Dormer, Noor Kalshekar, Paul Morgan, and Ian Weeks, and the secretaries, Chris Hullin, Dot Thomas, Dilys Marks, and Jeannie Burt. Goodness knows how they have managed to read my writing or cope with me over the years; their support has been much appreciated. I am also grateful to Andrew Newby, Alan Mcgregor, Ralph Marshall, Tony Jackson, and particularly Alastair Compston for their interest and collaboration, and the department of Medical Illustration and the Animal House for their invaluable support.

I owe many people thanks for material in this book, but particularly Peter Herring for some of the photographs of luminous animals, Woody (J. W.) Hastings for his most generous gift of one of my treasured possessions, a copy of Harvey's *A history of luminescence* (1957), Professor G. Pérès for the photograph and review of Dubois, and Professor George Reynolds for the photograph of E. Newton Harvey.

My research has been generously supported by the Medical Research Council, the Science and Engineering Research Council, the Arthritis and Rheumatism Council, the Multiple Sclerosis Society, the Department of Health and Social Security, The Welsh Office, the Browne and Maurice Hill Bequest Fund of The Royal Society and CLEAR Ltd. Without their financial support I would not have had the opportunity to discover and exploit living light.

I am extremely grateful to Ellis Horwood Limited, particularly Sue Horwood, for inviting me to write this book. I am a pencil and paper writer, so very special thanks go to Rosemary Harris for typing the manuscript so excellently and accurately, in spite of its apparent illegibility.

Finally I thank my family for all their support: my wife Stephanie, my sons David and Neil, and my daughter Georgina, whom I treasure greatly. I thank Stephanie particularly for her marvellous support. Families of scientists have a lot to put up with. Mine have chased me across the beach or up a mountain at night, as I excitedly search for a new luminous animal. They have put up with me escaping to our Anglesey retreat, where I find endless inspiration. I also thank Sue Campbell.

My mother and father first encouraged me to pursue my embryonic interest in Nature. Thank you for everything and for many valuable discussions since with my sister and collaborator, Caroline Sewry.

Go, view that House, amid the garden's bound,
Where tattered volumes strew the learned ground,
Where Novels, — Sermons in confusion lie,
Law, ethics, physics, school-divinity;
Yet did each author, with a parent's joy,

Survey the growing beauties of his boy,
Upon his new-born babe did fondly look,
And deem Eternity should claim his book,
Taste ever shifts, in half a score of years
A changeful public may alarm thy fears;
Who now reads Cowley? — The sad doom await,
Since such *as these are now* may be thy fate.

Thus wrote Gilbert White to his friends and relatives in 1788 whilst nervously awaiting the publication of his life's work *The natural history and antiquities of Selbourne*. Few of us can expect our contributions to stand the test of time as White's has done. Yet we all go through similar anguish and self-doubt. What then is the real purpose of the present book? As with Gilbert White, it is part of my own personal quest for a better understanding of the natural world. It is an attempt to marry the aesthetic appeal of a truly remarkable phenomenon, with the intellectual challenge of learning more about it and harnessing it to human good. Hopefully others will find pleasure as well as stimulation in some of its pages.

Anthony K. Campbell
Cartrefle
Ynys Môn

October 1987

The wonders of living light

This book is dedicated to three very special people:
the piddock hunters —
my children,
David, Neil, and Georgina

1

A natural history of chemiluminescence

The chilling night-dews fall: — away, retire!
For see, the glow-worm lights her amorous fire!
Thus, ere night's veil had half obscured the sky,
Th'impatient damsel hung her lamp on high:
True to the signal, by love's meteor led,
Leander hasten'd to his Hero's bed.

from 'The Naturalist's Summer-Evening Walk'
in *The natural history and antiquities of Selborne* by Gilbert White (1789)

1.1 ENCOUNTERS WITH NATURAL LIGHT

Imagine a warm summer's night and that you are camping on the beach. The danger of a thunderstorm has subsided after a few flashes of lightning in the distance, and the bluish wisps of St Elmo's fire which emanated from the tops of some of the ships' masts in the bay have now disappeared. The embers of your camp fire are still glowing red, and the light of your lantern has attracted a number of insects. As you saunter out towards the sea's edge you glance up. The flickering lights of the stars are visible now that the clouds have gone and a meteorite flashes across the sky. Just above the horizon there is a spectacular luminous arc, transparent white with touches of green, red, and violet. This is *aurora borealis*, the Northern Lights, if you are in the northern hemisphere, or *aurora australis*, if you are in the far latitudes of the southern hemisphere. As you walk closer to the sea you notice flashes of bluish-green light arising from something attached to some of the brown seaweeds, and sparked off by treading on them. The rock pools, and the sea itself, are also full of tiny organisms emitting flashes of blue light as you walk through the water. A dead fish, glowing bright blue, is lying on a rock.

These latter three phenomena are but a few of Nature's many examples of chemiluminescence, the others are not. What is it that distinguishes between them? What is the origin of these divers light emissions which you have seen? What is the cause of their striking colours and varying intensities? Has man succeeded in mimicking Nature in the laboratory? Has the scientific study of such phenomena lead to any important conceptual or technological advances in biology and medicine?

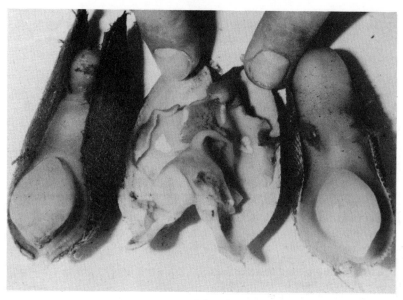

(a)

(b)

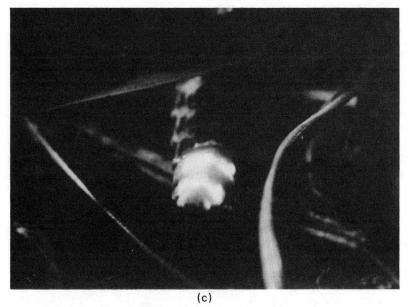

(c)

Fig. 1.1 — Living light. (a) The common piddock *Pholas dactylus* — white triangular organs in centre = luminous glands; (b) The hydroid *Obelia geniculata* — bright spots along stem = luminous cells; (c) The European glowworm *Lampyris noctiluca* — last three main segments contain the the luminous cells.

Wales is separated from England by the Severn estuary, which has one of the largest tidal falls in the world. Along its shores, at very low tide, you may be lucky enough to discover the workings of an extraordinary shellfish known as the common piddock *Pholas dactylus* (Fig. 1.1a). It lives all its life inside a rock. Tweak its siphon and the animal is likely to squirt a luminous cloud from out of its rock burrow. This will be a surprise to most people, yet perhaps even more surprising is the fact that the chemiluminescent system from this animal can be used to examine the workings of single human cells. The particular cells in question invade the heart both before and after a heart attack, the most common cause of death in the Western world. The same cells are also found in large numbers in the joints of patients with rheumatoid arthritis, a particularly common cause of invalidism. Another organism, the exquisite jelly fish *Obelia geniculata* (Fig. 1.1b), is one of the most abundant luminous species to be found off the British coasts during the summer. Its chemiluminescent system has enabled a link to be established between the role of calcium inside cells and the mechanism underlying another crippling disease, multiple sclerosis. It also sparked off an idea for using chemiluminescence in biotechnology to replace radioisotopes such as ^{125}I and [32]P. The glow-worm *Lampyris noctiluca* (Fig. 1.1c) can be found during the summer in a quiet Welsh valley displaying her 'amorous fire'. This beetle, and her American relative, the firefly *Photinus pyralis* provide a means of testing whether any material, terrestrial or extra-terrestrial, contains life.

If all of this 'living light' is chemiluminescence, what exactly is chemiluminescence? How have scientists been able to harness it to illuminate some of the major diseases of the 20th century, as well as providing new diagnostic aids for use at the patient's bedside?

1.2 WHAT IS CHEMILUMINESCENCE?

'Das bei chemischen Processen auftretende Leuchten würde Chemiluminescenz...'; or translated 'Light emission occurring as a result of chemical processes would then be referred to as
chemiluminescence'

Eilhard Wiedemann, 1888

Light is energy transmitted by photons (Fig. 1.2). If solids, liquids, or gases are to
emit light they therefore require a source of energy. This energy can be obtained
either externally, for example by absorbing heat through conduction or radiation, or
internally via physical or chemical changes such as nuclear transformations, electronic transitions, or chemical reactions. During the 1880s Eilhard Wiedemann was

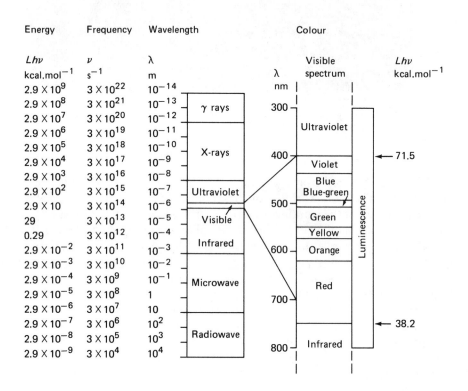

Fig. 1.2 — The electromagnetic spectrum and luminescence.

studying a variety of phenomena which produced light. He was the first, in 1888, to
use the term 'Chemiluminescenz' to describe chemical reactions which emit light. He
also distinguished two distinct mechanisms by which solids, liquids, or gases can
produce light; one required a rise in temperature, the other did not. To understand
how Wiedemann reached this conclusion we must go back some three hundred years
to the observations of Robert Boyle (Fig. 1.3a) and Isaac Newton.

(a)

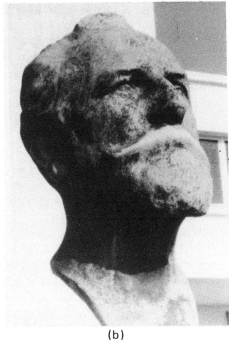

(b)

Fig. 1.3 — Some historic figures in chemiluminescence. (a) Robert Boyle FRS 25th Jan 1626–31st Dec 1691 with permission of The Royal Society; (b) Professor Horace Raphael Dubois, first director of the Laboratorie Maritime de Physiologie of the University of Lyon at Tarmaris-sur-Mer; (c) Professor E. Newton Harvey, Nov 25th 1887–July 21st 1959 Henry Fairfield Osborn Professor of Biology, University of Princeton, with thanks from Professor George Reynolds.

(c)

1.2.1 Luminescence or incandescence?

Everyone knows that when solids and liquids get very hot they emit light. This is incandescence, exemplified by the glowing embers in the fire of our mythical camper. Boyle, in 1668, carefully described five similarities between burning coal and shining wood, the latter now known to be caused by a luminous fungus. Both produced their own light, but only in the presence of air, and both could be extinguished by removal of air. However, whereas 'a live coal' was irreversibly extinguished by withdrawing air for a few minutes, 'shining wood', on the other hand, could easily regenerate its light, even if air was not re-admitted for half an hour. Furthermore, 'live coal was actually and vehemently hot'; whereas shining wood was 'not sensibly lukewarm'.

Newton in his *Opticks* of 1704 wrote 'And what else is a burning coal but red hot wood? Do not all fixed bodies, when heated beyond a certain degree, emit light and shine?' By 1794 J. Hutton had proposed the term 'incandescence' (Latin, *incandare* — to become white) to describe the emission of light by a body heated to high temperature. The limits of colour for the human eye are approximately 400 nm (violet) to 750 (red) (Fig. 1.2), though at extreme intensities this range can sometimes be extended to 350–900 nm. In incandescence a faint red glow can be detected at 525°C (798 °K). As the temperature rises the light becomes dark red then turns yellow, and becomes increasingly white as blue light contributes to the spectral emission (Table 1.1). Kirchhoff established in 1860 the laws governing this type of radiation, and in particular the dependence of the intensity and colour of light

Table 1.1 — Typical colours of incandescence

Approximate temperature (°C)	Colour observed
525	faint red
700	dark red
900	cherry red
1100	dark yellow
1200	bright yellow
1300	white
1400	blue white

Note: The colour temperature of an incandescent body is defined as the temperature of a black body whose colour matches the body being studied. This 'colour temperature' may be greater or less than the true temperature. Thus the colour temperature of an intense white carbon arc lamp is about 3500°C, whereas that of a candle flame, yellow owing to incandescent carbon particles, is about 1650°C. The light from flames is, however, not always incandescent. In many instances luminescence (chemi- and pyro-) plays a major role.

emission on the temperature of the emitting body. The Stefan–Boltzmann law established, not only that the total radiant energy emitted increases with temperature, but also that for a 'perfect black body radiator' the radiant energy varies as the fourth power of the absolute temperature. However, a major puzzle for 19th century scientists was, why does the *peak* in the emission spectrum shift to shorter wavelengths as the temperature increases? The problem of finding a law to explain this property of incandescence was solved by Planck in 1900 by considering the radiation to be emitted by oscillators, in discrete packets called 'quanta'. The equation enabling the energy (1.1) of these 'quanta' to be calculated was defined by Einstein in 1904 as a result of his studies on the photoelectric effect (see section 2.1):

Energy of radiation = the sum of the number of quanta $(N) \times$ the Planck constant $(h) \times$ frequency (v) (1.1)
$$= N h v$$

Of the emissions observed by our camper described in the opening section, not only were the glowing embers of the fire an example of incandescence, but also the glowing tungsten filament in the lantern as well as the meteorite flare, heated by friction as it entered the earth's atmosphere.

The discovery of the 'Bolognian Stone' in 1603, by an Italian cobbler and amateur alchemist called Vincenzo Cascariolo, was to lead to the characterisation of another group of phenomena, classified by Wiedemann as luminescence (Latin, *lumen* = light). The Bolognian stone, made of a heavy spar of barium sulphate, could 'imbibe' the light from the sun and re-emit in the dark. A large number of inorganic and organic compounds have now been identified which also luminesce after being irradiated by ultraviolet or by visible light. Wiedemann, in his original paper of 1888, proposed that a 'luminescent substance' was one which 'becomes luminous by the action of an external agency which does not involve an appropriate rise in temperature'. These examples of 'cold light' do not conform to Kirchhoff's laws of radiation. Wiedemann initially distinguished six types of luminescence designated by a prefix: *photo*luminescence caused by absorption of light, *electro*luminescence produced in gases by an electric discharge, *thermo*luminescence produced by *slight* heating, *tribo*luminescence as a result of friction, *crystallo*luminescence as a result of crystallisation, and *chemi*luminescence, the result of chemical processes. Others have since been added to the list (Table 1.2).

Returning to our camper on the beach at night, we can now see several phenomena as luminescence: St Elmo's fire, a form of electroluminescence; the aurora, a form of radioluminescence caused by the bombardment of the earth's atmosphere by particles from the sun; and chemiluminescence responsible for the light emanating from the various luminous organisms — hydroids on the seaweed, dinoflagellates in the water, and bacteria on the dead fish. Triboluminescence, on the other hand, can be readily observed at home by grinding crude sugar, cracking certain peppermints, and by stripping Sellotape or opening self-seal envelopes, in the dark. Somewhat paradoxically heat can generate luminescence, as in pyro- and thermo-luminescence (Table 1.2A).

What then is the physical basis of the difference between incandescence and luminescence? How much energy is needed, and where does it come from?

1.2.2 The electron and light
On Friday 30 April, 1897, J. J. Thomson announced to an audience at the Royal Institution in London the results of his experiments aimed at measuring accurately the ratio of charge to mass of the electron, named in 1891 by G. Johnstone Stoney as the elementary unit of electricity. Thomson found that the electron was many times smaller than the smallest known atom, hydrogen.

Table 1.2 — A classification of luminescence, based on E. N. Harvey

Type	Basis of light emission	Example
A. Associated with heating (distinct from incandescence)		
1. Candoluminescence	luminescence of incandescent solids emitting shorter λ than expected	heat ZnO
2. Pyroluminescence	luminescence of metal atoms in flames	yellow Na flame
3. Thermoluminescence	luminescence of solids and crystals on mild heating (i.e. well below that necessary to produce incandescence)	heat diamond
B. Associated with prior irradiation (fluorescence–short lived; phosphorescence — long lived)		
1. Photoluminescence	irradiation by UV or by visible light	Bologna stone ($BaSO_4$)
2. Cathodoluminescence	irdadiation by β particles (electrons)	television screen
3. Anodoluminescence	irradiation by α particles (the nuclei)	zinc sulphide phosphor
4. Radioluminescence	irradiation by γ or X-rays	luminous paint
C. Associated with electrical phenomena		
1. Electroluminescence and piezoluminescence	luminescence associated with electric discharges and fields	fluorescent strip light
2. Galvanoluminescence	luminescence during electrolysis	electrolysis of NaBr
3. Sonoluminescence	luminescence from intense sound waves in solution	ultrasonic probe in pure glycerol
D. Associated with structural rearrangements in solids		
1. Triboluminescence	luminescence on shaking, rubbing, or crushing crystals (material stressed to point of fracture)	gentle agitation of uranyl nitrate ($UO_2(NO_3)_2 \cdot 6H_2O$)
2. Crystallo(= tribo-)luminescence	luminescence on crystallisation	HCl or ethanol to saturation alkyl halide solutions (NaCl, KCl)
3. Lyo(= tribo-)luminescence	luminescence on dissolving crystals	LiCl or KCl coloured by irradiation by cathode rays
E. Associated with chemical reactions		
1. Chemiluminescence (oxyluminescence)	chemical reaction	luminol + H_2O_2 and peroxidase
2. Bioluminescence (organoluminescence)	luminous organisms	O_2 + luciferin–luciferase from sea firefly (*Vagula*)

See Harvey (1920, 1940, 1957) and Walton (1977) for details, history, and references

Light was conceived by Huygens in 1678 as waves, but by Newton in 1704 as particles, or corpuscles as he called them. Between 1830 and 1850 Faraday's experiments established that light had both magnetic and electrical components. This lead Maxwell in 1864 to propose an electromagnetic theory of light, light being an electromagnetic wave, essentially like any other wave form in nature. Thus, the velocity of light (c) = frequency of wave × distance between wave peaks (i.e. wavelength):

$$c = v \times \lambda. \tag{1.2}$$

In a vacuum $c = 3 \times 10^{10}$cm s^{-1}. From equation (1.2), the six intervals of the spectrum observed by the human eye — red, orange, yellow, green, blue, and violet — can be related to frequency and wavelength (Fig. 1.2), and to the other regions of the electromagnetic spectrum. Between 1900 and 1905 Planck and Einstein re-established the particle concept, later named the 'photon', to explain the spectral emission of black-body radiation and the photoelectric effect. The modern theory of light marries together the two theories, electromagnetic and quantal.

Thus, energy of a photon (E)

$$= hv = hc/\lambda \text{ in ergs or joules (SI unit)} \tag{1.3}$$

where h = The Planck constant.

A frequently used unit is the wave number, $\bar{v} = 1/\lambda$

$$\bar{v} \text{ cm}^{-1} = E/hc, \text{ the unit being a Kayser (K).} \tag{1.4}$$

Wave numbers are directly related, though not equal to, energy, whereas wavelength is inversely related to energy (see Chapter 2, section 2.1). They are strictly a more correct way of representing spectra, being more precise than using the 'wave theory' parameter, wavelength.

Current understanding of luminescence is based on the direct relationship between the energy of electrons within atoms and molecules and the energy of photons, either absorbed or emitted by these atoms and molecules.

From the Planck-Einstein equation (1.3) it is easy to calculate the energy necessary to generate a photon of blue ($\lambda = 450$ nm), green ($\lambda = 500$ nm) or orange-red ($\lambda = 600$ nm) light.

Note: a photon of wavelength 400 nm has a wavenumber $\bar{v} = 25\,000$ K or 25 kK.)

where Energy = hv ergs per molecule or Lhv ergs mol^{-1} (1.5)

L = The Avogadro constant = 6.023×10^{23}

h = The Planck constant = 6.63×10^{-27} ergs s or $\times 10^{-34}$ J s (SI unit)

v = frequency of light = c/λ

c = speed of light = 3×10^8 m s^{-1} in a vacuum (SI unit).

Energy of 1 photon = (kcal.mol^{-1}) $\times 6.95 \times 10^{-21}$J

Since 1 erg = 2.39×10^{-8} cal; 1 joule (J) = 10^7 erg; 1 kcal = 4.186 joule,

then Energy = 63.5 kcal mol^{-1} (2.75 eV) for blue light, 1 photon = 4.41×10^{-19}J

(1.6)

57.1 kcal mol^{-1} (2.48 eV) for green light, 1 photon = 3.97×10^{-19}J

and 47.6 kcal mol^{-1} (2.07 eV) for red light, 1 photon = 3.31×10^{-19}J

Note: 1 electron volt (eV) $= 1.602 \times 10^{-19}$J. The kcal is the common unit for enthalpy in a chemical reaction. Whether the phenomenon is incandescence or luminescence, this energy must be supplied if visible light is to be produced.

Energy provides an essential link between molecules, atoms, subatomic particles, and the fundamental particles of Nature. The energy of atoms and molecules can be subdivided into essentially five parts, three from movement (translation/vibration/rotation) and two 'potential' (electronic/nuclear). Absorption of energy from an external source, or from one generated internally from a chemical reaction or particle rearrangement, raises one or more of these components in the atom or molecule to a higher energy level. The stability of the molecule can be restored in one of three ways:

(1) A chemical reaction, resulting in bond cleavage or formation.
(2) Transfer of energy to another atom or molecule.
(3) Loss of energy by radiation.

Electromagnetic radiation in different parts of the spectrum is associated with distinct molecular and atomic events (Table 1.3). As the temperature of a solid or

Table 1.3 — Atomic and molecular events, and the electromagnetic spectrum

Wavelength	Region	Atomic or molecular response	Type of spectroscopy
3×10^8 m	alternating	none	none
3×10^5 m	electric power		
3×10^5 m	audio	none	none
300 m			
300–3 m	radiowave	molecular translation nuclear spin	NMR
3–10^{-4} m	microwave	molecular rotations	ESR
10^5–700 nm	infrared	inter- and intra-molecular vibrations	IR
700–400 nm	visible	electronic excitation	luminescence
400–10 nm	ultraviolet	electronic excitation	luminescence
10 nm–0.03 nm	X-ray	atomic diffraction electron transitions in inner shells	X-ray fluorescence
0.03–(3×10^{-4}) nm	γ ray	nuclear excitation	Mössbauer

liquid is increased, so the vibrational and rotational energy levels between the atoms and molecules increase. Below about 525°C the energy is sufficient for radiation in the infrared, above 525°C it is visible (Table 1.1). In gases, heating increases mainly the translational energy component. However, sufficient energy may be available to raise some of the electrons to new orbitals at higher energy levels than the ground state. The proportion of atoms or molecules in which this occurs is small and is determined by the Boltzmann distribution:

$$N_{ex}/N_0 = \exp\left(-E_g/kT\right) \tag{1.7}$$

where N_0 = original number of atoms at ground state energy

N_{ex} = number of atoms in the excited state
E_g = energy difference between the states
T = temperature in degrees absolute
k = The Boltzmann constant = 1.381×10^{-16} erg $°K^{-1}$.

The flame temperature of an air–coal gas flame is about 2000 °K. At this temperature the ratio of atoms in the excited state to those in the ground state is only 10^{-5} for sodium ($\lambda = 589$ nm). Thus the intensity of light depends on this figure, whereas the colour of the light emitted depends on E_g (1.1). This is pyroluminescence and is responsible for the yellow colour of flame when salt (NaCl) is dropped into it, or the bluish-green colour when copper wire is thrown into a fire. It is thus an apparent exception to our rule about the difference between incandescence and luminescence. The key difference between these is thus not so much heat, but rather whether the physical process necessary for light emission involves transitions in electronic energy levels *within* atoms or molecules in the case of luminescence, or transitions in energy levels *between* atoms or molecules in the case of incandescence. Both nuclei and electrons also have spins, with the consequence of another set of energy levels.

Different types of spectroscopy have been developed to study the various interactions within atoms and molecules as well as between them, by using the appropriate segment of the electromagnetic spectrum (Table 1.3). The lowest discrete energy levels are associated with commercially generated electric current at 50–60 cycles per second (50–60 Hz), power, and the audio region used in line telecommunications. The electromagnetic waves of lowest energy having direct effects on atoms or molecules are the radio waves (Fig. 1.2 and Table 1.3). These are used in nuclear magnetic resonance spectroscopy, involving nuclear spin. Those of highest energy, namely γ rays, on the other hand, raise nuclei to higher energy states and are used in Mössbauer spectroscopy. The visible region involves changes in electronic energy. This is not really very surprising since the physicochemical processes in the eye which respond to photons and enable us to see are, in principle, similar to those responsible for generating photons in this region of the spectrum.

The electronic energy itself can be further subdivided into three parts:

$$E_{electronic} = E_{orbital} + E_{vibrational} + E_{rotational} \tag{1.8}$$

In isolated atoms the energy transitions involve changes between orbitals, hence the sharpness of spectral lines. On the other hand, in molecules, vibrations and rotations of bonds formed through electrons between the individual atoms, superimpose a further set of fine energy levels for each orbital energy level. Only the orbitals having energy separations large enough will generate photons within the visible to ultraviolet regions (Fig. 1.2). Vibrational energy levels within each orbital state produce photons within the near infrared, whereas rotational energy separations are smaller still, producing photons in the far infrared to microwave regions. The energy levels can be displayed diagrammatically by an 'energy well' diagram, or more simply by the Jablonski diagram first introduced in the 1930s (Fig. 1.4).

Thus luminescence is concerned primarily with the emission of visible or near-visible radiation (200–1500 nm) when electrons in excited orbitals decay to ground state, the light arising from the *potential* energy of electronic transitions within atoms or molecules. Many types of luminescence have been identified, designated by a prefix which identifies the energy source responsible for generating or releasing the light (Table 1.2). Incandescence, on the other hand, is light arising from losses in kinetic energy between atoms or molecules. This energy is usually supplied initially as heat. As we have already seen, heat can generate electronically excited states in the gas phase giving rise to pyroluminescence. In contrast, in *thermoluminescence* gentle heating, usually 100–500°C, can provide the activation energy necessary to release electronic energy initially absorbed from ultraviolet and visible light or subatomic particles. This occurs, for example, in a crystal or liquid containing impurities. Absorption of a photon by the major component can excite an electron to a higher energy level. A 'hole' is left behind which, since it has lost a negatively

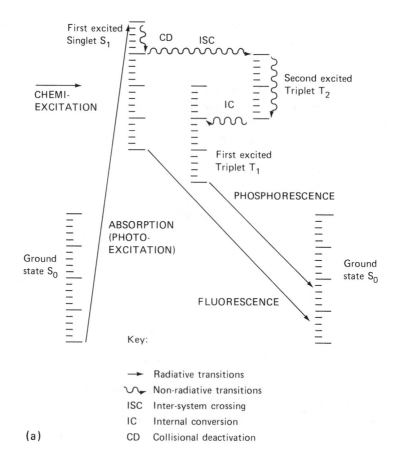

(a)

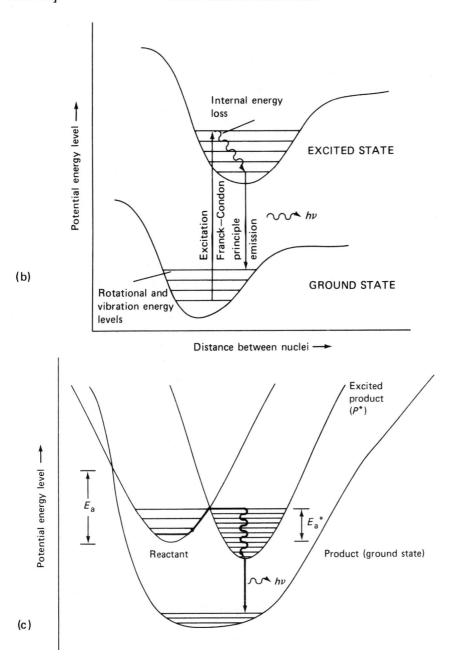

Fig. 1.4 — Excited state energetics. (a) Jablonski diagram; (b) Energy wells for ground and first electronically excited state between two atoms, for simple luminescence; (c) Energy pathway for chemiluminescence. For a reaction to occur activation energy must be supplied so that the reactant energy well crosses that of the product. E_a = activation energy for dark reaction (i.e. $R \rightarrow P$ with no photon emission), E_a^* = activation energy for chemiluminescence (heavy line = energy pathway). For good ϕ_{CL} $E_a^* < E_a$.

charged electron, behaves like a positive charge. The positive 'hole' attracts the excited electron, the association being known as an *exciton*. But the exciton can jump between molecules in the crystal. In a pure crystal the energy of the exciton on re-association of the 'hole' and the electron is dissipated as heat. However, if there are dislocations, structural boundaries, or chemical impurities within the crystal, the exciton may be trapped here for long enough to allow thermoluminescence to occur. Now, when the crystal is heated, the energy of the exciton can be emitted as light, the emission spectrum being characterised by the impurity rather than the main component. This phenomenon has interesting applications for problems as wide ranging as the dating of archaeological specimens and the study of structural changes in biological molecules.

In *chemiluminescence* the electronically excited state is generated by a chemical reaction. As a result the excited molecule, the product of the reaction, has a different atomic structure from the initial substrate(s). Here, too, gentle heating can initiate the phenomenon. Two such examples are the thermal degradation of dioxetans (see section 1.5), well known chemical intermediates in many organic chemiluminescent reactions, and the degradation of certain aromatic peroxides, e.g. rubrene, formed orginally by reaction with singlet oxygen.

The electronic transitions in chemi, and other types of, luminescence can occur within a wide range of vibrational and rotational energy levels. This is one reason for the broad spectrum of light emission from molecules compared with the sharp line spectra associated with excited atoms. Even though events associated with electrons are very fast compared with rates of chemical reactions it is important to realise that the time scale of different electronic events varies over several orders of magnitude (Table 1.4). Thus, the life time of a singlet excited state may be 1–10 ns, whereas the

Table 1.4 — Time scale of atomic and molecular events

Event	Typical half time in seconds[†]
Formation of electronically excited state	10^{-15}–10^{-12}
Nuclear and molecular vibration	10^{-13}–10^{-11}
Molecular re-orientation and rotation	10^{-13}–10^{-11}
Resonance energy transfer	10^{-12}
Intersystem crossing (singlet reactant to triplet state product)	10^{-12}–10^{-10} (aliphatic) 10^{-8} – 10^{-6} (aromatic)
Lifetime of singlet excited state (decay to singlet ground state)	10^{-9}–10^{-8}
Time for molecule to diffuse 10 nm in solution	10^{-9}
Time for 100 collisions in a gas at 1 atm	10^{-8}
Lifetime of triplet excited state (decay to singlet ground state)	10^{-5}–10^{-3} (solution or gas) 10^{-4} many seconds (rigid media)
Ligand–ligand binding	10^{-3}
Bond splitting	10^{-2}

†These are intended only as an illustration of the order of magnitude involved.

time scale for formation of the initial excited state, as well as vibrational and rotational changes within it, is two or three orders of magnitude faster than this. The result is that the emitting species is usually in the lowest vibrational and rotational energy level. Electrons have spin. Like a spinning top which can rotate clockwise or anticlockwise, so electrons can spin one way or the other. A pair of electrons with opposite spin in one molecule is known as a singlet spin state. However, if the spins are parallel the result is a triplet spin state. But the Pauli exclusion principle insists that the electrons are in different orbitals. If transition from singlet to triplet spin state occurs this will slow the emission process considerably, since the decay of triplet to singlet ground state can be anything from a hundred to a million times slower than the singlet–singlet transition.

1.2.3   The discovery of chemiluminescence
Unlike many natural phenomena it is difficult to credit any one individual with the discovery of chemiluminescence. However, it is possible to identify certain key observations and experiments in the history of luminescence (Table 1.5) which led up to the first definition of chemiluminescence by Wiedemann in 1888. These lead on to

Table 1.5 — The history of luminescence

Event	Discoverer	Date
Discovery of the Bolognian stone	Cascariolo	1603
First observations on luminescence vs incandescence	Boyle	1668
Discovery of phosphorus	Brandt	1669
First use of 'phosphorescence'	Elsholtz	1677
Light as a wave	Huygens	1678
Light as particles	Newton	1704
Definition of incandescence	Hutton	1794
First observations of fluorescence	Herschel	1845
First use of 'fluorescence'	Stokes	1853
First measurement of fluorescence lifetime	Becquerel	1858
Laws of electromagnetic radiation	Maxwell	1864
First use of 'electron'	Johnstone Stoney	1891
Discovery of the electron (e/m)	J. J. Thomson	1897
Energy in discrete packets	Planck	1900
Quantal theory of photoelectric effect: $E = h\nu$	Einstein	1904
Quantisation of the atom	Bohr	1912
Prediction of energy transfer	Franck	1922
Exclusion principle (electron spin)	Pauli	1925
Uncertainty principle	Heisenberg	1927
Energy transfer equation	Förster	1948

the realisation, during the early years of the 20th century, that chemiluminescence involves chemical reactions which produce electronically excited states in sufficient quantity, and sufficiently quickly, to allow emission of light to occur.

From Siberia to New Mexico, from Europe to China, the early writings of human civilisation contain references and mythology about luminous animals such as fireflies and glow-worms, of shining flesh and glowing wood. However, it was the Greeks and Romans who first reported their peculiar characteristics. Aristotle (384–322 BC) appears to have been one of the first writers to recognise 'cold' light. In his *De coloribus* he wrote 'some things, though they are not in their nature fire nor any species of fire, yet seem to produce light'. Pliny the Elder (AD 23–79) provides a description of several luminous organisms in *Historia naturalis* including a luminous jelly fish *Pulmo marinus* (thought now to be *Pelagia noctiluca*), a lantern fish *Lucerna piscis*, the luminous bivalve *Pholas dactylus*, and the glow-worm, as well as luminous fungus. 'The glow-worms', he wrote in Book IX Chapter 28, 'are named by the Greeks, Lampyrides, because they shine in the night like a spark of fire'. In spite of such vivid descriptions of bioluminescence it was not until the end of the 17th century that serious scientific investigations of 'cold light' began.

Hennig Brandt (or Brand) was a physician in Hamburg intent on making a fortune through alchemy. In 1669 he produced a material, by chemical reduction of the solids from distilled urine, which had the remarkable property of glowing with blue light in the dark. No previous exposure to light, no heating, and no mechanical treatment, were necessary. He called the substance *phosphorus* from Greek ϕως light, ϕορως bringing, without significant heat. His first discovery of artificial chemiluminescence was called appropriately 'phosphorus mirabilis' (miraculous light), and is now a well recognised property of phosphorus when exposed to oxygen in the air.

At the time of Brandt's discovery it was not realised that living organisms were responsible for 'shining wood' and 'shining flesh'. The proof that 'glowing or shining wood' was caused by a luminous fungus, and that luminous bacteria were reponsible for 'shining flesh', was definitively established by Johann Florian Heller, professor of medicine in Vienna. His observations were first reported in 1843 and published in 1853 as a series of papers entitled *Ueber das Leuchten in Pflanzen-und Tierreiche*. Nevertheless, in the year preceding Brandt's discovery of phosphorus, Robert Boyle (Fig. 1.3a), using his 'air-pump', had shown that complete removal of the air surrounding either 'shining wood' or 'shining flesh' caused a large reduction in the intensity of light emission, and in some cases extinguished it altogether. On re-admitting the air the material glowed vividly again. Although Boyle was not aware of the existence of oxygen, as it was discovered independently by Scheele and Priestley over 100 years later, this was the first experimental demonstration that oxygen, or one of its derivatives, is required in all known bioluminescent reactions and most artificial organic chemiluminescence. One of the problems one faces when trying to demonstrate such requirements for oxygen is the very high affinity for oxygen of some enzymes involved in bioluminescence. In order to fully extinguish the light it may be necessary to lower the oxygen concentration in solution to less than 10 nM, more than 20 000 times less than in air-saturated water at room temperature. This is difficult to achieve with the somewhat crude vacuum pump apparatus used by Boyle. This may be the reason why the initial experiments of Boyle and others on removing air from phosphorus were somewhat inconclusive. However, even this inorganic chemiluminescence has now also been shown to require oxygen.

Some puzzling exceptions to the oxygen-requiring rule in luminous organisms were apparently discovered by McCartney in 1810, who found that extracts of luminous jelly fish could glow without air. These observations were repeated more carefully and extended to luminous radiolarians by Harvey (Fig. 1.3c) in 1926, who used Pt/H$_2$ to remove the oxygen. This puzzle was resolved in 1962 by the discovery, by Shimomura and colleagues, of proteins which could be extracted from luminous jelly fish containing an organic prosthetic group with oxygen in the form of a hydroperoxide covalently attached to it. Addition of calcium to the protein triggers a chemiluminescent reaction within the prosthetic group (see Chapters 3 and 7). Peter Herring and I have found similar proteins in luminous radiolarians. In spite of the frequent involvement of oxygen in organic chemiluminescence, there are many examples of inorganic chemiluminescence in solution and in the gas phase which do not involve oxygen. Furthermore, there are many radical reactions capable of generating chemiluminescence which also do not require oxygen (see Chapter 9, sections 9.4 and 9.7).

What other critical requirements have been found for chemi- and bio-luminescence?

Spallanzani in the 1790s showed that water was necessary for luminous wood or jelly fish (medusae) to glow. However, for the first clear demonstration of chemiluminescence in solution we have to go back again to inorganic compounds. Johann Friedrich Meyer in 1764 described at length the whitish light observed on adding water to freshly slaked lime. This was followed by the studies of Heinrich and Davy in the first quarter of the 19th century who observed many examples of light emission from inorganic chemical reactions. The chemical basis of such inorganic chemiluminescence was irrefutable, yet arguments based on chemo- and organo-vitalist principles persisted throughout the 19th century when examining the physico-chemical basis of 'living light'.

In 1821 Macaire concluded that the luminous material in glow-worms was composed mainly of 'albumine', and required oxygen. Macaire's experiments lead away from the old idea that the light was caused by phosphorus. Between 1885 and 1887 Raphael Dubois (Fig. 1.3b) from Lyons reported a series of ingenious experiments, carried out using first the luminous beetle *Pyrophorus* and then the luminous rock-boring bivalve *Pholas*. The conclusions have had a major impact on the study of bioluminescence ever since. He showed that extraction of a luminous organ in cold water produced a suspension which glowed initially and then gradually faded away. On the other hand, extraction in hot water resulted, perhaps after an initial glow from the dissected organ, in no light emission. However, once this latter extract had cooled he added it back to the now dim 'cold water' extract. To his delight emission was restored. He therefore argued that luminescence was the result of a chemical reaction and must require a heat-stable factor which he called *luciférine*, and a heat-labile factor which he called *luciférase*. He correctly identified the luciférase as 'a potent albuminoid like. . .diastase, zymases etc' but incorrectly assumed that 'its phenomenon is not an oxidation'. Oxygen is also, of course, required. All bioluminescent reactions discovered since involve the oxidation by oxygen or one of its metabolites of small organic molecules catalysed by a protein (see

Chapter 3). E. Newton Harvey (Fig. 1.3c) spent much of his scientific career exploring Dubois' concept in a remarkable variety of luminous organisms.

Although the man-made substance luminol was discovered in the mid-19th century it was not until 1928 that it was reported by Albrecht, following an observation by W. Lommel of Leverkunsen, and found to be chemiluminescent. The first recognised synthetic chemiluminescent organic compound, lophine (Fig. 1.5a),

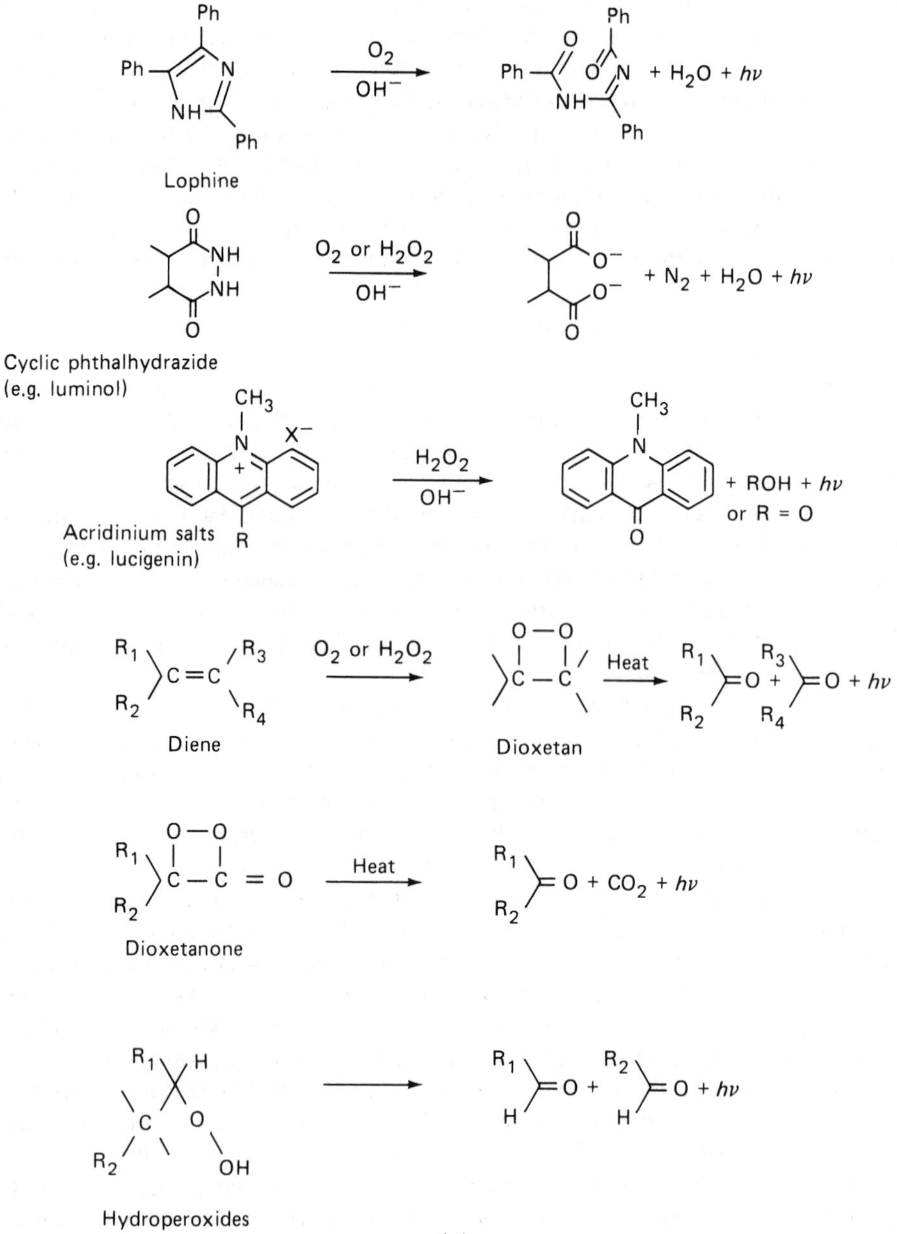
Lophine
Cyclic phthalhydrazide
(e.g. luminol)
Acridinium salts
(e.g. lucigenin)
Diene
Dioxetan
Dioxetanone
Hydroperoxides
O_2
OH^-
O_2 or H_2O_2
H_2O_2
Heat
+ H_2O + $h\nu$
+ N_2 + H_2O + $h\nu$
+ ROH + $h\nu$
or R = O
+ CO_2 + $h\nu$

(a)

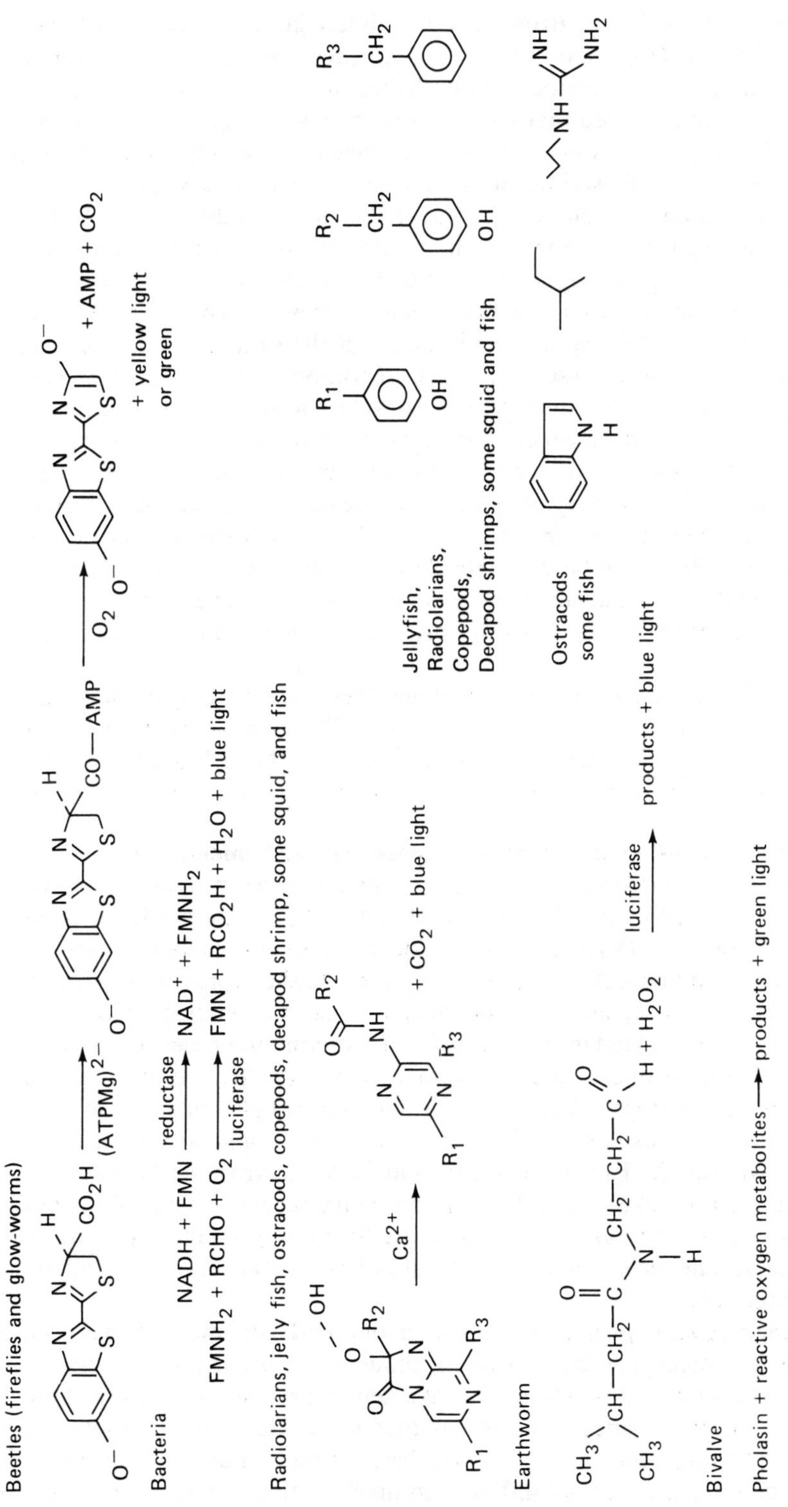

Fig. 1.5 — Some chemiluminescent reactions. (a) Synthetic; (b) Bioluminescence.

was prepared in 1877 by Bronislaus Radziszewski, Professor of Chemistry at Lemberg, Galicia. His first observations concerned the luminescence from lophine formed from hydrobenzamide, when shaken in air with an alkaline alcoholic solution. He also observed that the luminescence was brightest at the air–solution interface. Hydrogen, removing all traces of oxygen, extinguished the luminescence. His first paper was followed in the same year by another reporting light emission from various aldehydes and amides under the same conditions. By 1880 he had formulated a long list of synthetic chemiluminescent organic compounds, as well as compounds of biological origin including terpenes, cholates, and fatty acids. In the same year he characterised the first chemiluminescent spectrum of an organic compound, lophine, finding it to be brightest at the Fraunhofer E line, similar to some luminous animals. These important observations confirmed the belief reported in his original paper of 1877 that 'it was of considerable interest to establish the cause of this chemical phosphorescence of an organic substance, because there is no doubt that the phosphorescence of some mushrooms and animals is likely to be due to similar causes'. The discovery of hydrogen peroxide by Thernard in 1819 and that of ozone by van Marum in 1785, together with the characterisation of ozone by Schönbein in 1839–40 and Soret in 1866, led to the discovery of many other synthetic reactions capable of producing light, as well as to the discovery of enhancement of several extracts from luminous organisms, when oxygen or one of its derivatives was added.

In 1877 Radziszewski reported 'lophine does not emit a light when heated by itself'. This concluded the prelude, enabling Wiedemann in 1888 not only to distinguish chemiluminescence from incandescence but to establish it irrevocably as a fascinating natural phenomenon at the interface between chemistry and physics.

1.2.4 The relationship of chemiluminescence to other luminescence

Since Wiedemann first defined six types of luminescence in 1888, fifteen varieties have been observed (Table 1.2). Conventionally these different luminescent phenoma are distinguished by a prefix, which identifies the energy source responsible for initiating light emission. As we have already seen, heat can provoke luminescence as well as incandescence, including the very familiar pyroluminescence seen when metal salts are dissociated into atoms in flames, thermoluminescence, first observed by Robert Boyle in 1662 in diamonds when heated gently, and the chemiluminescence of dioxetans (Figs 1.5 and 1.10). Irradiation by light and subatomic particles gives rise to many examples of luminescence. We would not be able to watch television without it. Electric currents and local electrostatic fields give rise to electro-, galvano- and a certain type of sonoluminescence. Frictional forces during crystal formation or breakdown can give rise to tribo-, crystallo-, and lyo-luminescence. Finally, chemical reactions, artificial and naturally occurring, result in chemi- or bio-luminescence.

Although several puzzles still exist in our understanding of many types of luminescence, much has been learned about their physical basis since Newton Harvey expanded and consolidated Wiedemann's original classification. This presents a problem, since it turns out that there are many examples where a particular phenomenon classified on the basis of the initiation mechanism should really be incorporated within a different group if the original source of the electronic excitation is taken into account. For example, thermoluminescence requires previous irradiation of the material by ultraviolet or by visible light, subatomic particles, γ or X-rays. Sonoluminescence (20 kHz–500 mHz), occurring because of cavitation in minute gas bubbles, and leading to collapse of the bubble, in liquids may be caused by small electric discharges, but can also initiate or potentiate chemical reactions producing light. Flosdorf and co-workers in 1963 showed that intense sonic signals in luminol cause chemiluminescence as long as oxygen is present. The large electrical potential generated at the gas bubble–liquid interface is thought to provide electrons for reduction of O_2 to its metabolites, O_2^- or H_2O_2 (see section 1.4). The green or blue light easily observed when stripping Sellotape off a reel, or on opening self-sealing envelopes, is usually described as triboluminescence. Yet it appears to be the result of local electrostatic fields produced from the disruption of frictional forces. Should it be really called electroluminescence? Harvey showed in 1939 that the triboluminescence of chips of glass was red in a neon tube, suggesting that it was really electroluminescence. Triboluminescent substances often require previous irradiation by light or subatomic particles in order to luminesce. The luminescence during electrolysis — galvanoluminescence — is really chemiluminescence resulting from ion or free radical reactions, and is now usually described as electrochemically generated chemiluminescence (see Chapter 9, section 9.7). So can the old classification (Table 1.2), based on the energy source responsible for initiating the luminescence, be rationalised with modern concepts of the electronically excited state?

Three parameters have now emerged which are fundamental to distinguishing and reclassifying the various types of luminescence:

(1) The real source of energy responsible for generating the electronically excited state, necessary for eventual light emission, as opposed to that sometimes responsible for releasing it.
(2) The source of energy responsible for overcoming the activation energy of any metastable intermediate(s) that may be formed during the whole process.
(3) The interaction, or mechanism, responsible for generating the metastable intermediate(s).

Examining the first of these, there are essentially three physicochemical processes capable of primary generation of the electronically excited state.

(1) Bond cleavage, either inter- or intra-molecular.
(2) Bond formation, either intramolecular or via the collision and interaction of ions, radicals, atoms or molecules.
(3) Energy transfer from a primary energy source, which may be radiative or non-radiative, from an atom or molecule already in an excited state, to another without the direct transfer of a photon.

The first of these, namely bond cleavage, can be generated chemically, for example in an enzyme-catalysed reaction or degradation of a dioxetan (Fig. 1.5), electrically, as in electrolysis, or mechanically, as in crystal crushing. The second can involve ion pair formation, radical annihilation, molecular collisions, and intermolecular forces, such as those occurring in crystal formation. In the third, heat through conduction, electromagnetic radiation through infrared, visible light, ultraviolet, γ or X-rays, or energy from subatomic particles, provides radiative mechanisms for generating electronically-exicted states by energy transfer. Non-radiative energy transfer can occur through a variety of mechanisms including coulombic and dipole–dipole interactions, and electron transfer (see Chapter 9).

Metastable intermediates may be chemical, such as dioxetans or hydroperoxides, or physical, such as excitons involved in certain types of thermoluminescence. Breakdown of the metastable intermediate, thereby releasing the potential energy as a photon, can be induced by supplying energy through heat or through changes in molecular and atomic interactions, the latter being responsible for examples of triboluminescence in certain irradiated crystals.

Thus, chemiluminescence as a term for chemical reactions producing light remains valid and useful. However, as we have seen, phenomena classified as sono-, thermo-, or electro-luminescence may, in reality, be special examples of chemiluminescence. Many of the terms in Table 1.2 are now so well established in the literature that it would be foolish, if not impossible, to alter their phenomological basis. Nor is it very helpful to use unwieldy double-barrelled terms like sonochemiluminescence or thermochemiluminescence. Our increased understanding of luminescent phenomena over the hundred years since Wiedemann first defined them might warrant the invention of some new terms based on the parameters I have outlined. However, throughout this book I shall prefix the term luminescence to indicate the source of energy responsible for initiating the phenomena, as orginally conceived by Wiedemann.

1.2.5 Problems of nomenclature

Whilst the difference between incandescence and luminescence was a source of confusion for several centuries, more recently there has been even more confusion over the terms fluorescence and phosphorescence, together with their relationship to chemiluminescence.

The impressive monograph of Fortuno Liceti entitled *Litheosphorus sive de lapide Bononiensi lucem in se conceptam ab ambiente claro mox in tenebris mire conservante* , published in 1640, contained detailed accounts of many 'stony light bearers' including the Bolognian stone. However, the modern scientific usage of the term phosphorescence, as opposed to the 'phosphorus' of Brandt, is ascribed to Johann Elsholtz in 1677 being applied to substances or objects which, like the Bolognian stone, absorb light and reemit light, after a delay, in the dark. However, phosphorescence soon became a popular term for any kind of 'cold light', including that emitted by luminous organisms.

E. Becquerel carried out may experiments on luminescence in the mid-19th century, including the invention of a phosphoroscope for measuring the minimum duration of phosphorescence. This led to the later discovery by his son Henri of radioactivity, and to the distinction between true phosphorescence and fluorescence. The term fluorescence (Latin *fluo* = I flow) was first introduced by Sir George Stokes in 1853 to describe a phenomenon observed in 1845 by Sir John Herschel. Herschel found that certain substances, for example quinine sulphate or calcium fluoride minerals (known as fluorspar), when exposed to ultraviolet radiation gave off visible blue light. However, the phenomenon was of very short duration since unlike phosphorescent minerals, no light was observable immediately the UV lamp was switched off. Phosphorescence is most striking when it lasts for many minutes, though in 1858 Becquerel using his phosphoroscope reduced the time between irradiation and light emission in some phosphorescent substances to 0.1 ms. Yet for fluorescence, even this brief time interval is some four orders of magnitude less than the lifetime of an excited fluor (Table 1.4). The recognition of the electronically excited state and electron spin provided a more objective criterion for distinguishing these two types of *photo*luminescence: fluorescence involving singlet–singlet and phosphorescence triplet–singlet transitions.

For an interesting difference between chemi- and photo-luminescence we can look at an original discovery of Dewar who showed that at very low temperatures true fluorescence and phosphorescence are dramatically increased. Try it after thin-layer chromatography! However, chemiluminescence decreases with temperature.

Unfortunately, the *Oxford English Dictionary* still defines 'phosphorescent' as the 'property of shining in the dark'. Furthermore, phosphorescence is commonly used by the layman — and appears in the older scientific literature — to describe luminous organisms. A further confusion is that in the scientific literature fluorescence and phosphorescence are sometimes used to describe only the decay of the singlet or triplet excited state respectively, rather than the complete process beginning with absorption of radiation. At least one medical dictionary (Dorland's) used by the *British Medical Journal*, uses *chemo*- not chemi-luminescence. *Chemo*-luminescence is not the accepted scientific term for this phenomenon.

A supplement to the Oxford English Dictionary, *O-Scz* (1982) gives:
   phosphorescence. Add: In scientific use now distinguished from fluorescence on tech. grounds (see quotes.); (the various definitions are all broadly equivalent).

1.2.6   Definition guidelines
The following definitions will therefore be used throughout this book:

luminescence
— is the emission of electromagnetic radiation in UV, visible and IR (light) from atoms or molecules as a result of the transition of an electronically excited state to a lower energy state, usually the ground state.

chemiluminescence
— is luminescence as the result of a chemical reaction.

bioluminescence
— is visible chemiluminescence from living organisms.

photoluminescence
— is luminescence as a result of absorption of electromagnetic radiation in the ultraviolet, visible, or near infrared regions.

fluorescence
— is luminescence from a singlet electronically excited state, which is of very short duration after removal of the source of excitation.

phosphorescence
— is luminescence from a triplet electronic state which remains detectable, some-times for considerable periods, after the source of excitation is removed. Fluorescence and phosphorescence are usually, but not always, used to describe photoluminescence.

chemiluminescent reaction (synonymous with a luminescent chemical reaction)
— is a chemical reaction which gives rise to luminescence. The light intensity may or
   may not be bright enough to be visible. The reaction may be inter- or intramolecular, may involve radical reactions, or energy pooling.

chemiluminescent compound
— is a compound capable of participating in a chemiluminescent reaction and whose
   product is the molecule generated in an electronically excited state which
   produces the light. The product of the reaction thus has a different chemical
   structure from the original *chemiluminescent compound*, or substrate, but is not
   itself regarded as a *chemiluminescent compound* unless capable of participating in
   some other *chemiluminescent reaction*. The primary product of the reaction will
   be capable of *photoluminescence*, but only of course when light is shone on it.

chemical reaction
— is an interaction between two or more atoms giving rise to a new chemical species.
   These atoms may be isolated, within one molecule or in different molecules.
   Included in this definition are reactions involving the transfer of electrons from
   one atom or molecule to another, even when this is induced by radiation, for
   example in a photochemical reaction.

1.3 SOURCES OF CHEMILUMINESCENCE

Many hundreds of inorganic and organic chemical reactions have now been discovered which produce visible light (for examples see Tables 1.6 and 1.7, Figs 1.5 and
1.8). The chemical mechanism of light emission in several groups of luminous
organisms, 'living light', has also been established (see Chapter 3). Chemiluminescence can occur in gases, in liquids, and at the interface between solids and either a
gas or liquid phase. In theory chemiluminescence could occur within solids, for
example through production or release of excitons. However, little work has been
done on this. Furthermore, the opacity of most solids would preclude significant light
emission, except from the surface.

What, then, are the sorts of chemical reaction responsible for
chemiluminescence?

There are essentially four principal factors which characterise a particular
chemiluminescent reaction:

(1) The brightness of the light emission, in particular whether it is visible to the
    naked eye or not. Some very dim emissions are classified as ultraweak chemilu-
    minescence (see Chapter 6).
(2) The state in which the chemiluminescent reaction occurs, in particular whether
    the excited state emitter, and the reaction producing it, occur in a gas, in a liquid,
    or at a gas–liquid, gas–solid, liquid–liquid, or liquid–solid interface.
(3) Whether the chemiluminescent reaction is organic or inorganic. In particular, is
    the light-emitting species organic (i.e. a compound of carbon) or inorganic?
(4) The existence of, or indeed a requirement for, an acceptor substance which takes
    the energy from the initial excited product of the chemiluminescent reaction,
    and then becomes the actual light emitter. This is known as *energy transfer* or
    sensitised chemiluminescence.

Thus, chemiluminescence can be visible and *bright*, visible and *dim*, or invisible and
ultraweak. When visible chemiluminescence occurs in living organisms it is known as
bioluminescence. Any one of these can involve energy transfer (see Chapter 9).

Table 1.6 — Some examples of inorganic chemiluminescence

Reaction	Colour of light
Solid (on the surface)	
Oxidation of phosphorus in air (vapour just above surface)	blue
Oxidation of alkali metals in air	whitish
Liquid	
H_2O_2 or strong acids on slaked lime	white
Alkali metal hydroxides + acids	whitish
$HgNa_3$ or $PbAc_2 + H_2SO_4$	whitish
Siloxene $(Si_6H_6O_3) + H_2O_2$, oxidised by HNO_3, H_2O_2 or permanganate	red
$H_2O_2 + NaOCl$	red
Gas	
$NO + O \rightarrow NO_2^*$ (air afterglow)	yellow–green
$N + N(+M) \rightarrow N_2^*$ (nitrogen afterglow)	yellow
$^1O_2 \rightarrow {}^3O_2$	red-infrared
$2\,O_2 \rightarrow 2{}^3O_2$	red
$O_3 \rightarrow O_2$	ultraviolet
O_3 and atomic oxygen with other gases (e.g. nitrogen oxides, olefins, SO_2, H_2S)	various
Sodium + halogens	yellow
Alkali, metal vapours + N_2, CO, SO_2, hydrocarbons	various
Phosphine $(PH_3 + P_2H_4)$ oxidation	blue
$H + H + S_2 \rightarrow H_2 + (S_2)^*$	ultraviolet–violet

Table 1.7 — Some examples of organic chemiluminescence

Reaction	Colour of light
Solid	
Rubrene peroxide dissociation by heat	red
Liquid	
Organo magnesium halides (Grignard reagents), in ether + O_2	ultraviolet–violet
Oxidation of luminol in dimethyl sulphoxide	blue (λ_{max} 480–502 nm)
Oxidation of luminol in aqueous alkali	blue (λ_{max} 425 nm)
Oxidation of lucigenin by alkaline H_2O_2	blue–blue green (λ_{max} 500 nm)
Oxidation of lophine in alcoholic NaOH	yellow (λ_{max} 530 nm)
Pyrogallol in alkaline H_2O_2	reddish pink
Gas	
Carbon monoxide flame	blue
Ether flame	blue

1.3.1 Types and characteristics of chemiluminescence

Two main types of chemiluminescent reaction have been defined, direct and indirect (Fig. 1.6), designated Types I and II. The latter is usually known as sensitised, or energy transfer, chemiluminescence.

In type I, direct, the reaction generates the primary excited state molecule (Product in Fig. 1.6), and this is then the molecule responsible for light emission. Since the physics of the electronically excited state is the same for all types of luminescence, one would expect the product of the chemiluminescent reaction (Product in Fig. 1.6) to be photoluminescent, i.e. to be capable of fluorescence or phosphorescence after the chemiluminescent reaction has finished. Furthermore, one would have thought that the photoluminescent emission spectrum of the product from the chemiluminescent reaction, now excited by absorbing light, would be identical to that of the chemiluminescent reaction. However, there are four particular reasons why this may not always be so, and that it may be difficult or impossible to regenerate the exact conditions once the chemiluminescent reaction has occurred for AB* (Fig. 1.7):

(1) The product eventually isolated to test for photoluminescence may not be the actual emitter in the chemiluminescent reaction. A metastable intermediate can be the chemiluminescent emitter, which then reacts further to form a stable product. Thus, the products of the reaction in luminous bacteria are a non-fluorescent long-chain fatty acid, and FMN with a yellow fluorescence more than 40 nm from the blue peak of the emitting bacteria. Similarly, in luminous dinoflagellates the eventual product appears to have no fluorescence.

(2) The chemistry of molecules in an electronically-excited state is different from that in the ground state. For example, the binding constants of groups such as $-CO_2H$, $-NH_2$, and $-OH$ for H^+ or metal ions are different between excited and non-excited states. Thus the fraction of protonated, or metal-bound, species formed from the chemiluminescent reaction may be different from that of the product, in the ground state before excitation by light. Protonation of chromophores has major effects on the colour of several bioluminescent reactions, including the firefly and jelly fish systems.

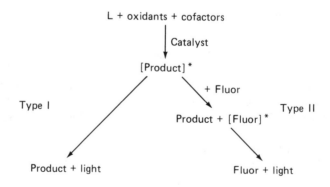

Fig. 1.6 — Two types of chemiluminescent reaction.

(3) The environment in which the electronically excited product is formed may be unique to the chemiluminescent reaction. Hydrophobicity, ligand–ligand interactions, and hydrogen bonding may all affect the colour. This is a striking feature when comparing the enzymes of bioluminescence, the luciferases, responsible for catalysing the same chemiluminescent reaction in different species. The fluorescence light yield of oxy-luciferins unbound to luciferase is much less than that on the luciferase, and the emission maxima are usually different.

(4) For one reason or another, in type II or indirect chemiluminescence it may be impossible to generate the same conditions for energy transfer when observing the photoluminescence of the products, compared with the chemiluminescent reaction itself. In which case the light yield and spectrum will be different.

In practice, the primary product can be isolated from many chemi- and bioluminescent reactions, and is photoluminescent, conditions being generated so that its fluorescence is very close to that of the chemiluminescence.

In contrast, in the indirect type (type II) the excited product of the reaction (AB*) is not the actual light emitter, but rather transfers its energy to an acceptor (CD) which then emits light. Several mechanisms for such energy transfer have been identified (see section 1.3.6 and Chapter 9). The actual spectrum may be entirely from Fluor* (Fig. 1.6) or a combination of that from Product* + Fluor*.

Four main parameters are characteristic of both types of chemiluminescence:

(1) Intensity.
(2) Colour.
(3) Speed of onset and decay of light intensity.
(4) Polarisation of light, if any.

In direct (type I) chemiluminescence these parameters are determined by the rate of the primary chemical reaction, together with the chemical structure and environment of the excited state product (Product*). To see the importance of the latter one only has to examine the enzyme-catalysed reactions involved in bioluminescence. The enzyme luciferase not only controls the rate of reaction, and thus light intensity, but can also dramatically affect the colour of the light. For example, fireflies and glow-worms, which are all really beetles, produce a range of colours from green to yellow depending on the individual species. The chemical reaction responsible is identical in all species. The colour differences are due entirely to differences in the environment at the active centre of the individual luciferase, each species having a unique amino acid sequence. Furthermore, *in vitro*, at acid pH or in the presence of heavy metal cations oxidation of firefly luciferin produces red light.

The intensity of chemiluminescence is dependent on both the rate of reaction and its efficiency at generating molecules in an excited state. The latter is expressed quantitatively as the quantum yield, which itself is made up of three components:

$$\text{overall quantum yield} = \phi_{CL}$$
$$= (\text{total number of photons emitted})/(\text{number of molecules reacting})$$
$$= \phi_C \, \phi_{EX} \, \phi_F \qquad (1.9)$$

where ϕ_C = chemical yield, i.e. fraction of molecules going through the chemilumi-
      nescent pathway,

ϕ_{EX} = yield of excited state molecules, i.e. fraction of those molecules going
      through the chemiluminescent pathway which actually produce an
      excited state product,

ϕ_F = excited state quantum yield, i.e. the fraction of excited state molecules
      which produce a photon.

There are three principal mechanisms by which the energy of an electronically
excited molecules can be dissipated (Fig. 1.7): a chemical reaction (molecular
dissociation, reaction with another molecule, or internal rearrangement), physical
(quenching to produce heat or intra- or inter-molecular energy transfer), and
luminescence. As with efficient photoluminescence, the loss of energy by non-
radiative processes must be low if significant light emission is to be observed. Overall
quantum yields (ϕ_{CL}) ranging as widely as 10^{-15} to nearly 1 have been observed.
However, many artificial chemiluminescent reactions (Fig. 1.5) have overall quantum yields in the range 1–10% ($\phi_{CL} = 0.01$–0.1), a few may be as high as 60%
($\phi_{CL} = 0.6$), whereas in bioluminescence yields in the range 10–30% (0.1–0.3) are
common. The firefly luciferin–luciferase reaction appears to be even more efficient
(80–100%; ($\phi_{CL} = 0.8$–1).

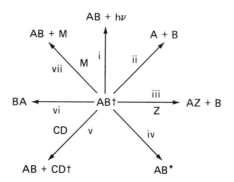

Fig. 1.7 — Seven ways for an electronically excited state to lose energy. (i) luminescence; (ii)
molecular dissociation; (iii) chemical reaction with another molecule; (iv) intramolecular
energy transfer; (v) intermolecular energy transfer; (vi) isomerization; (vii) physical quenching
(e.g. by collision with solvent or solute molecules).

Reactions with very low quantum yields ($\phi_{CL} < 10^{-4}$) are usually difficult or
impossible to observe with the naked eye. However, the chemiluminescence of
Grignard reactions can be seen in spite of the very low ϕ_{CL}, perhaps as low as
10^{-6}–10^{-8}, since reagent concentrations are usually very high. However, even
chemiluminescent reactions invisible to the naked eye, and there are many, can be
detected and quantified using sensitive modern detectors (see Chapter 2). In fact,
when using a highly sensitive photomultiplier tube, it is amazing how many reactions
can generate some chemiluminescence. This dim, or *ultraweak chemiluminescence* as

it is known (see section 1.3.5), has recently attracted much interest from biologists (see Chapter 6). The inorganic, organic, and bioluminescent reactions described in sections 1.3.2, 1.3.3, and 1.3.4, all produce sufficient light to be visible with the naked eye. Similarly the examples of energy transfer described in section 1.3.6 also can be seen, though the efficiency of light emission in the *absence* of an energy transfer acceptor molecule (type II, Fig. 1.6; CD) may be very low, as is the case with oxalate esters (Fig. 1.5a). The intensity and colour of energy transfer chemilumines- cence is thus critically dependent on the acceptor. In some reactions the acceptor may also affect the rate of reaction, particularly if it interacts with the catalyst, requires electron transfer, or interacts directly with the excited state molecule.

Efficient chemiluminescent reactions require up to five components:

(1) The chemiluminescent substrate or substrates, which react together or cleave to form the electronically-excited state molecule, responsible for light emission in type I or acting as the energy transfer donor in type II.

(2) An electron acceptor, such as oxygen, if an oxidation reaction is involved as is the case in a large number of synthetic chemiluminescent and all bioluminescent reactions.

(3) A catalyst, such as an enzyme or metal ion. The 'catalyst' has one or more of three functions: (a) to reduce the activation energy and thus initiate or speed up a reaction would either be very slow, or not occur at all, (b) to provide an environment for a high ϕ_{CL}, where ϕ_C, ϕ_{EX} and ϕ_F are all high, (c) to process an oxidant.

(4) Cofactors, necessary to convert one or more of the substrates into a form capable of reacting or interacting with the catalyst, or necessary to provide an efficient leaving group if bond cleavage is required to generate the excited emitter.

(5) An energy or electron acceptor, if energy transfer chemiluminescence is involved (type II).

I have divided the chemiluminescent reactions which follow into five classes: inorganic, organic (i.e. man-made), bioluminescence (all organic), ultraweak, and energy transfer.

1.3.2 Inorganic reactions

The inorganic elements which can participate in chemiluminescent reactions range from the alkali metals and heavy metals such as mercury and lead, to non-metals such as O, S, N, P, As and the halogens (Table 1.6). Compounds of silicon, the next element to carbon in group IV of the periodic table, also can produce chemilumines- cence. Whilst the earliest inorganic chemiluminescent reaction in solution, between water and calcium oxide, was discovered by J. F. Meyer in 1764, it is in the gas phase that we find three examples of chemiluminescence with which most people will be familiar. One is the 'airglow' in the upper atmosphere, observable from twilight to dawn, though actually occurring by day and night. The second is the long-lived yellow afterglow caused by active nitrogen (N·) formed by an electric discharge, and first discovered by J. F. A. Morren in 1865. The third is the famous blue Will o' the Wisp thought to be due to spontaneous ignition of phosphine ($PH_3 + P_2H_4$) or methane, formed from the decomposition of animal matter in marshy ground, but where a puzzle still exists (see reference [37]).

The first light in the night sky has four contributory factors: starlight, sunlight scattered by fine dust particles in space, aurora, and airglow. The aurora is an example of luminescence, but is mainly radioluminescence — the result of direct excitation of atoms and molecules in the atmosphere by energetic particles entering along the lines of force from the earth's magnetic field. On the other hand, 'airglow', or 'air afterglow', is the result of chemiluminescence. The process begins in the upper layers of the atmosphere from 90–300 km above the earth's surface. Ultraviolet radiation from the sun causes dissociation and ionisation of molecules and atoms. The recombination of these reactive species results in light emission from violet (370 nm) to deep red (660 nm). Line spectra corresponding to the red and green lines of atomic oxygen (O·), the yellow doublet of sodium (Na), the blue and red lines of hydrogen (H), as well as blue lines of calcium (Ca), potassium (K), and N_2^+, have all been observed. The spectrum extends far into the infrared and includes that of hydroxyl radical (·OH). These chemiluminescent reactions include:

$$O + O + O \rightarrow O_2 + O^* \rightarrow h\nu \tag{1.10}$$

$$O^+ + O_2 \rightarrow O + O_2^+ \xrightarrow{+e} 2O^* \rightarrow h\nu \tag{1.11}$$

$$NO + O \rightarrow NO_2^* \rightarrow h\nu \tag{1.12}$$

This latter reaction is well known in the laboratory, producing yellow–green air afterglow, e.g. following electrical discharges.

$$NaH + O \rightarrow OH + Na^* \rightarrow h\nu \tag{1.13}$$

The relative importance of these molecular reactions to the dayglow, twilight glow, and nightglow varies with altitude from 50–350 km. In the laboratory the NO + O reaction (1.12) has been extensively investigated, being known as the yellow–green air afterglow. This reaction has been used as a secondary standard for measuring quantum yield and as an assay for oxygen atoms. Gas phase chemiluminescent reactions involving metal vapours are also known, including the reaction of alkali metals with halogens. Chemiluminescence is also responsible for light emission in many flames including those of N_2O, H_2S, CS_2, CO, and CN in air or oxygen. It is important to distinguish the blue colour of many flames, including organic oxidations such as coal gas in air, the chemiluminescence, from the yellow colour, the incandescence of hot carbon particles. Another gas producing an afterglow under similar conditions to N_2, but of much shorter duration ($<$ 1s), is hydrogen; first observed by R. W. Wood in 1920. The discovery of ozone (O_3) led to the identification of many new gaseous chemiluminescent reactions, including its reaction with NO, S, I_2, Na, and other metals. In addition, the breakdown of another active oxygen species, hydrogen peroxide, in solution provoked by catalysts such as colloidal silver and platinum, lead dioxide (PbO_2), and manganese dioxide (MnO_2), produces light. The addition of H_2O_2 to sodium hypochlorite (NaOCl) produces a weak red flash, observed first by Mulliken in 1928 and now known to be caused by singlet oxygen (see section 1.4.3).

Finally, many solid metals and non-metals glow at their interface with gas or liquid phases. The famous glow of phosphorus has already been described (section 1.2.3). Humphry Davy, the discoverer of the alkali metals and alkaline earths as well as several non-metals in the early years of the 19th century, also noticed the chemiluminescence of sodium and potassium at a freshly cut surface. This can be enhanced by gentle heating to 60–70°C. The layer of Na_3N, K_3N, Na_2O, K_2O which forms in air eventually stops the chemiluminescence.

1.3.3 Organic reactions

Since the synthesis of lophine by Radziszewski in 1877, an enormous range of organic compounds have been produced which chemiluminesce (Figs. 1.5, 1.8, Table 1.7), emitting light from the ultraviolet to the infrared. The majority can be separated essentially into two main groups:

(1) Oxidation reactions requiring oxygen or one of its metabolites.
(2) Reactions requiring electron transfer, particularly those involving radicals, and their annihilation.

Apart from aliphatic compounds in flames, most are aromatic substances. Important exceptions are the aliphatic oxalate esters, which have a poor quantum yield on their own but which act readily as energy transfer donors in a sensitised chemiluminescent reaction (type II, Fig. 1.6), and stable aliphatic dioxetans (Fig. 1.8). Two particularly well studied reactions, the oxidation of luminol and acridinium salts, emit blue to blue–green light, whereas the oxidation of pyrogallol and the breakdown of rubrene peroxide produce red light. The latter compound has been used by Kasha and co-workers as an indicator of singlet oxygen. The colour and quantum yield can be affected markedly by the polarity of the solvent. For example, luminol oxidised in aqueous alkaline solution produces blue–violet light (λ_{max} 425 nm), whereas in dimethyl sulphoxide (DMSO) the colour is blue–green (λ_{max} 480–502 nm). In both cases the excited product is 3-aminophthalic acid (Fig. 1.5), the chemiluminescence spectrum corresponding closely to the fluorescence emission spectrum of this product in the particular solvent. But in DMSO ϕ_{CL} is 0.05 compared with 0.01 in H_2O.

Luminol (5-amino-phthalhydrazide, otherwise known as 5-amino-2,3-dihydro-1,4-phthalazine dione) chemiluminescence was first published by Albrecht in 1928, although it had been synthesised in 1853. It has been one of the most well studied compounds. Drew and co-workers in the 1930s showed that adding electron withdrawing groups (R_1 and R_2 in Fig. 1.5) onto the free amino group had dramatic effects on the quantum yield. Both steric hindrance and electron withdrawal reduce light yield. Luminol, and derivatives of isoluminol, where the amino group is shifted along the benzene ring (Fig. 1.8) to reduce steric hindrance, have several important biomedical applications (see Chapters 4–9). Surprisingly, the precise mechanism of the luminol reaction under aqueous conditions is still controversial (see section 1.5). In DMSO the reaction involves oxygen directly. However, in aqueous media, where the reaction proceeds best under alkaline conditions, hydrogen peroxide or another oxygen metabolite (see section 1.4) is the oxidant, the reaction being catalysed by a range of transition metals, metal complexes, and enzymes, particularly peroxidases.

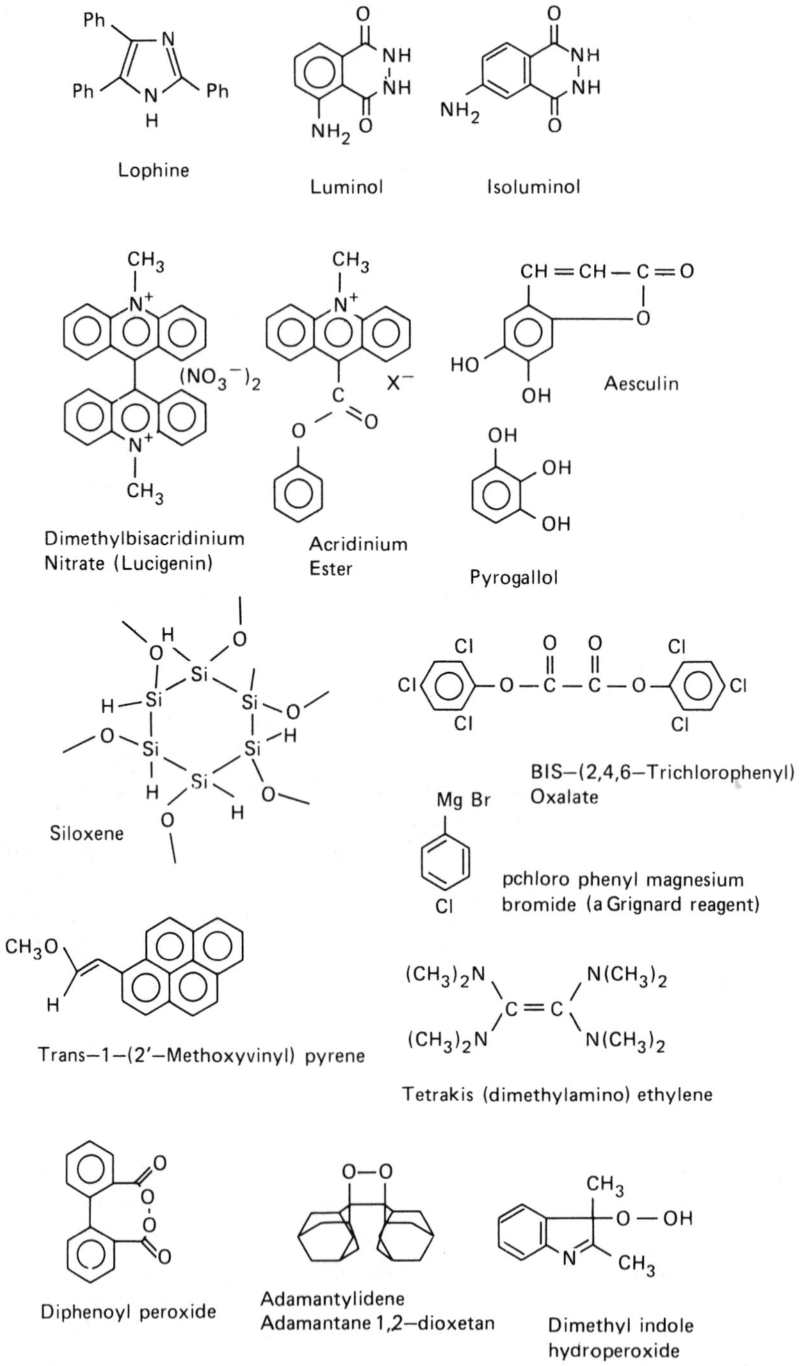

Fig. 1.8 — A range of chemiluminescent compounds.

Lucigenin (Fig. 1.8, bis-*N*-methyl acridinium nitrate, otherwise known as 10,10'-dimethyl-9, 9'-bisacridinium dinitrate) was first synthesised by Gleu and Petsch in 1935. Like luminol, oxidation in alkaline solution occurs with hydrogen peroxide and some other oxygen metabolites. Also like luminol the light emission lasts several minutes and is blue–green (λ_{max} *c*. 500 nm). Heating the solution converts the light to blue. In both cases the excited product of the reaction is *N*-methyl acridone (Fig. 1.5), the blue chemiluminescence emission in hot solution corresponding to the fluorescence emission spectrum of this latter compound. Unlike luminol, acridinium salts do not require a catalyst. At the concentrations of lucigenin usually employed in these experiments (*c*. 10 mM) the initial substrate, lucigenin, also fluorescent acts as an energy transfer acceptor from the *N*-methyl acridone (see section 1.3.6), thereby altering the spectrum. McCapra and Rauhut in the 1960s independently synthesised a group of acridinium esters (Fig. 1.8), based on one half of the lucigenin molecule. They too produce blue light when oxidised by alkaline H_2O_2, due to the product *N*-methyl acridone. The quantum yield of these compounds can be high ($> 10\%$) and the reaction very fast, more than 90% of the luminescence occurring within 1–10s. They have several important biomedical applications (see Chapters 4 and 8).

The metastable intermediate responsible for generating the excited carbonyl of *N*-methyl acridone following acridinium salt oxidation is a dioxetan for lucigenin, and a dioxetanone for acridinium esters (Fig. 1.8). Stable aliphatic and aromatic dioxetans have been synthesised (Fig. 1.8) which undergo thermal decomposition, thereby producing light. A particularly dramatic compound is tetrakis (dimethyl amino) ethylene. This is a liquid at room temperature which on pouring in air chemiluminesces, and can be used as a 'light' pencil for a nice demonstration in a lecture.

Whilst aromatic compounds dominate the direct (type I) chemiluminescence of organic molecules in solution, in flames many small aliphatic carbon molecules such as methane, CO, alcohols, and ethers burn and thereby chemiluminesce. The temperature of some of these flames is remarkably low. Many will be aware of the ease with which one can hold one's hand in an ether flame, for example.

Finally, in this group are compounds at the interface between organic and inorganic chemistry, the organometallic compounds. Chemiluminescence of Grignard compounds, such as the oxidation of phenyl magnesium iodide, has been known for nearly a century. Other organometallic compounds which chemiluminesce include porphyrin derivatives and compounds related to phthalocyanin (Fig. 1.8). Several of these generate red light.

1.3.4 Bioluminescence — living light
All living light is chemical light, or so we now believe. But Isaac Newton in 1704 thought that the light emission from shining, rotten wood, now known to be caused by luminous fungus, was the result of 'vibrating motions of bodies'. In other words, the origin of the light had some physical basis. What then is the evidence for the current dogma, that the origin of luminescence in all luminous organisms is chemical? There are three key experiments, begun by Robert Boyle at the end of the 17th century, and developed by E. Newton Harvey and others in our century:

(1) All living light requires oxygen in some form, either O_2 itself or one of its metabolites such as H_2O_2.

(2) A small organic moiety, the luciferin, can be isolated from luminous organisms, which when oxidised (Fig. 1.5), in the presence of the necessary cofactors, produces light.

(3) Proteins, the luciferases and apophotoproteins, have been isolated which are responsible for catalysing the bioluminescent reaction. Several different chemistries have been identified, and five major chemical families of luciferin, based on aldehydes, imidazolopyrazines, benzothiazoles, linear tetrapyrroles and flavins, have so far been characterised. At least half a dozen more chemistries remain to be unravelled (see Chapters 3, Fig. 3.11).

The evidence that bioluminescence is simply a living example of chemiluminescence has been obtained in literally hundreds of luminous species: bacteria, fungi, dinoflagellates, jelly fish, worms, shrimp, insects, some squid, and fish. The chemical reaction producing the light is either contained within or secreted by specialised cells in the organism. Yet a definitive case for chemiluminescence in many other luminous species remains unproven: for example, certain polychaete worms, starfish, certain squid, tunicates, and even some fish. Here we expect to find chemiluminescence, but the reaction is either so unstable that it rarely survives freezing, or is particulate, and has so far evaded the conventional 'grind and find' approach of the biochemist. Wouldn't it be fun if some forms of living light turned out to be examples of tribo- or electro-luminescence after all!

How common is living light, and where can we find it?

There are literally thousands of individual luminous species representing 16 phyla and almost 700 genera. Some will be familiar to most people, others a surprise. Living light has invaded the air, the land, and is very common in the sea (Tables 3.1–3.3). More than half of the earth's surface is covered by ocean over 1000 metres deep. In this cold, pressurised environment, hostile to man, most of the creatures are luminous. At these depths virtually all the light is 'living light', since most of the light from the surface will have been scattered or absorbed by the sea above. Prokaryotes are represented by several free-living marine bacteria, which also occur as symbionts in the organs of other marine animals such as some squid and fish. The famous angler fish, with its dangling and tempting lure containing luminous bacteria, is one such example. One freshwater species of luminous bacteria has also been identified. However, there are no known luminous cyanobacteria (blue–green algae) nor archaebacteria.

On land the most familiar luminous species are the fireflies and glow-worms, actually beetles, and the occasional earthworm. The first two have individuals which can fly. There are no luminous plants, but many luminous fungi. These include not only mushrooms growing in the ground, but also more amorphous species such as certain leaf moulds. The occasional fresh-water species, such as the limpet *Latia neritoides* in New Zealand, may be found. But by far the greatest variety of luminous species are found in the sea (Figs 3.1–3.4). A walk on the beach, like our mythical camper, will reveal many luminous species: hydroids, worms, starfish, dinoflagellates, and probably bacteria on some decaying organic material. Some, such as the extraordinary rock-boring bivalve *Pholas* (Fig. 1.1a), the common piddock, remain hidden from view. The surface of the oceans contains many similar and related species. Research ships such as RRS *Discovery*, with sophisticated nets (Appendix

III, Fig. 3.2), capable of being opened and closed at different depths by sonic signals from the ship, have established that luminous species exist from the surface right through the deep oceans to benthic animals on the ocean floor several thousand metres below the surface. Here we may find luminous starfish, and other luminous invertebrates in the surrounding waters such as decapod and euphausiid shrimps and the tunicate *Pyrosoma*. The latter can grow to several metres long, giving spectacular displays when ensnared by fishermens' nets or hit by ships when near the surface. There are also many species of luminous fish. One genus, *Cyclothone*, is so abundant throughout the oceans of the world that it can make a claim as the most common vertebrate in the world, either on land or sea. Another example illustrating the great quantities of luminous animals in the sea are the euphausiid shrimp. All but one are luminous. They constitute the krill eaten by whales. A study by Beebe in the 1930s showed that as many as 90% of the species of fish caught in the deep oceans can be luminous. In most catches, of small creatures in the oceans, from the surface down to the bottom, more than 50% may be luminous.

To the human observer, living light appears blue, blue–green, green, yellow, or very occasionally, red. Most luminous marine organisms emit blue or blue-green light (440–510 nm), though there are few notable exceptions. For example, some scale worms and starfish emit green light. Blue chemiluminescence's secreted by the earthworm *Diplocardia*. However, the glow-worms and fireflies emit green to yellow light. Luminous animals emitting red light are very rare (Fig. 3.10). Two fascinating examples have been discovered both of which have light organs emitting two colours. The females and larvae of the South American railroad worm, *Phrixothrix tiemani* have two red light organs on the head and eleven greenish-yellow pairs on the body. Certain deep sea stomiatoid fish, such as *Malacosteus niger* or *Aristostomias scintillans*, have one pair of large light organs below the eye, emitting deep red light (λ_{max} *c.* 705 nm) and one smaller pair behind the eye emitting blue light (λ_{max} 470–480 nm), as well as many smaller photophores distributed on the body.

Luminous organisms in nature were first described by Chinese writers from at least as far back as 1500 BC, and by the Greeks and Romans. Many hundreds of species were identified and classified during the 17th, 18th and 19th centuries. Yet new species are still being discovered. Until the discovery by Kymazawa and Haneda in Singapore in 1942 of *Dyakia striata*, no luminous land snails were known. More recently a land nematode has been found containing luminous bacteria, which have now been found in the soil. This is an exciting find since hitherto all luminous bacteria were thought to be marine. It may explain the puzzle of how common 'shining flesh', never apparently in contact with fish, was in the 17th and 18th centuries. There is scope for more finds in the tropical rain forests. Here, many species of fungi and beetle probably remain undiscovered by man. Recent discoveries of luminous animals in groups hitherto thought to be entirely non-luminous include octopods and holothurians.

Several luminous organisms are responsible for spectacular displays at night (Table 3.2): dinoflagellates, for example, light up a ship's wake and, on occasions, can cause whole bays to 'phosphoresce', whilst the synchronous flashing of fireflies in the tropics provides a living 'Christmas tree' display for the fortunate observer.

This widespread occurrence of living light presents us with many intriguing problems. How can some organisms produce a flash lasting less than a second whilst others glow for many minutes or even hours? What is the chemical basis of this and the colour of the light produced? What is its function? Why do some animals secrete luminous clouds and others use intracellular luminescence? What is the symbiotic relationship between luminous bacteria and those creatures which harbour them? Life began some 3500 million years ago. But when did bioluminescence first appear? What is the evolutionary significance of bioluminescence, and why is it that it may be common in one genus, yet rare or absent in another closely related to it? Why are there no luminous mammals; nor any luminous flatworms, higher plants, and cyanobacteria?

Living light provides the scientist with a fascinating and aesthetically pleasing phenomenon for probing some of the central problems of biology: the evolution and stabilisation of ecosystems, the differentiation and development of cells, and the origin and evolutionary selection of biochemical processes. Nature has evolved an excellent model for the chemist wishing to understand the electronically excited state and energy transfer. Clinical and research scientists have also been able to harness bioluminescence in the laboratory as a powerful analytical tool in biotechnology, providing insights into the processes underlying disease and as a new diagnostic aid with immediate impact on patient care.

1.3.5 Ultraweak chemiluminescence

There are no luminous mammals, yet many of our cells are capable of producing an 'ultraweak' chemiluminescence, invisible to the naked eye but detectable by a sensitive photomultiplier tube.

Perception of a light signal by the dark-adapted human eye needs 0.85×10^{-6} candle m^{-2} (6×10^{-10} watts sr^{-1} m^{-2}) or about 30 000 photons hitting a square centimetre of the retinal surface per second at 510 nm, the optimal wavelength for sensitivity. This is equivalent to 1 quanta hitting an area of some 500 retinal rods per second (1 rod = 2.6μ). Luminous bacteria can emit at least a thousand photons per cell per second. Eurkaryotic luminous cells emit at least a hundred times more than this. These light emissions are easily visible to the dark-adapted eye. All of the original discoveries of chemiluminescence, up to the first quarter of this century, involved initially observations by the naked eye. The advent of highly sensitive photomultiplier tubes capable of detecting just a few photons per second led to the discovery of the much weaker or 'ultraweak chemiluminescence' not visible to the naked eye, and which may be less than one photon per cell per second.

In the late 1920s A. G. Gurvich reported a weak light emission in the ultraviolet region from cells during mitosis. His studies were originally on dividing cells in the onion root, the regenerating cornea of the frog, and dividing cells of malt yeast, and blood cells. In subsequent experiments Gurvich claimed that the intensity of the UV radiation was only 10–10 000 photons cm^{-2}s^{-1}, and between either 190–280 nm or 320–350 nm. Since the emission originated from many hundreds if not thousands of cells, it was several orders of magnitude lower in intensity than that from luminous organisms. He called the radiation *mitogenetic*. He subsequently reported such ultraweak chemiluminescence from a wide range of plant and animal cells during mitosis, as well as from several test-tube reactions, particularly when amino acids and other biological molecules were oxidised by hydrogen peroxide. He claimed that the weak mitogenetic emission was not only necessary for mitosis, but was itself also responsible for stimulating mitosis, as well as causing irreversible morphological and permeability changes in neighbouring cells.

The measurement of UV in these experiments utilised either biological detectors, such as yeast and bacterial cultures or plant meristems, or physical detectors such as photographic plates or a UV-sensitive Geiger or photomultipler tube. These experiments were severely criticised at the time, and later, because of the underdevelopment of such detectors for weak luminescence, together with their susceptibility to extraneous factors. Furthermore, whilst Gurvich's experiments were followed by several hundred reports in the Russian and German literature, confirming the phenomenon, extensive investigations in the West during the 1930s failed to reproduce it. Work on mitogenetic luminescence subsequently ceased in the West, although it was still actively pursued in the Soviet Union. As recently as 1974, Quickenden and Que Hee re-examined the phenomenon in dividing yeast cells. They reported 'limited support for certain published claims relating to the existence of a so-called mitogenetic radiation from dividing cells' and 'the present results indicate that a case exists for re-investigation of some of the mitogenetic phenomena, using modern photon counting and cell counting equipment. . . '

In contrast to this controversy, the emission of a weak chemiluminescence from many cells in the plant and animal kingdom in the visible region is now firmly established (Tables 6.1, 6.2, 6.3). It was first observed in plants by Colli and co-workers in 1955, and then in mammalian organs and subcellular organelles by Tarusov and co-workers in 1961. The spectrum of the emission is broad (400–700 nm). The emission may be only 1–10 photons per second per cell, or even less, more than a thousand times less than the luminescence of bacteria visible to the naked eye.

The reactions responsible for this invisible ultraweak chemiluminescence are oxidations involving oxygen, oxygen metabolites, and radicals. Peroxidation of fatty acids within lipids often plays an important part. The oxygen metabolites, e.g. H_2O_2, generate excited hydroperoxides and carbonyls or, through singlet oxygen, emit light directly (see section 1.4.3). Excited carbonyls emit mainly in the blue region, whereas singlet oxygen emits in the red. Red ultraweak chemiluminescence has even been detected in human breath. The discovery by Allen in 1972 of ultraweak chemiluminescence in activated phagocytes, such as neutrophils, together with its enhancement by adding the synthetic compounds luminol and lucigenin , has led to the widespread use of chemiluminescence as an indicator of reactive oxygen metabolites and lipid peroxidation in cells and whole organs (see Chapter 6). This may be important in an understanding of the pathogenesis of several inflammatory and immune based diseases. The ultraweak chemiluminescence may thus be either indigenous to cells or a reaction, or may be 'indicator-dependent' when a compound such as luminol is added. In the latter case, the emission spectrum reflects that of the indicator.

The use of modern, highly sensitive photomultiplier tubes capable of detecting just a few photons has extended the known number of ultraweak chemiluminescent reactions to a wide range of cellular and acellular systems. The most well studied involve oxidation of organic compounds, but many examples of inorganic ultraweak chemiluminescence also exist. The question therefore arises, is the low intensity of these luminescent phenomena because of a low rate of reaction, or because of a low overall quantum yield (1.9)?

1.3.6   Energy transfer

A particularly intriguing feature of some luminous organisms is their ability to alter the colour of light emission from that of the isolated chemiluminescent reaction. The hatchet fish *Argyropeleus* (Fig. 3.4) passes its bluish–green light through a layer of pigment which absorbs much of the red end of the emission. The resulting bluer light is virtually identical in its spectrum to the dim ambient light at the depth at which this fish lives. This is consistent with the proposed biological function of its chemilumi-nescence, camouflage. Several other luminous organisms have the ability to alter the spectrum of chemiluminescence by using absorbing filters. However, there is another mechanism for altering colour used in some organisms, energy transfer. 'Trivial' transfer can, of course, occur via photons, but the more interesting process occurs in many luminous organisms without the direct transfer of a photon (see Chapter 9).

In 1963 Johnson and colleagues observed that the light emitted by the jelly fish *Aequorea forskalea* was slightly greener than that from the Ca^{2+}-activated photo-protein, aequorin, which can be isolated from it. Morin and Hastings found similar spectral shifts in several other luminous coelenterates, including *Obelia* and *Halis-taura*. Using the sea pansy, *Renilla*, Ward and Cormier showed that these luminous animals have a radiationless energy transfer process within the cells responsible for light emission. The energy from the excited-state molecule produced by the chemilu-minescent reaction is transferred, without the direct transfer of a photon, to a green fluorescent protein which becomes the actual light emitter. This is thus a natural example of sensitised or type II chemiluminescence (see section 1.3.1). Another type of non-radiative energy transfer occurs in the luminous marine bacterium *Photobac-terium phosphoreum*. In this case the fluor, instead of shifting the spectral emission towards the green or red end of the spectrum, shifts it the other way, towards the blue. The question now arises whether non-radiative energy transfer is responsible for some of the peculiar spectral changes in other luminous animals, where spectral emissions may be bimodal, may change with time, or, as in the railroad worm and the deep sea fish *Malacosteus*, results in different colours being emitted from separate light organs?

Energy transfer through electrons and radical annihilation is involved in many well known synthetic examples of chemiluminescence including siloxene, peryoxa-lates, and oxalate esters (see Chapter 9). The colour of the emission is dependent on the fluorescent acceptor. Oxalate ester chemiluminescence energy transfer is used in commercially available light sticks, sold widely as safe night lights and as decorative toys. Energy transfer chemiluminescence is also observable in the gas phase, for example between active hydrogen (H) or singlet oxygen and an acceptor such as a metal atom. Energy transfer (or pooling) is responsible for the luminescence of singlet oxygen in the visible region (see section 1.4.3).

Certain examples of non-radiative energy transfer occur only over very short distances of <0.1 nm, whereas others occur over distances as long as 10 nm. The result is a change in quantum yield, colour, and sometimes rate of reaction. Occasionally these changes can be dramatic. With oxalate esters (Fig. 1.5) the quantum yield is several orders of magnitude greater in the presence of a fluor than in its absence. Light emission is visible only with the fluor present. Energy transfer occurs in many luminous organisms as well as with several synthetic compounds. It is found with solids, liquids, and gases, and may be inter- or intra-molecular. There are six main types: exciton transfer, resonance, electron exchange (collisional), electron transfer, chemical reaction, and energy pooling (Table 9.6). The mechanism of transfer may be physical as with resonance transfer, or chemical, as is the case of singlet oxygen reacting with certain multicyclic aromatic compounds such as rubrene or anthracene.

Not only are the physicochemical basis and biology of energy transfer a challenge to the multidisciplinary approach in science; they also provide a potentially exciting approach to studying ligand–ligand interactions in living systems (see Chapters 8 and 9).

1.4 OXYGEN AND CHEMILUMINESCENCE

Oxygen is the most abundant element on the earth, and the existence and diversity of life depends on it. Yet paradoxically oxygen can also be very toxic to cells. Many of the molecules utilising oxygen in living organisms may have evolved orginally to combat this toxicity, at a time when anaerobes were the dominating presence. Oxygen, in some form, is required for the reactions responsible for light emission in all luminous organisms, as well as for the chemiluminescence of many inorganic and man-made organic compounds.

What then is so special about the chemistry of oxygen which can explain its intimate relationship with chemiluminescence?

1.4.1 The discovery of oxygen and its metabolites

On the first of August 1774, Joseph Priestley was experimenting at Bowood House in Wiltshire. Much to his surprise, he discovered that heating mercuric oxide produced mercury and a gas capable of causing a candle to burn with 'a remarkably vigorous flame'. Initially he thought the gas to be nitrous oxide, which he had discovered previously. He soon realised it was a new gas, and duly called it 'dephlogisticated air'. A year earlier, in fact, Scheele in Sweden had discovered the same gas on heating silver carbonate, calling it first 'vitriol air' and later 'fire air', but failed to publish his discovery until some years later. Unfortunately, both scientists believed in the phlogiston theory. It was therefore up to Lavoisier to give the new gas its modern name, oxygen. Initially, in 1778, Lavoisier called the gas 'principle oxygine' but changed it in 1789 to oxygène (from two Greek words oxys = acid and ginomae = I produce), believing incorrectly that it was an essential principle in the formation of acids. Nevertheless, in 1783, with some foresight, he wrote in his 'Reflections on Phlogiston' 'principle oxygène explains . . . . the chief difficulties of chemistry . . .', thereby hitting the final nail in the phlogiston coffin.

Thenard discovered, in 1818, that addition of dilute acid to an oxygen compound of barium (barium peroxide) produced a new substance, capable of oxidation or reduction, hydrogen peroxide. Ozone, O_3, (from Greek ozin = to smell) was identified by Schönbein, in 1840. Yet many aspects of the chemistry of oxygen remained a puzzle well into the 20th century. Two particular problems were:

(1) Molecular oxygen (dioxygen) was kinetically less reactive than would have been expected.
(2) Since four electrons are required to fully reduce O_2 to $2H_2O$ was it possible to reduce oxygen one electron at a time and thus generate intermediates?

The first of these was solved once the electron orbital occupancy and spin of the outer electrons in O_2 were known. Normal oxygen has its outer electron pair in the triplet spin state. Singlet oxygen ($^1O_2$) predicted by Mulliken in 1927 is considerably more reactive than $^3O_2$ and is chemiluminescent. It is responsible for the weak red chemiluminescence from H_2O_2 and NaOCl, a flash first observed independently by various workers including Mallet in 1927 and Groh in 1938 when trying to provoke the chemiluminescence of various organic compounds.

In 1932 Fritz Haber and Joseph Weiss reported that 'the catalytic decomposition of neutral solutions of hydrogen peroxide in the presence of ferrous salts, . . . were not explicable by the earlier theory that the reaction took place through the interaction of six-valent iron–oxygen compounds, but could be easily understood if the decomposition was actually a chain reaction . . .'. Two years later Professor Sir William Pope FRS revised a paper by the same authors given to him shortly before Haber's death. As a result of studying the rate of reaction under different conditions, particularly the concentration of Fe^{2+}, Fe^{3+}, and H_2O_2, they proposed a chain radical reaction involving a new oxygen metabolite, hydroxyl radical ·OH. The main stages of the reaction proposed were:

$$Fe^{2+} + H_2O_2 = Fe^{3+} + OH^- + \cdot OH \tag{1.14}$$
$$\cdot OH + H_2O_2 = H_2O + HO_2 \, (O_2^- + H^+) \tag{1.15}$$
$$HO_2 + H_2O_2 = O_2 + H_2O + \cdot OH \tag{1.16}$$
$$\cdot OH + Fe^{2+} = Fe^{3+} + OH^- \tag{1.17}$$

Equation (1.14) is now often known as the Fenton reaction. Fenton in 1894 had, in fact, discovered the reaction under alkaline conditions, but had not realised that it generated ·OH. Equations 1.15 and 1.16 were originally proposed by Haber and Willstätter. A fifth equation was required to explain catalysis by ferric salts:

$$Fe^{3+} + H_2O_2 = Fe^{2+} + HO_2 + H^+ \tag{1.18}$$

Similar reactions ocur with Cu^+/Cu^{2+}. The sum of the reactions is known as the Haber–Weiss reaction:

$$H_2O_2 + O_2^- \rightarrow O_2 + OH^- + \cdot OH \tag{1.19}$$

Oxido–reduction reactions involve transfer of electrons between atoms or molecules. Oxidation of a substance involves loss of electron(s), reduction a gain in electron(s). During the 1940s Michaelis argued that every oxidation (or reduction) could proceed only in steps of univalent oxidation (or reduction). Application of this principle to oxygen (O_2) led to the later discovery of superoxide anion (O_2^-). Although the possible existence of O_2^- had been speculated since the 1940s it was not until 1959 that Känzig and Cohen provided the first clear spectroscopic evidence for its existence. By the end of the 1960s O_2^- had been established as an intermediate of dioxygen in several enzyme catalysed reactions, including oxidases orginally thought to be enzymes reducing dioxygen directly to H_2O_2 and leaving the organic substrate oxidised. O_2^- can act either as a one-electron reductant or a one-electron oxidant, the latter being thermodynamically favoured. Since relatively few reactions are known where O_2^- acts as an oxidising agent, such reactions may be less kinetically favourable because of the faster, spontaneous dismutation of $2O_2^-$ to H_2O_2. This reaction is catalysed in cells by the enzyme superoxide dismutase, isolated by McCord and Fridovich in 1969. In this reaction O_2^- oxidises itself to O_2 and reduces itself to H_2O_2. Thus, the stepwise, four-electron reduction of O_2 to H_2O produces O_2^-, H_2O_2 and $\cdot OH$ together with protonation states of these species (Fig. 1.9, Table 1.8). The copper-containing cytochrome oxidase in mitochondria, first discovered by Mann and Keilin in 1938, has apparently evolved a mechanism at its active centre whereby the complete reduction of O_2 to H_2O can occur without significant release of the very reactive and potentially toxic intermediates. Other oxidases, for example xanthine oxidase or glucose oxidase, lead to the formation of O_2^- and H_2O_2 via one- or two-electron reduction respectively. Xanthine oxidase produces O_2^- whereas glucose oxidase mainly H_2O_2. Of course, O_2^- will spontaneously dismutate to H_2O_2 anyway. The list of known oxygen derivatives is daunting (Tables 1.8 and 1.9). Many occur in living systems, and several are required for chemiluminescence.

Oxygen features in the beginning of the radical concept, L. B. Guyton de Moreau (1737–1776) first using the word 'radical' for an 'acidifiable base' which combined with oxygen to give an acid. The term radical is now used to describe any atom or molecule with one or more unpaired electrons. Usually the radical has one unpaired electron on its own; as a result it is highly reactive, searching for another to make up the pair. Atomic oxygen ($O\cdot$), superoxide anion ($O_2^-\cdot$), and hydroxyl radical ($\cdot OH$) are true radicals. H_2O_2 and OCl^- are not. Thus the use of 'oxygen radicals' to describe O_2^-, $\cdot OH$, H_2O_2, and OCl^- in biological systems is strictly incorrect. I will try to use the term reactive oxygen metabolites. Naturally-occurring dioxygen (O_2) is itself, in a sense, a biradical as it has two outer electrons in separate orbitals. These radical features are a major reason for the high chemical reactivity of oxygen, and for its role in chemiluminescence.

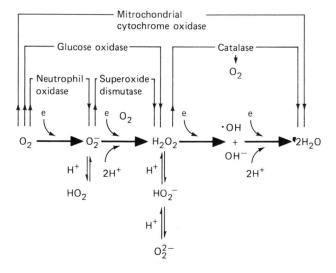

Fig. 1.9 — The four electron reduction of oxygen to water.

Table 1.8 — Protonation of oxygen metabolites

Metabolite	Approx % present in H_2O at pH 7.4
O_2	100
O_2^-	>99.9
HO_2	<0.1
O_2^{2-}	negligible
HO_2^-	<0.1
H_2O_2	>99
O^-	<0.1
OH	>99
OCl^-	>95
HOCl	<5

Hence major metabolite species at pH 7.4 are O_2^-, H_2O_2

$$HA \rightleftharpoons H^+ + A^-$$
$pK_a = pH + \log_{10} HA/A^-$, i.e. calculate $HA/(A^- + HA)$ or $A^-/(A^- + HA)$

Table 1.9 — Oxygen and some derivatives

Species	Name	Protonated form(s) (or other species)	Ntural example or source
$O.(^3P, ^1D^1S)$	atomic oxygen	—	electrical discharge
$O_2(^3O_2, ^1O_2)$	dioxygen	—	photosynthesis
O_2	superoxide anion	[a]HO_2	xanthine oxidase reaction
H_2O_2	hydrogen peroxide	[b]HO_2^-, [c]O_2^{2-}	glucose oxidase reaction
·OH	hydroxyl radical	[d]O^-	breakdown of H_2O_2 by Fe^{2+}
(O_2^{3-})	hypothetical precursor	—	reduction of H_2O_2
O_2^+	dioxygenyl cation	—	electrical, airglow
O_3	ozone	—	electrical discharge
O_3^-	ozonide	HO_3	?
OCl^-	hypochlorite anion	[e]HOCl	myeloperoxidase reaction
OBr^-	hypobromite anion	HOBr	eosinophil peroxidase
$OSCN^-$	hypothiocyanite anion	HOSCN	lactoperoxidase reaction
$OOCl^-$	chloroperoxy anion	HOOCl	$H_2O_2 + OCl^-$
ROOH	organic hydroperoxide	ROO^-	reaction of oxygen radicals with fatty acids in biological membranes
$M^{n+} - O$	metal ion-oxygen complex	—	peroxidase active centre

a: hydroperoxy, hydroperoxyl or perhydroxyl radical; b: hydroperoxyl anion; c: peroxide or peroxy anion; d: oxene anion; e: hypochlorous acid (many other oxy-acids exist with halogens and sulphur)

1.4.2 The role of oxygen in light-emitting reactions

Oxygen, or any one of its derivatives, can participate in both inorganic or organic chemiluminesence in solids, liquids, or gases (Table 1.10). Atomic oxygen (O.) is mainly responsible for the red or yellow–green of aurora in the atmosphere. Ozone

Table 1.10 — Some chemiluminescent reactions involving oxygen and its derivatives

Reaction	Colour	Approx λ_{max} (nm)
$NO + O \rightarrow NO_2^*$	yellow–green	530–570
$O_2 + e \rightarrow 2\,O^*$	red and green	630,571
Vargula luciferin + O_2		
$\rightarrow$ oxyluciferin*	blue	460
Pholas luciferin + O_2^-		
$\rightarrow$ oxyluciferin*	blue–green	490
acridinium ester + HO_2^-		
$\rightarrow$ N methyl acridone*	blue	429
firefly luciferin + O_2		
$\rightarrow$ oxyluciferin*	yellow	565
$H_2O_2 + OCl^- \rightarrow OOCl^- \rightarrow {}^1O_2$		
${}^1O_2 + {}^1O_2 \rightarrow 2\,{}^3O_2$	red	634,703
$R - O - O - H \rightarrow\, >C = O^*$	blue	460

(O_3) chemiluminesces, and oxidises several synthetic organic and biological molecules resulting in chemiluminescence. It enhances light emission from many bioluminescent reactions, for example the firefly. Singlet oxygen produces red light, and is a very reactive oxidant in several chemiluminescent reactions. O_2 is the source of oxidant for luminol chemiluminescence in DMSO. It is easy to demonstrate this visually by leaving a flask of luminol in DMSO + t-butyl alcohol + KOH pellets. Chemiluminescence, after a while, is visible only at the surface exposed to the air. In aqueous media, O_2^-, OCl^-, and H_2O_2 plus a peroxidase are all good oxidants for luminol chemiluminescence. Similarly peroxide is necessary for acridinium salt and oxalate ester chemiluminescence (Fig. 1.5).

Dioxygen is the primary oxidant for most bioluminescence. However, the fact that O_2^- can provoke chemiluminescence from several isolated luciferins, such as that from the bivalve *Pholas* (Fig. 1a), suggests that O_2^- may be an intermediate in the natural reaction. *Pholas* luciferase is a peroxidase, thus H_2O_2 + peroxidase and OCl^- also generate chemiluminescence from the luciferin. However, the quantum yield and kinetics of the latter may not be the same as with the luciferin–luciferase reaction. In a few luminous species, for example earthworms and the unrelated acorn worm *Balanoglossus*, H_2O_2 is the primary oxidant. The organism generates light by releasing the appropriate oxidase and substrate to generate H_2O_2 in the gut of the animal.

How then does one identify the naturally occurring oxidant source in bioluminescence, and how can the molecular mechanism of luciferin oxidation be established? The important experiments are:

(1) The relationship between $(O_2)^n$, $(H_2O_2)^n$ or $(OCl^-)^n$, and the intensity and quantum yield of the reaction.
(2) The labelling of the product from either ${}^{18}O_2$ or $H_2^{18}O_2$, the light emitter being usually an excited carbonyl ($>C = O$) or hydroperoxide (R–O–O–H).

Transition metals such as Fe^{2+}, Fe^{3+} or Cu^{2+}, either isolated or within enzymes such as the haem-containing enzyme myeloperoxidase, are potent catalysts of several reactions involving H_2O_2, possibly through metal-O–OH intermediates.

1.4.3 Singlet oxygen

R. S. Mulliken in 1928 provided an interpretation of the atmospheric oxygen absorption bands, thereby establishing the electronic states of dioxygen, and described it as $1S^2$, $2S^2$, $2P^4$ (singlet or triplet). In the previous year, Mallet had observed a weak red flash when H_2O_2 was added to sodium hypochlorite. By 1960 Seliger had identified one narrow emission band at 633 nm. Stauff and Schmidburg in 1962 suggested that this emission was due to a dimol $2\ ^1\Delta_gO_2 \rightarrow (O_2 \ldots O_2)^* \rightarrow hv$; (see Chapter 9). A year later Khan and Kasha described two emission bands, one at 633.4 nm and a stronger one at 703.2 nm, and identified singlet oxygen as the source of several examples of red chemiluminescence because of its sharp bands at these wavelengths. Individual molecules of singlet oxygen give rise to chemiluminescence in the infrared (λ_{max} 1268 nm). The overall chemiluminescent quantum yield of $^1O_2$ is however very low ($c.\ 10^{-6}$). Singlet oxygen has been reported in biological systems including the myeloperoxidase reaction ($H_2O_2 + Cl^- \rightarrow OCl^- + {}^1O_2$), a number of cells which generate reactive oxygen metabolites including phagocytes, and even human breath. It is also generated in several photochemical reactions, including those involving porphyrins. Whilst red chemiluminescence is observed in all of these cases, the speed of quenching $^1O_2$, together with the lack of sharp lines in the biological spectra, make its existence in living systems still controversial. The infrared emission at 1268 nm is more definitive than the visible emission at 633 or 703 nm, as emission at these latter wavelengths can often be caused by oxidation of endogenous organic substances.

Singlet oxygen can thus give rise to chemiluminescence in any one of four ways:

(1) Unimolecular emission.
(2) Energy pooling with itself (dimol emission).
(3) Energy transfer with a luminescent acceptor, usually in the gas phase.
(4) Chemical reaction with a chemiluminescent substrate, usually organic, e.g. rubrene or vinyl pyrenes.

But what is so special about the reaction of a particular substance with oxygen, or one of its derivatives, which makes it chemiluminescent?

1.5 HOW CAN A CHEMICAL REACTION GENERATE LIGHT?

If a chemical reaction is to generate light, there must first be sufficient energy available from it to produce a photon; and secondly, the formation of the excited product from a reaction intermediate together with its energy loss through light emission must obey certain rules of quantum mechanics.

However, the credibility of any scientific explanation inevitably depends on the depth and breadth of perspectives held by the beholder. The biologist perceives real images of mechanisms, whilst the physicist or chemist often views Nature in terms of abstract ideas. So with chemiluminescence. The physics and chemistry of molecules in electronically-excited states can appear to the uninitiated a jungle of jargon, esoteric concepts, and obscure equations. Ultimately, we shall have to delve into the strange world of quantum mechanics, if we are to understand fully what makes chemiluminescent reactions so special. But first, let us see if it is possible to probe the special nature of chemiluminescent reactions at a qualitative level. Two key questions need to be answered:

(1) Why can some chemical reactions produce light, whereas others apparently cannot?
(2) What physicochemical factors determine the most obvious characteristics of a chemiluminescent reaction, namely intensity, kinetics, colour, and polarisation, if any?

As we have already seen (section 1.2.2) if a reaction is to be chemiluminescent, molecules must not only be generated in electronically-excited states but also these excited molecules must be capable of either emitting photons directly, or of transferring their energy to molecules which can. A chemiluminescent reaction therefore has three essential features:

(1) There must be sufficient energy generated by the reaction for formation of the electronically-excited state. In other words, the reaction must be exothermic.
(2) There must be a pathway by which this energy can be channelled to form an electronically-excited state. If all the chemical energy is lost as heat, e.g. via vibrational and rotational energy, the reaction will not be chemiluminescent.
(3) The excited product must be capable of losing its energy as a photon, or be capable of energy transfer to a fluor. If all of the energy is lost or transferred via non-radiative processes (Fig. 1.7), then the reaction will not be chemiluminescent.

A full explanation of the conditions to satisfy the second of these requires examination of the rules of quantum mechanics. To examine the first we need to known how much chemical energy is required and in what form. Critical to all three is the identification of the excited product and light emitter, together with the final step in chemical pathway responsible for generating it.

Normally, we are used to thinking of the free energy (ΔG) determining whether a reaction occurs or not:

$$\Delta G = \Delta H - T\Delta S. \tag{1.20}$$

This concept is still valid for the chemical reaction itself in chemiluminescence. However, it is the enthalpy, ΔH, which is the actual energy source for generating the electronically-excited state product. Fortunately, in many chemiluminescent reactions the entropy change (ΔS) is small, so ΔH and ΔG are very similar in magnitude. Energy will also be absorbed to initiate the chemical reaction, the activation energy (ΔH_A). Most chemiluminescent reactions produce photons in the range 400 (violet)–750 (red) nm. From equations (1.5) and (1.6) we can see that the energy required from the chemical reaction will be in the range 71–38 kcal mol^{-1} (blue to red). Thus:

$$\text{energy available} = \Delta H_A - \Delta H_R \geqslant \Delta E_{EX} \qquad (1.21)$$

where

$$\Delta H_R \quad = \text{reaction energy}$$
$$\Delta E_{EX} \quad = \text{energy required to generate the excited state.}$$

Note that it is ($-\Delta H_R$) since for an exothermic reaction ΔH is negative. In practice ΔE_{EX} will be greater than the energy required to produce a photon since energy will be lost during the complete process, but as a rule of thumb $\Delta E_{EX} \geqslant 70$ kcal mol^{-1}. Furthermore, if the chemiluminescent reaction is to proceed at an observable rate, the activation energy cannot be too high. For biomolecular reactions $\Delta H_A \leqslant 35$ kcal mol^{-1}, and for unimolecular ones $\Delta H_A \leqslant 25$ kcal mol^{-1}. This means that ΔH_R, the reaction energy, is the major contributor to the excitation energy in a chemilumines-cent reaction. The heats of formation of $H_2O(H_2 + \frac{1}{2}O_2)$ and $CaO(Ca + \frac{1}{2}O_2)$ are 68 and 151 kcal.mol^{-1} respectively, and for the complete oxidation of glucose $(C_6H_{12}O_6 + 6O_2 \rightarrow 6CO_2 \rightarrow 6H_2O)$ 673 kcal.mol^{-1}. All therefore have enough enthalpy for chemiluminescence. In contrast formation of $NH_3(\frac{1}{2}N_2 + \frac{3}{2}H_2)$ only generates 11 kcal.mol^{-1} and ATP hydrolysis less than 10 kcal.mol^{-1}. Not enough!

The two most common mechanisms with sufficient energy for generating an excited state are:

(1) Electron transfer.
(2) Cleavage of linear or cyclic peroxides.

The former is well illustrated by many gas phase chemiluminescent reactions and by radical annihilation reactions in solution (see Chapter 9). Flames are rich sources of chemiluminescence involving CH, OH, C_2, CO, CO_2, HCO, and SO_2, S_2 as the emitting species. Electron transfer can generate such species by recombination of radicals, by bond cleavages, or by transfer of one atom to another (metathesis). For example:

Recombination:

$$2Cl \rightarrow Cl_2^* \rightarrow h\nu \qquad (1.22)$$
$$N + O \rightarrow NO^* \rightarrow h\nu \qquad (1.23)$$
$$O + O + SO_2 \rightarrow O_2 + SO_2^* \rightarrow h\nu \qquad (1.24)$$
$$SO_2 + 2H_2 \rightarrow 2H_2O + S, \; S + S \rightarrow S_2^* \rightarrow h\nu(\lambda_{max} \; 300\text{–}425 \text{ nm}) \qquad (1.25)$$
$$NO + O_3 \rightarrow O_2 + NO_2^* \rightarrow h\nu(\lambda_{max} \; 600\text{–}2800 \text{ nm}) \qquad (1.26)$$

Metathesis:

$$Ba + NO_2 \rightarrow NO + BaO^* \rightarrow h\nu \qquad (1.27)$$

Radical ion annihilation:

$$R_1^-. + R_2^+. \rightarrow R_2 + R_1^* \rightarrow h\nu \qquad (1.28)$$

Radical ion annihilation can also generate excited molecular complexes which become the light emitters. Complexes made up of pairs of like molecules are known as excimers, or exciplexes if the pair is of unlike molecules. Aromatic radical ions can be generated chemically (chemically-initiated electron-exchange chemiluminesence

— CIEEL) or electrochemically (electrochemiluminescence — ECL). Both can be considered as a form of energy transfer (see Chapter 9).

The most common excited product in organic chemiluminescence, where the oxidant is oxygen or one of its derivatives, is the carbonyl of a ketone ($\begin{smallmatrix} R_1 \\ R_2 \end{smallmatrix}$ = O). In solution, excited aliphatic ketones are quenched rapidly, so, in the absence of an energy transfer acceptor, either R_1 or R_2 is usually a fluor, which is inevitably aromatic. However, in the gas phase, particularly at low pressure where the time between collisions is long relative to the lifetime of the excited ketone, aliphatic ketones can readily be observed as the products of a visible chemiluminescent reaction. The ketones, be they aliphatic or aromatic, can be formed by cleavage of either a linear hydroperoxide ($\begin{smallmatrix} R_1 \\ R_2 \end{smallmatrix}$–O–OH) or a cyclic peroxide. The latter appears to be the favoured route, since linear peroxides usually first convert to the cyclic peroxy derivative. This latter intermediate consists typically of a four or six-membered ring involving an –O–O– bridge, though bigger rings can occur. The four-membered ring is known as a dioxetan (or dioxetane in the USA) when of the form

$\overset{\text{O–O}}{\underset{\text{–C–C–}}{\mid\quad\mid}}$, and a dioxetanone when it contains a carbonyl, $\overset{\text{O–O}}{\underset{\text{O}}{\mid\quad\mid \atop -C-C \diagdown}}$. Dioxetandione

($\overset{\text{O–O}}{\underset{\text{O} \diagup \quad \diagdown \text{O}}{\diagup C-C \diagdown}}$) is thought to be the key intermediate in oxalate ester chemilumines-

cence. Reaction of O_2, or a derivative, with a carbon–carbon double bond, can form these cyclic rings directly, or they can form after formation of a hydroperoxide attack on one carbon. Formation of the six-membered endoperoxide is illustrated by the reaction of luminol with O_2 (Fig. 6.4) or of rubrene with singlet oxygen. Cleavage to the excited ketone then occurs directly, or after rearrangement of the ring, to a four-membered dioxetan or dioxetanone.

Although attention has focused on the role of 1,2 dioxetans in organic chemiluminescence and bioluminescence only over the past 20 years, the first report of such a compound was made by Blitz in 1897. During the 1960s White, McCapra, Rauhut, and others predicted that they were the critical intermediates in the chemiluminescent reactions of lophine, acridinium salts, indoles, peroxylates, and several luciferins (Fig. 1.10). Ring strain in the –O–O– bond would make the intermediate prone to spontaneous cleavage in a highly exothermic reaction to form the carbonyl products, one of which would be in an electronically-excited state. Originally it was thought that they would be too unstable to isolate, and the evidence for their existence would have to be based on spectroscopic and [18]O labelling data. However, in 1969, Kopecky and Mumford synthesised a relatively simple dioxetan, trimethyl-1,2-dioxetan which was quite stable ($t_{\frac{1}{2}}$ at 60°C = 20 min). Hundreds of analogous alkyl and aromatic dioxetans have now been synthesised. All are chemiluminescent, as was discussed previously; those not coupled to an aromatic ring or a fluor are readily quenched in solution and are thus best studied either in the gas phase, or in solution in the presence of an energy transfer acceptor. Dioxetanones are inherently less

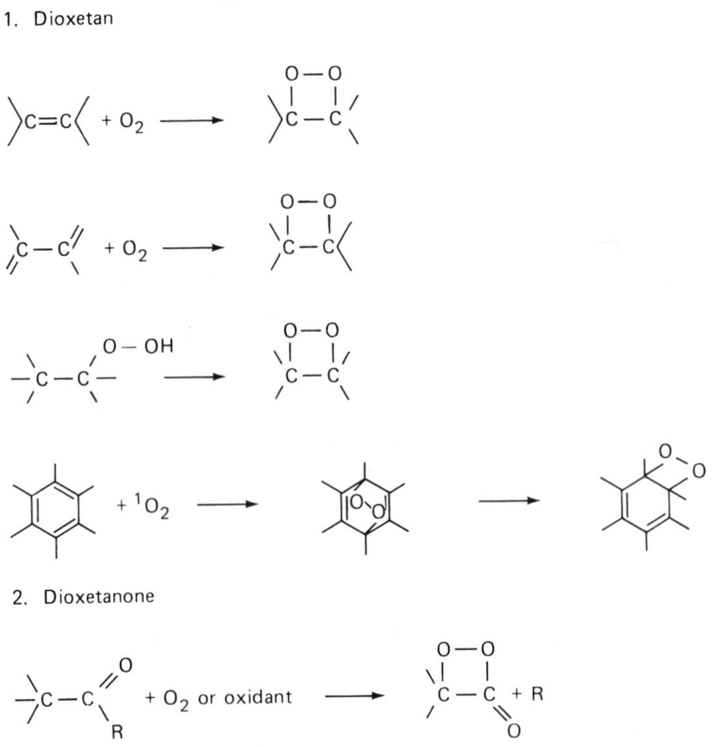

1. Dioxetan

2. Dioxetanone

Fig. 1.10 — Mechanisms for generating dioxetans or dioxetanones.

stable, but of particular interest because of their role as intermediates in several bioluminescent reactions (Fig. 1.10). There are four main ways to form a dioxetan:

(1) Reaction of oxygen ($^3O_2$, $^1O_2$, O_2^-, O_3) with a carbon–carbon double bond or aromatic ring.
(2) Rearrangement of a six-membered ring endoperoxide.
(3) Cyclisation of a hydroperoxide.
(4) Cleavage of esters to form a dioxetandione.

A typical cyclisation reaction is the silver-catalysed reaction:

$$\overset{OOH}{\underset{|}{-C}} - \overset{|}{\underset{|}{C}} - X \xrightarrow{Ag} \overset{O-O}{\underset{|}{-C-C-}} + HX \tag{1.29}$$

where X is a halogen.

For this reaction to be efficient X in Eq. 1.29 must be a good 'leaving group' (i.e. electron acceptor). In the beetle luciferin (benzothiazole) reaction, the leaving group is AMP^{2-}, in the ostracod and coelenterate luciferin (imidazolopyrazine) reaction the electron acceptor is the imidazolopyrazine ring itself (Fig. 1.10), thereby forming the dioxetanone derivatives. Thus any fluorescent ketone with a good leaving group will be susceptible to an oxidation reaction which is chemiluminescent (1.30):

$$\text{Ⓕ} -\overset{\overset{O}{\|}}{\underset{|}{C}}-\overset{\overset{O}{\|}}{C}-X+(O_2)\to \text{Ⓕ} -\overset{\overset{O}{\|}}{\underset{\underset{\underset{O-}{|}}{\overset{|}{O}}}{C}}-\overset{\overset{O}{\|}}{C}-X\to X^- + \text{Ⓕ} -\overset{\overset{O}{\|}}{\underset{\underset{O-O}{|}}{C}}-\overset{\overset{O}{\|}}{C}\to CO_2+(F-C=O)^*\to h\nu \qquad 1.29)$$

Ⓕ =fluor

So much for the chemistry. What about the physics, and the factors determining light intensity, kinetics, colour, and polarisation, if any?

In general, chemical events such as ligand–ligand binding, bond cleavage and formation are much slower than electronic events such as excitation or decay to new orbital or intramolecular vibrations (Table 1.4). Thus the kinetics of a chemilumines-cent reaction are determined primarily by the activation energy, the concentration of the reagents and any catalyst involved, and temperature. The brightness of the light emission at any one time is dependent on the number of excited state emitters at that point. This in turn depends on two factors:

(1) The chemical kinetics and reagent concentrations.
(2) The overall quantum yield, ϕ_{CL} (1.9).

As we have already seen $\phi_{CL} = \phi_C \cdot \phi_{EX} \cdot \phi_F$. The chemical yield ($\phi_C$), will be reduced below one if there are dark reactions competing with chemiluminescent ones. ϕ_{EX} and ϕ_F are themselves reduced to below one, the maximum possible, by two factors (Fig. 1.7):

(1) Pathways tapping energy away from the final step in the reaction, i.e. the intermediate to excited product formation.
(2) Non-radiative pathways taking energy from the excited state product.

To follow this further we have to understand the nature of the excited electron and the way in which quantum mechanics describes its energy levels.

Excitation of an electron in an atom raises it into a higher energy orbital, be it s, p, d, or f. Likewise in a molecule, excitation of an electron raises the electron to another orbital, in this case from a bonding to an anti-bonding orbital (designated $n \to \pi^*$ (common) $\sigma \to \sigma^*$ (rare), or $\pi \to \pi^*$ (anisotropic). This effectively breaks half of the chemical bond between two atoms. The result is structural re-arrangement and a molecule with different physicochemical properties from the ground state, in terms of the pKs of ionisable groups and chemical reactivity. The simplest molecule consists of two atoms (A–B). The energy levels within such a molecule are determined by the distance separating the two atoms. This can be represented graphically (Fig. 1.4b), the energy well or minimum being the most stable form of the molecule. The electronically excited state will be represented by another 'energy well' of higher energy. If there is no minimum then the molecule is inherently unstable and liable to dissociate. Since the time of nuclear vibrations is picoseconds and the excitation event takes femtoseconds (Table 1.4), 1000 times faster, the transition from ground to excited state is represented by a vertical line (Fig. 1.4b). This is the Franck–Condon principle, i.e. the positions of the nuclei are the same before and immediately after photoexcitation, and enables the events to be simplified as the Jablonski diagram (Fig. 1.4a).

However, the Franck–Condon principle is not really applicable, since the excited substance in chemiluminescence is structurally different from the orginal substrate. Furthermore, the reverse event, i.e. excited to ground state, is much slower, typically about 1–10 ns. Thus, the excited molecule will normally lose energy and fall to the bottom of its energy well before photon emission occurs. But the electron may not return to the lowest vibrational level in the ground state. Since the photon energy is equal to the energy jump the electron takes between excited and ground state, these energy losses affect the colour of the light actually emitted, it being 'redder' than might otherwise have been expected. Three complicating factors must now be considered when applying this diagramatic representation of molecular energy to chemiluminescence:

(1) The route from ground state to excited state will be different when provoked by a chemical reaction compared with absorption of a photon (photoluminescence).
(2) The spin state.
(3) The chemiluminescent molecule may contain considerably more than two atoms.

Each additional atom adds another dimension to the graphical representation of a molecules energy. A molecule with 100 atoms requires 100 dimensions! This is of course impossible to represent in graphic form. Nevertheless, it has meaning, and is known as a 'hypersurface'. Each spin state, singlet (S) or triplet (T), has a separate surface. Since the triplet state is usually of lower energy than the singlet, it might be thought that triplet excited-state emission would be favoured in chemiluminescence. However, the slowness of intersystem crossing (Table 1.4), i.e. $T \rightarrow S$ or $S \rightarrow T$, redresses the balance. Both triplet and singlet emissions are found. Structural modification has predictable effects on the ratio between the two from any particular chemiluminescent molecule. For example, alkyl, phenyl, and alkoxy groups tend to have a high proportion of triplet state product, whereas substitution with aromatic, electron donating groups, as in several luciferins, tends to favour singlet excited state formation. If the decay to ground state also involves intersystem crossing then this will also be slow, increasing the chance of quenching by solvent and other molecules. O_2 is a particularly efficient triplet state quencher, since its ground state is itself triplet. In fact, it is possible to estimate the average distance an oxygen molecule would move if the decay to ground state took 10 ns:

Einstein's equation states:
The average distance moved $= (\Delta_x^2)^{\frac{1}{2}} = 2D\tau$ (1.31)
where $D =$ diffusion coefficient (for O_2 $D = 2.5 \times 10^{-5}$ cm^{2}s^{-1})
$\tau =$ time

It turns out to be as much as 7 nm, equivalent to the thickness of a cell membrane. Thus if the decay of the excited state from $T \rightarrow S$ is much slower than normal $S \rightarrow S$ of

10 ns then this considerably increases the chance of collision with $^3O_2$. The question arises whether the active centre of a luciferase might protect a long-lived triplet state from quenching. An increased lifetime of an excited state will also increase the chance of energy transfer via collisions or electron transfer.

There is one final point about the potential energy hypersurfaces describing the energy levels of a multi-atomic molecule. Where a ground state and electronically excited state surface cross, the geometry of a molecule in either state is identical. It can thus cross over. This provides a radiationless route from the excited state to ground state. This is what happens in the many chemical reactions which are sufficiently exothermic to produce energy for photon emission, yet are non-chemiluminescent. In summary, for a reaction to be chemiluminescent at all, there must be enough energy available, and at least some of this energy must be available to form an electronically state without all the energy required to form it, or to take it back to ground state, being lost by quenching, internal vibrations, subsidiary chemical reactions, or energy transfer to a non-fluorescent molecule.

An ideal chemiluminescent reaction has no subsidiary dark reactions ($\phi_C = 1$), obeys the spin-coupling rules, and channels sufficient energy into an excited state ($\phi_{EX} = 1$) which can only release it as a photon ($\phi_F = 1$). Overall chemiluminescent quantum yields can vary from 10^{-8} in the Grignard reactions, still visible, to virtually 1 in the firefly luciferin–luciferase reaction. Reactions with $\phi_{CL} < 10^{-12}$ are known.

Since the speed of rotation of molecules is fast relative to excited state decay (Table 1.4), and since molecules are randomly distributed in free solution or in a gas, it would be surprising if the light arising from the chemiluminescence of such molecules was polarised. However, in a regular solid or in a biological membrane things may be very different. Nature has much to teach us about the chemistry of the excited state.

1.6   THE REST OF THE BOOK
Next time you are in the country or on the beach at night I hope that you will now be able to distinguish any chemiluminescence you see, from the diverse sources of light observed by our camper who introduced this chapter. Perhaps I have enthused you to ask a few more questions:

(1) How do we study the phenomenon, quantitatively?
(2) What are the chemistry, function, and evolutionary significance of the chemiluminescence found in living organisms?
(3) Does chemiluminescence have any use, other than as an aesthetically pleasing and fascinating natural phenomenon?

Scientists observe, and then measure. We begin the voyage of discovery about chemical light by first identifying the parameters of a chemiluminescent reaction we ought to measure, and how this can be achieved.

1.7   RECOMMENDED READING

Historical
  (1) Aristotle (1908, 1923 editions) *De amina, De sensu, Historia animalium, meteorologica.* Oxford.

(2a) Boyle, R. (1668) *Phil. Trans. Roy. Soc.* **2** 211–214. Experiments concerning the relation between light and air in shining wood and fish.

(2b) Boyle, R. (1668) *Phil. Trans. Roy. Soc.* **2** 215–216. Comparison between burning coal and shining wood.

(3) Dubois, R. (1884, 1885) *Comptes Rend. Seanc. Soc. Biol.* (ser 8) **1** 661–664; **2** 559–562. Note sure la physiologie des Pyrophores.

(4) Dubois, R. (1887) *Comptes Rend. Seanc. Soc. Biol.* (Ser 8) **4** 564–565. Fonction photogenique chez le *Pholas dactylus*.

(5) Harvey, E. N. (1920) *The nature of animal light.* Philadelphia.

(6) Harvey, E. N. (1940) *Living Light.* Princeton University Press, Princeton.

(7) Harvey, E. N. (1957) *A history of luminescence.* From the earliest times until 1900. American Philosophical Society, Philadelphia.

(8) Jablonski, A. (1935) *Z. Phys.* **94** 38–46. Über den Mechanismus des Photolumineszenz von Farbstoff-phosphoren.

(9) Macaire, J. (1821) Annales de Chimie et de Physique **17**, 251–267. Mémoire sur la Phosphorescence des lampyres.

(10) Pliny, C. *Historia naturalis*, translated (1855) by Bostock, J. and Riley, H. T. Bohn, London.

(11) Radziszewski, Br. (1877) *Beriche d Chemischen Gesellschaft* **10** 70–75. Untersuchungen über Hydrobenzamid, Amarin und Lophin (translation available on request).

(12) Wiedemann, E. (1888) *Ann. d. Physik u. Chemie* **34** 446–463. Ueber fluorescenz und phosphorescenz. I. Abhandlung. (translation available on request).

(13) Weidemann, E. & Schmidt, G. L. (1895) *Ann. d. Physik u Chemie* **54** 604–625. Ueber luminescencz.

Light
(14) Brill, T. B. (1981) *Light: its interaction with art and antiquities.* Plenum Press, New York and London.

Chemical reactions producing light
(15) Adam, W. & Cilento, G. (eds) (1982) *Chemical and biological generation of excited states.* Academic Press, New York.

(16) Cormier, M. J., Hercules, D. M. & Lee, J. (eds) (1973) *Chemiluminescence and bioluminescence.* Plenum Press, New York and London.

(17) De Luca, M. A. & McElroy, W. D. (eds) (1981) *Bioluminescence and chemiluminescence.* Basic chemistry and analytical applications. Academic Press, New York.

(18) McCapra, F. (1973) *Prog. Org. Chem.* **8** 231–277. Chemiluminescence of organic compounds.

(19) McCapra, F. (1982) *Proc. Roy. Soc. B.* **215** 247–272. The chemistry of bioluminescence.

Bioluminescence
(20) Beebe, W. M. (1933) *Zoologica N.Y.* **16** 5–11, 15–91. Deep sea fishes of the Bermuda oceanographic expeditions.

(21) Harvey, E. N. (1952) *Bioluminescence.* Academic Press, New York.

(22) Herring, P. J. (ed) (1978) Bioluminescence in action. Academic Press, London and New York.

(23) Herring, P. J. (1984) *New Scientist* 45–48. Lights in the night sea.

Ultraweak chemiluminescence

(24) Barenboim, G. M., Domanskii, A. N. & Turoverov, K. K. (eds) (1969) *Luminescence of biopolymers and cells*. Plenum Press, New York and London. pp. 114–142. Trans. Chen. R.F.

(25) Boveris, A., Cadenas, E. & Chance, B. (1981) *Fed. Proc.* **40** 195–198. Ultraweak chemiluminescence: a sensitive assay for oxidative radical reactions.

(26) Gurwitch, A. & Gurwitch, L. D. (1959) *Die mitogenetische Strahlung*. Gustav Fischer Verlag, Jena.

(27) Zhuravlev, A. I. (ed). (1972) *Trans. Moscow Society of naturalists* **39** Ultraweak luminescence in biology (in Russian).

Energy transfer

(28) Ward, W. W. (1979) *Photochem Photobiol. Rev.* **4** 1–58. Energy transfer processes in bioluminescence.

Oxygen

(29) Priestley, J. (1775) *Phil. Trans. Roy. Soc.* **65** 384–394. An account of further discoveries in air.

(30) Gilbert, D. L. (ed) (1983) *Oxygen and living processes: an interdisciplinary approach*.

(31) Gorman, A. A. & Rodgers, M. A. J. (1981) *Q. Rev. Chem. Soc.* **10** 205–231. Singlet molecular oxygen.

(32) Frimer, A. A. (ed) (1985) *Singlet O_2*. CRC Press, Florida, including article by Kahn A. U.

Biomedical applications

(33) Campbell, A. K., Hallett, M. B. & Weeks, I. (1985) *Methods Biochem. Anal.* **31** 317–416. Chemiluminescence as an analytical tool in cell biology and medicine.

(34) DeLuca, M. A. (ed) (1978) *Methods in Enzymology* **57**. Bioluminescence and chemiluminescence.

(35) DeLuca, M. A. & McElroy, W. D. (1987) *Methods in Enzymology* **33**. Bioluminescence and chemiluminescence part B.

Triboluminescence and Will-o'-the-wisp

(36) Walton, A. J. (1977) *Adv. Phys.* **26** 887–948. Triboluminescence.

(37) Mills, A. A. (1980) *Chemistry in Britain* **16** 69–72. Will-o'-the-wisp.

2

Detection and quantification of chemiluminescence

2.1 THE NATURE OF LIGHT

2.1.1 The problem

'What is light dad?' Such a simple question asked of me some years ago by one of my sons. Yet it lays bare what has been, and still is, one of Nature's great puzzles.

Ever since the 16th and 17th centuries, when the early physicists began to consider this as a serious scientific problem, it has provoked much argument and controversy. But how fundamental a question really is it? Our perception of the world around us is critically dependent on our ability to receive, process, and interpret light, its brightness, its colour, and the direction from which it came. We may be but a speck in the universe, but life on earth depends, ultimately, on the energy of light from the sun. Just consider the number of physical or chemical processes involving absorption or emission of visible light and other parts of the electromagnetic spectrum (Table 1.3), or the number of the laboratory techniques using it to determine quantities and structures — nmr, esr, luminescence, X-ray diffraction, Mössbauer spectroscopy, and radioisotopes, for example. Clearly light does play a fundamental role in the physics of the universe. One reason why this role has been so difficult to elucidate is that it travels so fast–so fast that it is difficult to relate to any normal experience.

Galileo, in the 17th century, attempted to measure the speed of light and concluded: 'if not instantaneous it is extremely rapid'. However, from several astronomical observations over the next century or so, it was possible to deduce that light must travel at about $2\text{–}3\times10^8$ metres per second (ms^{-1}). The first non-astronomical value was obtained by the French physicist Armand Hippolyte Louis Fizeau in 1849. By using mirrors 8630 m apart, and a spinning, toothed wheel he was able to measure the speed necessary to stop an image which would otherwise travel to and fro between the mirrors. He obtained a value of 3.13×10^8 ms^{-1}. The accepted value is now 2.997929×10^8 ms^{-1}. The accepted value is now 2.997929×10^8 ms^{-1}, to six decimal figures. This means that it takes about 33 picosecond $(3.3\times10^{-11}$ s$)$ for light to travel one centimetre, or, since the radius of the earth is just less than 4000 miles, light can travel around the world seven and a half times in one second.

Relating this to the electronic events in generating the excited state of atoms, and in their luminescence emission, we see that light travels about 30 cm in the time it takes for an electron in the singlet excited state to decay to ground state (Table 1.4), namely 1 nanosecond (10^{-9} s). However, the excitation process or energy transitions can take only a femtosecond (10^{-15} s) when light would travel only 300 nanometres, less than the wavelength of visible light (Fig. 1.2)! Here we reach the dilemma: you cannot go on reducing the distance light moves, nor can the energy of a light beam be subdivided, for ever.

Light can be bright or dim, and it can be coloured. This colour can change if the emitting body travels very fast, close to the speed of light (the Doppler effect). Light can be refracted or reflected, it can be polarised, and it can be bent by electric, magnetic, or gravitational fields. Light also has energy since it can generate heat, or can itself be generated by thermal or chemical energy sources. The problem is that in order to relate light to a particular physical or chemical phenomenon sometimes we consider it as an electromagnetic wave, but at other times as a stream of particles of energy photons. Which one should be used to understand the nature of chemiluminescence, to study and measure this phenomenon, or to use it as an analytical tool?

2.1.2 Wave or particle?

The Hay Wain, painted by John Constable in 1821, will be familiar to the many people who have visited the National Gallery in London. From a distance one can see the reflections of many objects on the surface of the water. However, closer examination will reveal that, in common with many artists, Constable has failed to observe a phenomenon well known to many children — refraction. Place a stick into clear water and it appears to bend at the air–water interface. As we saw in Chapter 1 (section 1.2.2) Christian Huygens in 1678, initiated the wave theory of light which explains quite adequately phenomena such as this. For a wave the distance between each peak or trough, the wavelength of light, is inversely proportional to the frequency which the wave takes to return to the same point each second (see equation (1.2)):

$$\text{wavelength } (\lambda) = \text{velocity } (c)/\text{frequency}(\nu)$$

The wavelength spectrum of visible light, embracing all chemiluminescence spectra, goes from violet ($ca\ \lambda = 400$ nm) to red ($ca\ \lambda = 750$ nm) (Fig. 1.2). The electromagnetic wave theory of light developed by Clerk Maxwell in 1872 adequately explains reflection, refraction, diffraction, polarisation, and the Doppler effect. However, this theory does not explain the relationship between colour (i.e. the emission spectrum) and the temperature of an incandescent body. Nor can it explain the relationship between the colour of chemiluminescence to the chemical energy generated by the reaction. For this we need a form of the 'corpuscular' treatment of light, first introduced by Newton, but correctly developed by Einstein using Planck's quantum principle to explain the photoelectric effect, discovered by Wilhelm Hallwachs in 1890 following earlier observations of Heinrich Hertz.

The energy levels of electrons in atoms or molecules are discrete, and not continuous. When an electron in an excited atom or molecule drops down to a lower vibrational or rotational state, or drops back to ground state, the energy has to be released as a packet, a quantum. If this energy is released as light, rather than being transferred to another molecule, or released as heat, then the energy difference determines the colour of the photon, the quantum of light, according to Einstein's equation (1.3):

$$\text{Energy difference } (E) = E_2 - E_1 = h\nu \tag{2.1}$$

where h = Planck's constant and ν = frequency of the light.

However, since the energy levels in atoms and molecules are discrete, and not a continuum, $h\nu$ can only have a series of finite values for any particular excited state. Thus, strictly, the description of spectra should take account of this. This is why spectra are often presented as luminous intensity plotted against wave number $(\bar{\nu})$ rather than wavelength (λ); however, since energy levels within excited states are large in number, and the wave number of photons of blue light is many tens of thousands greater than that of photons of red light. So, for practical purposes:

$$\bar{\nu} = 1/\lambda \tag{2.2}$$

and

$$E = hc/\lambda = hc\bar{\nu} \tag{2.3}$$

Presentation of spectra as an *apparent* continuum of intensity versus wavelength will only be misleading if we forget that in reality the abscissa can only have discrete values.

There are some situations, however, where we must use the appropriate quantum parameter, for example, when relating chemical reaction energies to chemiluminescence spectra, when trying to detect very low light intensities, or when trying to describe the workings of a light detector dependent on the photoelectric effect, such as a photomultiplier tube. Otherwise our conceptual understanding may be distorted, or even wrong, if we continue always to try to use the parameters of wave theory. In contrast, properties such as the depth of focus of an image, or plane and circular polarisation of light, are easy to explain with wave theory, but rather difficult when considering individual photons. The modern theory of light, therefore, tries to marry the two historical concepts of the nature of light. A photon is considered as a discrete wave packet of energy.

What then do we need to measure if we are to study chemiluminescence, to understand how chemical energy can be converted to light and to use it as an analytical tool?

2.1.3   What do we measure?

A bewildering array of parameters have been used to describe, quantitatively, the emission of light from a particular source, no matter what its origin (Table 2.1, Fig. 2.1). Four of these stand out, and must be measured if the chemistry, physics, and biology of chemiluminescence are to be investigated comprehensively, and if its unique potential as an analytical tool in chemistry, biology, and medicine is to be exploited:

(1) Light flux (Φ) i.e. the number of photons emitted per unit time

Table 2.1 — Emission and detection of light from chemiluminescence

Term	Symbol	Derivation
Light yield (total or partial)	Q	$\int \Phi \cdot dt$
Radiant or luminous flux	Φ	dQ/dt
Radiant or luminous intensity	I	$d\Phi/d\Omega$
		(Ω = solid angle)
Radiant or luminous exitance	M	$d\Phi/dA$
		($A = Q/dV$)
		(V = volume)
Radiant density	W	dQ/dV
Radiance or luminance	L	$dM/d\Omega$
Irradiance or illuminance	E	dI/dA
Radiant or luminous exposure	H	$\int E dt$
Spectral intensity	$I(\lambda)$	$dI/d\lambda$
Spectral irradiance or illuminance	$E(\lambda)$	$dE/d\lambda$

Note: (1) Radiant = light emitted designated by subscript R.
      (2) Luminous = light detected designated by subscript L.
      (3) Light measured as energy (e) or as photons (p).
      (4) What is measured by a chemiluminometer is luminous intensity.
See Appendix I for definitions.

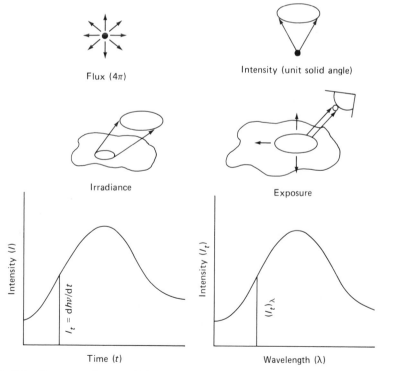

Fig. 2.1 — Parameters for quantifying light emission: radiant = emission by source: luminant = detected by observer.

(2) Light integral $= \int_0^\infty \Phi_t dt$, i.e. total number of photons emitted during the complete reaction

(3) Colour (Φ_λ) i.e. the spectrum of light intensity versus wave number (or wavelength).

(4) Polarisation, plane or circular, if any.

Actual measurement and expression of these parameters can inevitably only be carried out to a first approximation.

Consider a simple chemiluminescent reaction:

$$LH_2 + O_2 \xrightarrow{k_1} H_2O + L = O^* \xrightarrow{k_2} L = O + h\nu \qquad (2.4)$$

e.g. $LH_2 = $ a luciferin.

The rate of formation of $L = O^*$, the excited state emitter,

$$= -d[LH_2]_t/dt = k_1 [[LH_2]_t] [O_2] \qquad (2.5)$$

$[LH_2]_t = $ concentration of LH_2 at time t where $k_1 = $ second order rate constant in $M^{-1}s^{-1}$.

Note that there is a negative sign since the concentration of the chemiluminescent substrate, LH_2, is *decreasing* with time.

If $[LH_2] \gg [O_2]$, or if $[O_2]$ remains constant as a result of continuous gassing, then the reaction is 'first order' with respect to $[LH_2]$, and the loss of $[LH_2]$ follows a pure exponential:

$$\int_{[LH_2]_0}^{[LH_2]_t} d[LH_2]_t/[LH_2]_t = -\int_0^t k_1[O_2] \, dt \qquad (2.6)$$

where $[LH_2]_0 = $ concentration of LH_2 at the start of the reaction, i.e. time 0, therefore

$$[LH_2]_t = [LH_2]_0 \exp(-k_1[O_2]t) \qquad (2.7)$$

and

$$d[LH_2]_t/dt = -k_1[O_2] [LH_2]_0 \exp(-k_1[O_2]t) \qquad (2.8)$$

But how does this relate to light intensity? There are two important points to be noted. Firstly, we measure total light emission from a reaction cuvette, and not a light emission per unit volume. Secondly, since the luminescence decay of the excited state $L = O^*$ (2.4) occurs within 1–10 ns, i.e. $k_2 \gg k$, the second stage of the reaction has a negligible effect on the measured kinetics. Thus, the rate of consumption of the substrate is directly proportional to light flux:

$$(dh\nu/dt)_t = -\phi_{CL} (d(LH_2)/dt) \qquad (2.9)$$

$$= \phi_{CL}(LH_2)_0 k_1[O_2] \exp(-k_1[O_2]t) \qquad (2.10)$$

Note $\Phi_0 = \Phi_{max} = \phi_{CL}(LH_2)_0 k_1[O_2] = $ peak height

$\phi_{CL} = $ overall quantum yield, i.e. fraction of molecules LH_2 reacting generating a photon

and $(LH_2)_0 = $ number of molecules (*not* concentration) of LH_2 at time 0

$$(2.11)$$

i.e. $(LH_2)_0 = LV[LH_2]_0$

where L = the Avogadro constant = 6.023×10^{23}

$[LH_2]_0$ = molarity (or molality) of LH_2 at time 0

V = volume of reacting solution

$$\text{light flux} = \Phi_t = dhv/dt \tag{2.12}$$

$$\text{but in reality measures } \Phi_t = \Delta hv/\Delta t \tag{2.13}$$

where Δ_t is finite. For Φ_t, the flux, to be used meaningfully Φ_t must, to a first approximation, be constant over the time interval required for measurement, Δ_t; i.e. Δ_t is much less than the time constant of the decay in the chemiluminescence caused by the consumption of the primary substrate. In practice Δ_t should be 1–10 ms for half times for chemiluminescence exponential decays of hundreds of milliseconds to seconds. For slower decays, lasting minutes, Δ_t can be as much as 1–10 seconds. Thus for a typical chemiluminescence reaction where the *oxidant* concentration remains constant throughout, the reaction is apparently first order: that is,

$$k_{app} = k_1[O_2] \text{ s}^{-1} \tag{2.14}$$

The measured flux

$$\Phi = \Delta hv/\Delta t \tag{2.15}$$

reaches a maximum rapidly and then decays exponentially. A semilog plot ($\log_e \Phi$ vs t) will thus be linear, the slope being $-k_{app}$ (Fig. 2.2). As with radioactivity a convenient way of expressing the uniformity of the decay rate constant (k_{app}) with time is through $t_{\frac{1}{2}}$:

$$t_{\frac{1}{2}} = \text{time to half maximum (or time to } \Phi_{t/2} \text{ from } \Phi_t) \tag{2.16}$$

From (2.10)

$$\log_e (\Delta hv/dt)_t = \log_e(\Phi)_t = \log_e(\phi_{CL}(LH_2)_0 k_1[O_2]) - k_{app}t \tag{2.17}$$

$$\log_e(\Phi)_{t_{\frac{1}{2}}} = \log_e(\phi_{CL} \cdot (LH_2)_0 k_1(O_2)) - k_{app}(t + t_{\frac{1}{2}}) \tag{2.18}$$

Subtracting (2.17) from (2.18)

$$\log_e(\Phi_t/\Phi_{t_{\frac{1}{2}}}) = k_{app}t_{\frac{1}{2}} \tag{2.19}$$

$$\ln 2 = k_{app} \, t_{\frac{1}{2}} \tag{2.20}$$

$$t_{\frac{1}{2}} = 0.693/k_{app} \tag{2.21}$$

Another parameter sometimes used is the time constant τ = time to $1/e$ of maximum intensity.

Remember that from (2.14) the $t_{\frac{1}{2}}$ will depend on oxidant concentration, unless, like jelly-fish photoproteins, the chemiluminescent substrate has already been precharged with oxygen. The O_2 concentration in air saturated H_2O is approx 220 μM, and decreases with temperature (Graham's Law).

Two practical problems arise when trying to measure the total photon emission from the complete reaction ($\int_0^\infty \Phi_t dt$) in order to estimate the chemiluminescence quantum yield (ϕ_{CL}).

$$\phi_{CL} = \text{total photon emission}/(LH_2). \tag{2.22}$$

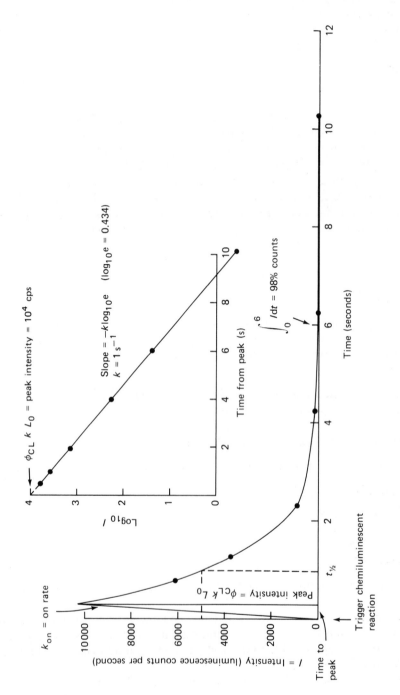

Fig. 2.2 — A first order, 'exponential', chemiluminescent reaction. $I = \phi_{CL} k \, L_0 \exp(-kt)$; ϕ_{CL} = overall chemiluminescence quantum yield; k = decay rate constant; L_0 = total chemiluminescent substrate (in molecules). As an illustration $L_0 = 10^6 = 1.66$ attomol (1.66×10^{-18} mol); $k = 1 s^{-1}$; $t_{\frac{1}{2}} = 0.69$ s after peak; time to peak = 0.25 s.

The first relates to the fact that one actually measures light intensity (I), a flux per solid angle, and not the total flux (Φ), which can only be measured by collecting the light emitted in all directions. Thus, to estimate real photon emission the measured intensity or total integral must be multiplied by a factor to correct for geometry. The measured light emission may be as low as 1% of true emission, even in a good chemiluminometer! The second problem relates to the difficulties of measuring the photon emission for long enough to collect all light from the reaction. Since the flux, or intensity, is in theory a continuous exponential, total photon emission can only be measured approximately, as the integration time is finite. For a pure exponential the flux, or measured intensity (I_n), after n half times will be $(\frac{1}{2})^n$ times the initial peak in light intensity. So, for example, the Ca^{2+}-activated photoprotein obelin from the hydroid *Obelia geniculata* has a decay rate constant of approximately $4\,s^{-1}$ in a saturating Ca^{2+} solution, equivalent to a $t_{\frac{1}{2}}$ of 0.18 s. Thus after six half times, i.e. 1.1 seconds, the intensity will be $\frac{1}{64}$ of that at time 0. What about total light emission? From equations (2.10) and (2.14):

$$\int_0^T dh\nu = \int_0^T \Phi_0 \cdot \exp(-k_{app}t) \cdot dt \qquad (2.23)$$

where Φ_0 = intensity at time 0 = $k_{app}\phi_{CL} \cdot [LH_2]_0$

$$h\nu_T = \Phi_0/k_{app}(1 - \exp(-k_{app}T)) \qquad (2.24)$$

$$\text{but}\,\Phi_T = (dh\nu/dt)_T = \Phi_0/k_{app}(\exp(-k_{app}T)) \qquad (2.25)$$

$$h\nu_T = (\Phi_0 - \Phi_T)/k_{app} \qquad (2.26)$$

$$\text{if } T = n\, t_{\frac{1}{2}} \text{ then } \Phi_T = (\tfrac{1}{2})^n\, \Phi_0 \qquad (2.27)$$

$$\text{therefore } h\nu_T = \Phi_0(1 - (\tfrac{1}{2})^n)/k_{app} \qquad (2.28)$$

but if $T = \infty$, then

$$h\nu_T = \Phi_0/k_{app} = \phi_{CL} \cdot (LH_2)_0 = \text{true total,} \qquad (2.29)$$

$$\text{but } h\nu_T = \text{measured total light yield} = (1 - (\tfrac{1}{2})^n)\, \text{true total,} \qquad (2.30)$$

i.e. after five half times

$$h\nu_T = (63/64) \times \text{true total} \qquad (2.31)$$

or 96.9% of true total, or after six half times 98.4%.

The result of all these calculations is that to measure $>98\%$ of the photons emitted in an exponential chemiluminescent reaction the integration time must be at least six half times. Most people would probably accept this 2% error.

Very bright, visible emissions produce thousands of millions of photons per second and are usually expressed as energy emitted per unit time. However, a single cell from a luminous organism may emit only 10^3–10^6 photons per second, and in dim or ultraweak chemiluminescence the emission may be less than 1 photon per second, and its chemiluminescence expressed in photons. Thus intensity (I), a fraction of the total real flux, and the total light yield can be related to the kinetics of a chemiluminescent reaction and its quantum yield (see section 2.7). The spectrum can tell us about the energetics of the reaction and provide clues about the electronic mechanisms involved. The existence of polarisation is, however, poorly documented in chemi- and bio-luminescence. If it occurs, then this is evidence for structural organisation of the molecules responsible for the chemiluminescent reaction, or the medium through which the light passed before being detected. Some of the organic molecules producing the excited states as the light emitters in chemiluminescence are initially anisotropic, i.e. they are asymmetrical with no plane of symmetry running through them. Others with planar rings are not anisotropic. If the chemiluminescence product is also anisotropic, it is potentially a source of plane or circularly polarised light. Excitation of these molecules by polarised light will result in polarised fluorescence emission with the plane of polarisation rotated, the degree of rotation being related to the rotation rate of the molecules. However, in free solution chemiluminescent molecules will have random orientation and thus would not be expected to produce polarised light. However, membrane structures such as the photosome in scale worms, coupled with an anisotropic precursor of the excited product such as firefly oxyluciferin should give a polarised emission. This can be detected by using the effect of a plane or circularly polarised filter, which will change the measured light intensity as it is rotated, if the incident light is itself originally polarised. The only evidence so far is in the firefly. Some thought needs to be given to the concept of polaristation when measuring individual photons.

In addition to the four key parameters, light yield, light flux, spectrum, and polarisation, several other factors (Table 2.1, Fig. 2.1) determine whether the light is visible to an animate observer, or detectable by an artificial light sensor. The former is essential if we are to understand the control and function of bioluminescence, the latter essential if we are to study chemiluminescence scientifically at all. If we cannot detect and quantify it with our experimental equipment, we cannot study it. Since the terminology embodied in Table 2.1 can be confusing, it might be useful to define the terms and their significance (see Appendix I).

Light yield (Q) is a measure of light emitted in the complete reaction (total) or partial reaction time (*t*). *Flux (Φ)* is a measure of the rate of light emission per unit time. In the old terminology, this was called either *radiant* flux if related to absolute energy units emitted or *luminous* flux if measured according to its ability to be visualised by a human observer. In the latter case a point light source, the candle, was used as a standard. Flux is often confused with intensity, which again could be radiant or luminous in the old terminology. Every light source can emit light in any direction. If the emission is uniform in all directions then it has *4π geometry*. Furthermore, the source, i.e. the chemiluminescence, can be, to all intents and purposes, a point, e.g. a photophore of a euphausiid shrimp; a line, e.g. a thin reagent tube or luminous organ; a surface, e.g. a reaction at a solid–liquid interface or a large light organ; or three-dimensional, e.g. a test-tube or flask. Whether any one of these sources can be seen by the observer or detector depends on how much light is actually received. This, in turn, depends on two geometrical factors, in addition to the innate brightness and colour of the light source. Firstly, how far away the detector is from the source, and secondly the distribution of the chemiluminescence within the source. The former can be described in one of two ways, either as flux emitted per unit solid angle, this being the strict definition of *intensity* (Table 2.2, Fig. 2.1), or as flux or light yield received per unit area of detector surface, the *exposure*. Qualitatively,

Table 2.2 — Units of light measurement

Parameter	SI unit name	Symbol	Photon unit
A. Radiant (emitted)			
Quantity (Q_R)	joule	J	Einstein
Flux (Φ_R)	watt = joule per second	W	Einstein s^{-1}
Intensity (I_R)	watt per steradian	W sr^{-1}	Einstein sr^{-1}s^{-1}
Exitance (R)	watt per square metre	Wm^{-2}	Einstein ,$^{-1}$s^{-1}
Density (W_R)	joule per cubic metre	J m^{-3}	Einstein m^{-3}s^{-1}
Radiance (L_R)	watt per steradian per sq metre	W sr^{-1}m^{-2}	Einstein sr^{-1}m^{-2}s^{-1}
Irradiance (E_R)	watt per sq metre	W m^{-2}	Einstein m^{-2}s^{-1}
Exposure (H_R)	joule per sq metre	J m^{-2}	Einstein m^{-2}
Spectral intensity ($I_R(\lambda)$)	intensity per unit λ	W sr^{-1} nm^{-1}	Einstein sr^{-1}-vs^{-1}
Spectral irradiance ($E_R(\lambda)$)	irradiance per unit λ	W m^{-2} nm^{-1}	Einstein m^{-2}-vs^{-1}
B. Luminous (received)			
Quantity (Q_L)	candle	cand	Einstein
Flux (Φ_L)	lumen	lm	Einstein s^{-1}
Intensity (I_L)	candela	cd	Einstein sr^{-1}s^{-1}
Exitance (M_L)	lux	lx = lm m^{-2},cd sr m^{-2}	Einstein m^{-1}s^{-1}
Luminance (L_L)	candela per sq metre	cd m^{-2}	Einstein sr^{-1}m^{-2}s^{-1}
Illuminance (E_L)	lux	lx	Einstein m^{-2}s^{-1}
Exposure (H_L)	lux second	lx s = cd sr m^{-2}s	Einstein m^{-2}
Spectral intensity ($I_L(\lambda)$)	candela per unit λ	cd nm^{-1}	Einstein sr^{-1}v̄ s^{-1}
Spectral illuminance ($E_L(\lambda)$)	lux per unit λ	lx nm^{-1}	Einstein m^{-2}v̄ s^{-1}

Note: (1) Old units include stilb (sb) = cd cm^{-2}; apostilb = $(1/10^4\pi)$ sb; lambert (L) = $(10^4/\pi)$ cd m^{-2}; foot lambert (ft$_L$) = π^{-1} cd ft^{-2}; phot = lm cm^{-2}, 1 candle emits 4π lumens.
(2) Constants: speed of light = $c = 2.99793 \times 10^8$ m s^{-1}, 1 joule = 10^7 erg = 4.126 kcal, 1 electron volt = 1.60206×10^{12} erg, The Planck constant (h) = 6.62617×10^{-34} joule s, The Avogadro constant (L) = 6.023 or 6.025×10^{23}.
(3) Photons or Einsteins, 1 Einstein = L photons.

however, light intensity is used to describe the observed brightness of a luminous source without defining precisely the solid angle captured. The total flux received by the detector can be estimated either from the overall solid angle it sees or from the surface area of receiver. The latter, the exposure, is particularly useful when describing light emission from chemiluminescent sources with complex geometries, e.g. luminous euphausiid shrimp, squid or fish with multiple light organs. Exposures are usually quoted at a standard distance, e.g. 1 m or 1 cm from the source. The geometry of a single source can be taken into account through the *irradiance* (E), the flux emitted per unit area of source. Alternatively the *density* (W), is light yield per unit volume of the emitter, the *exitance* (M), the light yield per unit area of emitter.

The spectral characteristics of chemiluminescence are described by two terms, spectral intensity ($I(\lambda)$) the flux per unit solid angle, i.e. the intensity, at a defined wavelength; and the spectral irradiance ($E_R(\lambda)$), the flux per unit area of source, i.e. the irradiance, at a defined wavelength. Chemiluminescence spectra are usually plots of $I(\lambda)$ against λ, or strictly $I(\bar{v})$ against $\bar{v}$.

These parameters will enable us to describe the brightness and colour of a chemiluminescent image. But if it is to be quantitative we have to have some units to describe them.

2.1.4 Units

The quantitative description of light emissions involves a bewildering array of parameters, and even more units of measurement (Tables 2.1 and 2.2). As we have just seen, the parameters (Table 2.1) are defined as *radiant* when measuring, or calculating directly, the light *emitted* from the source, e.g. using chemiluminescence quantum yield×molecules reacting; or as *luminous* when measuring the light *received* from the source. What we want to know is how much light is emitted or received per unit time, per unit area, and how this is spectrally distributed. The amount of light can be expressed either as an energy or as photons, and then related to time, area, and wavelength using cgs, MKS, or Imperial units. When relating the light to the number of molecules reacting these can be expressed either per individual atom or molecule, or per mole from millimols (10^{-3}) to tipomols (10^{-21}).

In the early days light standards were required to provide uniformity of units for luminosity within the scientific community. By 1918 an *International candle* was established as a standard point source emitting a uniform flux (Φ) in all directions. One firefly flash equals 1/400–1/50 candle, 6400 glow-worms one candle. The unit of solid angle to describe the intensity (I) of this source is the *steradian*, 4π steradians covering the whole emission. The detected intensity unit of the candle is the *lumen*. Thus the total radiant flux from one candle is 4π lumens, i.e. a spherical *candle* emits 12.57 lumens. A 100 W filament is about 1200 lumens. The *luminous* flux, together with *luminous* intensity in candelas, are old terms developed to relate the power of a light source to that required to produce a visual sensation in the observer. The situation then becomes confused between metric (cgs or MKS) and British imperial units. The irradiance emitted from a surface used to be expressed in *stilbs* or *lamberts* (1 lambert=1 lumen emitted per cm^2) in the metric system, or in lumens emitted per ft^2 (1 lumen ft^{-2}=1.076 millilamberts). Alternatively, using the area of the light receiver, 1 lux=1 lumen hitting the detector (incident) per m^2, 1 phot=1 lumen incident per cm^2 (1 phot=10^4 lux), and 1 foot candle=1 lumen incident per ft^2=1.076 milliphots=10.76 lux.

If the solid angle of the emitter was not taken into account, then the units of lux per unit area of source were *candles*. The problem with these units was that each country originally had its own standard light source, Britain the Pentane lamp, Germany the Hefner lamp, France the Carcell lamp, and the USA both the Pentane and Hefner lamps.

1 International candle=1 Pentane or 1 American candle=1.11 Hefner

units=0.104 Carcell units. Once accurate methods, became available for measuring the energy of emission, e.g. in ergs, joules, or kcalories, then the *radiant* flux, expressed as energy per second, gradually superseded the old units.

The historical, and somewhat anthropomorphic, units are not of any real help in studying chemiluminescence. The fact that a luminous bacterium emits 2×10^{-14} candles per second is of far less significance than the total number of photons emitted. Two types of unit are therefore now used to quantify chemiluminescence emission, or the illumination of a detector, as well as other classes of luminescence (Table 2.2). One, based on the particle nature of light, expresses light yield and flux in photons. A 'mole' of photons is known as an Einstein. The other is based on the energy of the emission, expressing light yield as ergs (cgs), joules (MKS), or kcal (British), and sometimes in electron volts. Flux is then expressed in erg s^{-1} (cgs) or watts (MKS). Emissions from many luminous organisms are expressed in μ watts. To convert energy units to photons we need to know the wavelength (or wave number), or the spectrum of the chemiluminescence.

number of photons=total energy released (Q_R)/energy of each photon   (2.32)

where h=the Planck constant, $\nu=(c/\lambda)$ Hz,

L=the Avogadro constant=6.023×10^{23}.   (2.33)

Individual luminous cells, a point source, emit some 10^3-10^6 photons per second. Pirenne & Denton in the 1950s showed that the dark-adapted human eye needs about 30 000 photons hitting a square centimetre of retinal surface per second to perceive an image. This is equivalent to about 1 photon per retinal rod. Dinoflagellates emit about 2×10^{-9} J per flash, other luminous invertebrates $10^{-9}-10^{-3}$ μW (or $10^{-15}-10^{-9}$ W) per cm^2 of receiver surface at 1 metre from the source.

Since $Q_R=h\nu=hc/\lambda$
$h=6.62617 \times 10^{-27}$ erg s$=6.62617 \times 10^{-34}$ joule
$c=2.99793 \times 10^{10}$ cm s^{-1}
λ=wavelength in nm

therefore 1 watt (1 joules s^{-1})=$(\lambda$ (in nm)$\times 10^{16}/1.986)$ photon s^{-1} (2.34)

or photon s^{-1}=$1.986 \times 10^{-16}/\lambda$ (in nm) watts, thus for a blue emitter, e.g. a luminous dinoflagellate with λ_{max}=479 nm and the receiver detecting 10^{-13} watts, this is equivalent to 241 188 photons per second, easily detectable.

The firefly emitting at 565 nm, and being detected as 10^{-10} watts per sq cm of receptor surface is equivalent to about 2.84×10^8 photon s^{-1} cm^{-2}. Bright! But dinoflagellates also do very well, one cell being able to flash some $1-10 \times 10^9$ photons. These calculations are, of course, only approximate, but illustrate the fun you can have with such numbers. The instrument we need to measure these numbers, to relate light emission to the chemical reaction, is called a *chemiluminometer,* or *luminometer* for short.

2.2 WHAT IS NEEDED OF A CHEMILUMINOMETER?

2.2.1 Quantification of a Flash or Glow

There are essentially three ways to express the light output from a chemiluminescent reaction: total output in all directions, light power emitted from the source, light power incident on a surface some distance away. In practical terms the last is the basis of a chemiluminometer, and is related to the time course, spectrum and polarisation. A typical chemiluminescent reaction results in rise in light flux (or intensity) which reaches a peak and then decays (Fig. 2.3A). This decay is often, but by no means always, a true exponential (Fig. 2.2). In spectrophotometry or fluorimetry the accumulation of the product, or consumption of the substrate, can be measured until the reaction is complete. A typical graphic representation of the increase in concentration is shown in Fig. 2.3B. However in chemiluminescence the light output is transient. Unlike the absorbing or fluorescent substance, light is not accumulated in the reaction cuvette. That is why the graphic representation (Fig. 2.3A) is a differential of Fig. 2.3B. The total light which has been emitted after time T can only be estimated by summing all the photons emitted up to this point (Fig. 2.3C): that is,

$$\text{light yield at time } T = \sum_{0}^{T} \Phi \cdot dt \qquad (2.35)$$

If one of the substrates of a chemiluminescent reaction is increasing with time, e.g. in an enzyme assay, then the light intensity will increase with time. On the other hand a constant *rate* of the chemiluminescence reaction produces a plateau trace in light intensity, which may be increased to another plateau if the reaction is 'spiked' with a standard concentration of substrate. If the chemiluminescence reaction rate is decreasing, e.g. as a result of consumption of one of the substrates, then the light intensity will also decrease (Fig. 2.3A). However, this decay will only be a true exponentional when just one of the substrates is consumed. Consumption of more than one, e.g. chemiluminescent substrate and oxidant, results in a second order reaction and non-exponential intensity traces.

From the chemiluminescence trace, we need to measure several temporal parameters: the lag time preceding light emission, the time to peak height, the half time of the rise and decay (rate constant, $k=0.693/(t_{\frac{1}{2}})$), if the up or down phases contain major exponential components. We also want to measure the peak flux (or

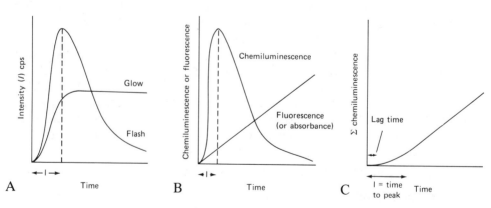

Fig. 2.3 — Types of luminous recordings: A. Flash versus glow; B. Fluorescence versus chemiluminescence intensty; C. Integration (Σ) of chemiluminescence ($\Sigma = \int_{0}^{T} I dt$).

intensity), as well as the total light yield, both as it accumulates and for the complete reaction, and we will also want to relate flux and light yields to the chemiluminescence spectrum.

The 'up' phase of a chemiluminescence flash may be as short as 10–100 ms, the 'down' phase 100 ms to many minutes or hours. The dimmest emission may be only 1 photon per second, the brightest many thousand millions of photons per second. So for maximum flexibility the chemiluminometer needs to be able to quantify as efficiently as possible a wide range of fluxes from dim to bright, linearly, over time intervals of just a few milliseconds to many minutes or even hours. The spectral sensitivity should ideally cover the entire visible range 380–750 nm or wider for UV or IR chemiluminescent reactions, though in practice most chemiluminescence spectra studied by biologists are within 400–700 nm.

How then can these requirements be satisfied?

2.2.2 The principal components
The sensitivity and flexibility of a chemiluminometer are dependent on the specification of its five principal component units (Fig. 2.4):

(1) The sample housing.
(2) The monochromator, if present.
(3) The light detector.
(4) The signal processor.
(5) The recorder and data processor.

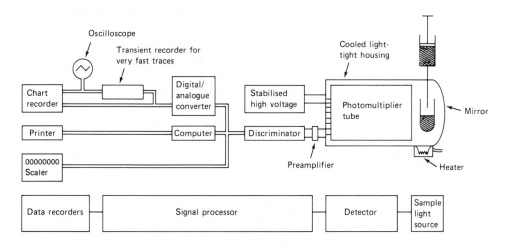

Fig. 2.4 — The components of a chemiluminometer.

2.2.3 The sample housing
The volume of a sample for chemiluminescence analysis can vary from just a few μl to hundreds of ml, from one cell to a whole organ, or even to a whole animal such as a squid or a fish many centimetres long. The sample holder may be a test tube, a flask, a

beaker, a microtitre plate, or a gel. The function of the housing is to maintain the sample in a temperature, ideally, pressure controlled, and completely dark environment, to which substances can be added at will, and from which as much as possible of the light produced reaches the light detector. The design of the housing must take into account six features:

(1) It must be light-tight.
(2) The sample volume should be adjustable, without affecting the fraction of light reaching the detector.
(3) It must be possible to add, sequently to the sample, various substances, e.g. cell stimuli or the final component of the reaction, with rapid and efficient mixing (at least within 50–100 ms).
(4) It must have temperature regulation.
(5) There must be efficient reflection of light from the chamber to the detector.
(6) It must handle the frequency of samples and analysis, and be suitable for automation.

Obviously, we must exclude ambient light from the detector. It might seem simple to make a completely dark housing for a chemiluminometer. It is not. A million millionth of a strip light can give 10^7 photons per second, the *maximum* permissible for a photo-multiplier tube! The first problem is the injector port which can, if one is not careful, act as a light guide. In a darkened room a narrow syringe needle above the sample may be adequate, resulting in an increase in background of less than 1–10 photon counts per second (Fig. 2.5A). However, on the open bench in the laboratory a shield over the injector port will be required. The other major problem is combating internal sources of light. Many paints, silicone grease and silica gel, glass and plastic tubes, either phosphoresce or chemiluminesce. This should be tested for and thus avoided. Contaminants in the reagents can give rise to large chemical chemiluminescence blanks. The purity of the water is paramount; for example, when using alkaline H_2O_2+peroxidase as the trigger. This can produce tens of thousands of photons per second, depending on the water source. Equally, someone handling grams or even milligrams of a chemiluminescence compound on one bench should be aware that one speck of it, equivalent to just a few μg, dropping into a stock reagent can generate as many as 10^{10}–10^{12} photons.

Measurements in solution will normally be carried out in a few μl to 2 ml volumes. The tube has to be inserted without allowing high light levels to leak into the housing. Otherwise the detector can be irreversibly damaged. For experiments with large cells, e.g. invertebrate muscle or nerve, whole organs and animals, or microtitre plates, dedicated housings are usually constructed. Microtitre plate readers are controlled by computer operated stepper motors which can read as many as 96 wells in a few seconds (Fig. 2.5B). There must be no light carried over from one well to another. An alternative for multisample analysis is to use a belt containing 100–1000 tubes (Fig. 2.6B). Sometimes another piece of apparatus, e.g. an oxygen electrode (Fig. 2.5C), is incorporated in the housing in order to relate directly chemiluminescence and another parameter, in this case O_2. Dual detector housings (Fig. 2.5D) are required for energy transfer and spectral measurements. Manual injection, or injection from a spring or solenoid operated syringe, of 10–500 μl volumes, usually provides adequate mixing, within 100–200 ms, in a test tube.

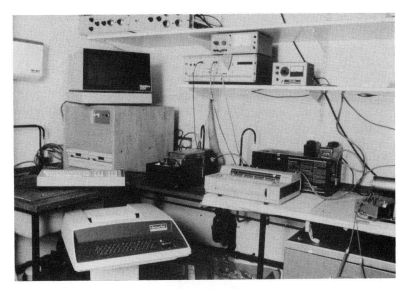

Fig. 2.5A

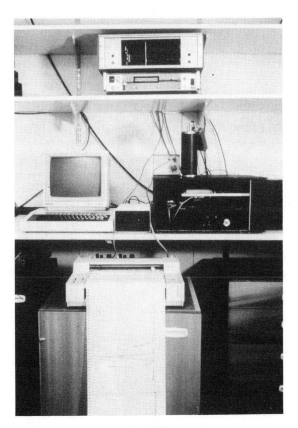

Fig. 2.5B

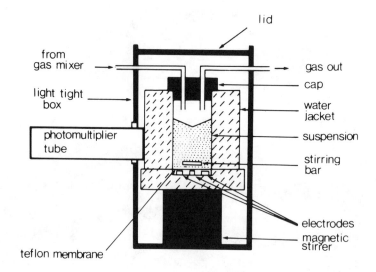

Apparatus for simultaneous oxygen/
chemiluminescence measurements

Fig. 2.5C

Fig. 2.5D

Fig. 2.5E

Fig. 2.5 — Home built chemiluminometers: A. Computerised and digital; B. Microtitre plate reader; C. O$_2$ electrode; D. Dual wavelength (PMT's opposite to each other); E. Calibration apparatus.

However, spinning the tube or use of a 'flea' may improve mixing, but magnetic stirrers and 'fleas' should be avoided as they often generate electrical noise in the photon detector. For rapid kinetics, i.e., <10 ms, a stopped-flow injector with a special housing is required.

Biological experiments are usually carried out over the range +4°C to 45°C, the temperature of mammalian blood being within 35°C to 45°C. Ambient temperature can, of course, vary from below freezing to over 30°C. The temperature of the deep oceans is around +5°C. The full range can be achieved electrically by using a peltier cooler+heater thermostat, or by using a jacket connected to a regulated bath. However, it can be instructive to study a chemiluminescent reaction over a much wider range of temperature, e.g. −20 to +100°C, or in the solid or gas phase from 77 K (liquid N$_2$) to several hundreds of degrees Celsius. This requires a specially designed sample housing with emphasis on protecting the light detector from effects of temperature.

It is essential that a high proportion of the light generated within the housing actually reaches the detector. This is determined by six factors:

(1) The geometry of the sample.
(2) The geometry of the housing.
(3) The reflectiveness of the housing.
(4) The presence of light absorbers in the sample or within housing.
(5) The distance from the sample to the detector.
(6) The surface area of the detector.

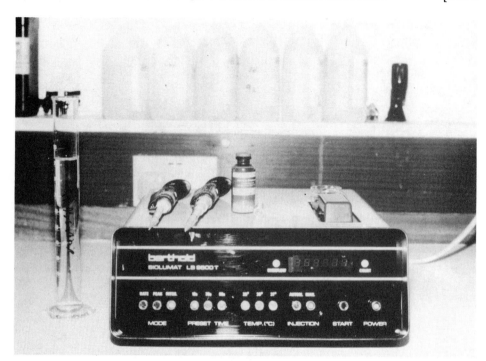

A

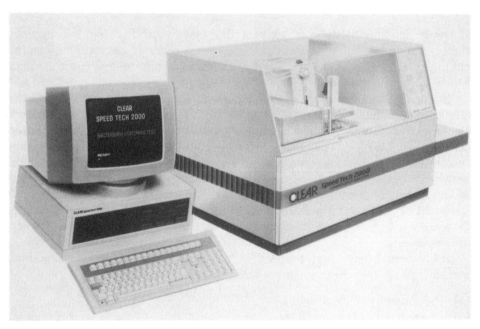

B

Fig. 2.6 — Two commercial digitalised chemiluminometers: A. Single tube — Berthold Biolumat LB9500T; B. Multitube — CLEAR Speed Tech 2000.

Light can be trapped by internal reflections and refraction. The latter worsens with increasing refractive index (*n*). Values of *n*<1.2 are usually acceptable. Spherical containers allow virtually all the light to escape, whilst cylinders (e.g. tubes) and parallel walled cuvettes are adequate. Mirrors, silvering, or white paint behind the sample will help to reflect light out of the housing on to the detector.

The closer the light source is to the detector the better. In most chemiluminometers this is usually a few cm. Longer distances may require lenses or light guides. However, these may result in large losses of light through reflection or scattering. The end surfaces of fibre optics require specialised polishing for low light levels. Remember the inverse square law!

In practice the amount of light escaping from the housing and then reaching the detector is usually 10–50%. It is worthwhile checking this with a light standard (see section 2.7). One important requirement is that this geometric efficiency should be independent of sample size, e.g., a fixed amount of chemiluminescent compound triggered in 10 μl should produce the same detected signal as if it is triggered in 2 ml.

2.2.4 The monochromator

Many chemiluminometers are not able to separate the light intensity into different colours; they monitor the whole spectrum simultaneously. However, if graphic representation of a chemiluminescence spectrum, or the flux (intensity) at a particular wavelength, are required, then the light from the sample housing must pass through a device which either disperses the spectrum or selects a designated part of it. Dispersion can be achieved by using either a prism or a diffraction grating. It is impossible to select a single wave number, the narrowness of the spectrum monitored at each point of the curve (Fig. 2.7) is determined by a slit; the wider the slit the larger the bandwidth measured, but the more the light available to the detector.

Selection of a particular band in the spectrum can be achieved by using an absorption filter or an interference filter, usually composed of a thin metal film protected by a glass envelope. These can be anything from a few mm or several cm across, round or square. Absorption filters have broad bandwidths, the part of the spectrum they let through. They are thus not 100% selective, and often have poor transmission even over the spectral region of their maximum transmittance. Interference filters, on the other hand, can have very narrow bandwidths, e.g. 5–10 nm (defined as the wavelength separation between the points where transmission is one half that at the maximum). Interference filters have high transmission (e.g. 10–50%) at the desired wavelength. This can easily be checked in a spectrophotometer.

Prisms have the disadvantage of non-linear dispersion, losses due to reflection and scattering; they are often cumbersome and thus slow to move, making spectra from fast flashes very difficult to obtain, and they can be expensive. Diffraction gratings, on the other hand, can be rotated very rapidly, enabling complete spectra to be obtained every 50–100 ms with some instruments (Appendix VA). However, they are not without problems. They reflect light, giving rise to stray light which may bypass the grating and reach the detector. Also they disperse spectra in several 'orders', which overlap. If the instrument is set up to look at the first grating order then the second order will overlap from a wavelength of λ/2. However, if we are only looking at 400–700 nm, then the second order of the low end, i.e. 800 nm, will not significantly overlap with the first order at the top end. Thus, this is not a serious

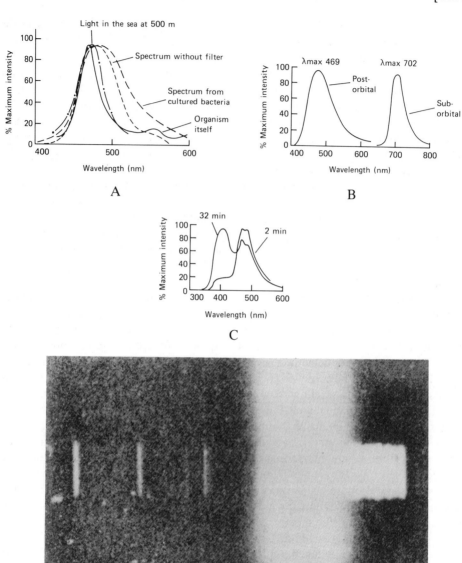

Fig. 2.7 — Chemiluminescence spectra: A. The bacterial photophores of the deep sea fish *Opisthoproctus* showing effects of filter to match emission with ambient light at these depths, for counter-illumination camouflage. Adapted from Denton *et al*. (1985). B. The deep sea fish *Malacosteus niger*, showing blue and red emissions from its two main pairs of light organs. Adapted from Widder *et al*. (1984). C. The deep fish *Searsi koefoedi*; showing shift in emission spectrum with time. Luminescence from exudate stimulated with H_2O_2. Adapted from Widder *et al*. (1986) with permission of Pergamon Press. D. Spectrum of the glowworm *Lampyris noctiluca* obtained using a prism and photographic film from Ramdas & Venkiteshwaran (1931). Reprinted by permission from *Nature* **128** 726–727, copyright © 1931 Macmillan Magazines Ltd.

problem in most chemi- or bio-luminescent spectra. A further problem is that the transmission of the grating varies with wavelength, so to obtain absolute spectra the instrument must be calibrated. The 'blaze angle' determines the most efficient wavelength (λ_B). For the first order spectrum the efficiency decreases to 50% of this at $2/3\lambda_B$ and $2\lambda_B$. Thus a grating blazed with $\lambda_B=500$ nm will have a satisfactory efficiency over the range 333 nm to 1000 mm, encompassing anything we may wish to study with chemi- or bio-luminescence. The flux ($\Phi(\lambda)$) through a monochromator is described by the equation:

$$\Phi(\lambda)=E_g(\lambda) \cdot E_0(\lambda) \cdot s\ d/(f) \tag{2.36}$$

where Φ is a measure of the overall efficiency of the grating

$E_g(\lambda)$	=grating efficiency at λ
$E_0(\lambda)$	=optical efficiency at λ
s	=entrance slit height
d	=linear dispersion
f	=optical aperture—the reciprocal f-number

High efficiency, therefore, requires high E_g and E_0, a large grating and entrance slit, and a small f-number, with the blaze angle selected to give the best λ in the middle of the spectral range required.

The main sources of error in determining spectra are stray and scattered light, and spectral calibration of the instrument. Since chemiluminescent glows or flashes are decaying, this must also be compensated for if the monochromator cannot move fast enough. This can be done by using a reference detector with no monochromator (Φ), but usually with a neutral density filter. The ratio (R) from the two detectors is then estimated and plotted versus λ:

$$\text{Ratio }(R)=\Phi(\lambda)/\Phi \quad . \tag{2.37}$$

To check that the spectrum remains unchanged over the minute or so required for a complete scan, the spectrum is then rescanned in reverse to show that it superimposes on the initial trace. Fast spectral analysers have the advantage that temporal changes in spectra can be plotted in three dimensions. Remember that a monochromator will probably reduce the signal to the detector to <1% of the light intensity without it.

2.2.5 The detector
The detector is the centrepiece and most critical component in any chemiluminometer. Incorrect selection will limit the experiments or analyses you are about to carry out. There are four main requirements:

(1) The detector must be able to detect a light signal over several orders of magnitude of intensity, from just a few photons per second to tens of millions of photons per second.
(2) The detector must be sensitive at least over the spectral range 400–600 nm, and ideally over the complete range of the visible spectrum (i.e. 380–750 nm) and even into the UV and IR. If its sensitivity varies with wavelength, as is usually the case, then it must be possible to correct the data for this.

(3) The signal output presented *by* the detector should be directly related to the light intensity hitting it, ideally linearly over the entire sensitivity range required. The signal from the detector should be produced in a form which can be displayed, recorded, and easily analysed.

(4) The speed of response of the detector must be much faster than the rate of the chemiluminescent reaction, if the signal output is not to be a distorted version of the true signal. This usually means that for the fastest reactions with initial time constants (time constant=time to 1/e maximum intensity) of the order of 1–10 ms, the detector should be capable of responding within 1 ns–1 μs. The time constant of the detector and signal processor, resulting from their response times and capacitances, should be negligble. Neither the detector nor the signal processor should exhibit any significant hysteresis. This means that both recover fully between pulses, i.e. for pulse rates of 10^7 protons s^{-1}, the recovery time must be in the ns range.

As we shall see, the most popular detectors are photographic film, photodiodes, or photomultiplier tubes (section 2.3). Some, for example photomultipliers, are better at detecting light in the blue region of the spectrum than in the red. Others, like photodiodes, are better in the red. Some can detect a few photons, others require bright, visible signals. The range of the most sensitive detectors can be extended to quantify bright emissions where they would otherwise become saturated, by using a neutral density filter.

A further requirement may be the ability to visualise the signal, for example to localise and record continuously where in an animal or where within a cell the chemiluminescence is being generated. Monoclonal antibody production, immunoassay, single cell analysis, cloning, and recombinant DNA technology may require an image of an entire microtitre plate, nitrocellulose sheet or slab gel for rapid quantification. To do this a special imaging detector is required, e.g. a fast photographic film, a TV camera, or an image intensifier (section 2.5).

2.2.6   Signal processing

The aim of the signal processor (Fig. 2.4) is to amplify the output from the detector and to present this to a recording device, for a permanent record (Fig. 2.8A), and for analysis. The original light signal hitting the detector is in digital form, i.e. photons. At high light fluxes the output pulses from an electronic detector will merge into a current, producing an analogue signal. At low light levels the output should still be in pulses so that one has the opportunity of processing it digitally. Which is best, analogue or digital?

At high light fluxes there is little to choose between analogue and digital processing. In fact, many commercial chemiluminometers, which are really based on analogue processing (see Appendix VA), provide a digital read-out of the accumulated light yield over a fixed period, e.g. 10 or 30 s. This is simply an integration of the analogue signal. In contrast, when digital processing is used, in order to produce a graphic representation (Figs. 2.3 and 2.8), rather than go through a digital computer, it is simpler to use a digital-to-analogue converter and plot the trace on a chart recorder or an oscilloscope (Fig. 2.8B). It is in the detection of low light levels where the real argument lies between analogue and digital.

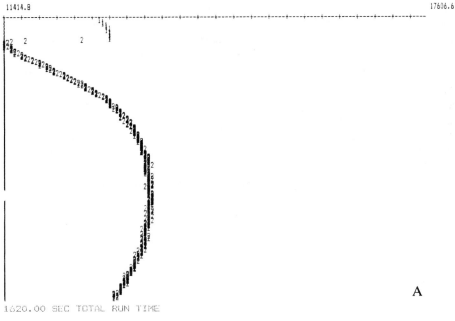

```
EXPERIMENT DOSE RESPONSE LATEX   PHOLASIN V LUMINOL
PHASE STATISTICS FOR RUN   5      5^7BEADS+PHOLASIN

1/ MEAN  :   15707.5    S.DEV :      147.3    N =  12
INTEGRAL :    188490    MIN :      15399    MAX :     15865

2/ MEAN  :   16164.3    S.DEV :     1679.8    N = 300
INTEGRAL :   4849290    MIN :      11415    MAX :     17607

PLOT EXPERIMENT              DOSE RESPONSE LATEX   PHOLASIN V LUMINOL
RUN NUMBER   5      5^7BEADS+PHOLASIN
```

```
 11414.8                                                                    17606.6
```

```
1620.00 SEC TOTAL RUN TIME
```

A

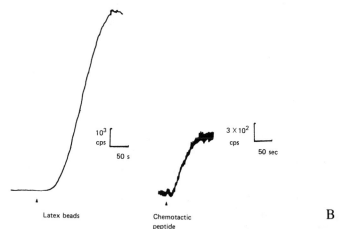

10^3
cps

50 s

3×10^2
cps

50 sec

Latex beads

Chemotactic
peptide

B

Fig. 2.8 — Chemiluminometer data output: A. Computer print out from Fig. 2.5 A Abscissa = time (sec); Ordinate = counts. From Davies *et al* unpublished. B. Analogue output from chart recorder in Fig. 2.5 A, showing luminol chemiluminescence from rat neutrophils stimulated by latex beads or chemotactic peptide, N formyl met leu phe. The dimmer emission with chemotactic peptide means more amplification of the signal, and hence more noise on the trace.

The ability to detect a signal from any instrument, or photograph for that matter, depends on being able to distinguish it from the 'noise' generated by the detector and processor. A simple way of comparing the sensibility of different devices is to measure the 'noise', the background from the instrument, and then the signal produced by a light standard. A radioactive luminescent standard, quenched so that it is a single photon emitter, is ideal. Strictly the detection limit of an instrument is related to (signal)^2/noise. However, for a simple rule of thumb, the sensitivity of any chemiluminometer can be optimised by minimising the noise, and producing the highest signal-to-noise ratio. Digital chemiluminometers have both a more stable instrument noise and a better signal : noise than analogue devices. Thus a digital processor is more sensitive for low level chemiluminescence than an analogue one.

The background noise escaping from the digital processor is minimised by a *discriminator*, which looks at the energy level of each pulse as it leaves the detector, and only records pulses above a set level. This level is chosen to produce the best signal : noise, rather than simply the largest signal (Fig. 2.9). The response of the discriminator must be fast. If pulses up to 10^7 per second are to be processed without loss or distortion then the response time must be of the order of nanoseconds. The discriminator converts pulses of varied energy levels into uniform, square wave, pulses amplified 10–1000 times, which can then be recorded on a scalar, or be transformed into an analogue signal for acceptance by a chart recorder. In an analogue device the 'noise' is subtracted either electronically or manually by the operator.

A good photon counting detector, using a discriminator, has a pulse noise of 10–50 pulses s^{-1}. Rates as low as 1 pulse s^{-1} are possible. But beware, some commercial luminometers 'doctor' the noise to produce an artificial low background count!

One way of estimating, statistically, the lowest possible signal detectable is to accept a signal as being real if, as recorded, it is more than two standard deviations above background. Thus, if the noise is 10 pulse s^{-1}, the standard deviation, assuming a Poisson distribution, is $\sqrt{N}$, i.e. approximately 3. Since the efficiency of the detector at responding to photons is likely to be no better than 25%, a noise of 10 pulse s^{-1} means that the detection limit will be equivalent to about 24 photons per second hitting the surface of the detector, assuming that all pulses are accepted by the discriminator, or approx. 6 recorded pulses s^{-1} above the background. In practice, this may not quite be the case. However, it is impossible even to approach this level of sensitivity when using an analogue processor. The principal reason is the instability of the background noise when using analogue processing. Thus compared with an analogue device digital processing is theoretically more valid, practically more sensitive, and is more flexible and easier to interface to a digital computer.

2.2.7 Recording and data processing

The point of going through all this rigmarole is to end up with a permanent, quantitative record, which can be analysed and related to the kinetics, the chemistry, or biology of the chemiluminescent reaction being studied. Pulses, or integrals, are conventionally displayed on a scalar. A signal in analogue form is best recorded on a flat-bed chart recorder, or displayed on the screen of a storage oscilloscope and then photographed. Most flat-bed recorders have a time constant of some 100–200 ms.

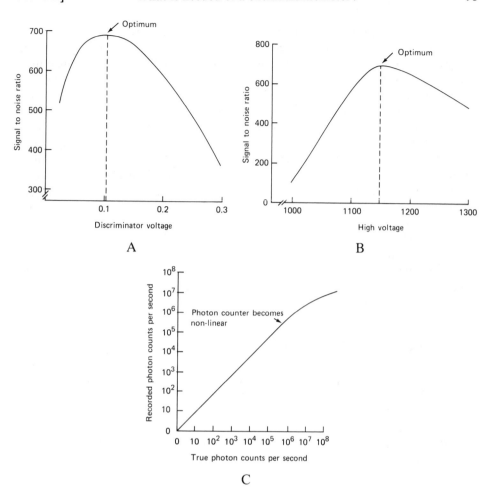

Fig. 2.9 — Optimisation of signal to noise in a chemiluminometer. A. Effect of discriminator.
B. Effect of high voltage. C. Saturation of optimised chemiluminometer at high photon counts.
Set optimal discriminator then adjust high voltage. The optimal settings will not be where the
recorded counts are highest. Luminescent standard = quenched ^{14}C hexadecane or $^{3}H_2O$ in
toluene scintillant (PPO/POPOP). Saturation prevented by reducing the first dynode — photocathode voltage or a neutral density filter.

This will lead to distortion of fast traces. For example, the Ca^{2+}-activated photoprotein obelin has a maximum decay constant of 4 s^{-1}, equivalent to a $t_{\frac{1}{2}}$ of 180 ms. The
distortion of such fast traces can be circumvented either by using a transient recorder
(Fig. 2.5D) which has a ms response time but can play back the record over many
seconds, or by interfacing the discriminator output to a computer.

Many of us have, I am sure, spent hours measuring peaks off graphic traces, or
have laboriously replotted them as log y against time. Thank goodness it is now quite
simple to interface a discriminator output directly to a microcomputer, or to a main
frame if you wish. The appropriate software will then estimate for you the rate
constant, $t_{\frac{1}{2}}$, the integral over the complete or partial experiment. It will also convert

the signal to the parameter being measured by comparison with a stored standard curve. All of the results can be stored on disc, and plotted graphically (Fig. 2.8). The computer software can also be written to control the chemiluminometer, to optimise the instrument, to measure the background noise, to subtract this from the signal, and to start and stop the experimental count.

Most laboratories have equipment capable of detecting light, for example spectrophotometers, fluorimeters, and scintillation counters. Spectrophotometers are insufficiently sensitive for studying chemiluminescence. Fluorimeters tend to be a little better, but are still incapable of detecting dim chemiluminescence invisible to the naked eye. Scintillation counters can easily detect low level chemiluminescence but they are not really designed for this purpose. You cannot, for example, inject anything into a sample tube whilst it is in front of the detector. Furthermore, scintillation counters are designed to quantify radioactive particles, which generate bursts of photons from the scintillant, and not single photons as in chemilumines- cence. If you want a good chemiluminometer, build your own (Fig. 2.5) or select one from the many commercial ones now available (Fig. 2.6, Appendix VC). But whichever you go for the light detector is crucial. So, what criteria does one use to select this vital component?

2.3 SELECTION OF THE DETECTOR

The detector in a chemiluminometer has to respond to light, producing an effect which can then either be directly seen and recorded by a human observer, or be converted to an electrical or chemical signal. This signal is then processed through another device to provide a permanent record. Four principal methods have been used to detect and quantify changes in light intensity emanating from various physical and chemical phenomena, including chemiluminescence (Table 2.3). What are their advantages and disadvantages for studying chemi- and bio-luminescence?

Table 2.3 — Detectors of light from chemiluminescence

Principle	Example
(1) Biological	the human eye photoreceptors from other organisms photosynthetic and photosensitive pigments
(2) Chemical	photographic paper and film photosensitive substances in solution
(3) Thermal	pneumatic device bolometer thermopile
(4) Electrical	photoelectric device (photomultiplier) photovoltaic device (light meter) photoconductor device (photodiode)

2.3.1 The choice
A fluorescent strip light can generate 10^{19} photons per second, but as we have already seen, ideally the detector of chemiluminescence should be able to respond to a range of light intensities from just a few photons per second up to bright sources emitting thousands of millions of photons per second, over the entire visible spectrum. Most emit in the 400–600 nm region. However, one group of luminous fish, the stomiatoids (see Chapter 3), have a pair of light organs emitting around 705 nm, and some ultraviolet and infrared solution, gas and ultraweak chemiluminescence has been reported. The ideal detector optimises seven characteristics:

(1) High sensitivity, down to 1 photon s^{-1}, with low noise.
(2) Sensitivity range, at least six orders of magnitude, ideally linearly.
(3) Wide spectral response, ideally unchanged over the visible spectrum of 380–750 nm.
(4) Quantitative, the signal from the detector can be easily converted to a number and not simply judged by the naked eye.
(5) Precise and reproducible.
(6) Stable and hardy.
(7) Cheap and available.

Biological detectors
The human eye cannot, of course, detect in the ultraviolet or infrared. The first method developed to measure light in the visible region was to compare the experimental sample with a standard light source. The eye is extremely sensitive, and when trained can be surprisingly accurate. However, it can only really be used to judge constant glows. Rates of decay, particularly when fast, cannot be quantified accurately. In theory, linking a highly sensitive photoreceptor from a deep sea organism to an electrical recording device should produce a very sensitive detector. This has yet to be achieved. Nor have photosynthetic and photosensitive substances, such as the chlorophylls, phycobiliproteins, rhodopsin and porphyrins, yet found application as quantitative light detectors.

Chemical detectors and photography
Many photosensitive substances have been discovered during the past two centuries. Yet none has been as useful as the silver halides which, as Carl Wilhelm Scheele, co-discoverer of oxygen, discovered in 1777, decompose to silver and thus darken when exposed to light. But the basis of the photographic process, established during the nineteenth century, is to form first an invisible 'latent' image which can be developed into a visible one. Surprisingly, the precise chemistry of this process is still not fully understood.

The photographic emulsion consists of grains of gelatin containing silver bromide ($AgBr$), and minute amounts of AgS essential for the formation of the 'latent image'. Modern emulsions may also contain acid, basic, or neutral sensitising dyes, and in colour film colour-sensitive dyes. The initial reactions when a grain is hit by, and absorbs, a photon are:

$$Br^- + h\nu \rightarrow Br + e^- \qquad\qquad (2.38)$$

$$Ag^+ + e^- \rightarrow Ag \qquad\qquad (2.39)$$

These two reactions do not occur at exactly the same spot. It appears that the electron released from the Br moves through the crystal lattice until it hits a fault in the crystal, or an impurity such as AgS or Ag. The electron then becomes 'trapped'. AgS is not itself the initial light absorber, but silver nuclei build up at these sensitivity points, by electron trapping, causing neutralisation of Ag^+ to Ag. These 'Ag nuclei', which form the latent image, act as catalytic centres for the developer, e.g. hydroxylamine,

$$2AgBr + 2NH_2O^- \xrightarrow[\text{latent image}]{Ag} 2Ag + N_2 + 2H_2O + 2Br^- \tag{2.40}$$

providing an amplification of some 10^9. Unreacted AgBr is converted by the fixative (hypo) into salts which are then washed away. If the latent image is to form at all, the Ag and Br first form within the grain (2.38) and (2.39) but must not recombine. Gelatin can act as a halogen acceptor, thereby preventing recombination:

$$2Br + H_2O \rightarrow OBr^- + HBr + H^+ \tag{2.41}$$

$$OBr^- + \text{gelatin} \rightarrow \text{gelatin } Br + OH^- \quad . \tag{2.42}$$

Similarly, sensitising dyes and AgS may also act as a Br receptor. The process looks, on the face of it, extremely sensitive. Yet photographic film cannot detect low intensity chemiluminescence, invisible to the naked eye. Why?

The 'reciprocity law' of Bunsen and Roscoe (1862) states that the photochemical reaction depends on the total light energy absorbed over a given area (development density), and not the rate of absorption: that is,

$$\text{effective exposure} = \int_0^T I \, dt \tag{2.43}$$

The characteristic curve for any particular emulsion is a plot of development density versus log effective exposure. As long as $I \times T$ is constant, then the effective exposure will be the same. However, this law breaks down at both very high and low light intensities. This is called 'reciprocity failure',

The sensitivity or 'speed' of photographic film usually increases with grain size and is indicated by the ASA(ISO) or DIN numbers. The bigger the ASA(ISO) number the more sensitive. Most people use 100–400 ASA(ISO) in their cameras. Film of 20 000 ASA(ISO) is more sensitive and can be used to detect chemiluminescence assays in microtitre plates, for example (Fig. 2.10). The quantum efficiency (ϕ_p) of a photographic film, i.e. the number of photons required, on average, to form one 'latent image' grain, can be estimated by relating the effective exposure in total photons to the number of grains formed after development. The figure produced using visible light is:

$$\phi_p = 0.1 \quad . \tag{2.44}$$

However, this is misleading. It is not that, on average, 10 $h\nu$ must hit a grain to form an Ag nucleus in the latent image. Rather, some 3–4 photons must hit *one* grain within a very short time interval, if recombination of Ag + Br is to be prevented. The overall quantum efficiency ($\phi_p \doteqdot 0.1$) of the emulsion is thus a combination of this,

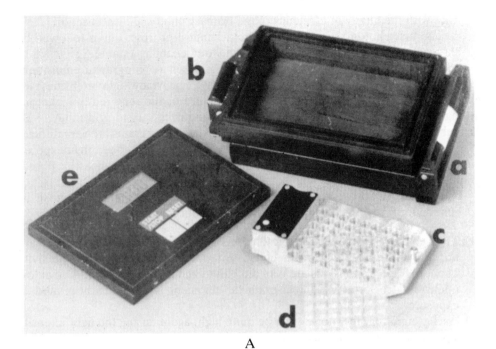

A

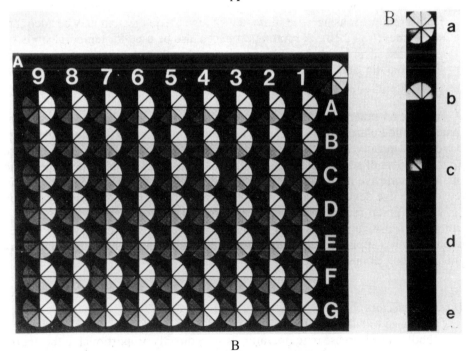

B

Fig. 2.10 — Photographing chemiluminescence: A. Camera luminometer (a) Polaroid film back; (b) shutter; (c) mask; (d) microtitre plate; (e) lid and timer. B. A range of neutral density filters (A) for quantifying microtitre plate chemiluminescence luminol (B) (a) 50 pmol, (b) 5 pmol, (c) 0.5 pmol, (d) 50 fmol, (e) 5 fmol. From Bunce *et al.* (1985) *Analyst* **110** 657 with permission from The Royal Society of Chemistry.

together with the quantum efficiency of the primary photochemical reaction (2.38, 2.39). The first photon thus starts the photochemical process; this is followed by similar reactions induced by hits two, three, and four, enabling a stable Ag nucleus to form. Therein lies the problem. The number of grains in a photographic emulsion is very large, tens of thousands to millions per square cm. At low light intensities, for example a few 100 photons per second, the chance of the required 3–4 photons hitting the same grain with the fraction of a second necessary to prevent Ag + Br recombination, is negligible. Furthermore, a few developed grains are very difficult to find, particularly since there is always a background fog, the 'noise', in unexposed film. However long exposures of weak chemiluminescence enhance the photographic image without increasing 'noise'.

In spite of its apparent convenience photographic film is thus limited in its application for studying chemiluminescence. The four main problems are:

(1) Threshold sensitivity — low photon emissions invisible to the naked eye are impossible to detect. Unfortunately quantitative data on the absolute sensitivity of photographic emulsions are very difficult to find in the literature. Also at low intensities (reciprocity failure) grain development is non-linearly related to exposure.
(2) Background fog, the 'noise', can be quite high, again raising the light intensity necessary for a detectable signal.
(3) Accurate quantification requires a 'grain' counter. Often this is done manually from a screen, or using a densitometer, though eye judgement may be adequate sometimes (Fig. 2.10). A rotating drum can also be used for temporal studies.
(4) Untreated silver halides are blue sensitive with very poor sensitivity in the red. Panchromatic and colour film contain sensitising dyes to circumvent this problem.

There is no inherent difference in the photographic process between conventional film and Polaroid film. The latter uses the ingenuity of its thin film technology to produce a negative and positive together, coupled with rapid development. There are three layers of silver halide grains in Polaroid colour film, blue-sensitive at the top, green-sensitive in the middle, and red-sensitive at the bottom. Sandwiched in between are dye developer layers with dyes complementary to the primary colour, e.g. yellow (green-red) next to the blue sensitive layer. None of these colour films is sufficiently sensitive to detect low intensity chemiluminescence. However photographic film has application in detecting southern and western blots, immuno assay and expressing clones.

Thermal detectors

Any substance will absorb light energy and warm up. The most efficient is a 'perfect black body'. The increase in temperature will be directly proportional to the energy absorbed ($\mu W\, cm^{-2}$), and can be quantified by an increase in pressure or volume if it is gas (pneumatic), or an increase in the electrical resistance of a wire, the bolmeter invented by Langley in 1880. Alternatively, using a thermocouple, the temperature increase can be detected by the development of a potential at the junction of two different metals (the Peltier effect) or two semiconductors (the Seebeck effect). The use of such a thermoelectric effect to measure light using a Cu and Fe junction was invented by Mellons in 1830. It forms the basis of the thermopile, a device commonly used by Dubois (Fig. 1.3b) and others in the nineteenth century studying lumines-cence. The potential change developed by the thermopile is recorded, using a galvanometer. C. V. Boys combined the two by suspending a small coil, on a quartz fibre, between the poles of a magnet. The coil was in the same circuit as an Sb/Bi thermojunction. Light caused the coil to move. An alternative method was deve-loped by Sir William Crookes, who allowed light to fall on microvanes which moved as convections occurred. The vanes had a mirror connected to them reflecting a light spot enabling the movement to be detected. The problems with many of these early thermally based light detectors were poor sensitivity, lack of precision, and some-times problems with linearity.

The thermopile still has application as a device for calibrating bright light sources, e.g. those used in fluorimetry. The noise of a very good thermopile, about 5×10^{-9} W cm^{-2}, is equivalent to some 10^{10} photons s^{-1} cm^{-2} of blue (400 nm) light or 1.8×10^{11} photons s^{-1} cm^{-2} of red (700 nm) light. (see Eqn 1.6)! This is a long way away from the detection limit of the best photoelectric device, the photomultiplier tube.

Photodiodes and photomultipliers

In 1887 Hertz observed that the length of a spark generated between electrodes was decreased if the secondary electrode was shielded from the light generated by the spark. He showed that this, the first description of a 'photoelectric' effect, was caused by UV radiation. The following year, Hallwachs showed that ultraviolet light caused a Zn electrode to lose its charge if originally negative, but not if positive. By using Na and K as electrodes instead of Zn, Eister and Geitel, in 1889, generated the first photoelectric cell capable of producing a current from visible light. J. J. Thomson's discovery, or definition, of the electron in 1897 resulted in the demonstration that the photoelectric effect was the result of electrons being emitted following the absorp-tion of electromagnetic radiation. Einstein, in 1905, finally established the theoreti-cal framework necessary to relate the energy absorbed to the electrons emitted. As a result he defined the photon, using Planck's quantum concept proposed some five years earlier.

There are three ways in which light can be measured using its ability to generate, or alter, electrical signals:

(1) Photoelectric cells and tubes, where absorption of a light photon releases an electron from a photosensitive material.
(2) Photovoltaic cells, where absorption of light generates a voltage.
(3) Photoconductive cells, where absorption of light changes the resistance (or conductance) of the material it hits.

The first of these is the primary basis of electron generation in a photomultiplier tube, the second is used in photographic exposure meters (Se) and solar batteries (Si), and the third is the basis of photodiodes and phototransistors.

Orthicon TV cameras use emission of photoelectrons from the light target, where they produce a localised charge. This charge is scanned by an electron beam, thereby providing an amplification of the original signal. In contrast, when light is received by a Vidicon TV camera tube it alters the electrical conductivity of the target, rather than inducing photoelectron emission. This is less sensitive than the orthicon but less noisy. This brief interlude on TV cameras highlights the three key features to be examined when selecting an electronic photon detector:

(1) Quantum yield—i.e. absolute sensitivity.
(2) Electron noise.
(3) Method of multiplication and its effect on noise.

Modern photovoltaic and photoconductive cells use semiconductors. Like the silver halides of photographic emulsions the activity of semiconductors depends on the presence of impurities. Germanium (Ge) and silicon (Si) have valency four, whereas arsenic (As) has valency five. Thus Ge contaminated with a bit of As results in excess of conducting electrons, resulting in positive 'holes' on the As within the Ge matrix. This is an 'n' type semiconductor. In contrast, with boron (B) as an impurity, with valency three, there are not enough electrons to satisfy Ge in a matrix, resulting in B^- $-Ge^+$, a 'p' type semiconductor. Formation of a pn junction results in a resistance of a few hundred ohms in one direction, but more than a million ohms in the other. Current can thus pass only in one direction, i.e. alternating current is rectified. Shining light on a pn junction will generate a voltage, thereby producing a photovoltaic cell. However, these are relatively insensitive to light, and extremely non-linear. They are therefore only of limited use in chemiluminescence analysis. But light also alters the resistance of a pn junction. This is the basic of the photodiode. The response is very rapid, i.e. ns. Sensitivity is better than in a photovoltaic cell, though still really only suitable for photon emissions visible to the human eye, i.e. a minimum of tens of thousands of photons per second. Furthermore, such solid state devices are red sensitive (Fig. 2.11), where they can have quantum efficiency of 50%. However, they are very poor in the blue region. Modern photodiodes are much better than the original ones based on cadmium sulphide or selenide. These old devices had very slow responses and variable spectral sensitivity. Photodiodes, and their pnp derivative, the phototransistor, are small, cheap, use a battery as electrical supply, and thus can be used to construct a pocket chemiluminometer for use in the field. They have dark currents, the noise, of just a few picoamps. If energy transfer (see Chapter 9) could be used, in an analogous way to that found in red and blue-green algae, then photodiodes may one day compete with what are currently the most sensitive detectors of light in the visible region of the spectrum, photomultiplier tubes. Increasing the voltage across a pn junction can result in the formation of an avalanche photodiode (APD). Cooling to -10 to $-20°C$ results in a very high quantum efficiency ($\phi E=60-80\%$) detector with very low noise. Single photons may eventually be detectable using this device which is ten times smaller, cheaper and efficient than a photomultiplier tube.

The two special features of a photo-tube designed to detect photons are, firstly, that it uses the photoelectric effect to generate electrons at one surface, the photocathode, and secondly, that it amplifies the initial electrons produced by the cathode, through a cascade, generating a measurable current or pulse of charge at the other end. *Gas filled* photo-tubes work by the photoelectrons generated at the

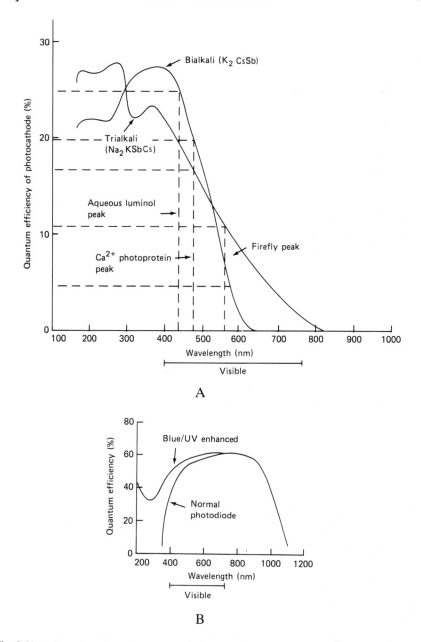

Fig. 2.11 — Spectral sensitivity of electronic light detectors: A. Efficiency of trialkali versus bialkali photomultiplier tube (from Thorn EMI). B. Efficiency of two photovoltaic diodes (from Thorn EMI).

cathode provoking an ionisation cascade in the gas, thereby generating a current. The response times of these tubes are slow, because of the slow movements of the relatively large positive ions of the gas towards the cathode. In contrast, *vacuum* photo-tubes use primary photoelectron generation at the cathode followed by secondary electron generation and electron amplification through a chain of 'dynodes'. This produces a device capable of detecting just a few photons per second. This is the photomultiplier tube.

2.3.2 The photomultiplier — how it works
The photomultiplier is one of a handful of electronic devices which have had a major impact on scientific discovery in this century, and on biology in particular. Just walk into any chemistry or biology laboratory and you will see spectrophotometers, fluorimeters, scintillation counters, and maybe even a chemiluminometer, all using a photomultiplier tube as the light detector. Yet how many biologists know how this key piece of equipment works, or how it compares with its major rival, the photodiode? How do these two devices, potential detectors in a chemiluminometer, compare in sensitivity, precision, reproducibility, spectral response, and range? Indeed the commercial catalogues and brochures contain an impressive array of different photomultipliers and photodiodes. How can one select the right one in order to construct a good chemiluminometer?

A photomultiplier tube (Fig. 2.12) is a photosensitive device which generates electrons at the photocathode, which are focused and multiplied by an electrode chain, eventually being collected at the anode. Several complicated-sounding terms are used to describe the key components and characteristics of a photomultiplier (Table 2.4). The tube can be conveniently divided into four main parts:

(1) The glass envelope, which maintains a vacuum within it and has a transparent glass window through which photons enter.
(2) The photocathode, the light receiver and generator of the primary electrons.
(3) The dynode chain, the electron focusing and multiplication device which generates secondary electrons.
(4) The anode, the electron collector.

The photomultiplier tube receives a photon at the photocathode which, because of the photoelectric effect, emits an electron. This is *primary emission*. The electron is then electrostatically accelerated, and focused so that it hits the first dynode. This is made of a material different from that of the photocathode and emits several electrons as a result of *secondary emission*. Each of these electrons is then accelerated, and focused to hit the second dynode, where a further multiplication of electrons occurs. This procedure may be repeated as many as 13 times, before the resulting electrons are collected as a pulse of charge at the anode. The anode then delivers the pulse, via a preamplifier, to the discriminator. This then decides whether it is big enough to be considered as a 'count'.

What determines whether the photocathode emits an electron at all?

The maximum kinetic energy (K_e) of the primary photoelectron can be estimated from the Einstein equation:

$$K_e = (hc/\lambda) - P_w \tag{2.45}$$

where h =the Planck constant
      c =speed of light
where λ =wavelength
where P_w =photoelectric work function.

P_w is a property of the photocathode, and is a measure of how difficult it is to knock out an electron from it. 'Blue' photons have more energy, and are of shorter wavelength, than 'red' photons. Thus K_e is greater and the photoelectron finds it easier to escape from the photocathode surface. This is why photomultiplier tubes, in

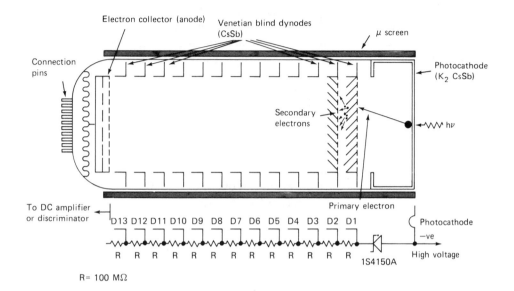

A

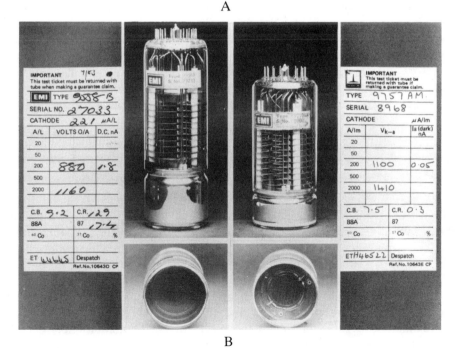

B

Fig. 2.12 — The Photomultiplier tube: A. The structure of a typical tube for chemiluminescence. B. Two photomultiplier tubes, 9558 = S20 trialkali, 9757 AM = bialkali.

Table 2.4 — Some photomultiplier (PMT) terminology

Term	Function or meaning
A. Component	
envelope	sealed glass container, under vacuum
window (end or side)	aperture through which light enters
photocathode (PC)	photosensitive material producing primary electrons
dynode (D)	multiplier producing secondary electrons
anode (A)	electrode collector (+ ve *relative* to cathode)
pin holder	external electrical connectors to PC, D and A
voltage divider network	resistors maintaining voltages across dynodes
end bell assembly	PMT holder including resistors for dynode chain
preamplifier	first amplifier of signal pulse from anode
μ shield	diamagnetic shield surrounding PMT, against cosmic noise
magnetic focusing ring	external magnet over PC to reduce effective size of PC and thus decrease noise
B. Characteristic	
bialkali, trialkali, multi-alkali	composition of photocathode
photocathode sensitivity	current (in nA) at anode with PMT in dark
gain †(G ideally = $(\delta)^n$)	multiplication factor from 1 electron at PC to output at A
response time — rise time	time to 1/e of time for anode pulse to reach maximum
response time — transit time	time for primary e at PC to give a pulse at anode
spectral sensitivity	relationship of PC sensitivity against λ
test ticket	manufacturer's ticket with key characteristics for optimal use

†δ = average secondary emission at dynode
n = number of dynodes. In fact, $G = f\,(g\delta)^n$ where f and $g < 1$, f = collection efficiency, g = transfer efficiency

contrast to photodiodes, are blue-sensitive (Fig. 2.11). At the optimal wavelength the quantum efficiency of a photocathode can be as high as 25%, i.e. on average a photoelectron is generated every fourth photon. Whether this electron reaches the first dynode depends on the negative voltage pushing it there. Then, if the efficiency of each dynode for secondary emission is four, four electrons will be omitted for every electron hit. The amplification factor of the complete tube, the *gain*, will therefore be 4^n, where n = number of dynodes. If $n = 13$ then the overall gain will be approximately 7×10^7. Herein lies the power of the photomultiplier tube. The complete process is also fast, usually <100 ns.

The problem is that a glance at any catalogue of photomultipliers reveals a plethora of different types of tube, whose characteristics (Tables 2.5, 2.6, and 2.7)

Table 2.5 — Component options in a photomultiplier tube

Component	Option
(1) Envelope	*borosilicate glass 300 nm–IR UV glass cut off below 185 nm fused silica (quartz) ideal for 175–350 nm MgF_2 specialised for UV soda lime glass no good below 350 nm rugged sapphire metal ceramic for high temperatures
(2) Window	side or *end-on
(3) Photocathode (PC)	
type	*bi or *multialkali
size	2 mm–5 cm
polarity (negative relative to anode) potential relative to anode	*negative or *earthed 200–2000 V (*1000–2000)
(4) Dynodes	
type	*venetian blind (CsSb), compact focussing, linear focused, box and grid
voltage divider network number	*voltage between PC and first dynode *ca* 300 V 6–13 (*10–13)
(5) Anode (positive relative to PC)	earthed or positive
(6) Accessories	*μ shield magnetic focusing ring *cooling

*signifies option for chemiluminometer PMT

Table 2.6 — Typical photocathode compositions

Description	Composition	Spectral peak (nm)
Bialkali (ambient)	K_2CsSb or $RbCsSb$	380
(high temperature)	Na_2KSb	360
S	Cs_3Sb	380
S1	Ag-OCs	(360) 800
S10	BiAg-OCs	420
S11	Cs_3Sb–O	390
Super S11	Cs_3Sb–O	410
S20 (tri- or multi-alkali)	Na_2KSbCs	380
Others	GaAs, GaCs, CsTl, CsI	flat <320, <200

NB: Sb = antimony, a non-metal

Table 2.7 — Typical characteristics of a photomultiplier tube specially selected for chemiluminescence

Characteristic	Description
window of glass envelope	borosilicate
photocathode	1–5 cm bialkali, or multialkali for better red sensitivity
dynodes	10–13 of CsSb
operating voltage	900 to 1200 V (some now 1500–1800 V)
gain	10^6 to 10^7
dark current at maximum sensitivity	0.1–10 nA
saturation current	many μA cm^{-2}
rise time	2–20 ns
transit time	20–80 ns

are described in bewildering detail. But which to choose? Let us consider the options. These relate to the composition and/or the electronic circuitry associated with each of the four main components, i.e. the window, the photocathode, the dynode chain, and the anode (Table 2.5). For chemiluminescence we only need an end window, transparent to visible light. The photocathode (Table 2.6) will be bialkali (e.g. K_2CsSb), or multialkali, sometimes called trialkali, designated S20 (e.g. Na_2KSbCs) for increased sensitivity in the red (Fig. 2.11). The dynodes will be Cs Sb, and number 11–13. The more dynodes the greater the multiplication, but also the greater the noise. Thus 13 dynodes may not necessarily be better in terms of signal-to-noise than 11. The controversy really starts when the circuitry is set up. There are three main decisions to be taken:

(1) Is the photocathode to be negative and the anode earthed, or the photocathode earthed and the anode positive?
(2) What is to be the overall voltage across the tube, and across each dynode, particularly between the photocathode and the first dynode?
(3) What is to be the size of the photocathode?

We are trying to achieve the maximum output signal from the anode, but with minimum noise. There are many potential sources of noise in a photomultiplier tube: ionisation of residual gases or those permeating through the glass envelope; light generated in the tube by ionisation; electrons hitting the glass (this can be minimised by a +ve anode and an earthed photocathode); electron leakage at the output, e.g. caused by dirt, fingerprints, or moisture at the base of the tube; primary electron emission from the dynodes when at high potentials; radioactive elements in the glass, e.g. ^{40}K, ^{238}U, ^{232}Th; and cosmic rays. In addition, tubes may inexplicably 'burst', particularly if damaged or if near other apparatus, e.g. vortex mixers, causing sparks, radio signals, or mains surges. However, the most common source of noise is thermal (thermionic), i.e. electrons emitted spontaneously from the photocathode and the dynodes. Here digital counting wins over analogue processing, since pulses from the dynodes will mostly be removed by the discriminator. To get the best out of your photomultiplier:

(1) Ask for one specially selected for low dark current, and have it spectrally calibrated.
(2) Optimise the high voltage and discrimination settings for the best signal:noise, using a chemiluminescent standard (Fig. 2.9).
(3) Use a magnetic μ shield to minimise magnetic interference, particularly when the photocathode is negative.
(4) Cool the photomultiplier tube (e.g. to $-20°C$), particularly if the sample is to be heated above room temperature. This is particularly important in the multialkali (S20) tubes. However, even bialkali tubes increase their dark current some 2-fold between 0 and 20°C, and a further 5 to 10-fold between 20°C and 37°C!

Having selected your tube, there are a number of points to watch:

(1) Do *not* expose the tube to daylight with the high voltage on! This will irrevocably 'burn' the photocathode, with the result that the tube will never return to its original low dark current, and will probably 'burst'. Exposure to light should be avoided even with the high voltage off, because when the voltage is switched on the tube will take longer to stabilise a low dark current compared with a dark-adapted tube. Beware also of bright chemiluminescent flashes when the photomultiplier is operating at maximum sensitivity, for the same reason. My rule of thumb is, if you can see it with the non-dark adapted eye, it is too bright for a tube set at optimal sensitivity.
(2) Saturation. If the light intensity becomes too high, e.g. at 10^7–10^8 photons s^{-1}, the photon rate becomes greater than the rate of transit of the electrons down the tube. Furthermore, if the current between the photocathode and the first dynode becomes high there will be a voltage drop between them. It may be possible to circumvent saturation of the photocathode by reducing the high voltage, but this is dependent on the voltage divider network across the dynodes. Sometimes this is designed to maintain the photocathode to first dynode voltage at about 300 V, even as the overall voltage drops. To increase the light intensity at which saturation occurs, the first dynode voltage must also be decreased. Alternatively, a neutral density filter can be used to reduce the intensity hitting the photocathode.
(3) Environmental factors to be wary of include exposure of the photomultiplier tube to daylight, to high temperature, to magnetic fields, to vibration, to shock, and to helium, which permeates glass. Normal bialkali tubes should not be used above 60°C, whereas at very low temperatures, they may lose sensitivity. As we have already seen, the dark current increases some 10-fold between 20 and 37°C.
(4) UV radiation — remember most chemiluminometers have photomultiplier tubes with envelopes made of borosilicate glass. These are essentially opaque to light below 350 nm. Hence you will never record UV chemiluminescence with these detectors.
(5) Safety — the high voltage across a photomultiplier tube is dangerous!

Which then is best, photomultiplier tube or photodiode? From the practical point of view, the answer to this question is determined by two factors:

(1) Sensitivity — ultimate, range and spectrum.
(2) Convenience — size, cost, and the need for expensive, complicated, or bulky accessories such a coolers and stabilised EHT (extreme high tension) supplies.

There is no doubt that silicon photodiodes win hands down on the second of these. They are small, but can still have a light catchment similar to that of a 1–2 cm diameter photomultiplier tube. Photodiodes can work off a small battery, at tens as opposed to thousands, of volts, and are cheap. If you need a pocket chemilumin-ometer, in the field, for quantifying visible reactions, e.g. luminous animals, a photodiode detector is highly suitable. Although the quantum efficiency of the best photodiodes may apparently range from 10–30% in the blue to as much as 60% in the infrared, and their noise is in picoamps. A substantial number of photons are needed to build up a charge detectable even over this low background. There is no built-in amplification. Although there is a class of photodiode, known as an avalanche photodiode, with a gain of some 10^3, this is only suitable for the red to infrared region. For light intensities below 10^3–10^5 photons per second then a bialkali photomultiplier tube in photon counting mode will be required. Quite small ones are now available, some with photocathode surfaces of just a few mm. But it will need a stabilised high voltage supply giving at least 1000–1500 volts. Cooling is not essential for bialkali tubes, but probably will be if you need to use a multi- or trialkali tube for better sensitivity in the red (Figs. 2.11 and 2.12). The noise of a photomultiplier will be reduced by cooling, particularly down to -20 to $40°C$, and also by reducing the effective area of the photocathode by using a magnetic focusing ring. Background counts as low as 20–$50 \, s^{-1}$ at high sensitivity are relatively easy to obtain, and with some effort one can approach a background of 1 count s^{-1}, enabling just a few photons per second to be detected.

The difference between a photodiode and photomultiplier tube is reflected in their specifications of quantum efficiency, noise, and gain. The optimal wavelength for photodiodes is usually around 900 nm, well into the infrared. Their sensitivity can decrease some 5- to 10-fold between 650 nm (red) and 400 nm (blue). In contrast, photomultiplier tubes are blue sensitive. A good bialkali tube has a maximum quantum efficiency of some 25–30% at 350–400 nm, which drops off markedly to only 1% at 600 nm. Thus a firefly flash at 565 nm will need some 5–10 times as many photons to generate the same signal in the chemiluminometer, as a flash from a luminous dinoflagellate or jelly fish (460–510 nm). The multi- or trialkali (S20) tube improves sensitivity in the red to some 8% quantum efficiency (Fig. 2.11). The gain of a photomultiplier can be as high as 10^8, but is typically 10^5–10^7 for the best signal : noise in photon counting. Its working range at optimal sensitivity is between 10 photons s^{-1} and 10^6 photons s^{-1}, when it saturates. This range can be extended to bright light levels by using a neutral density filter or by reducing the high voltage, provided that the dynode chain is set up appropriately. In contrast the best photodiodes have a working range of 10^{-8}–10^{-4} watts, equivalent to approximately 10^{10}–10^{14} photons s^{-1}.

One final technical point concerning the response time. The mean time taken for a pulse to appear at the anode of a photomultiplier after the photocathode has been hit by a photon is called the 'transit time'. This is usually about 30–100 ns and has a spread of 2–50 ns. In contrast, photodiodes, with no amplification chain, respond more rapidly to light. Light causes electrons to move from the valence to the conduction band, leaving behind charged carriers. The fast initial response, response time 1–10 ns, of a typical silicon photodiode (spd) reflects the transit time of these carriers. But there is also a much slower component in the spd response of some 1–10 μs, corresponding to charged carriers formed outside the region of the electric field moving in to it by diffusion. An spd also has a significant capacitance, around 1 nanofarad resulting in a decay phase of μs after the light is switched off.

2.4 THE IDEAL CHEMILUMINOMETER
Ideally, our chemiluminometer needs to be able to quantify chemiluminescent reactions emitting as little as 1 photon per second up to visible, bright emissions of some 10^{12} photons s^{-1}, over the whole visible spectrum (380–750 nm), over a wide temperature range (usually 0–50°C), and capable of detecting flashes of just a few ms in duration to glows lasting many minutes or even hours. To achieve this we need a highly sensitive bialkali (or multialkali) photomultiplier tube with a stabilised, but variable, high voltage supply set up in digital mode (Table 2.7). Analogue devices cannot detect low pulse rates. Their noise is very variable, and unlike the discriminator in a digital system, cannot distinguish low energy noise pulses emanating from the dynode chain from those originating at the photocathode. The photomultiplier will require some cooling if the reaction is above room temperature. In summary, the main components will therefore be:

(1) A thermostatable light tight housing, with an injector port, having the flexibility to cope with tubes, animals, gels, microtitre plates, as well as being able to automate hundreds of tubes for multiple assays. It should also be possible to insert an oxygen electrode when required. The best way to achieve this flexibility is to construct the housing in modular form, if necessary using fibre optics as light guides.
(2) A specially selected bialkali photomultiplier tube with a dark current <0.1 nA if 11-stage, or <1 nA if 13-stage, placed as close to the light source as possible. It should be cooled at least to +4°C, or even to −20°C.
(3) A stabilised high voltage 500–2000 V setting the photocathode highly negative (some people argue that the noise is lower with the anode highly positive and the photocathode at earth potential).
(4) A preamplifier to provide the discriminator with the pulse in an appropriate form.
(5) An eight-digit scalar to display the accumulated photon counts, a chart recorder output, and a computer interface to process the data, print and plot it on a printer and VDU.

I, and others, have tried to meet these requirements by building our own chemiluminometers (Fig. 2.5). However, many commercial instruments are now available. Only a few are set in true pulse counting, digital mode (see Fig. 2.6, A,B and Appendix VC), in spite of their presenting the data in digital form. Several are designed to handle sequentially, or occasionally simultaneously, many samples. The problem with commercial instruments is that they do not have the flexibility, particularly in the sample housing, required by the imaginative experimenter. However, at least two can detect biochemical events in single eukaryotic cells.

So far we have assumed that the detector needs simply to capture as many of the photons emitted by the reaction as possible. However, there is also a need to identify from where within the sample the light is originating. For example, what is the time sequence of flashing of individual organs in a luminous fish or squid with multiple photophores? From where is the light coming within a single luminous cell, or one injected with a chemiluminescent indicator? Would it not be simpler and quicker to visualise a complete microtitre plate of 96 wells, a whole gel, or nitrocellulose sheet, rather than have to scan it? To solve these problems we need to record and quantify the time sequence of a chemiluminescence image.

2.5 IMAGING OF CHEMILUMINESCENCE
2.5.1 The problem
Bright chemi- and bioluminescent reactions and organisms can be recorded with a still or cine camera, preferably using a large-grained, high-sensitivity film (high ASA number). The ability of photographic emulsions to produce a permanent image of a wide area, dissectable in minute detail because of the many thousands of grains on each square cm of film, many times per second, is unique amongst imaging dievieces. However, anyone who has waited, frustratingly, whilst watching down a microscope a low intensity fluorescent image photobleach before their eyes will be well aware of how difficult it is to record faint images using photographic film. As we saw in section 2.3, in technical terms, the photographic process suffers from three main problems: non-linearity of response over the wide range of light intensities required to observe chemiluminescence, restricted dynamic range, reciprocity failure and adjacency effects. At the light levels required for visualising low intensity chemiluminescence, i.e. $1–10^4$ photon s^{-1}, photographic emulsions are hopeless. What is needed is a means of intensifying the original image so that it is bright enough to be photographed, or a way of converting it to a computerised image which will enable temporal changes in intensity, and colour if possible, to be analysed. The problem of sensitivity will be compounded if the light emitted by the object being observed has to pass first through a microscope. Light losses through reflections and scattering at optical interfaces in a microscope can reduce the light signal leaving the microscope by more than 90%.

2.5.2 Some solutions
An imaging device for chemiluminescence has five requirements:

(1) Sensitivity: it must be able to amplify an original image ranging from just a few photons per second per unit area, up to many millions.
(2) Response time: it should be capable of millisecond response, when studying fast biological phenomena such as muscle contraction.
(3) Quantification: the light intensity per unit time per unit area should be quantifiable to produce a temporal, intensity contour map, or a pseudo-image where different colours represent intensity thresholds (red=bright, blue=dim).
(4) Spatial resolution: the shape of the image should be undistorted with high spatial resolution, down to 0.1–0.5 μm when studying individual living cells.
(5) Spectrum: the spectral response should cover the entire visible spectrum (380–750 nm), with the capacity for quantifying changes in colour at particular locations within the image.

These can be achieved by converting the photons to an electrical signal, which can be amplified whilst retaining the image. In 1934 Kiepenhauser proposed that the focused electrons could produce an image directly on a photographic emulsion, and in 1936 Lallemand independently began to put this idea into practice — the Camerà Electronique. Modern electrographic cameras use nuclear, rather than photo-graphic, emulsions. These are very fine grained, have low background fog, and have good electron stopping power because of their high silver halide-to-gelatin ratio. Importantly, density of the image is directly proportional to exposure ($\int I\, dt$). There is no reciprocity failure. These 'cameras' have found application in astronomy, but have yet to be used by biologists. Imaging devices currently used in biology convert the electron image to a visible image by using a phosphor screen or by using computer processing (Tables 2.8 and 2.9). The former can be photographed and the dots counted, or quantified by using a densitometer, or recorded on videotape using a TV camera. Ideally the imaging device first converts photons to electrons, then amplifies the electron signal with minimal amplification of the noise (i.e. in the absence of light), and finally displays the electron amplified signal visually and digitally. There are three ways of converting a light signal to an electrical one:

(1) A television camera.
(2) A charge-coupled device.
(3) A photocathode.

TV cameras

A vidicon TV camera scans a light sensitive, conductive target with an electron beam. This beam initially charges the rear side of the light target. When light hits the front of the target, which is now positive with respect to the rear, its conductivity increases, and as a result the rear side becomes more positive. The electron scanning beam then reads this signal by depositing electrons on the positively charged areas. The resulting change in capacitance in the photoconductive target thus serves as the signal electrode. In contrast in an orthicon TV camera the electron beam scans the light target, deposits electrons on positive areas, and then returns to an electron multiplier which surrounds the electron gun, the source of the scanning beam. TV cameras are the only devices available for recording a true colour image. However,

Table 2.8 — Abbreviations associated with imaging

CCTV	capacity coupled television
SIT	silicon-intensifier target
SIV	silicon-intensifier vidicon
SID	silicon imaging device
CCD	charge-coupled device
MCP	microchannel plate
IPD	imaging photon detector
MOS	metal oxide semiconductor
ISIT	intensifier silicon intensified target

Table 2.9 — Imaging of chemiluminescence

Feature	Example	Instrument
A. Light detector	silicon target	orthicon, Vidicon TV camera
	silicon diode array	Ultricon
	SbS_3	sulphide Vidicon
	SeAsTl	SATICON Vidicon
	lead oxide	Vistacon
	pn junction	photodiode array
	photocathode	
	(bialkali or S20)	silicon intensifier target tube (SIT)
		image orthicon
		image photon detector (IPD)
		image intensifier
	charge-coupled device (CCD)	image Isocon
	image intensifier	
B. Electron multiplier	phosphor-dynode	image intensifier
	microchannel plate	IPD
C. Electron detector (+ image producer)	phosphor screen (direct)	image intensifier
	TV set	TV camera
	computer + TV set	IPD
D. Electron focusing	electrostatic	image intensifier
	magnetic or electro-magnetic	image intensifier
	direct on to solid state array	CCD
	microchannel plate	IPD

they are not capable of detecting low level chemiluminescence. To improve the sensitivity of a TV camera a photocathode is incorporated as the initial light detector. The photocathode may be bialkali or multialkali (Table 2.9). The photo-electron then hits a target which is scanned by the electron beam in orthicon (image-orthicon and image Isocon) or vidicon (SIT) mode. The best SIT vidicons have gains of some 10^4 or better, with a resolution of about 15 μm, and a noise of 100–1000 events cm^{-2} which can be improved by cooling.

Charge-coupled device (CCD)
A charge-coupled device is a solid state two-dimensional array of semiconductors containing as many as 2.5×10^5 units, or 'pixels' as they are called, on an area of just a

Table 2.10 — Television camera tubes

Type	Light target (photoconductor)
A. electron beam scans change in target charge	
orthicon	Si
Isocon	bialkali or multi-alkali photocathode
image orthicon	S-10 photocathode
B. electron beam scans change in target conductivity	
Vidicon	Si
sulphide Vidicon	SbS_3
Plumbicon	Si
Ultricon	silicon diode array
Vistacon	lead oxide
SATICON Vidicon	Se-As-T1
SIT Vidicon	multi-alkali photocathode
ISIT Vidicon	image intensifier coupled to a SIT

few square centimetres. These units are very well insulated from each other. Photons hitting a pixel build up a charge. The array is then read off in sequence to see which pixels are charged. An image is then reconstructed by computer. Another type of solid state imaging device detects a change in voltage or capacitance depending on the position of the photon hit. This information is stored in a computer and the image then reconstituted.

Image intensifiers

Whilst TV cameras and CCDs may be able to record a chemiluminescence image, they do not strictly 'intensify' it. The electrical signal is, of course, amplified but this amplification also increases the noise, necessary to record the signal but without enhancing the signal (S) to noise (N) (or strictly S^2/N or $S/\sqrt{N}$). To amplify the image, electrons produced by a photocathode are 'multiplied', using either secondary emission or a phosphor-photocathode sandwich. The former is used in 'microchannel plates' which are composed of a bundle of a large number of glass fibres (capillaries $10–20\mu$ in diameter) coated with secondary electron emitting material (eg Cs Sb). The electrons are accelerated by several kV and are reflected internally, leading to an avalanche at the end with a gain of some 10^7. The fibre bundle retains the 'electron image' which is reconverted to light by a phosphor at the far end. Alternatively, the photoelectrons can be focused electrostatically or magnetically to hit secondary, internal phosphors sandwiched with a simple photocathode, e.g. composed of KCl, which generates more electrons. Each stage can produce a gain of 50 to 100-fold. Thus the best electromagnetically focused image intensifier tubes have four such stages, resulting in a gain of $10^6–10^7$ when operating at maximum voltage (40 kV) and a dark current equivalent to that of a very good photomultiplier tube. Electrostatically focused tubes are more compact, work at lower voltages, but are less sensitive.

The question arises, which is the best device for imaging chemiluminescence? Many of the electronic advances necessary to develop these imaging devices have been made as a result of military programmes and the needs of astronomers. The specifications of absolute sensitivity and fast time resolution required by biologists are different. Experimental data are now needed to assess the value of these various devices for biology.

The most sensitive and flexible systems for imaging in biology use a combination of light detectors and electron multipliers (Tables 2.8 and 2.9). In other words, use a TV camera or CCD to record and quantify the image from an intensifier. Popular combinations include:

(1) Intensifier — silicon intensified — TV tubes, a SIT Vidicon optically coupled by fibres to the phosphor screen of a two stage, electrostatically focused image intensifier.
(2) Intensifier-Ultricon TV camera, an Ultricon (a more sensitive Vidicon) TV camera optically coupled to the phosphor screen of a photocathode–microchannel plate image intensifier.
(3) Image photon counting system, a Vidicon TV camera coupled by lenses to the phosphor screen of a four-stage magnetically focused intensifier tube.
(4) Image photon detector, a charge-coupled device optically coupled to the phosphor screen of a photocathode-microchannel plate image intensifier.

The relative merits of these four systems for imaging chemiluminescence depend on three main factors:

(1) Sensitivity in detecting small numbers of photons per second over the visible range.
(2) Time resolution.
(3) Ease of data processing.

Semiconducting solid state devices tend to be red sensitive, with poor sensitivity in the blue, though this can be improved by cooling. In contrast photocathodes are blue sensitive. Multialkali photocathodes are slightly more sensitive in the red than bialkali ones, but must be cooled to produce equivalent low noise in the absence of light. Until relatively recently the outstanding imaging system for visualising bioluminescence and locating chemiluminescence indicators inside cells was the four stage electromagnetically focused intensifier tube, with a maximum gain of 10^6–10^7, and a noise of only a few events per cm^2 per second on the phosphor screen, enabling single photoevents to be detected. The image can be recorded by a photographic still or cine camera, or a TV camera (Fig. 2.13). The elegant studies of Reynolds and his collaborators have shown what a powerful tool this device is for localising bioluminescent microsources in single cells and waves of intracellular Ca^{2+}, for example in fertilised eggs (Fig. 2.14). The resolution when using a microscope is 0.5 μm, which is perfectly adequate for most cells, and the response time is dependent on the chemistry of the indicator, not that of the detector. However, these devices are cumbersome, fragile, and work at very high voltages, up to 40 kV.

In contrast the recent introduction of image photon detectors using S20 multialkali photocathodes with quantum efficiencies of 5–25% (650–450 nm) coupled with a microchannel plate and CCD provides a compact device, working at just a few hundred volts, with a dead time of only 15 μs, of great potential for imaging chemiluminescence. The quoted intensity range of these devices is from a few photons s^{-1} to some 10^5 photons s^{-1}, when it saturates. Experimental data are now required to see whether this type of device is a real alternative to the four stage, electromagnetic image intensifier tube.

CCDs have the great advantage of generating a signal which can be easily processed digitally by a computer. Their spatial resolution is limited by the number of pixels per unit area. Their time resolution is, at present, relatively slow, as it can take several seconds to charge up and several seconds to read out. Their potential in chemiluminescence is, as yet, unproven. Simple TV cameras and even SITs are too insensitive for low intensity chemiluminescence, though they have application for fluorescence. Intensified TV cameras are more sensitive, but their ultimate sensitivity as single photon detectors also remains unproven in biological systems. Their spatial and temporal resolution is limited by the 525 lines, but is usually quite adequate for most biological experiments. A large memory is required for full computer analysis of the images with time. However, a simple software can be developed to scan, and then plot, the intensity distribution of each line, enabling, for example, a single cell to be scanned semi-automatically.

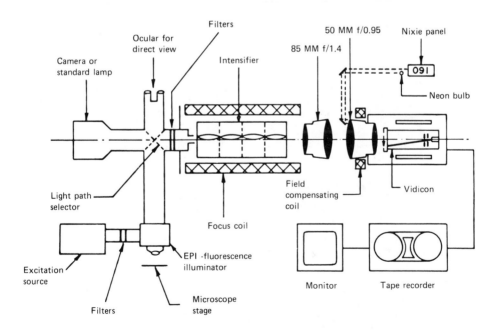

Fig. 2.13 — An imaging system for bioluminescence. From Reynolds (1980) *Microscopica Acta* **83** 55–62, with permission of S. Hirzel Verlag.

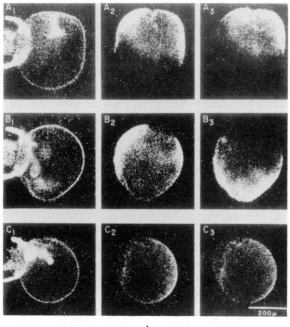

A

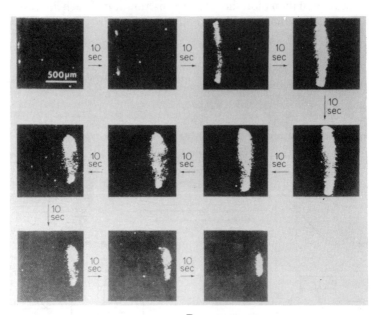

B

Fig. 2.14 — Examples of imaging chemiluminescence in living cells. A. Microsources of dinoflagellate bioluminescence (*Noctiluca miliaris*) from Eckert & Reynolds (1967) *J. Gen. Physiol.* **50** 1429–1450 with permission of Rockefeller University Press. Object to left = pipette holding cell. Spots outside cell = noise. Three specimens A, B, C under different conditions. B. Free Ca^{2+} wave in a *Medaka* fish egg detected using aequorin, from Campbell (1983) *Intracellular calcium* copyright (1983). Reprinted by permission of John Wiley & Sons Ltd © 1983 adapted from Gilkey *et al.* (1978) *J. Cell. Biol.* **76** 448–466.

Image intensification of luminescent indicators in living cells is one of the most exciting developments in contemporary cell biology. More experimental evidence is required to enable newcomers to select the appropriate device for their particular application.

2.6 CHEMILUMINESCENCE SPECTRA

The emission spectrum is a fundamental property of any chemiluminescent reaction. Firstly, it relates directly light emission to the energetics of the reaction through Einstein's equation $E=hv$. Secondly, comparison of the chemiluminescence spectrum with fluorescent spectra of potential reaction products plays a key role in identifying the structure of the excited-state molecule generated by the reaction. In bioluminescence this also provides clues to the structure of the initial substrate. Thirdly, the colour and quantum yield of many synthetic reactions, as well as those from several luminous organisms, is changed through energy transfer to a 'fluorescent' acceptor (see Chapter 10). The discovery and understanding of this fascinating aspect of chemiluminescence requires careful measurement of spectra. Fourthly, environmental factors, such as pH, ionic strength, and hydrophobicity, can modify the colour of a chemiluminescent reaction. These factors can even modify a reaction within the active centre of a luciferase. The firefly *Photinus pyralis* emits yellow light (λ_{max} 565 nm), whereas its European relative, the glow-worm *Lampyris noctiluca*, emits green light (λ_{max} 550 nm). Yet the structure of the chromophore and the chemical reaction are identical. Chemiluminescence spectra are often not quite identical to the fluorescence spectrum of the reaction product. Furthermore, many luminous organisms show peculiar spectral changes during a single flash or glow. To investigate the cause and function of these spectral perturbations we need to measure temporal changes in the chemiluminescence spectrum. Could it be possible to identify a particular species in the deep ocean by the temporal pattern of its chemiluminescent spectrum?

Ideally, we would like to measure the true spectrum, designated $T(\lambda,t)$, i.e. the absolute photon intensity at each wave number as it changes with time. This would require a chemiluminescence spectrometer calibrated to record absolute photons (see section 2.7) and capable of the finest possible resolution of the spectrum, using a detector having a uniform response over the full spectrum. In practice $I(\lambda,)_t$ is measured with an instrument which has a discrete spectral and time resolution $S(t)$, an efficiency of trapping photons $\ll 100\%$, and using a detector with a non-uniform spectral response (Fig. 2.11). The desire to correct for the spectral sensitivity of the detector is no esoteric whim. Without it, even qualitative assessment of the spectral shape can be wrong or misleading. However, correction for the quantum efficiency of the instrument, reduced by geometric and electronic factors, is not essential, provided that these factors remain the same over the whole spectrum. Thus obtaining the true temporal chemiluminescence spectrum ($T(\lambda,t)$) requires two stages, firstly a measurement of $I(\lambda,)_t$ to produce $S(\lambda, t)$ over the visible range 380–750 nm, perhaps selecting a narrower region once the main spectral region has been identified, and secondly correction for spectral perturbations as a result of non-uniformity within the instrument to produce $S(\lambda, t)$ 'corrected'. If necessary, the efficiency of an instrument can be taken into account and $S(\lambda, t)$ 'corrected' and converted to $T(\lambda, t)$.

To achieve the first of these objectives the light from the sample is split into discrete spectral components ($\Delta\lambda$), and the intensity of each is measured simultaneously or sequentially. As we saw earlier (section 2.2.4) the monochromator is usually a diffraction grating, the discrete part of the spectrum ($\Delta\lambda$) being selected by positioning a slit. Rotation of the grating or the slit enables the detector to scan the spectrum. Alternatively a discontinuous spectrum can be obtained by using a series of interference filters with narrow bandwidths (e.g. 7–10 nm). The problem with these two approaches is that obtaining a spectrum can take many minutes. If the chemiluminescent reaction decays significantly over this time, as is usually the case, then this can be corrected for by having a reference detector (R_D) placed at 90° or 180°C to the spectral detector (Fig. 2.5D)), monitoring the complete spectrum, i.e. without the monochromator. Thus

$$S(\lambda,t)=I(\lambda,t)/R_D(t) \tag{2.46}$$

At any point in time the intensity of the light seen by the reference detector will be at least 10–100 times that of the narrow bandwidth seen by the spectral detector, at lower sensitivity than the spectral detector. This will correct for temporal changes in intensity, but temporal changes in the chemiluminescence spectrum will still be distorted unless the reference detector is also corrected for spectral sensitivity.

If the temporal changes in intensity or spectra are relatively slow, i.e. several minutes, then a scan from 400–600 nm and back again should produce two superimposable spectra for each measurement. Alternatively the grating, or discs containing several interference filters, can be rotated very fast. However, to record spectra from very short flashes, for example those seen in the hydroid *Obelia,* it may be necessary to record a complete spectrum within 100 ms. This requires the grating, or disc of filters, to be rotated at some 600 rpm. The data are then processed by computer. A more elegant way of recording a spectrum within 50–100 ms is to display the whole spectrum onto an imaging device as suggested by Reynolds.

Many of the first chemi- and bioluminescence spectra were obtained by splitting the spectrum with a prism, and allowing the light to fall on the naked eye, using a hand spectroscope, or on to photographic film. Thus in 1935 the glow-worm was shown to have a peak emission at about 550 nm (Fig. 2.7C). A plot of intensity versus wavelength can be obtained simply by placing an absorbing 'step' wedge at 90° to the spectrum (thinnest point on the axis), so that the brighter the light the higher up will be the intensity threshold, below which no latent image forms. This method does not take into account the spectral sensitivity of the photographic emulsion, and can be used only for bright, visible emissions. A more modern approach is to display the spectrum from a diffraction grating onto the photocathode of an image intensifier tube, an Ultricon or SIT TV camera, a charge-coupled device, or an image photon detector with a microchannel plate. A linear array of 700 silicon photodiodes coupled by a bundle of fibre optics to a microchannel plate image intensifier has enabled unique recordings of bioluminescence spectra and of single photophores from hundreds of luminous animals to be obtained (Fig. 2.7B and C see Appendix VA). This particular system may, however, be limited in sensitivity to emissions producing visible light.

Correction of chemiluminescence spectra for spectral perturbations in the apparatus is much easier if a computer is interfaced to the chemiluminescence spectrometer. There are two main sources of such perturbation, independent of the experimental sample or animal being studied:

(1) Non-uniformity of spectral transmission through the monochromator.
(2) Non-uniformity in spectral sensitivity of the light detector.

The percentage transmission of a diffraction grating at different wavelengths, or that of a narrow band width interference filter, can be measured easily, by placing it in the beam of a spectrophotometer. There are three ways of estimating the spectral perturbation of the complete instrument, incorporating both the monochromator and the light detector distortions:

(1) Standard lamp
    Compare the spectrum obtained when using a standard lamp source with the spectral irradiance obtained from physical tables with a knowledge of the colour temperature. The latter can be measured with a meter based on energy output. Typical standards are a 1000 watt tungsten filament having a colour temperature at 2342°K or a 45 watt quartz iodine lamp. The correction factor (F_λ) at each wavelength is the ratio of the measured signal to the calculated lamp intensity, based on the laws of black body radiation, and is available from standard tables.

$$S(\lambda, t) \text{ corrected} = S(\lambda, t) \text{ measured}/F(\lambda) \qquad . \qquad (2.47)$$

    The problem with this method when used to calibrate a chemiluminometer is that the lamp is extremely bright for single photon counting instruments. It may emit as many as 10^{16} photons s^{-1}. Hence it has to be placed many metres away and a pin hole must be used. Furthermore, the high voltage has to be drastically reduced if the photomultiplier tube is not to be saturated.
(2) Fluorescent standard
    The true fluorescent spectra of many fluors are well established. Incorporating an exciting light source into the chemiluminometer enables the measured fluorescence spectrum to be compared with the true spectrum. A further correction is required for the spectral emission of the exciting source, which will be non-uniform.
(3) Chemiluminescence standard
    Thanks to painstaking work by many enthusiasts, the true chemiluminescence spectra of many synthetic compounds and bioluminescent systems are now known. A highly efficient energy transfer system, e.g. oxalate esters, can also be useful. The chemiluminescent spectrum of these should be identical to the fluorescence of the acceptor. However, take care to ensure that environmental factors within the sample do not perturb the chemiluminescent spectrum of these standards.

Having obtained a true, i.e. corrected, chemiluminescence spectrum, we are now in a position to assess a number of important qualitative and quantitative characteristics. What is the shape of the spectrum? Is it unimodal with a single peak, or does it have more than one peak or a shoulder? Does the shape of the spectrum change with time (Fig. 2.7C)? What are the wavelengths of the peak emission(s) (I_{max})? How broad is the spectrum? This is described quantitatively as the bandwidth between the two wavelengths producing $I_{max}/2$ (see chapter 3)? How does the chemiluminesce spectrum compare with the fluorescence spectrum of the product of the reaction, or with that of a potential energy transfer acceptor? How does the *in vivo* bioluminescence spectrum compare with that measured in broken cells or with that of the purified system? The answers to all of these questions provide fundamental insights into the physics, chemistry, and biology of the phenomenon. The other crucial characteristic of a chemiluminescent reaction is its quantum yield. To measure this, we need to calibrate not only the relative spectral sensitivity of the instrument, but also its absolute sensitivity.

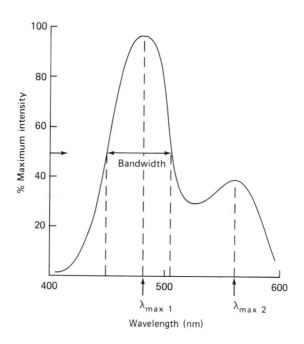

Fig. 2.15 — Parameters to measure from a chemiluminescence spectrum.

2.7   QUANTUM YIELD AND INSTRUMENT EFFICIENCY

The overall quantum yield (ϕ_{CL}) of a chemiluminescent reaction, as we have already seen (see Chapter 1), is the number of photons emitted divided by the number of molecules reacting. A reaction with $\phi_{CL}=1$ is 100% efficient. In most chemi- and bioluminescent reactions, capable of producing light bright enough to be seen by the human eye, ϕ_{CL} is 0.01–1 (Table 2.11). However, several dim emissions, and some ultraweak chemiluminescence invisible to the naked eye, may have $\phi_{CL}\ll0.001$. To

Table 2.11 — Some chemiluminescence quantum yields

Chemiluminescent substrate	λ_{max} (nm)	approx. quantum yield
A. Synthetic compounds		
luminol ($OH^-/H_2O_2/Fe(CN)_6^{3-}$)	424	0.01
luminol (t butoxide/DMSO 25°C)	485	0.015
isoluminol ($OH^-/H_2O_2/Fe(CN)_6^{3-}$)	ca 425	0.001
aminobutylethyl isoluminol (OH^-/H_2O_2/peroxidase)	ca 425	0.01
lophine	530	—
lucigenin (OH^-/H_2O_2)	500	0.01
oxalate ester + fluor	max. for fluor	0.05–0.5
$^1O_2-^1O_2$ dimol	634,703	10^{-6}
Grignard reagent p Cl phenyl MgBr	475	$10^{-6}-10^{-8}$
benzylamine + benzoyl Cl	—	10^{-15}
B. Bioluminescent reactions *in vitro* [†]		
Vargula luciferin/luciferase	460	0.3
bacterial luciferase/oxidoreductase	480–490	0.1–0.2
Renilla luciferin/luciferase	480	0.05
aequorin	469	0.15–0.20
pholasin	495	0.09
firefly luciferin–luciferase	565	0.88[‡]

[†]Some values estimated only once.
[‡]One estimate as low as 0.3

remind ourselves, the overall quantum yield is made up of three components (see Chapter 1):

$$\phi_{CL} = \phi_C \cdot \phi_{EX} \cdot \phi_F$$

where ϕ_C = fraction of molecules going through the necessary reaction (i.e. excluding 'dark' side reactions)

ϕ_{EX} = fraction of those molecules which react in the chemiluminescence pathway which become electronically excited

ϕ_F = fluorescence quantum yield of the excited product, i.e. the number of photons emitted divided by the number of molecules actually excited.

ϕ_{CL} is therefore susceptible to modification by at least eight chemical and physical factors, many of which also affect the spectrum of the chemiluminescence:

(1) Chemical structure of the primary substrate, not only the centre piece containing the electronically excited group, but also side chains. The latter is well illustrated by the range of quantum yields obtained with different phthalazinedione derivatives, the former by the difference between the efficiency of beetle and jelly fish chemiluminescence (Table 2.11).

(2) The nature and concentration of other substrates, e.g., the oxidant(s), affecting competition of the chemiluminescent substrate between different pathways, including dark reactions. Luminol, for example, can be oxidised through at least two pathways, of widely differing quantum yields (see Chapter 8).

(3) The nature of the catalyst. Without the luciferase, ϕ_{CL} in bioluminescent reactions is much reduced.

(4) Metal ions, particularly transition metals involved in processing the oxidant.

(5) pH.

(6) Temperature.

(7) Hydrophobicity of the solvent, e.g. luminol in aqueous versus organic solvents (Table 2.11).

(8) The presence of energy transfer acceptors.

The three parameters making up ϕ_{CL}, ϕ_C, ϕ_{EX}, ϕ_F can be estimated independently. ϕ_C can be measured chemically by estimating substrate loss and product formation:

$$\phi_C = \text{product via chemiluminescent pathway/total substrate loss} \qquad (2.48)$$

ϕ_{EX} is more difficult to measure, but can be calculated if we can measure ϕ_{CL} and ϕ_F. ϕ_F of the final product can be measured from its photoluminescence, but may not be the true value because the precise chemical structure, the ionisation state, and environment of the excited state emitter formed in the chemiluminescent reaction may be different from that used to measure ϕ_F. To measure ϕ_{CL}, the number of total substrate molecules which react are measured, chemically or by absorbance or fluorescence change. The total photon emission is estimated from the efficiency of the chemiluminometer (f) and the counts recorded over the same time interval, usually when the reaction is >98% complete, i.e. more than $6 \times t_{\frac{1}{2}}$ (section 2.1.3, Eqn 2.31.)

$$\phi_{CL} = \text{counts recorded}/(f \times \text{substrate molecules reacted}) \qquad (2.49)$$

The efficiency factor (f) is dependent on four factors:

(1) The quantum efficiency of the light detector, which is always <1. For the best photomultiplier tube at the optimal wavelength this is not better than 0.25–0.3.

(2) The quantum efficiency of the detector varies with the wavelength of the light. For a yellow emitter this can reduce the recorded counts to a fifth of what they would be for blue light.

(3) The light received by the detector is always less than that emitted because of imperfect geometry. This is because of trapped and absorbed light in the housing, asymmetrical emission, and imperfect optical coupling between the sample and the detector.

(4) The high voltage and discriminator settings of the chemiluminometer for optimal signal to noise will, almost certainly, not be those for maximum yield of counts (Fig. 2.11).

All in all, the apparent ϕ_{CL} based on recorded counts can never be better than 10–20% of the true ϕ_{CL}, and may be as little as 0.1–1% of the true ϕ_{CL}.

There are, in principle, three ways of obtaining the absolute calibration of a chemiluminometer:

(1) Chemical actinometry — the use of a light *sensitive* chemical reaction.
(2) Radiometry — the use of a standard incandescent light source.
(3) Luminescent source, previously standardised.

An actinometer measures the total amount of incident radiation falling on it. Chemical actinometry is based on surrounding the light source, the chemiluminescent reaction, with a solution of a light sensitive substance which changes chemically when it absorbs light. By making sure that all the light is absorbed and by measuring the amount of product formed, and with a knowledge of the quantum efficiency of the photosensitive reaction, the light yield from the chemiluminescence source can be estimated. An aqueous solution of potassium ferrioxalate has been used to determine quantum yields of gas phase, radical ion chemiluminescence, and electroluminescence. However, the method is cumbersome and insensitive. It can require as many as 10^{15}–10^{16} photons and is therefore only applicable to relatively large quantities of high quantum yield chemiluminescent reactions.

The basis of the other two methods is to use a standard light source to calibrate the geometry of the instrument, and its overall efficiency. The light source can be a standard incandescent lamp, whose spectral and absolute emissions can be obtained from physical tables based on the laws of black body radiation. However, as we have already seen, it is difficult to make these of sufficiently low intensity to prevent saturation of a chemiluminometer. Two other popular light sources for calibrating a sensitive chemiluminometer have been either a chemiluminescent compound with known spectrum and quantum yield, e.g. luminol, or a radioluminescent standard produced by adding a known amount of ^3H or ^{14}C to a scintillant, in liquid or solid form. These radioluminescent standards produce a constant emission over the time course required for most biological experiments.

The number of disintegrations per second can be calculated, taking into account loss as a result of ^3H decay in old standards. The problem is that there is sufficient energy in a β particle to produce tens or hundreds of photons within the nanosecond resolution time of the photomultiplier. These photons are emitted in all directions, but only a few have to hit the photocathode to register a 'count'. Thus, even with poor geometry (e.g. <10%) ^{14}C will be counted in a chemiluminometer with 95–100% efficiency. To circumvent this it is necessary to quench the radioactive sample, liquid or solid, e.g. with $CHCl_3$, to reduce it to a single photon emitter. The relative geometrical efficiences of different instruments can be compared easily by using this standard, detected with the same photomultiplier tube.

If a chemiluminescent reaction has, to a first approximation, the same spectrum as the luminescent standard, then its ϕ_{CL} is easy to estimate. If for the chemiluminescence standard:

$$R = (\text{counts recorded/molecules reacting}) \times 1\phi_{CL}^{standard} \qquad (2.50)$$

or for the radioluminescent standard, $R = \text{counts recorded/nuclear disintegrations}$, then for the unknown:

$$\phi_{CL}^{unknown} = (\text{counts recorded/molecules reacting}) \times 1/R \qquad (2.51)$$

Luminol in dimethyl sulphoxide at 25°C has a ϕ_{CL} of 0.015, and provides a useful standard for many other blue chemiluminescent reactions. However, yellow and red standards are more difficult to obtain. Radioluminescent standards emit in the 380–420 nm region, some way from the spectral range of most chemiluminescent reactions. A further problem is that the spectral calibrations of photomultiplier tubes obtained using current (i.e. analogue) outputs, and available from the manufacturer, do not always match the spectral calibration obtained using digital processing:

$$(I_\lambda)_{\text{true}} = (I_\lambda)_{\text{measured}}/Q_{E\lambda} \tag{2.52}$$

where $Q_{E\lambda}$ = quantum efficiency of the photomultiplier tube at wavelength λ.

$$\tag{2.53}$$

A chemi- or radioluminescent standard can be used to measure $Q_{E\lambda}$, and thus ϕ_{CL} of an unknown can be estimated from equation (2.52).

$$\phi_{CL} = \left(\int_{\lambda_1}^{\lambda_2} \int_0^\infty I(\lambda,t)_{\text{true}} \, d\lambda dt \right) / \frac{\text{molecules}}{\text{reacting}} \tag{2.54}$$

In practice integration is not ∞ but at least 6 half times (Eqn 2.31).

From this discourse the reader will have gathered that calibrating accurately the absolute efficiency and spectral response of a chemiluminometer is not as easy as it might at first appear. This is why there is some variation in the published quantum yields and spectra of many reactions. The problems are highlighted by a controversy concerning the use of luminol chemiluminescence as a standard, compared to a ^{14}C radioluminescent standard. The latter produces values two to three times higher from those using luminol as a standard.

2.8 THE FUTURE

Accurate and precise measurement wisely used is the touchstone of the scientist. But the power of any form of scientific analysis is related directly to the questions being asked, and to the concepts being developed by the user. We have seen in this chapter how to construct a chemiluminometer capable of detecting reactions and their spectra, emitting but a few photons per second. I have tried to explain how, in principle, the key components work and for what the instrument can be used. Future advances in microelectronics will offer exciting opportunities for us to develop spectrometers for ultraweak chemiluminescence, intensifiers producing a true colour image, as well as photodiodes and solid state devices capable of detecting single photon events at the same or better levels of sensitivity and time resolution as a photomultiplier tube, thereby enabling miniature instruments to be constructed. These future developments will have unique potential in cell biology and biomedicine for probing the chemistry of reactions within healthy and sick cells, as well as having application in multi-assay systems and gel techniques required to exploit recombinant DNA technology safely in biology and medicine (Chapters 4–9). Polarisation of chemiluminescence emissions provides a challenge both for the basic and applied scientist.

The remainder of this book will attempt to highlight how quantification of light, and its colour, emanating from synthetic or living sources of chemiluminescence can open the door to the discovery of the wonders of this phenomenon and its exploitation for human good.

2.9 RECOMMENDED READING

Light and units of measurement

Brill, T. B. (1980) Light: its interaction with art and antiquities. Plenum Press, New York and London.
Murray, J. J., Nicodemus, F. E., & Wunderman, I. (1971) Proposed supplement to SI nomenclature for radiometry and photometry, *Physical tables of the Smithsonian Institute,* Washington, USA.

Photography

Bever, J. E. (1981) Bioluminescence and Chemiluminescence. pp 583–589. Eds De Luca, M. & McElroy, W. D., Academic Press, New York. *Autolumography: Using bioluminescence to 'stain' for enzyme activities resolved on polyacrylamide gels.*
Haist, G. (1979) Modern photographic processing. vols 1 & 2. John Wiley, New York & Chichester.
Mees, K. C. E. (1954) The theory of the photographic process. Macmillan, New York.
Thorpe, G. H. G., Whitehead, T. P. & Kricka, L. J. (1984) *Clin. Chem.* **30**, 806–807.

Photographic monitoring of enhancement luminescent immunoassays
Photomultipliers and photodiodes

Birks, J. B. (1964) *The theory and practice of scintillation counting.* Pergamon Press, Oxford.
Papers available from EMI:
    Thorn EMI photomultiplier manual
    A comparison of current-measurement with photon-counting in the use of photomultiplier tubes, by C. J. Oliver.
    Basic physics and statistics of photomultipliers, by J. C. Barton.
    Dark current in photomultiplier tubes, by J. Sharpe.
    Design of photomultiplier output circuits for optimum amplitude or time response.
    Image intensification systems.
    Photomultipliers (and accessories).
    Photomultipliers — space charge effects and transit time spread, by A. J. Parsons.
    Reducing noise from photomultipliers, by H. A. W. Tothill.
    Rubidium caesium photomultipliers, EMI catalogue.
    Sources of noise in photomultipliers, by A. G. Wright.
    Voltage divider design, EMI catalogue.
    Photomultiplier manual RCA, Harrison, N.J., USA.
    Solid state silicon photodiodes, RCA, Harrison, N. J., USA.
    Photomultiplier catalogue, Hamamatsu, Japan.

Instrumentation

Anderson, J. M., Faini, G. J., & Wampler, J. E. (1978) *Methods Enzymol* **57** 529–539. Construction of Instrumentation for Bioluminescence and Chemiluminescence Assays.

Seliger, H. H. (1978) *Methods Enzymol* **57** 560–600. Excited states and absolute calibrations in bioluminescence.

Stanley, P. E. (1982) In: *Clinical and biochemical luminescence,* pp. 219–260. Eds Kricka, L. J. & Carter, T. J. N. Marcel Dekker, New York. Instrumentation.

Stanley, P. E. (1987) *Methods Enzymol.* **133**.

Wampler, J. E. (1978) In: *Bioluminescence in action,* pp. 1–48. Ed. Herring, P. J. Academic Press, New York and London. Measurements and physical characteristics of luminescence.

Image intensification

Hayakawa, T., Kinoshita, K., Miyaki, S., Fujiwake, H. & Ohsuka, S. (1986) *Photochem. Photobiol.* **43**, 45–47. Ultra-low-light level camera for photon counting imaging.

Mackay, C. D. (1986) *Ann. Rev. Astron. Astrophys.* **24** 255–283. Charge-coupled devices in astronomy.

Reynolds, G. T. (1972) *Q. Reo. Biophys.* **5** 295–347. Image intensification applied to biological problems.

Reynolds, G. T. (1978) *Photochem. Photobiol.* **27** 405–421. Application of photosensitive devices to bioluminescence studies.

Imaging devices from EMI, RCA and Hamamatsu.

Chemiluminescence spectra

Harvey, E. N. (1952) *Bioluminescence.* Academic Press, New York.

Herring, P. J. (1983) *Proc. Roy. Soc. Lond. B* **220** 183–217. The spectral characteristics of luminous marine organisms.

Widder, E., Latz. M. F., & Herring, P. J. (1984) *Science* **225** 512–514. Far red bioluminescence of two deep-sea fishes.

Widder, E., Latz, M. F., & Case, J. F. (1983) *Biol. Bull.* **165** 791–810. Marine bioluminescence spectra measured with an optical multichannel detection system.

Quantum yield

Hastings, J. W. & Weber, G. (1963) *J. Opt. Soc. Am.* **53** 1410–1415. Total quantum flux of isotopic sources.

Hastings, J. W. & Reynolds, G. T. (1966) In: *Bioluminescence in progress.* pp 45–50. Eds Johnson, F. H. & Haneda, Y. Princeton University Press, Princeton. The preparation and standardization of different methods of liquid light sources.

Nakamura, T. (1972) *J. Biochem. (Tokyo)* **72**, 173–177. Standardization of an aqueous light source containing luminol: application to measurement of quantum yields of bioluminescent reactions.

3

Bioluminescence: living light

It is apparent . . . that no clear development of luminosity along evolutionary lines is to be detected but rather a cropping up of luminescence here and there, as if a handful of damp sand has been cast over the names of various groups written on a blackboard, with luminous species appearing wherever a mass of sand struck.

E. Newton Harvey, 1952

3.1 WHAT IS LIVING LIGHT?

Bioluminescence, or 'living light', is the emission of light by living organisms. The light is generated by the organism itself, is visible to the naked eye, and plays some role in the survival or reproduction of the species. It is found both in pro- and eu-karyotes, though not in man or any other mammal. Luminous organisms of one kind or another have invaded virtually all the major habitats on the earth's surface, the air, the land, and are particularly common in the sea (Table 3.1). Some 700 genera from 16 major phyla are luminous. There are thus many thousands of individual species of living light (see Appendix II).

3.1.1 Chemiluminescence versus phosphorescence

All living light is chemical light, chemiluminescence (Chapter 1, sections 1.2.6 and 1.3.4) — so the current dogma goes. This has been supported by literally thousands of experiments over the one hundred years since Dubois first identified the luciferin-luciferase reaction in the click beetle *Pyrophorus*, and in the common piddock, *Pholas dactylus* (Fig. 3.1), a boring, or to some a very interesting bivalve mollusc.

The *Oxford English Dictionary* will tell you that 'phosphorescence' is the word in common usage to describe living light. This is confirmed by the writings of many naturalists and scientists over the past three centuries, where the term has frequently been used to describe this and a wide range of examples of chemiluminescence. However, as we saw in Chapter 1 (section 1.2.6), this is, scientifically, strictly incorrect. Phosphorescence is the emission of light from triplet excited states decaying to ground state. This in itself is rare in bioluminescence, where singlet excited emission is the norm. Furthermore, the term phosphorescence is usually restricted to phenomena where light itself is responsible for generating the triplet excited state, i.e. it is a type of photoluminescence. Nor, when we describe

Table 3.1 — The invasion of different habitats by living light.

Habitat		Example Species (common name)
A. *Terrestrial*		
Soil		*Diplocardia longa* (earthworm)
		Xenorhabdus luminescens (bacterium)
		Heterorhabditis bacteriophora[†] (nematode)
Surface	— on vegetation	*Pleurotus olearius* (fungus)
	— free moving flight	*Quantula striata* (land snail)
	— capacity for flight	*Photinus pyralis* (firefly)
B. *Freshwater*		
Stones in lakes, rivers, and		*Latia neritoides* (freshwater limpet)
fast streams		*Vibrio cholerae* (*albensis*)
		(freshwater bacterium)
C. *Marine*		
Mud		*Achloë astericola* (scale worm)
Rock-boring		*Pholas dactylus* (piddock)
On algae		*Obelia geniculata* (sea fir)
On decaying fish		*Vibrio fisheri* (bacterium)
Under or on rocks		*Clytia johnstoni* (sea fir)
		Acronida brachiata (brittle star)
Free floating	— surface	*Noctiluca miliaris* (dinoflagellate)
	shallow	*Photoblepheron* (flashlight fish)
	ca 1000 m	*Acanthophyra* (decapod shrimp)
		Argyropelecus (hatchet fish)
Benthic	— shallow	*Renilla reniformis* (sea pansy)
	deep	*Peniagone théeli* (sea cucumber)
	(> 1000 m)	

This is only intended to illustrate the diversity of habitats for luminous animals. The interested reader may expand this list as the chapter progresses.

†The cause of the luminescence is the bacterium *Xenorhabdus*.

bioluminescence, are we referring to the fluorescent pigments which are responsible for the colours of many fish, such as those in a home aquarium, and even to be found in cells responsible for the chemiluminescence of living light.

Newton believed that the light from living creatures had a physical origin. The spectacular displays of luminous dinoflagellates often seen in a ship's wake have caused many sailors to worry, mistakenly, that electric sparks emanating from it might cause an oil fire. In principle, the crystalline structures observed within the luminous cells of certain species could give rise to triboluminescence, or the electric organ of fish to electroluminescence. However, until it is refuted, we shall assume here that living light, *bioluminescence*, is the result of a chemical reaction from substances produced within, or secreted by, the luminous organism producing the visible light. It is distinguished from ultraweak and from dim chemiluminescence, which can only be observed by using a sensitive photomultiplier tube and are invisible to the naked eye (see Chapter 6).

Fig. 3.1(a)

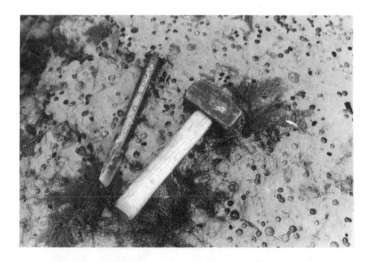

Fig. 3.1(b)

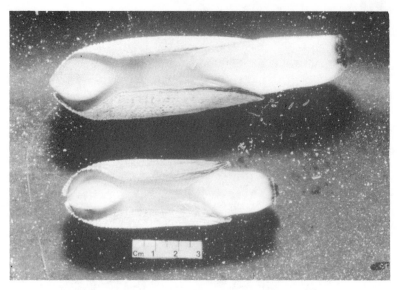

Fig. 3.1(c)

Fig. 3.1(d)

Fig. 3.1 — The common piddock *Pholas dactylus*. (a) Piddock hunters on the Severn estuary; (b) rock burrows, with the biochemist's typical weapons; (c) two *Pholas* out of their burrow. (d) internal light organs photographed by their own bioluminescence.

3.1.2   One phenomenon?

Striking as bioluminescence may be when we see it on a midnight swim or from a veranda in the tropics, is it really justifiable to deal with it as one phenomenon? Were it like muscle contraction and the cytoskeleton, where the physiology and bio-chemistry, together with the molecular biology of the structural and regulatory proteins, follow clear patterns of development throughout the animal kingdom, then there would be no question of not treating it so. However, five quite distinct chemical families have already been discovered (section 3.4). Further divisions can be identified when the oxidant and biochemistry of the protein catalysts are examined, and related to control and function. These are responsible for the bioluminescence of organisms as diverse as jellyfish and bacteria. Several others await further clarifica-tion or discovery. It is possible that chemiluminescence may have arisen indepen-dently more than 30 times during evolution. The oxygen carrying pigment in human blood is red, as is the pigment of the 'shell' or carapace of several deep sea decapod shrimp when examined on the surface. Yet the chemistry and biochemistry of these two pigments would hardly be considered together. There are four particular reasons why, as scientists, we should consider bioluminescence as one phenomenon:

(1)   Identification of common chemistry
      Only by studying the whole phenomenon can the similarities and differences in the chemical reactions responsible for generating excited states in Nature be identified. The wide-ranging importance in bioluminescence of excited carbo-nyls generated by dioxetan or dioxetanone decomposition (Fig. 1.10) has been discovered as a result of this approach. Common aromatic rings at the centre of the excited state molecule, but with different side chains, have been identified in luminous coelenterates on the one hand and in luminous ostracods on the other (section 3.4). An example of convergent evolution? This approach also provides valuable information about ecosystems and food chains. For example, the midshipman fish *Porichthys* is apparently luminous everywhere except in regions where the luminous ostracod *Vargula*, the sea firefly, is not found. Injecting these non-luminous fish with *Vargula* luciferin makes them luminous.

(2)   Common features in protein chemistry
      Protein families with common amino acid sequences provide information essential for understanding their catalytic role, and whether they have evolved via convergent or divergent evolutionary routes. The increasing identification of protein sequences, using recombinant DNA technology, provides sufficient information for evolutionary hypotheses based on gene transfer and exon sorting.

(3)   Ethology
      The common occurrence of bioluminescence in the deep oceans, the fact that most of the creatures in the depths have retained efficient eyes during evolution, even though the sunlight from above does not penetrate significantly below about 1000 metres, suggests that living light plays a major role in animal life here. Only by comparing and contrasting all the various groups can the role of bioluminescence in the behaviour and ecology of deep sea animals be fully elucidated. In contrast, distinguishing particular groups of organisms which produce light, from close relatives which do not, enables the biologists to

rationalise different behavioural mechanisms fulfilling, in different species, the same function, for example, sexual attraction.

(4) Evolution

As Harvey's quotation at the beginning of this chapter highlights, the apparent spasmodic occurrence of bioluminescence between, and within, different phyla has long been a puzzle. How is it, given the range of chemical processes necessary for producing and controlling the chemiluminescent reaction, that even within the same family one species may be luminous, yet another not. A typical example of this problem is to be found in the coelenterates. Two sea firs (hydroids) *Obelia geniculata* and *Laomedea* (*Campanularia*) *flexuosa* are both common in rock pools, growing on seaweeds and on the rock surface. They are closely related phylogenetically, and are often mistaken for each other. In a dark room, however, there should be no trouble in distinguishing them since *Obelia* is luminous, but *Laomedea* is not. Yet for *Obelia* to be luminous it must produce the protein and chromophore for light emission, the control mechanism requiring an increase in intracellular calcium to trigger the bioluminescence protein, and a green fluorescent protein required to shift the colour of light emission, all contained within one specialised cell. Rarely is such a dramatic discrepancy found in Nature between two such closely related species. Yet surely this presents us not just with a challenge but a great opportunity to discover new things about the biochemistry of evolutionary processes (see Chapter 10, section 10.3.1), still far from compatible from the modern Darwinian–Mendelian concepts applicable at the level of whole species and groups. If only we could ask the right questions!

3.1.3 Some questions

How often have we been sitting with a group of friends or students and been put on the spot by a simple, yet fundamental, question about our work? Many of the most obvious questions about living light posed over the past three centuries by laymen, by naturalists, and by professional scientists, have been answered, but several fundamental questions remain unsolved. To prevent the description of bioluminescence turning into a vast list of anecdotes and unconnected facts, we need to find a way of presenting the questions, not only in ways which lend themselves to scientific enquiry, but also which enable the chemistry and physics of the phenomenon to be rationalised with its biology. May I suggest six particular questions:

(1) How do we identify definitely bioluminescence, and in particular, distinguish it from other sources of light emanating from natural objects?

(2) How common is it? Indeed, where can we find living light and in what groups or organisms does it occur?

(3) What characteristics are important for its function? To answer this we need the answers to four supplementary questions, (a) what are the key physical characteristics, e.g. intensity, duration, colour, and polarisation? (b) how and why are these controlled? (c) where is the light generated? and (d) what are the components required for light production and how are these brought together?

(4) What is it for? In fact, can we use a knowledge of the physical and chemical characteristics, together with the cellular and molecular biology, to help us answer this question?

(5) What is its evolutionary origin and significance?

(6) Can it have any direct benefit to man, apart from the aesthetic satisfaction of studying it? As if that weren't enough!

3.2 ABUNDANCE AND OCCURRENCE

3.2.1 Is it or isn't it?

Two popular sources of confusion must be cleared up before we examine bioluminescence in detail. The first relates to the problem of distinguishing bioluminescence from other sources of light in or around living organisms. The second relates once again to the confusion between photoluminescence (fluorescence and true phosphorescence) and chemiluminescence.

Throughout the world, large blooms of dinoflagellates are sometimes found on the sea's surface during the summer. These can sparkle so brightly when the water is disturbed that the bioluminescence is dramatic, even in daylight. However, living light is most obvious at night. The problem is that once the human eye has dark adapted every pale stone seems to glow, and every pool of water flash when disturbed. It is thus necessary to distinguish glowing or flashing luminous organisms from other, non-living, sources of light. Common cases of mistaken identity are reflections from particles, pale solid objects, and water, especially on a moonlit night, as well as electrical phenomena such as St Elmo's Fire. On the beach at night the eye can soon learn to distinguish the distinctive blue emission of glowing bacteria or the blue and green flashes of various invertebrates, from non-living sources of light. On land, green or yellow emitters, such as glow-worms and fireflies, are even more distinctive. However only rarely, are we likely to encounter an example of 'living red light'.

Definitive proof that one is truly observing living light requires isolation of the organism(s) responsible, followed by observation of visible light emission in a completely dark room. This will eliminate reflections, or fluorescent emissions from brightly coloured specimens. Familiar fish in home aquaria, such as the elegantly red and blue striped neons, are not self-luminous. A microscope may be required to identify unicellular organisms such as bacteria or dinoflagellates. Having confirmed that a source of light is emanating from a live organism, and perhaps been able to demonstrate light emission from extracts of particular parts in the specimen, a more subtle problem arises. How do you know that this bioluminescence originates from the organism itself, as opposed to a contaminant?

Decaying flesh, including human, fish, and invertebrate tissue, provides an ideal culture medium for luminous bacteria. Luminous bacteria are also quite capable of infecting live invertebrates. 'Shining flesh' from animals and fish has been known of for centuries. Human cadavers on battlefields have often been seen to glow. Yet, as we have already seen, there are no mammals which are visibly self luminous. Land nematodes and various insects can also be infected with luminous bacteria from the soil or fresh water. I have often found live crabs infected in this way by marine bacteria, though they may be on their last legs. Similarly, dead or dying plant material, including vegetables, fruit, and wood, provide an ideal substrate for many luminous fungi.

The first clue that the 'phosphorescence' of dead animal matter was truly living light can be traced back to 1742 when Baker, in his *Microscope made easy*, suggested that the light might come from 'animalcules'. He had previously observed that similar light emissions from the sea were caused by 'tiny insects'. Many workers over the next century contributed to the now established dogma that the light emitted from decaying animal matter is caused by luminous bacteria. Two key experiments were those of Heller published in 1853 and Pflüger in 1875 who showed, firstly that fresh meat could be innoculated with luminous material and then become luminous itself, and secondly that bacteria could be filtered off from glowing flesh and then be grown in culture.

Another complication in identifying the origin of living 'light' is that several luminous fish and squid deliberately maintain symbiotic cultures of luminous bacteria. However, these are usually easy to distinguish from organisms simply contaminated with luminous bacteria since the former have clearly identifiable light organs, often with ancillary structures such as reflectors, lenses, filters, and shutters. The ubiquitous occurrence of luminous bacteria in the sea means that many extracts of marine specimens are contaminated with bacterial luciferase, albeit in small amounts.

Bacterial and fungal contamination can therefore be distinguished from other sources of bioluminescence, either by culturing the microorganism from a sample of the glowing organism, taking care to optimise culture conditions for induction of the enzymes responsible for light emission, or chemically and immunologically by identification of particular luciferases. Recombinant DNA probes combined with Southern and Northern blotting should be even more certainly definitive in the future. Contamination with other small organisms such as dinoflagellates, tiny jelly fish, or worms can usually be ruled out by careful inspection under the microscope. This is particularly important when examining organisms with crevices, such as sponges.

Two other sources of luminous contamination have confused, or fooled, searchers of living light. Firstly, non-luminous species can be spattered by the secretion of a luminous organism, or by fragments from it. Secondly, several luminous organisms are preyed on by larger organisms, their remains luminescing in the stomach of the predator. Thus, luminous toads have been observed having feasted on fireflies, whilst some translucent invertebrates have internally reflecting or absorbing surfaces in their stomachs to minimise such light being emitted, which would otherwise attract predators to themselves. The chemiluminescent substrate, the luciferin, or its precursor, may of course be obtained from the diet in this way, as apparently in the midshipman fish *Porichthys*. However, there are no established examples of the protein catalyst, the luciferase, being absorbed from the gut. Therefore a self-luminous organism must contain in its own DNA the gene(s) coding for the proteins responsible for light emission.

Thus, before investigating the chemistry and biology of a luminous animal or plant, it is essential to check, firstly that it is truly chemiluminescent, secondly that this is not due to a contaminant such as bacteria, or other small organism, and thirdly that the phenomenon can be observed reproducibly in other specimens of the same species, i.e. that your particular specimen is truly self-luminous and not luminous because it has eaten a luminous species or been spattered by one.

Conversely, there may be several reasons why a likely looking structure for bioluminescence within an isolated organism is not apparently luminous. Firstly, it may be spent out. Secondly, the life cycle or circadian rhythm may be such that the luminous system has not been induced, a particular feature in luminous bacteria and dinoflagellates. Thirdly, the diet may be deficient, particularly if the luciferin or its precursor originates from the organism's prey. This is particularly relevant if such luminous organisms are to be reared and bred successfully in the laboratory. Fourthly, it is feasible that the light emission is outside the spectral sensitivity of the human eye, i.e. approximately far violet (400 nm) to far red (750 nm). No established examples of this have been identified. Some ultraweak chemiluminescence has been reported to be in ultraviolet (see Chapter 6), whereas the red emission from the dissected suborbital organs of the deep sea fish *Malacosteus* is detectable by a photodiode array, but difficult to observe even by the dark adapted human eye.

3.2.2 Geography

'Living light' has invaded all of the major habitats on this earth (Table 3.1). Luminous creatures can be found from the Arctic to the Antarctic, on land and in the sea, from the surface to many thousands of metres down. A rare millipede can even be found some 1500–2000 m up in the Sierra Nevada mountains. Some can fly, some live in caves, some burrow into mud, others can swim. Living light is relatively rare on land, but common in the sea. Nearly three quarters of the earth's surface is covered by ocean, and of this, more than two thirds is greater than 1000 metres deep. The numbers of luminous organisms here are huge, but few people will have been lucky enough to have access to the plethora of luminous species available from the ocean depths. Research ships with specialised nets (Fig. 3.2) are required to catch them.

The results of bioluminescence can be spectacular or even frightening. A boat moving through a sea thick with dinoflagellates produces a brilliant trail of light. A few bays in certain parts of the world have become particularly famous for this (Table 3.2), although mini-displays can be seen all over the world. Sailors have sometimes had a rude shock when flushing the toilet during a passage through a bloom of dinoflagellates! Look over the bow of a large ocean-going vessel, at night, and you will see flashes of light as the ship hits a variety of larger luminous organisms. Dolphins, playing in the bow's wake, send off streaks of light. On land, shining wood, caused by luminous fungi, and the occasional glow-worm or firefly, are familiar sights to campers. More spectacular is the synchronous flashing of hundreds of fireflies in one tree seen in the tropics, for example New Guinea.

The phenomenon has scared many an ignorant observer. The tunicate *Pyrosoma* can grow to several metres in length. A fishing boat or its net disturbing this organism, causing it to glow, has worried many an unsuspecting sailor. A Chinese quartermaster has been known to leave the helm on witnessing 'phosphorescent wheels'. These banks of light, either in parallel or as spokes of a hub up to 100 metres in diameter, can move at several metres per second and are thought to be caused by shock waves stimulating luminous plankton. A particularly amusing anecdote was reported by a group of Israeli soldiers during the Six-Day War off the Sinai

(a)

(b)

(c)

(d)

Fig. 3.2 — The collection of deep sea luminous animals, kind gift from Dr Peter J. Herring with permission of Institute of Oceanographic Sciences. (a) RRS Discovery with permission of IOS, Wormley and Dr P. J. Herring; (b) neuston net; (c) plankton net; (d) rectangular midwater trawl (multiple net). Other nets include a benthic sledge.

Table 3.2 — Some famous displays of living light

Phenomenon	Where seen	Cause
'Christmas tree' flashing	Forests of New Guinea	Synchronous fireflies
Luminous 'fishing lines' (reflection from glow-worm lantern)	Waitomo caves of North Island (Wellington) of New Zealand	Larvae of the New Zealand glow-worm
Phosphorescent bays	Oyster Bay, Jamaica Baia Fosforescente, Puerto Rico	Dinoflagellates (*Pyrodinium*)
Red tides	many sites in tropics, and sometimes temperate waters	Dinoflagellates (particulary *Noctiluca Gymnodium* and *Gonyaulax*)
Milky sea	Western Arabian Sea	Bacteria
Phosphorescent wheels	Indian Ocean South China Sea	Shock waves on luminous plankton

Peninsula. On observing a greenish glow at night in the coral reefs they fired off explosives into the water. No frogman appeared, only large numbers of dead flashlight fish on the beach!

How common throughout the world then is bioluminescence? What habitats have been invaded most successfully by luminous organisms? In which major groups in pro- and eu-karyotic organisms do we find living light, in particular which groups contain either the largest number of species or the largest number of individuals?

In spite of the familiarity people may have with luminous beetles, i.e. fireflies and glow-worms, which have worldwide distribution, the variety and quantity of terrestrial luminous animals is relatively small (Fig. 3.3a, Table 3.3). Freshwater species are even rarer, being restricted to one or two species of bacteria and a limpet, *Latia neritoides*, found only in Australasia. The 'firefly shrimp' in Lake Suwa, Japan is infected by luminous bacteria which kill it. Similarly species of the fungus gnat, genus *Arachnocampa* (formerly *Bolitophilia*) (Order Diptera), seem to be restricted to Australasia. In striking contrast the sea has provided a habitat for an enormous variety of species and individuals (Fig. 3.3b, Table 3.3). There can hardly be a beach in the world which does not contain at least a few luminous species. Luminous marine bacteria and dinoflagellates are found in rock pools and near the surface of the sea all over the world. Mid-water trawls from a few metres below the surface to several thousand metres in the deepest oceans, or dredges from the ocean bed, always contain many invertebrate luminous species as well as varieties of luminous fish.

Precise quantification of the frequency of occurrence of particular luminous species or individuals is not easy. However, it has been estimated that as many as 90% of the individual small fish, and 60–80% of all the fish species, in the deep oceans are luminous. A genus of small luminous fish, *Cyclothone* (Fig. 3.3b), can make a claim to be the commonest vertebrate in the world, the particular species depending on the locality! Not only are some genera of worldwide distribution but also the particular species. Thus, several luminous bacteria and dinoflagellates are found in all oceans, and the hydroid *Obelia geniculata* is common both in the United Kingdom and in the inshore waters of New Zealand.

There are three key factors in a habitat which affect the abundance of luminous organisms: physical, e.g. temperature, pressure, light, and discontinuities (such as water layering); chemical, e.g. oxygen and ionic composition; and biological, e.g. food and predators. These factors not only vary in different parts of the world, but also with the time of day and season. The habitat may also be controlled through symbiosis. Many marine organisms move upwards at night, or have circadian rhythms such that their luminescence is most active at night. The reproduction of many species during the summer means that they are particularly abundant then. *Obelia* jelly fish (Fig. 7.1) are virtually absent from British waters during the winter, yet between June and October the plankton is full of them, and the hydroids can then be found growing in thick carpets on large brown seaweeds. Several organisms use their luminescence to attract or court the opposite sex. In the United Kingdom adult female glow-worms (Fig. 3.4a) only light up regularly during the summer months of July and August. The larvae, also luminous, are mainly responsible for maintaining a colony throughout the winter until the next mating season.

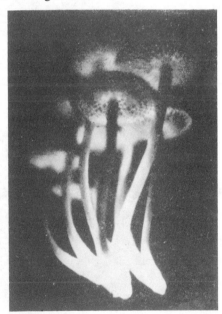

Fig. 3.3A(i)

Fig. 3.3 — Some luminous organisms.
(A) On land; (i) Mushroom *Polyporus*, from Harvey (1952) by permission of Academic Press; (ii) Glow-worm *Lampyris*; (iii) firefly *Photinus* copyright Natural History Photographic Agency, with permission.
(B) In the sea. (i) some luminous jelly fish and ctenophores; (ii) some small luminous creatures, dinoflagellates (Pyrocystis), ostracods and copepods (*Disseta and Lucicutia*); (iii) the luminous polychaete worm *Chaetopterus*; (iv) a luminous octopus and an angler fish; (v) the tunicate *Pyrosoma*; (vi) luminous decapods; (vii) luminous euphausiids; (viii) some luminous fish.
B(i),(ii),(v),(vi),(vii),(viii) = fixed specimens

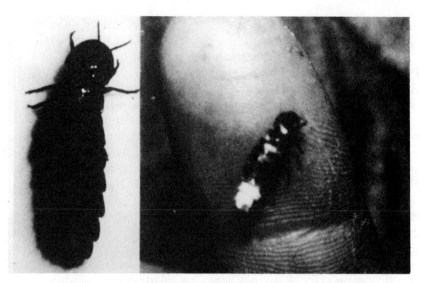

Fig. 3.3A(ii)

Fig. 3.3A(iii)

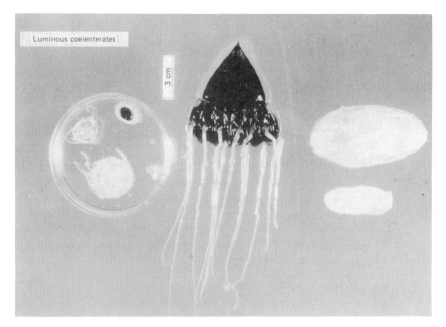

Fig. 3.3B(i)

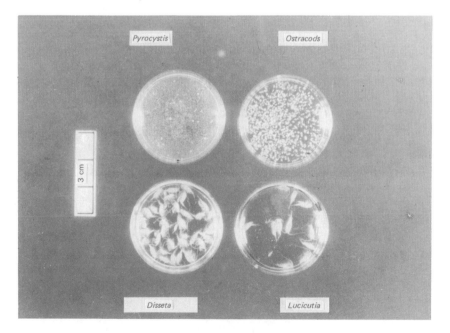

Fig. 3.3B(ii)

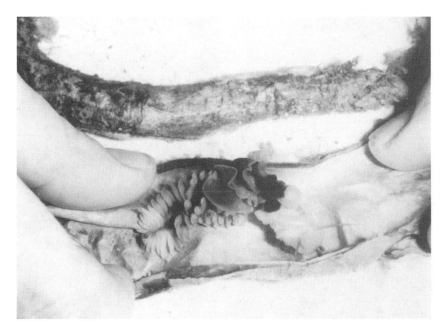

Fig. 3.3B(iii)

Fig. 3.3B(iv)

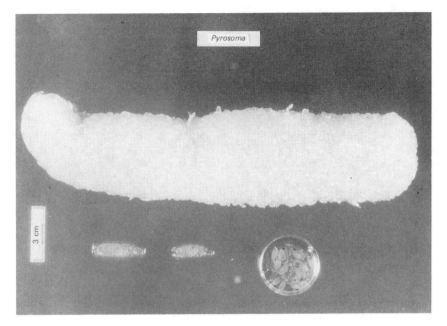

Fig. 3.3B(v)

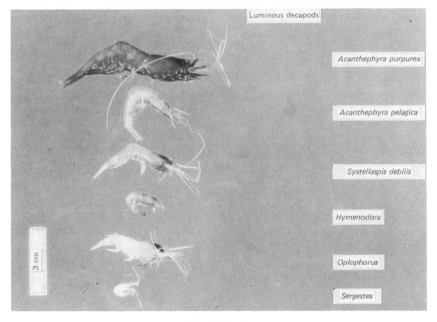

Fig. 3.3B(vi)

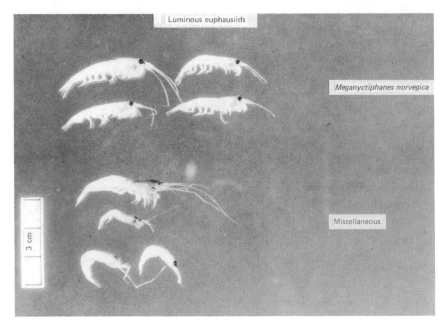

Fig. 3.3B(vii)

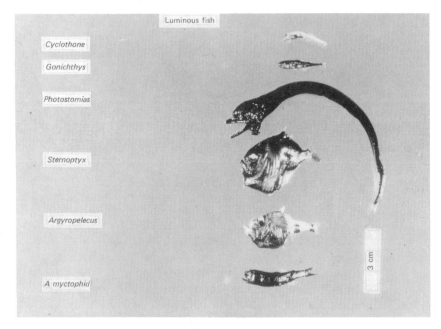

Fig. 3.3B(viii)

Table 3.3 — Some common and rare examples of bioluminescence

Habitat	Luminous group or species	Distribution
Underground	Earthworms	Worldwide, but local
Earth's surface	Fireflies and glow-worms	Worldwide, but local
	Millipede (*Luminodesmus*)	*USA* and just a few other sites
	Land snail (*Quantula*)	Malaysia and Cambodia only
	Fungi	Worldwide
	Fungus gnat (*Arachnocampa*)	New Zealand and Australia
Fresh water	Limpet (*Latia*)	New Zealand and Australia only
Beach and rock pools at low water	Bacteria on decaying flesh	Worldwide
	Dinoflagellates	Worldwide
	Hydroids	Worldwide
	Polychaetes	Worldwide
Rock	Bivalve	Europe (*Pholas*), Palau W. Pacific (*Gastrochaena* formerly *Rocellaria*)
Sea surface or coastal (< 50 m)	Free floating bacteria (non-luminous)	Worldwide
	Dinoflagellates	Worldwide
	Jelly fish	Worldwide, local
	Ostracods (*Vargula*)	Seas of Japan and North America
	Flashlight fish (bacterial organ)	Red Sea and Indian Ocean
Midwater (100–3000 m)	*Cyclothone* (a fish)	Abundant worldwide
	Pachystomias (a fish)	Very rare
	Symbiotic bacteria in fish and squid	Worldwide
	Decapod shrimp	Worldwide, local
	Euphausiid shrimp	Worldwide, local, very large shoals
	Ostracods (Cochaecia)	Worldwide
Benthic	Starfish and sea cucumbers	Worldwide

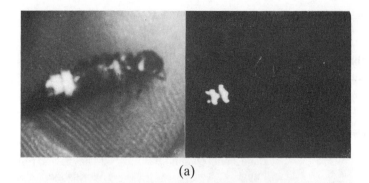

(a)

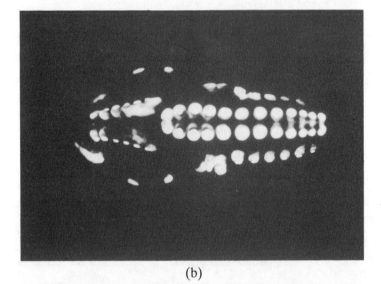

(b)

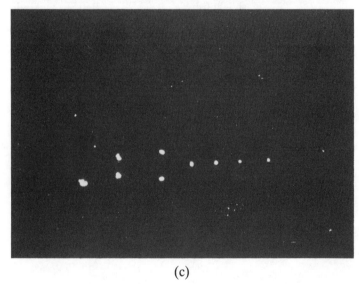

(c)

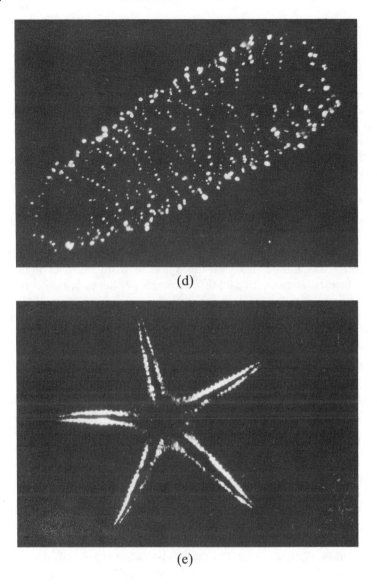

(d)

(e)

Fig. 3.4 — Some luminous organisms photographed by their own light. (a) glow-worm, *Lampyris noctiluca*; (b) the hatchet fish *Argyropelecus*; (c) the euphausiid *Meganyctiphanes*; (d) the tunicate *Pyrosoma*; (e) starfish *Plutonaster*. (b), (c), (d), (e), kind gift from Dr Peter J. Herring.

A puzzle with a few species is that in one part of the world they may be luminous, in another not. This peculiarity is well illustrated by certain fungi such as *Pleurotus* and the honey fungus *Armillaria*, and the midshipman fish *Porichthys*. The latter is luminous and found off the west coast of North America down to Mexico. Yet specimens caught off Washington State, in Puget sound, are non-luminous. However, such dichotomy is not a common occurrence with most luminous organisms.

What variety of *luminous* organisms can we find in particular habitats? Biolumi-nescence in the soil is restricted to strains of one species of bacteria, *Xenorhabdus* (previously called *Achromobacter*) *luminescens,* a nematode containing these bac-teria, a millipede living high up in the mountains, and at least 33 species of earthworm representing 16 genera. Luminous earthworms are found in Europe, Africa, Asia, India, the Far East, North and South America, and Australasia. Particular genera appear to be restricted to specific regions. For example, *Diplocar-dia* is found in North America, whilst *Spenceriella* lives in Australia.

On the surface we can find fungi, a variety of insects, some centipedes, and a few millipedes which are luminous. Some luminous fungi look like mushrooms, others are more amorphous. The 'mushroom', or *fruiting body* as it is known, contains the spores and develops from the *mycelium,* which grows unseen below the surface. They grow on rotting wood, on leaves, or as mould. They are found all over the world, from the forests of Europe to the rain forests of Ecuador and Borneo. More than 40 luminous species have been described, mainly from the Class Hymenomy-cetes (formerly Basidiomycetes), but there may be many more to be discovered in the tropical forests. A distinctive feature is whether either or both the mycelium and fruiting body are luminous. *Armillaria mellea* is very common in Europe. Its mycelium is luminous, but its fruiting body is not. At one time fungi with luminous fruiting bodies were thought to be rare, but several have now been found in the tropics and the far east. Some luminous fungi are associated with a particular habitat. They are well known to woodmen, foresters, miners, and wine merchants. *Pleurotus olearius* is common in the Mediterranean region growing on olive trees. Some have a wide distribution, others a restricted geography. *Panus strypticus* and *Armillaria mellea* are found in luminous and non-luminous forms which are separated geogra-phically. Glow-worms and fireflies also have essentially a worldwide distribution, though particular genera and species tend to be restricted to a geographical location. However, they are rare on the Pacific coast of the USA and in desert countries, and none are found in the Arctic and Antarctic. Fireflies are plentiful in North and South America, the West Indies, Southern Europe, Africa, and New Guinea. In Burma, Thailand, the Phillipines, and Indonesia, trees can be found with hundreds of fireflies flashing synchronously 100 times a second. In Europe a solitary female glow-worm clings to a blade of grass, (Fig. 3.4a) one of perhaps less than twenty, in a particular locality. Both sexes can fly in some species, in others only one. All stages of the life cycle, egg, larvae, pupae, and adult, are usually luminous.

There are a few other luminous insects, apart from beetles: the springtails, glow-worm flies, and lantern flies, though whether the latter are truly self-luminous has been much debated for nearly three centuries. One of the most celebrated luminous flies is the fungus gnat *Arachnocampa luminosa* living in caves in New Zealand, and called the New Zealand glow-worm. The larva dangle sticky, 'fishing lines' which transmit light from the luminous organ to attract and ensnare prey. The Australian species is called *Arachnocampa richardsae*. Some of these insect colonies are thought to be very old. The beetle 'glow-worm' seems to have remained in particular valleys of Snowdonia for several centuries, but the famous Waitomo 'glow-worm' cave in New Zealand is thought to be at least one million years old!

Littoral luminous organisms to be found on the beach and in rock pools include bacteria on decaying flesh or a dying invertebrate, dinoflagellates, starfish, scale worms under rocks, hydroids on rocks and algae, and polychaete worms on algae or buried in the mud and sand. Some may be well concealed, and only easy to find at very low tide. Only the neat holes drilled by the piddock *Pholas* are visible from the surface (Fig. 3.1). The scaleworm *Acholoë* can be found living in the same burrow as a larger, but non-luminous, polychaete worm. Occasionally a brilliant ctenophore can be found in a rock pool, but luminous shrimp, squid, and fish are not usually found on the beach. However, in the Red Sea or the Indian Ocean flashlight and pony fish can be found amongst the coral by scuba diving below about 30 metres. By using the appropriate nets (Fig. 3.2) the great variety of oceanic luminous organisms living in the sea, from the surface down to thousands of metres, can be collected (Fig. 3.3b).

Dinoflagellates and radiolarians some 0.1–1 mm in diameter are usually found within the top 10 cm, and are trapped by using a Neuston net. In 'phosphorescent' bays (Table 3.2) *Pyrodinium bahamense* can reach a density of some 10^4–10^7 per litre. Dinoflagellates responsible for the 'red tides' in different parts of the world are usually *Noctiluca*, *Gymnodinium*, or *Gonyaulax*, and off the East Indies *Pyrocystis*, though this species is also pelagic. Midwater trawls will yield nets flashing with jelly fish, ctenophores, decapod and euphausiid shrimp, tunicates, squid, the occasional octopus, and fish, as well as smaller luminous animals such as ostracods and copepods (Figs. 3.3b and 3.4). The benthic sledge will scoop up luminous hydroids, echinoderms (starfish and holothurians), and a few free-swimming luminous fish and shrimp. Some of the most spectacular sights include clusters of separated dinoflagellates producing blue luminous water, bright blue-green ctenophores, large shoals of euphausiids (krill) which can block engine filters, squid such as *Lycotheuthis* (formerly *Thaumatolampus*) with five bright organs on each eye ball, eight on the under surface of the body and four on each pair of tentacles, or the squid *Ommastrephes* with a blue emitting organ more than 10 cm long (Fig. 3.8a), a large specimen of *Pyrosoma* up to several metres long, or fish synchronously flashing their multiple photophores. These are indeed some of the wonders of the deep.

The geographical location of a particular species may not remain stationary either during a 24-hour period or at different seasons. For example, at night many marine organisms move upwards, sometimes several hundreds of metres. The firefly squid *Watasenia scintillans*, seen spectacularly off the coasts of Japan, is a deep water species but comes inshore in millions to breed.

Search at night, at the right time of year, in any region of the world and you have a good chance of finding living light. But can we make any sense of its phylogenetic distribution now that we know something about its geography?

3.2.3 Phylogeny
Currently nearly 700 genera from 16 phyla contain luminous species (see Appendix II and Herring, 1987). These include many familiar species such as mushrooms, jelly fish, squid, starfish, worms, fish and beetles, yet other equally familiar groups such as spiders, ants, crabs, lobsters, sea urchins, flatworms, amphibians, reptiles, birds and mammals are not represented. To all but the purist, taxonomy is an intellectual mine field. One enters it at one's peril, not only because of its complexity but also because the rules do not appear to be universally agreed on and seem to be changing continuously. Some of us, being more inclined towards cell biology, will divide living

organisms into three main groups, prokaryotes, archaebacteria, and eukaryotes. Eukaroytes can then be divided into three major kingdoms: the plants, the protists to include fungi, protozoa, and dinoflagellates, and the animals. Others, with a more conventional bent, will simply say that there are five kingdoms: Virus; Monera, to include bacteria, archaebacteria and blue-green algae; Fungi; Plantae, to include not only higher plants and algae but also dinoflagellates, containing chlorophyll and in some classifications Fungi; and Animalia, which can move. We were once told that there were only three genera of bacteria containing luminous species, *Photobacterium*, *Beneckea*, and *Achromobacter*. Now we are told that there are four or five and that *Achromobacter*, found in the soil, should be *Xenorhabdus*, *Beneckea* should be *Vibrio*, and that some *Photobacterium* species should be moved to this genus. Some call the luminous jelly fish providing the Ca^{2+}-activated photoprotein aequorin (see Chapter 7) *Aequorea aequorea*, others call it *Aequorea forskalea*, and more recently the cloning of aequorin has apparently used *Aequorea victoria*, which may or may not be the same species. The sea firefly was once in the genus *Cypridina*; now we are told to call in *Vargula*. Similarly the New Zealand glow-worm was originally placed in the genus *Bolitophilia*; now it is called *Arachnocampa*. There are many other such examples of name changes, without deed poll! Boundaries between kingdoms and

Table 3.4 — Relationship between common and systematic names of luminous organisms

Common name	Phylum	Systematic name
Moonnight mushroom	Eumycota	*Pleurotus japonicus*
Foxfire	Eumycota	*Armillaria mellea*
Jack-my-lantern	Eumycota	*Clitocybe illudens*
Firefly squid	Mollusca	*Watasenia scintillans*
Oceanic fireworm	Annelida	*Odontosyllis enopla*
Sea firefly	Arthropoda	*Vargula* (formerly *Cypridina*) *hilgendorfi*
Fire beetle	Arthropoda	*Pyrophorus* and others
European glow worm	Arthropoda	*Lampyris noctiluca*
Firefly	Arthropoda	*Photinus*, and others
Lightning bug	Arthropoda	*Photophorus* and others
Railroad worm	Arthropoda	*Phrixothrix tiemani*
Glow-worm fly	Arthropoda	*Keroplatus sesoides* and others
New Zealand glow worm	Arthropoda	*Arachnocampa luminosa*
Fire cylinder	Chordata	*Pyrosoma*
Hatchet fish	Chordata	*Argyropelecus* (and others)
Lantern fish	Chordata	*Myctophum spinosum* and others
Midshipman fish	Chordata	*Porichthys notatus* and others
Angler fish	Chordata	*Melanocetus peligicus* and others
Flashlight fish	Chordata	*Photoblepharon palpebratus* and others

See Appendix IIC for other examples.

phyla can change, new super- or sub-classes, orders or families appear, genera change their name apparently at the whim of the taxonomist. The names of individual species may be altered as knowledge and prejudice change. Did someone mention DNA? The confusion for the non-expert, who prefers common names (Table 3.4), is compounded by the tongue-twisting nomenclature, virtually imposs-ible to remember (Table 3.8).

Is it worth bothering then to examine the phylogeny of luminous organisms? I believe it is, and would argue that some important generalisations can be gleaned from such a taxonomical list (Appendix II). Firstly, it is important that scientists wishing to study bioluminescence, or wishing to use it to study other biological phenomena, should be able to find out whether luminous organisms in their vicinity are the same, or closely related, to species whose biology and chemistry have been well characterised elsewhere. Secondly, comparison of the chemistry, cell biology, and function, in relation to the chemistry and physics of the environment, of related luminous and non-luminous species, together with comparison between unrelated luminous species, provides important insights into the evolutionary significance and mechanisms responsible for selecting bioluminescence. Why is it that luminescence is very common in some phyla (Table 3.5), for example Ctenophora (comb jellies), but apparently absent in other major phyla (Table 3.6), for example Platyhelminthes (flatworms)? Why is it that bioluminescence plays a major role in the ecology of deep sea life? It is essentially the only source of light below 1000 metres. In spite of the

Table 3.5 — The occurrence of living light in nature

Kingdom	Phylum	Approx. no. of genera	Approx. no. of known species (i.e. luminous and non-luminous)	Approx. no. of luminous genera
1. MONERA	Bacteria	many 1000	many 1000	5
2. FUNGI	Eumycota	17 ?	500000	9
3. PLANTAE				
(Chromophycota)	Dinophyta	13	25000	11
4. ANIMALIA	Protozoa — Sarcomastigophora	140	50000	9
	Porifera	many	10000	1
	Cnidaria	80	9100	66
	Ctenophora	10	200	15
	Nemertea (Rhynchocoela = Nemertini)	11	650	1
	Nemata	100's	30000	1
	Mollusca	210	80000	74
	Annelida	150	8700	40
	Arthropoda	many 1000	1000000	207
	Ectoprocta (Bryozoa)	26	4000	1
	Hemichordata	10	10	3
	Echinodermata	60	5300	47
	Chordata	100's	46000	208

A very approximate guide to the abundance of genera containing luminous species; see Appendix II and Harvey (1952), Herring (1978, 1987)

Table 3.6 — Phyla with no known luminous species

Kingdom	Phylum (common name and description)[†]
VIRUS	None (except man made!)
MONERA	Cyanophycota (blue–green algae), Archaebacteria
FUNGI	Myoxomycota
PLANTAE	All algae and higher plants
ANIMALIA	Mesozoa (minute marine parasites)
	Platyhelmintha (flat worms)
	Rotifera (wheel animalcules)
	Gastrotricha (aschelminths)
	Kinorhyncha (pseudocoelomates)
	Nematoda (round worms)[‡]
	Acanthocephala (wormlike pseudocoelomates)
	Endoprocta (very similar to sea mats) = Entoprocta
	Priapulida (cucumber shaped sand or mud dwellers)
	Sipuncula (peanut worms)
	Echiura (marine worms living in burrows and rocks)
	Tardigrada (water bears)
	Pentatastomida (tongue worms)
	Onychophora (slug-like with legs)
	Phoronida (worm-like animals)
	Gnathostomulida (leeches) = Gnathobdellida
	Loricifera
	Brachiopoda (lampshells)
	Chaetognatha (arrow worms)
	Pogonophora (deep water, worm-like in chitinous tube = beard worms)

†Some earlier accounts may have suggested a luminous species that may have been due to contaminants.
‡A soil nematode with parasitic luminous bacteria is known. See Herring (1978, 1987) for further details.

dark environment here most organisms living at these depths still have eyes or photoreceptors to detect living light!

There are now five recognised genera containing luminous bacteria: *Vibrio*, *Photobacterium*, *Lucibacterium*, *Alteromonas*, and *Xenorhabdus*, though some classifications do not recognise *Lucibacterium*. Nine species have been well characterised, more may come to light using criteria based on recombinant DNA technology rather than conventional culturing, staining and morphological criteria, particularly since some symbionts in fish and squid have proved difficult to culture. Of the nine, seven require high salt (*ca* 0.09–0.5 M NaCl): *Vibrio* (formerly *Photobacterium*) *fisheri*, *Vibrio* (formally *Beneckea*) *spendida*, *Photobacterium phosphoreum*, *Photobacterium logei*, *Photobacterium leiognathi*, *Lucibacterium* (formerly *Beneckea*) *harveyi*, and *Altermonas hanadei*. One freshwater species has been clearly identified, *Vibrio albensis* (formerly *cholerae*), and one *Xenorhabdus* (formerly

Achromobacter) *luminecens* is found in the soil, a land nematode *Heterorhabditis bacteriophaga*, and infects wounds and meat.

Luminous bacteria are found worldwide in any of four conditions: free floating; saprophytic, living on dead organic matter such as decaying animal flesh or rotting plants; as parasites; and as symbionts within luminous glands. Luminous bacteria can infect a range of animals not normally luminous, and thus be parasitic. They are often found on insects and other invertebrates, both on their surface and in their guts, and this has often led to spurious assumptions that the host organism is self-luminous. Symbiotic bacteria form a true relationship with the host, which is assumed to benefit both species. Such symbiosis only occurs in certain squid and fish. Infection of the symbiotic glands in the developing adult occurs from environmental bacteria. All individuals of such animal species are usually luminous. In contrast, parasitic bacteria do not infect all individuals in a particular species. It is not clear whether the infection of the nematode *Heterorhabditis* by the soil bacterium *Xenorhabdus* should be regarded as symbiotic or parasitic, since it does appear to benefit the nematode by aiding its geographical spread. Bacterial growth and luminescence are affected by O_2 concentration, salts and nutrients, temperature, and auto-inducers produced by the bacteria in themselves (Fig. 3.5) required to switch on the genes coding for the luminescent properties, and catabolite repression. These parameters control the particular bacterial species which predominates in a given environment. A litre of sea

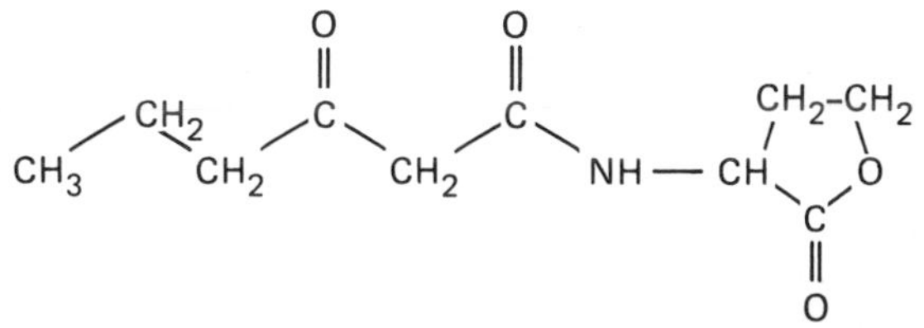

N-(3-oxohexanoyl) -3 aminohydro -2(3H)-furanone
(N- (β-ketocaproyl) homoserine lactone)

Fig. 3.5 — Auto-inducer of bacterial genes in *Vibrio fisheri*.

water may contain several hundred bacteria. but they will not be luminous since the auto-inducer is not present above the critical concentration necessary for gene expression. This provides a mechanism for switching on their light only when the population of cells is large enough to be visible. Arctic and Antarctic sea water will contain mainly *P. phosphoreum* and *logei*, whereas mid-water Atlantic samples will most likely contain only *P. phosphoreum*. Below 1000 m there are few, if any, free floating luminous bacteria. However, they can still be found at these depths in the light organs of certain fish and squid. There is also a seasonal and geographical variation of free floating luminous bacteria, reflecting the sensitivity to physical and chemical conditions of different species (see section 3.4.5). For example, in the surface water in California *Lucibacterium harveyi* predominates in the summer, but is virtually absent in the winter. *P. phosphoreum* is only found in significant numbers in the winter, whilst *V. fisheri* is present all the year round. In contrast, in the Mediterranean *L. harveyi* is found all the year round and *V. fisheri* only in the winter, and in the Gulf of Elat and in the Arctic Sea *P. leiognathi* is a major species, being replaced by *L. harveyi* in coastal waters during the summer.

All the known marine bacteria can be saprophytic on animal flesh, usually a dead or dying invertebrate or fish. However, they do not apparently grow naturally on plant material. It is not clear whether the marine, fresh water, or soil species were responsible for the celebrated historical accounts of shining flesh. Parasitic luminous bacteria, up to 10^5–10^7 per ml, are found in the guts of many surface and mid-water fishes. The bacteria appear to produce a chitinase which aids digestion. They can also infect live invertebrates such as sand fleas, shrimp, and crabs. The soil or freshwater species are presumably responsible for occasional reports of luminous caterpillars, may-flies, mole-crickets, ants, and woodlice. Bacteria, parasitic on the nematode *Heterorhabditis*, colour it red by day and make it luminous by night, thereby attracting predators, day or night. This kamikaze behaviour seems to provide a means by which the nematodes can be spread over a wide area. These bacteria also produce a potent antibiotic which may account for the unconfirmed anecdotes of the benefits of 'glowing' wounds in wars, before the availability of penicillin.

Of the 208 or so genera containing luminous fish, in about a third the luminescence is caused by symbiotic luminous bacteria living within special glands (Table 3.7). They are restricted to the teleosts (bony fish). The species include famous ones (Fig. 3.6), such as the anglers with their luminous lures, flashlight fish with the living torch below their eyes, and the pony fish. The bacteria in all these light organs are packed very densely, e.g. 1–5×10^{10} bacteria per g, equivalent to 10^7–10^9 cells per total organ, depending on the size of the fish. The bacterial culture is a pure cell line, and often but not always species specific for the host. There is much to be learned here from recombinant DNA technology. Three main species have been identified so far: *V. fischeri*, *P. phosphoreum* and *P. leiognathi*, the latter being the species found in pony fish. The only other established organisms with symbiotic luminous bacterial organs (Table 3.6c) are the squid (Phylum *Mollusca*; Order Cephalopoda). Of the 106 squid and cuttle fish genera, about 70% contain luminous species. In many cases, the whole genus is luminous. However, the species with bacterial light organs are mainly the shallow water sepiolid cuttle fish and myopsid squids, rather than the oceanic oegopsids. In the main, the latter produce their own photophores. It has been proposed that the tunicate *Pyrosoma* contains intracellular bacteria, but the evidence for this is weak.

All the known luminous fungi belong to the class Hymenomycetes (formerly Basidiomycetes). Nine of its 12 genera contain the 42 well established luminous species. Some look like familiar mushrooms, for example *Pleurotus olearius* often found at the base of olive trees in the Mediterranean. Others, e.g. the mycelium or rhizomorph of *Armillaria* infecting rotting wood, are more amorphous. Several are distributed widely throughout the world, for example *Pleurotus olearius* and *Armillaria mellea*, whilst some such as *Panus stypticus* and *Armillaria mellea*, apparently have luminous and non-luminous forms or forms which may be separated geographically.

Table 3.7 — The occurrence of luminous bacteria

Mode	Where found	Species
Free living	Freshwater	*V. cholerae*
	Soil	*X. luminescens*
	Seawater	all *P.* (*phosphoreum*, *logei*, *leiognathi*) and all *V.* (*harveyi*, *fischeri*)
Saprophytic	Flesh of dead marine animals, human and animal wounds, or meat	All *P.* and *V.* species *X.*
Parasitic	Outer surface of marine animals (e.g. crabs)	All *P.* and *V.* species
	Digestive tracts of fish and marine invertebrates	All *P.* and *V.* species
	Terrestrial and freshwater animals	*X.*
Symbiotic exo-	Land nematode and caterpillar	*X.*
	Teleost (bony) fish (19 families, 62 genera)	*P. phosphoreum*, *P. leiognathi*, *P. logei*, *V. fischeri* and others not identified or difficult to culture
	Sepiolid cuttlefish and myopsid squid (2 families)	*V. fischeri* and others not yet cultured
endo-	Tunicates	Not yet cultured

V = *Vibrio*; *X* = *Xenorhabdus*; *P* = *Photobacterium*.

There are eleven genera of dinoflagellates containing luminous species. Four species are particularly common: *Noctiluca scintillans* and *Gonyaulax polyhedra* are responsible for red tides in many parts of the world at greater than 10^6 cells per litre, *Pyrodinium bahamense* is responsible for the luminescence of Oyster Bay, Jamaica, where it can reach 10^4–10^7 per litre, and *Pyrocystis* spp is responsible for many 'milky seas' away from coasts. Dinoflagellates are large, unicellular organisms several 100 μm in diameter and easily visible to the naked eye. Collecting them in a net provides spectacular displays of blue luminescence when poured from one vessel to another. Even larger, and the only other known group of unicellular luminous organisms, are the luminous radiolarians (Phylum: *Protozoa*) of which there are nine luminous genera, some with cells several millimetres in diameter like *Thalassicolla* (Fig. 3.3c), others forming multinucleate colonies. They occur particularly but not exclusively in the top few centimetres of the ocean.

The phylum *Coelenterata* is now considered as two separate phyla: *Cnidaria* and *Ctenophora*. Luminous cnidarians are abundant in all three classes, *Hydrozoa*, *Scyphozoa*, *Anthozoa*, and are found littorally in rock pools, in shallow and in deep water. A high percentage of all ctenophores are luminous, although not all are, as was once thought. In contrast, bioluminescence is rare in most of the mulluscan

(a)

(b)

Fig. 3.6 — Two luminous organisms with symbiotic bacteria. (a) the angler fish; (b) flashlight fish *Photoblepharon* from Harvey (1952) with permission of Academic Press.

classes, except for the cephalopods (squid and octopods), where as many as 70% of all the 100 + genera contain luminous species. Of these more than half of the species in any one family can be luminous. Some have bacterial symbionts, others their own photophores. The latter can number several hundred scattered over the body (Fig. 3.8a). As recently as 1952 E. Newton Harvey (Fig. 1.3) wrote 'It is highly probable that no true octopus is luminous'. Yet there are now three known luminous octapods and just two bivalve genera. Luminous annelids (worms) are found all over the world, both on land and in the sea. They occur only in the polychaete and oligochaete classes, containing 26 and 12 luminous genera, respectively. They usually lie hidden from view in the soil or in mud. Here they can form their own tubes. For example *Chaetopterus* (Fig. 3.3c), are like some scale worms, they can establish a symbiotic relationship with another polychaete.

Arthropods represent more than three quarters of all the animal species on the Earth. The class Insecta alone has more than three quarters of a million documented species. There are some 400 000 species of beetle known, and maybe as many as 3–4 million awaiting discovery. Yet evolution has been highly selective within the six out of the fourteen main classes of arthropods containing luminous species. The greatest abundance of luminous species and individuals are to be found in two classes: *Crustacea*, particularly marine copopods, ostracods, euphausiid, and decapod shrimp, and *Insecta*, particularly coleoptera (beetles), which are entirely terrestrial. There are at least 2000 individual species of fireflies. One important, and fascinating, feature of Arthropod bioluminescence is that there are at least five quite distinct established chemical reactions responsible for light emission, in: beetle; decapod, mysid shrimp and copepods; ostracods; millipedes; euphausiid shrimp. (Fig. 3.11, Table 3.21).

The occurrence of luminous echinoderms is much commoner than was once thought. Harvey in 1952 reported that of the five classes of echinoderms, only the ophiuroids (brittle stars and snake stars), and possibly the asteroids (starfish), contained luminous species. Thanks particularly to the studies of Peter Herring, we now know that at the bottom of the deep oceans there are many luminous crinoids (sea lilies and feather stars), asteroids (starfish), and holothurians (sea cucumbers), although there are no known luminous sea urchins. To date, 47 luminous genera of echinoderms have been identified.

The number and variety of luminous fish is truly amazing. All are marine. Of the 4032 fish genera, representing some 450 families of fishes, 5% of the former and 10% of the latter are luminous. These occur from shallow water down to the deepest oceans. Some have only one light organ, others multiple organs, or photophores. A few can even emit two colours, for example, the stomiatoids, *Malacosteus* (see Chapter 9, Fig. 3.10b), *Aristostomias* and *Pachystomias*. The luminous organs are found in a remarkable variety of structural forms, ranging from large orbital light organs to tiny photophores, as well as in appendages such as the lures and barbels of the angler fish. Most of the luminous fish which are caught are relatively small, perhaps less than 15–30 cm in length, although much larger anglers sometimes appear in fish markets. Bacterial symbionts have been identified in some 19 families, the remainder producing their own photophores. As with the arthropods, several chemistries exist. For example, the stomiatoids are different from the myctophids (see section 3.3). There are certainly more discoveries to come!

The striking appearance of many luminous organisms means that humans must have been familiar with examples of living light for thousands of years. Ancient Chinese writers described glow-worms, Aristotle wrote about glowing wood and shining flesh, whilst Pliny in his *Natural history* described the luminous bivalve *Pholas* and a luminous 'lung', now presumed to be the jelly fish *Pelagia*. Yet the phenomenon of living light cannot be said to have been 'discovered' until its origins in *living* cells had been both identified and described. Whilst many distinguished scientists, such as Boyle and Spallanzini, studied the phenomenon in the 16th and 17th centuries, it was not until the work of Baker, in the mid-18th century, that scientists realised that they were truly observing living light. Subsequently, many discoveries of previously unknown luminous organisms were made in the 19th century, particularly as a result of the voyages to parts of the world hitherto unexplored by civilised man. Thus, *Pyrosoma* (Fig. 3.3 and 3.4) was 'discovered' by Peron in 1800 on an expedition to Australasia. Darwin described many luminous organisms in his *Beagle* accounts. The New Zealand glow-worm was described by Meyrick in 1886, and by the end of the 19th century the development of nets for trawling the deep oceans had lead to the identification of several new deep sea luminous species. Yet many of the luminous fish and squid were not discovered until this century. As recently as 1946 Haneda found the luminous land snail *Dyakia* (now called *Quantula*); and as we have seen with the holothurians and octopods, discoveries are still being made in the sea. On land, parts of South America remains to be fully explored. Furthermore, the taxonomy of many species is likely to be revolutionised as a result of applying recombinant DNA technology.

Now that we know something of the distribution of living light in Nature we are in a position to examine some of the cellular, chemical, and physical properties which characterise the light emission from particular groups. This will then provide a basis for understanding how the light is produced, how it is controlled, what it is for, what its role is in the ecology of a particular environment, and how it evolved in the first place.

3.3   THE SCIENCE OF LIVING LIGHT

The particular fascination of bioluminescence to the natural scientist is that in order to comprehend fully the whole phenomenon, not only is a synthesis of the three disciplines of physics, chemistry, and biology required, but also a marriage between the mechanistic and holistic philosophies is essential if we are to follow the path from ecosystem to individual molecule and back again. The physical characteristics of light emission, together with the luminous cells and their ultrastructure, need to be described before we can discover the nature of the chemical reactions responsible, or what its function might be. A full understanding of the chemistry and biochemistry of living light, together with a definition of the similarities and differences within and between groups of luminous organisms, is required before we can find out how the organism controls its light emission. Both the cell biology and chemistry are needed if the evolutionary significances of bioluminescence is to be understood.

In order to discover how luminous organisms produce and control their own light, and to discover what function it plays in their survival and reproduction, we need to know the answers to six questions:

(1) What are the important physical parameters which characterise light emission?
(2) Where is the light produced?
(3) What are the components, and the reactions, responsible for generating the light?
(4) Is the light emanating from the initial chemiluminescent reaction processed in any way before leaving the organism? If so, how and why?
(5) How are the components of the reaction brought together to initiate chemiluminescence, and how is this process controlled?
(6) What is the biochemical basis of the main physical similarities and differences between species? Are these of any significance to the function and evolution of the phenomenon?

Table 3.8 — Examples of the phylogeny of luminous animals

Class	Order	Family	Species
1. *Sub-kingdom: Protozoa, Phylum Sarcomastigophora*			
Actinopoda	Radiolaria (Spumellarida)	Thalassicollidae	*Thalassicolla* spp.
2. *Phylum: Cnidaria*			
Hydrozoa	Hydroida (Thecata)	Campunulariidae	*Obelia geniculata*
Scyphozoa	Semaeostomea	Pelagiidae	*Pelagia noctiluca*
Anthozoa	Pennatulacea	Renillidae	*Renilla reniformis*
3. *Phylum: Ctenophora*			
Tentaculata	Lobata	Mnemiopsidae (Bolinopsidae)	*Mnemiopsis leidyi*
4. *Phylum: Mollusca*			
Gastropoda	Basommatophora	Lattidae	*Latia neritoides*
Bivalvia	Myoida (Adepodonta)	Pholadidae	*Pholas dactylus*
Cephalopoda	Teuthoidea	Enoploteuthidae	*Watasenia scintillans*
5. *Phylum: Annelida*			
Polychaeta	Chaetopterida (Spioniformia)	Chaetopteridae	*Chaetopterus variopedatus*
Oligochaeta	Haplotaxida (Opisthopora)	Megascolecidae	*Dilpocardia longa*
6. *Phylum: Arthropoda*			
Crustacea (sub--phylum)-Ostracoda	Myodocopida (Myodocopa)	Cypridinidae	*Vargula (Cypridina) hilgendorfi*
Insecta	Coleoptera	Lampyridae	*Photinus pyralis*
7. *Phylum: Hemichordata*			
Enteropneusta		Ptychoderidae	*Balanoglossus* spp.
8. *Phylum: Echinodermata*			
Stelleroidea (Asteroidea)	Paxillosida	Astropectinidae	*Plutonaster* spp.
9. *Phylum: Chordata* subphylum Tunicata			
Thaliacea	Pyrosomida (Pyrosomatida)	Pyrosomitidae	*Pyrosoma* spp.
subphylum Vertebrata			
Osteichthyses	Salmoniformes	Sternoptychidae	*Argyropelecus* spp.

This list is intended only to illustrate the diversity of luminous species in nine major animal phyla, see Harvey (1952), Herring (1987) and Appendix II for further details. In brackets = former name.

Let us begin by examining the physical properties of living light and where it is produced in the organism. These determine whether the luminescence is visible or not.

3.3.1 Physical characteristics

All luminous cells are subject to control. In other words, sometimes they produce light, sometimes they do not. Of course, the distinction between visible and invisible might be thought rather subjective, particularly since most bioluminescence systems, in the absence of the trigger, produce a 'resting flow' invisible to the human eye, but detectable by a sensitive photomultiplier tube. Strictly we should define 'visibility' of a particular luminous organism with respect to the sensitivity of the eye or photo-receptor of the organism in which it has evolved to evoke a behavioural response. Below about 900 metres in the sea the ambient light is invisible to the human eye. Yet it may still be bright enough for organisms which live there, with photoreceptors some 5–10 times more sensitive than us humans, to perceive images.

Five physical parameters characterise the light emission from a particular luminous species: its direction, its timing, its intensity, its colour, and its state of polarisation.

Direction

The direction of the light emitted from a luminous organism depends both on the geometry of the source and its position in the organism (Fig. 3.8). Light emission can be regarded as emanating from a point, or a discrete or a diffuse source, depending on the size of the light emitting unit (Figs. 3.3, 3.4, 3.6, 3.7, and 3.8). For example, an isolated individual bacterium or a dinoflagellate would appear to be a point source to

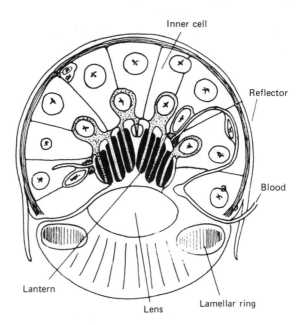

Fig. 3.7 — Ultrastructure of a euphausiid photophore. Adapted from Herring (1978) with permission of Academic Press.

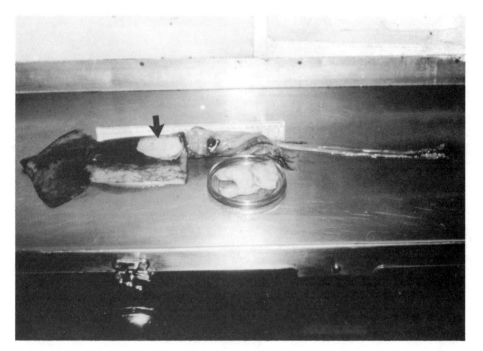

Fig. 3.8A(i)

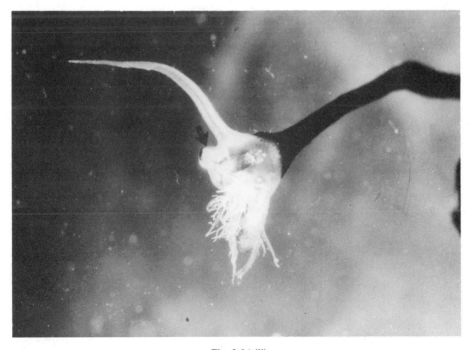

Fig. 3.8A(ii)

Fig. 3.8A(iii)

Fig. 3.8A(iv)

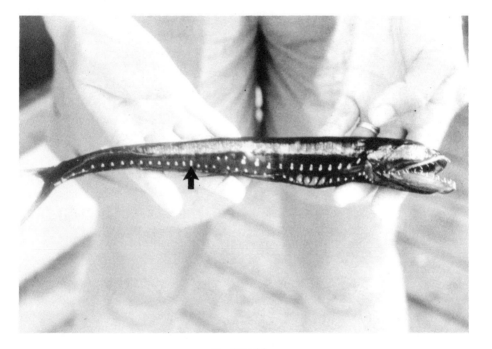

Fig. 3.8A(v)

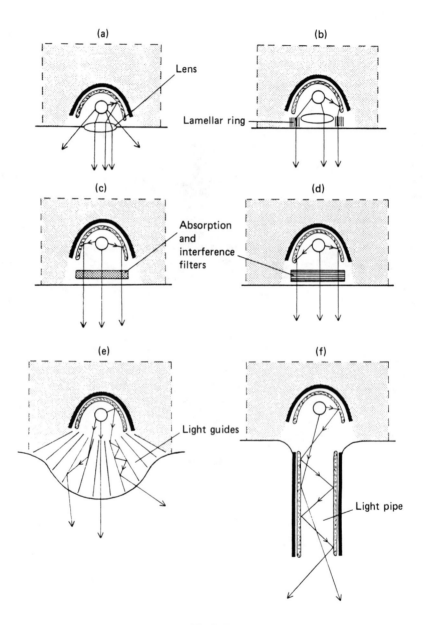

Fig. 3.8B

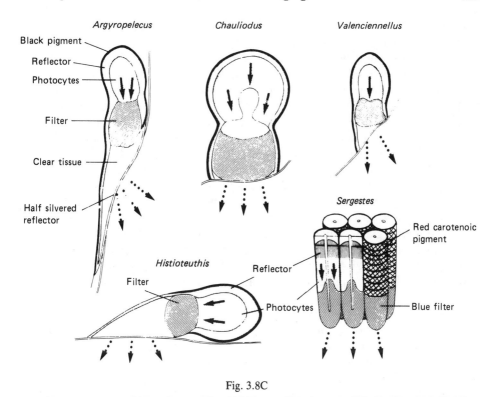

Fig. 3.8C

Fig. 3.8 — The position, detection, and colour modification of living light. (a) sites of photophores or light organs (↑); (i) single large organ on mantle of Ommastrephes + dissected organs from several individuals; (ii) lure of angler fish; (iii) mouth of octopus Japatella; (iv) below eye in Ecciostomias; (v) under (ventral) surface in Gonostoma. (b) Direction, from Herring (1985) with permission of Cambridge University Press. a. Lens alone e.g. myctophid fish. b. Lens plus lamellar ring e.g. euphausiid shrimp. c. Absorbing filter or plug e.g. fish Opisthoproctus. d. Interference filter e.g. enoploteuthid squid. e. Diffusing light guides e.g. fish Opisthoproctus. f. Light pipe e.g. ceratoid angler fish; (C) Filter pigments modifying colour of emission in three fish (Argyropeleus, Chauliodus, Valencienellus), a decapod (Sergestes), and a squid (Histioteuthis), from Denton *et al.* (1985) with permission SEB Symp. **38,** 323–49. How to survive in the dark: bioluminescence in the deep sea.

virtually any animal, yet both of these usually occur in large colonies emitting clouds of diffuse light. Euphausiid shrimp and many luminous squid and fish have many point sources distributed over their bodies. In contrast, other squid and fish have substantial discrete light organs. The light organs can be a few millimetres in diameter to several centimetres long. The squid *Ommastrephes* has on its body a spectacular light organ some 5–15 cm long. In contrast, some squid, as well as decapod shrimp, squirt a diffuse luminous cloud into the surrounding sea water. Obviously any cell, group of cells, or luminous cloud, will emit light in all directions, i.e. with 4π geometry. However, in the case of a point source, or a discrete light organ, not only does the position of the organ on the animal's body determine the direction in which the light is emitted, but also the organ may contain intricate reflectors and lenses in order to produce the light with the required angular distribution. For example, the hatchet fish, of the genera *Argyropelecus* and *Sternoptyx*, have light organs on the undersurface (ventral) with reflectors containing guanine crystals. These provide multiple reflections between a silvered outer surface to produce the required angular distribution of light emission. Euphausiid shrimp, and many other luminous fish, have a reflecting surface behind the cells producing the light. Euphausiid photophores, tiny as they are, also have a lens to direct the light emission downwards (Fig. 3.7). One group of luminous fish, the pony fish (family *Leignathidae*), have symbiotic luminous bacteria in their gut. The light reaches the surface and is directed by a series of light guides, 'living fibre optics'.

Most luminous fish, starfish, and euphausiids direct their light downwards. Some with organs below the eye, e.g. flashlight fish and stomiatoids such as *Malacosteus*, direct it sideways, whilst angler fish hold the lure in front of their mouths to attract prey. In common with many luminous marine creatures, fireflies and glow-worms have their light organs on their under-surface. However, some are able to direct the light from their lanterns upwards by curling the abdomen whilst clinging to a piece of foliage or grass (Fig. 3.4). Click beetles have their extremely bright light organs positioned so the light can be seen from above and from a side view.

Timing

To the casual observer, seeing a luminous animal at night, a precise quantitative description of the flash or glow may not apparently add anything to a good qualitative observation. Yet the kinetics and intensity of light emission provide two of the key parameters which the biochemistry of the light emitting reaction, and the molecular basis of cell activation, must explain. Furthermore, the precision in the timing of several particular species is critical for its function to be fulfilled. For example, the flashing or glowing patterns of fireflies are not only different for each species but also crucial for species recognition. Thus, the male European glow-worm *Lampyris noctiluca* recognises the female organ of the same species but not that of the female of the firefly *Phausis splendida*. An extraordinary adaptation has evolved in the female firefly of one species of *Photuris*, which can trick the male of the other genus, *Photinus*, by flashing just like its own female. It then kills and devours the unsuspecting smaller male! Thus, although there is much still to be learned about the relationships between the flashing patterns of luminous organisms and their chemistry and behaviour, we do at least need to define the kinetics carefully in order to begin to investigate scientifically its biological and evolutionary significance. So how do we describe, quantitatively, the time of a luminous glow or flash?

The first problem is that the kinetics depend on whether the measurement is made using the whole organism, one organ, or a cluster of luminous cells (photophores), a single cell (photocyte), or even a microlight source within one cell. What matters to the organism responding to the light is the pattern emitted from the whole organism. This may be a continuous glow as in luminous bacteria infecting a dead fish, or blinking as with flashlight fish (Table 3.9). Light emission may consist of synchronous flashing of multiple photophores, e.g. in the lantern fish. On the other hand, it may consist of luminous waves traversing over the organism or cell. In the case of hydroid colonies, sea pansies, sea pens, and tunicates like *Pyrosoma*, these waves may be mm–cm long moving at several metres per second. Alternatively, individual organs, photophores, or photocytes may be under independent control.

Table 3.9 — Examples of the diversity of timing of bioluminescence

Organism	Source of light	Lag or latent period	Interval between flashes	Flash duration
Noctiluca (dinoflagellate)	Microsource	2–3 ms	1–10 ms	1 ms
	Whole cell	2–3 ms	—	100–200 ms
Photoblepharon (flashlight fish)	Organism (blinking)	type 1 (75 bpm)[†]	640 ms	160 ms
		type 2 (29 bpm)	205 ms	260 ms
		type 3 (37 bpm)	800 ms	800 ms
Pyrosoma (tunicate)	Single zooid	600 ms	—	2 s
Renilla (sea pen)	Organism (wave)	120 ms	120 ms	275 ms
Porichthys (midshipman fish)	Organism	7–10 s	—	2–5 min
Lampyris (glow-worm)		—	—	> 30 min, sometimes hours
Watasenia (squid)	Brachial organ	—	—	up to 30 s
	Mantle organ	—	—	≥ 20 min
Photobacterium (bacterium)	Organism	—	—	Continuous glow for hours

†bpm = blinks per minute.

Furthermore, within clusters of luminous cells, or even within one cell, a luminous flash may be composed of individual spikes from microsources, quantifiable using image intensification. Thus, the smallest flashing unit can be an intracellular organelle, one cell, a cell cluster, or it may be a large organ, or groups of organs, in the organism. The time scale of their light emissions can vary from a few milliseconds to many minutes or even hours, depending on the species (Fig. 3.9, Table 3.9).

The dynamics of these acute responses can be divided into up to five phases (Fig. 3.9a):

Phase 1 — A lag between stimulation of the organism and stimulation of the cell(s) containing the components of the chemiluminescent reaction.

Phase 2 — A lag (or latent period) between receipt of a signal by the cell and the appearance of visible light.

Phase 3 — The emission of visible light with an increase in light intensity to a peak, or sometimes a plateau.

Phase 4 — A decay back to invisibility.

Phase 5 — A latent period before the photophore, the cell, or microsource recovers and can flash again. In some cases the source may only be able to light up once.

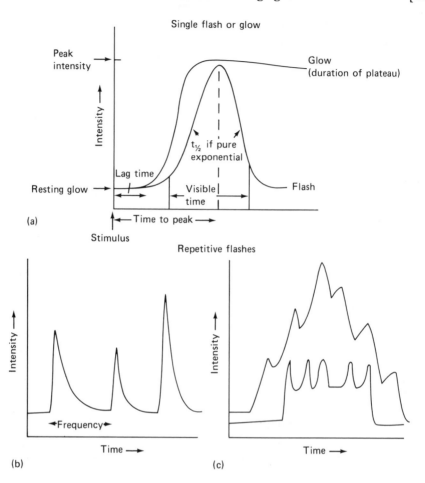

Fig. 3.9 — Types of flashes and glows.

A simple bioluminescent reaction can be described as:

$$E + L_{red} + O_2 \xrightarrow{\text{'on rate'}} E - L\begin{smallmatrix}O\\ \diagdown\\ O\end{smallmatrix} \xrightarrow{\text{'off rate'}} E + L_{ox} + h\nu \qquad (3.1)$$

where E = protein catalyst and L_{red} = the chemiluminescent substrate.

Light emission rises as the amount of the key intermediate ($E - L\begin{smallmatrix}O\\ \diagdown\\ O\end{smallmatrix}$), the precursor of the excited state emitter, increases (phase 3). This is so because the lifetime of the emitter (L_{ox}^*) is so short, usually just a few nanoseconds, and because its production from $E - L\begin{smallmatrix}O\\ \diagdown\\ O\end{smallmatrix}$ is very fast relative to the rates of the preceding steps in the chemiluminescent reaction. As L_{red} decreases, the amount of $E -$ L$\overset{\text{O}}{\underset{\text{O}}{\diagup\diagdown}}$ decreases and thus the light intensity (I) begins to decay. In this simple case the bulk of phase 3, up to peak, can be described by an exponential 'on' rate (k_{on}), or time constant (time to $1/e$ of the peak). Similarly the decay, phase 4, will be a exponential (i.e. $\log_e I$ versus time will be the linear with a slope $= -k_{off}$, $t_{\frac{1}{2}} = 0.693/k$). The literature is full of such constants, decay rate constants, and times to half maximum intensity. However, in luminous organisms not only are these often not true exponential flashes, but they may be composed of complex multiple flashes (Fig. 3.9).

The simplest way to describe a flash is to measure the latency time, the time to peak, and the duration of the complete flash. If the latter is more than 0.5–1 s it is usually considered to be a glow. A further complication arises if the flash does not decay truly to the resting level. The results (Table 3.9, Fig. 3.9) indicate the wide variation between different groups and species. Even different organs within one individual may have different kinetics. This is well illustrated in several squid. The same organ may flash differently depending on the behavioural response. Thus, the flashlight fish *Photoblepharon* is able to blink because the continuous glow from their symbiotic bacteria only reaches the outside world when the fish moves the shutter in front of its light organ. At least three types of blink have been identified (Table 3.9). These blinking patterns relate to three different functions. The fish use the light as a torch to search, as signalling lamp to communicate with neighbours, or as an alarm to distract predators.

It is possible to estimate the total light emission from a flash or glow in quanta, from the equation:

$$\text{total quanta} = \int_0^T I \, dt \tag{3.2}$$

where I = intensity at time $t = dhv/dt$, T = duration of flash.

A microsource in a dinoflagellate can emit a flash of some 10^4 quanta lasting 1–2 ms. From a knowledge of the chemiluminescence quantum yield (section 1.2.2; equation (1.9)) the number of molecules reacting can then be estimated. Bacteria only glow when the genes responsible for producing the components of the light-emitting reaction are switched on. Fully induced bacteria can emit some 10^3–10^4 quanta per second per cell. The total light output, or intensity, can be related to the number of chemiluminescent chromophore molecules (L) reacting.

$$dL/dt = I/\phi_{CL} \tag{3.3}$$

$$L_T = \int_0^T (I/\phi_{CL}) \, dt \tag{3.4}$$

where ϕ_{CL} = quantum yield of the chemiluminescent reaction.

It is surprising how full are the stores, or replenishable, in many luminous species. One warning about interpretation of kinetics: many data have been reported using experimental stimulation, e.g. mechanical, electrical, H_2O_2. This may give rise to abnormal kinetics or may damage the cells, thereby releasing the components externally. The light decay will then behave as if it were as secretory system, such as the worm *Chaetopterus* (Fig. 3.3c).

Intensity

The amount of light emanating from a luminous organism can be related either to the quantity and rate of emission (radiance, radiant flux, and radiant intensity) or to the quantity and intensity received by an observer (luminance, luminous flux, luminous intensity) (see Chapter 2, section 2.1.4). Radiant parameters are truly related to the chemistry of the light emitting reaction, whereas luminous parameters are what matter if the light is to fulfil its function.

Against a dark background, the perception of a luminous object is dependent not on the total light emitted during a flash but on the amount of light falling on a given area of the retina per unit time, i.e. intensity per unit area relative to the ambient light, together with the spectral sensitivity and quantum efficiency of the rods and cones in the eye. Thus the luminosity of living light is usually expressed as an intensity per unit area of receiver, e.g. watts cm^{-2} (1 watt = 1 joule per second) or photon s^{-1} cm^{-2} (1 photon s^{-1} = 4×10^{-19} watts for a photon at 500 nm since $E = h\nu$). Whilst the luminosity of literally hundreds of examples of living light has been recorded, variations in experimental conditions, together with differences in detector sensitivity, make accurate comparisons difficult. A further problem is that many luminous organisms monitored in the laboratory do not behave exactly as they do in the wild. Some, brought up from the deep oceans, are already exhausted or damaged. Naturally, they live in water at 4–6°C under hundreds of atmospheres of pressure. A far cry from conditions in the laboratory! A further problem is non-uniformity of distance between the organism and the recorder, varying from a centimetre to a metre or more. Nevertheless, some important qualitative observations have been made, particularly when compared with the ambient illumination of the natural habitat, together with the sensitivity of eye or photoreceptor or the observing creatures for which the luminescence has evolved to evoke a response.

Recordings made using photomultipliers under water, where the distance between the light source and the detector are difficult to define, have detected intensities of flashing from 10^{-13}–10^{-8} watts cm^{-2}. Under more controlled conditions, comparison of emissions from organisms studied in the laboratory, detected as flux or intensity one metre away, shows a similar range covering at least five orders of magnitude (Table 3.10). Among the brightest to the human eye are sea combs (ctenophores), deep sea shrimp (decapods and euphausiids), tunicates, and a large number of fish. Yet high concentrations of dinoflagellates (e.g. 10^9 per litre), emitting about 2 joule per flash equivalent to 7×10^{18} photons (see Chapter 2, eqn. (2.23)), are dramatic at night. In 'phosphorescent bays' this can produce very bright light indeed. Infection of decaying flesh with bacteria is easily visible, but dimmer than the previous examples. Some of the weakest sources of living light are the luminous fungi, though even they can often be seen shining brightly many metres away.

Table 3.10 — The intensity of living light

Organism	Estimated flux at 1 m in watts cm^{-2}	λ_{max} (nm)	photon s^{-1}cm^{-2}
Cytocladus (a radiolarian)	10^{-15}	ca. 450	2.27 × 10^3
Vibrio (a bacterium)	10^{-15}per cell per second	489	2.46 × 10^4 s^{-1}
Aeginura (a jelly fish)	10^{-14}	ca. 470	10^4
Mnemiopsis (a sea comb)	10^{-13}	485	2.44 × 10^5
Pyrosoma (a tunicate)	10^{-12}	490	2.47 × 10^6
Lampyris (glow-worm)	10^{-11}	ca. 550	2.77 × 10^7
Photoblepharon (a flashlight fish)	10^{-10}	490	2.47 × 10^8
Gonyaulax (a dinoflagellate)	10^{-9} per flash	479	2.41 × 10^9

These figures are only intended to illustrate the breadth of intensities. They may vary over one or two orders of magnitude from this, depending on conditions. Dinoflagellates, for example, may go up to 10^{10}–10^{11} quanta per flash per cell. Number of photons per second=(energy in watts×λ in mm× 10^{16})/1.986.

How much light is emitted per luminous cell? A 100 W bulb emits some 10^{18} photons per second, a fluorescent light strip 3–5 times more than this. The smallest luminous cells are the bacteria, only 1 micrometre long, but packed with luminous material. Some 5% of their soluble protein is the luciferasae, when maximally induced. One bacterium emits some 10^3–10^4 photons per second, equivalent to some 10^4–10^5 molecules reacting per cell per second. Bigger, eukaryotic cells, up to 10–30 μm in diameter, may emit > 10^6 photons s^{-1} per cell, although accurate data on this in a range of cells are lacking. One unfatigued dinoflagellate can emit a flash of some 10^9 photons!

A luminous cell or organism looks brighter when close to than it does when far away. This obvious fact is described quantitatively for a point source by the inverse square law. A luminous source emitting its photon capacity in a few milliseconds looks brighter than one producing the same total emission over several minutes. But what really matters biologically is, can this emission be detected by the animal in which the phenomenon has been involved to evoke a response? Alternatively, does the light help the organism emitting it to merge into the background, thereby using the luminescence as a camouflage? The human eye can perceive some 30 000 photons per square centimetre of retinal surface, equivalent to 10^{-16} watts cm^{-2} from a small light source. The eyes of some deep sea fish may be 10–100 times more sensitive than this, allowing them to 'see' living light many metres further away than would the human eye. Yet the intensity alone may not be sufficient for it to be visible. The colour of the light emitted is also critical.

Colour
How much beauty can we behold in the world because of our colour vision? However, though the human eye can see from violet to red, other organisms have a much narrower colour vision. Some insects can see violet, blue and green-orange light. But deep sea fish and invertebrates can only see blue light, apart from storniatoids which have red sensitive visual pigments. The sensitivity of marine visual pigments ranges from around 450 to 515 nm, coastal species being better in the blue-green region. Living light embraces all the colours of the rainbow from violet to red,

and occasionally beyond (Table 3.11, Fig.2.7). Some emissions appear purer than others. This depends on the sharpness of the spectrum and the spectral region in which it falls. There are four ways by which the spectra of different species can be rationalised with the chemistry of the reaction and the biology:

(1) Differences in spectra from reactions having the same chemistry require a biochemical explanation. For example, how can a firefly produce a different colour from the European glow-worm, yet both have the same chemiluminescent reaction?

(2) Comparison of spectra within, and between, major groups of luminous species provide clues about the function of the luminescence.

(3) Comparison of spectra with the natural ambient light can also provide information relevant to discovering the function of the luminescence.

(4) Relating spectra to function provides information necessary for understanding the evolutionary mechanisms responsible for selecting bioluminescence.

Universal generalisations are difficult to find. However, some traits can be identified.

Table 3.11 — The spectrum of living light

Colour to human eye	λ_{max} (nm)	Bandwidth in nm (range)[‡]	Genus (common name)
Violet	395	138 (378–516)	*Thourella* (sea coral)
Blue	408[†]	119 (391–510)	*Searsia* (a fish)
Blue	450	80 (417–497)	*Thalassicolla* (a radiolarian)
Blue	471	71 (444–515)	Postorbital of *Malacosteus* (a fish)
Blue	470	97 (440–537)	*Periphylla* (a jelly fish)
Blue	479	—	*Gonyaulax* (a dinoflagellate)
Blue	488	113 (451–564)	*Pyrosoma* (a tunicate)
Blue–green	508	23 (500–523)	*Aequorea* (a jelly fish)
Green	530	—	*Omphalia* (a fungus)
Green	550	—	*Lampyris* (glow-worm)
Yellow	565	60(530–590)	*Photinus* (firefly)
Red	705	47 (682–729)	Suborbital of *Malacosteus* (a fish)

These are just a few examples to illustrate the breadth of the spectrum of living light. See Herring (1983) for full documentation.
†One of the peaks.
‡Width of spectrum of $\frac{1}{2}$ peak emission (λ_{max}); range = range of bandwidth at $\frac{1}{2}$ maximum emission.

Species emitting blue light are particularly common in the sea. These include bacteria, dinoflagellates, radiolarians, many fish such as midshipman, hatchet, lantern and angler fish, decapods and euphausiid shrimp, some starfish, and tunicates. Blue–green emitters are found in coastal and pelagic species, for example, the coelenterates *Obelia*, *Aequorea*, *Renilla*, and several polychaete worms. The luciferin–luciferase reaction extracted from the bivalve mollusc, *Pholas*, also has a peak in the blue–green region (λ_{max} 490 nm). Green and yellow emitters are rare in the sea, but include the occasional ophiuroid starfish and a few squid. However, there are many green and yellow emitting species on land, thanks particularly to the large variety of luminous beetles. The true glow-worms and fireflies (family: *Lampyridae*) emit green or yellow light, whilst beetles from other families (*Elateridae, Homalisidae, Telegeusidae*, and *Phengodidae*) emit blue–green, green or yellow–green light. Some blue-emitting species are found on land, including earthworms and fungi. However, some fungi are green emitters; others appear whitish.

Red emitters are very rare, both on land and in the sea. In fact, there are only two well-documented examples (Fig. 3.10), the larvae of the beetle *Phrixothrix*, the railroad worm, and three stomiatoid deep sea fish, *Malacosteus, Aristostomias*, and *Pachystomias*. These two groups also illustrate another, sometimes puzzling, aspect of the colour of living light; different colours can be emitted from different organs in the same animal — in the case of the railroad worm red and green–yellow, with the stomiatoids red and blue. And the peculiarities do not stop here (Table 3.12). Why, and how, is it that closely related species with the same chemiluminescent reaction, such as fireflies and glow-worms or different species of jelly fish, emit different colours? How can colours vary between cells in the same species, or even with the same individual? How can the purity of the colour vary with time, between species, or between the intact animal and the extracted chemiluminescent reaction?

The colour of the light emitted by a luminous organism is determined by three parameters:

(1) The chemiluminescent reaction itself (see section 3.4).
(2) The environment in which the reaction occurs, being particularly susceptible to pH, ions, and the hydrophobicity of the active centre of protein catalysts (luciferase) i.e. 'the solvent' (see section 3.4).
(3) Mechanisms within the organism which can bring about a change in the colour eventually emitted.

There are two established ways in which luminous organsism can alter the colour of light emission from that of the primary chemiluminescent reaction; either by passing the light through an absorbing or interference filter, or by energy transfer to a fluor, which then becomes the actual emitter. The former involves passage of light from the luminous cells through another structure or cell layer, and occurs in several luminous fish and squid. The latter occurs within the same cells as the chemiluminescent reaction without the direct transfer of a photon, and includes several luminous coelenterates and bacteria (see Chapter 9).

Many other peculiar spectral characteristics remain unexplained (Table 3.12). One possibility is that the environment within individual luminous cells in the same organism is different. At acid pH or in the presence of heavy metals such as Pb, firefly luciferin–luciferase emits red light, instead of yellow light. It would be interesting to know whether this explains the red emission from the organs on the head of the railroad worm, in contrast to the yellowish-green ones on the body.

To examine differences in chemiluminescence spectra between organisms, and between reactions *in vivo* and *in vitro*, we need to define the four major parameters enabling these comparisons to be made quantitatively:

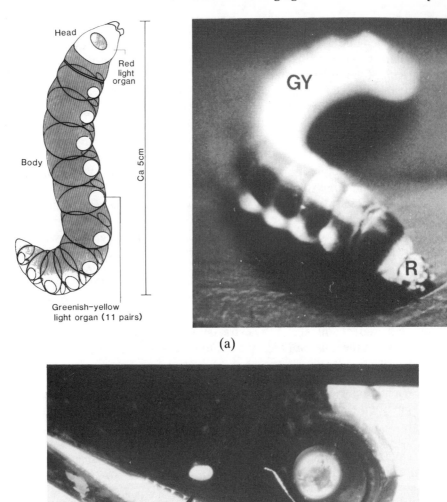

(a)

(b)

Fig. 3.10 — Two rare emitters of red living light. (a) the railroad worm *Phrixothrix tiemanni*
→ g–y = greenish–yellow organs → r = red organs; by Robert F. Sisson © 1970 National
Geographic Society (b) the deep sea fish *Malacosteus niger* → r = red organ, − b = blue
organ.

Table 3.12 — Some peculiarities in the spectra of living light

	Peculiarity	Group	Example
1.	Cells or organs in the same organism or colony emitting different colours	sea pen stomiatoid fish railroad worm some squid	*Umbellula* *Malacosteus niger* *Phrixothrix tiemani* *Heteroteuthis*
2.	Different individuals of the same species emit different colours (often dependent on geography)	bacteria sea pen jelly fish	*Vibrio fischeri* normal versus yellow Y1 stain *Parazoanthus* *Phialidium*
3.	Colour varies in closely related species	fireflies and glow-worms bacteria jelly fish	*Photinus v Lampyris* *Vibrio vs* *Photobacterium* *Aequorea vs Halis-taura*
4.	Spectrum bimodal	some amphipods	*Parapronoë crustulum*
5.	Spectrum changes with time	searsiid fish	*Searsia*
6.	*In vivo* spectrum different from *in vitro* extracts	some bacteria several cnidarians several fish	*Vibrio fischeri* *Obelia geniculata* *Argyropelecus*
7.	Polarisation	firefly	*Photinus pyralis*

(1) Spectrum uni-, bi- or multi-model (i.e. with one or more than one peak).
(2) λ_{max} of the peak(s).
(3) Bandwidth of the peak(s) = wavelength difference at half maximum intensity.
(4) Temporal changes in parameters (1), (2) and (3).

Very few organisms have peaks in their emission below 400 nm (violet) or above 600 nm (red). A large number have emission peaks in the range of 450–510 nm (blue to blue–green), and some in the range 550–580 nm (green to yellow). The bandwidth can be as narrow as 20–30 nm, as in the jelly fish *Aequorea*, or as wide as 150–200 nm, as in the squid *Abraliopsis*. Most bandwidths are in the range 50–100 nm (see Chapter 2 for definitions). In general, the narrower the bandwidth the purer the colour appears to our eyes, and presumably to the real beholder. The significance of this is not clear, except that a pure colour can stand out particularly from the colour of its surroundings.

Spectra can be related not only to biological function but also to the chemical energetics of the chemiluminescent reaction via Einstein's equation $E = h\nu$.

Polarisation
A poorly investigated physical characteristic of bioluminescence is whether the light emitted is either plane or circularly polarised. There are two probable reasons for

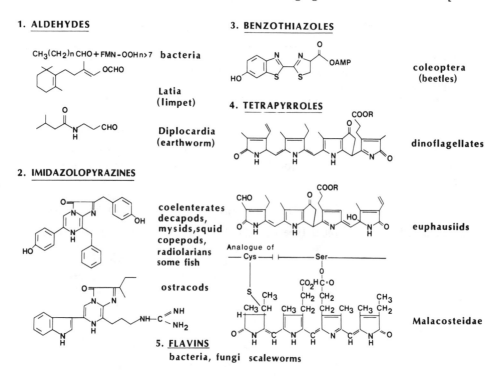

Fig. 3.11 — Chemical families in living light. Structure and preliminary biochemical characterisation of the bioluminescent system of *Ommastrephes pteropus*.

this. Firstly, polarisation is one of the most difficult conceptual aspects to grasp, particularly when considering light as a stream of photons. Secondly, it is, at first sight, difficult to see how an apparently diffuse bioluminescent reaction, where the excited state emitter has a plane of symmetry passing through it (i.e. it is not apparently anisotropic, and would thus not be expected to be optically active), could produce polarised light. Yet there is some evidence that fireflies produce circularly polarised light. Other possible candidates requiring investigation include polynoid worms and euphausiid shrimp, with paracrystalline structures within their luminous cells, and squid and searsiid fish with particulate or membrane associated bioluminescence. A further possibility is the polarisation of the emittted light by ancillary structures such as reflectors and filters.

Accurate descriptions of the four main physical parameters characterising bioluminescence — timing, intensity, colour, and state of polarisation — are essential if we are to relate the chemistry responsible for living light to its function and evolutionary significance. Yet to forge these relationships we also need to know from where the light is emanating and what chemical reactions are responsible.

3.3.2 Location of light emission

Where is the light produced? Is it intra- or extra-cellular? Where are the components of the chemiluminescent reaction stored? How are these components brought together to produce light? To answer these questions, we must first identify the cells responsible for generating the chemiluminescence, and then find out which organelles within these cells are involved. Before doing this, it is perhaps as well to clarify a few terms associated with the cell biology of living light (see Appendix I).

The cells responsible for living light are known as *photocytes* (Fig. 3.12a). These can occur individually, in small clusters when they form part of a *photophore* (Fig. 3.12b), or in large, dissectable structures when they are known as *light organs* (Fig. 3.12c). Two distinct types of photophore, or light organ, are identifiable. Those which produce their light discretely within the organ or photophore, and those which release a light cloud or slime into the surrounding fluid (Table 3.13). The presumption is that the former is the result of chemiluminescence within the photocytes, whereas the latter results from the secretion of the reactive components from cells within the photophore or organ. In most secretory luminous organisms the components are released in free solution. However, in a few, for example, the ctenophore *Eurhamphea* and searsiid fish, and some angler fish and squid squirting out luminous bacteria, the secretion is particulate. Although more than one cell type has been identified in several secretory light organs, including the bivalve *Pholas*, the polychaete worm *Chaetopterus*, several luminous squid, copopods, ostracods and the freshwater limpet *Latia*, it has yet to be established whether these different cell types contain separate components of the chemiluminescent reaction. Most secreting luminous organisms also release mucus, and one of the cell types may be responsible for this. This mucus can result in the secretion having a particulate appearance. Tap a rock containing *Pholas* and the entrances to many holes will glow blue–green, sending globules and streams of luminescence into the surrounding seawater.

The organisms producing light intracellularly, from within their light organ or photophores, can themselves be divided into two groups: those which generate their own photocytes and those which contain luminous bacteria.

Examples of the three types of location, namely bacterial, intrinsic within photophores, and intrinsic but secretory, can be found within quite closely related organisms. For example, intrinsic luminescence is found in oceanic squid of the suborder Teuthoidea (the oegopsids), some produce the light within the photophores, others squirt out the luminescence. In contrast luminous bacteria are found in some but not all the sepiolids and myopsid squid. A few organisms contain both symbiotic bacteria and their own photocytes. For example, the lures of all angler fish (Fig. 3.6a) contain luminous bacteria, but the luminous barbel found in some species contains its own photocytes. Bacterial light organs are usually found close to the surface of the animal, e.g. the flashlight fish (Fig. 3.6b), so that the light can shine out easily. However, in the pony fish the organ is internal, part of the gut. Symbiotic bacteria in fish and squid can leak out slowly into the surrounding water, but a few angler fish and sepiolid squid squirt out a cloud of luminous bacteria. In one squid the bacterial secretion is released like a toothpaste and then fragments into lots of tiny, luminous, globules. Unlike most other secretory systems the bacterial luminescence remains *intra*cellular. A striking exception to this rule are searsiid fish which secrete a cloud of cellular luminous acini.

It is usually assumed that light produced within an individual photophore is exclusively *intra*cellular. However, in several cases, this has yet to be proven. Intracellular luminescence can be observed emanating from within individual

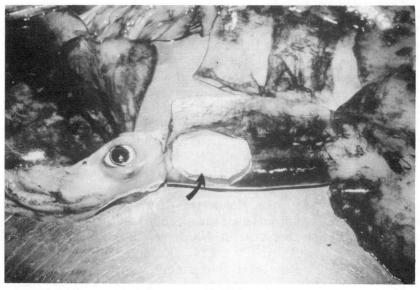

(a)

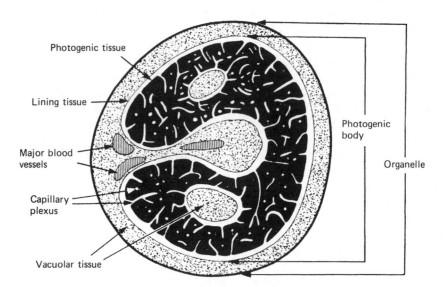

Schematic diagram of a single organelle.

(b)

Fig. 3.12 — Structural levels in bioluminescence. (a) A light organ of a squid (Ommastrephes)
↑ ; (b) a light unit (organelle) from (a) from Girsch *et al.* (1976) *J. Mar. Biol. Assoc.* **56** 707–722
with permission of Cambridge University Press; (c) Single luminous dinoflagellate from Eckert
and Reynolds (1976) *J. Gen. Physiol.* **50** 1424–1458 with permission of Rockefeller University
Press; (d) Bacteria in the cephalopod Sepiola. b = bacteria, bv = blood vessel, c = cilia,
er = endoplasmic reticulum, jc = junctional complex, n = nucleus, from Herring *et al.* (1981) *J.
Mar. Biol. Assoc.* **61** 901–916 The light organs of *Sopiola atlantica* and *Spirula spirula* with
permission of Cambridge University Press.

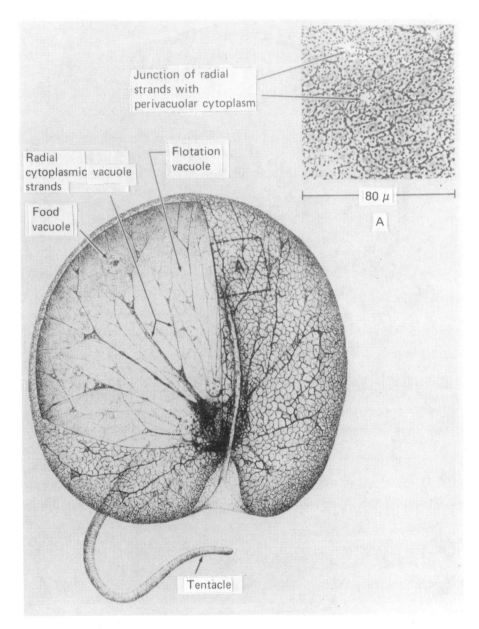

Junction of radial
strands with
perivacuolar cytoplasm

Radial
cytoplasmic vacuole
strands

Flotation
vacuole

80 μ

A

Food
vacuole

Tentacle

(c)

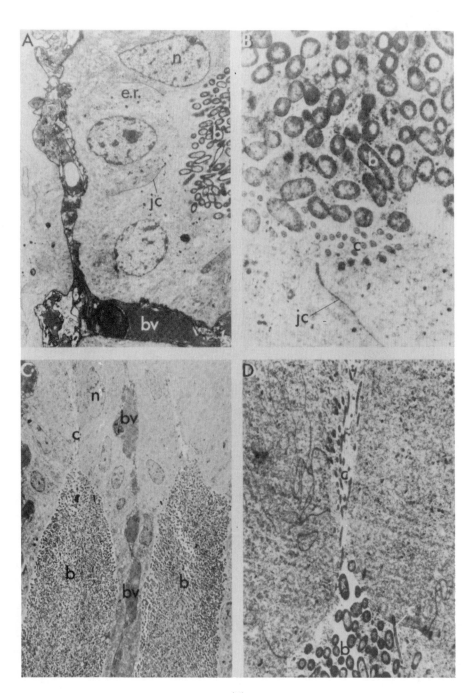

(d)

Table 3.13 — Intra- and extra-photophore luminescence

Taxonomic group	Example (common name)
A. *Light emission generated exclusively within photophore or light organ*	
1. All unicellular species	*Bacteria, dinoflagellates, radiolarians*
2. *All fungi*	*Pleurotus olearius*
3. Most cnidarians, except *Pelagia* and *Poralia*	*Obelia geniculata* (sea fir)
4. All ctenophores, except *Eurhamphea*	*Mnemiopsis leidyi* (comb jelly)
5. Many molluscs, including nudibranchs, most squid, octopus, and a land snail	*Watasenia scintillans* (firefly squid)
6. Certain annelid worms, including polynoids and syllids	*Acholoë astericola* (scale worm)
7. Certain arthropods, including most amphipods, a few decapod shrimp, all euphausiid shrimp, all insects, millipedes	*Photinus pyralis* (firefly)
8. Most echinoderms, except *Hymenaster*	*Plutonaster* (starfish)
9. Most chordates, including *Pyrosoma* and fish, except searsiids	*Argyropelecus* (hatchet fish)
B. *Photophores or organs release a luminous secretion outside the animal*	
1. A few cnidarians and ctenophores	*Pelagia noctiluca* (jelly fish)
2. Some molluscs including some myopsid and sepiolid squid, bivalves and gastropods	*Heteroteuthis dispar* (squid) *Pheolas dactylus* (piddock) *Latia neritoides*
3. Some annelid worms including most oligochaetes, and several polychaetes including Chaetopteridae, Cirratulids, terebelids and with gametes in some syllids	*Chaetopterus variopedatus* (worm), *Ddontosyllis* (fire-worm)
4. Several crustacean arthropods including ostracods, copepods, most mysid and decapod shrimp	*Vargula* (formerly *Cypridia*) *hilgendorfi* (sea firefly)
5. Centipedes	*Orya barbarica*
6. Most hemichordates	*Balanoglossus* (a corn worm)
7. A few echinoderms	*Hymenaster* (starfish)
8. A few fish	*Searsia* (particulate secretion)

This list is not complete, particularly with respect to exceptions, and is intended to point towards questions required to understand the control mechanisms for intracellular, intra-photophore, and secretory bioluminescence. Note that intraphotophore may not necessarily be intracellular, though it usually is.

photocytes in many of these species (Table 3.13a) using the naked eye aided by the light microscope, or with the aid of an image intensifier. However, Bassot has examined carefully the structure of several fish photophores, finding three types containing two morphologically distinctive cells, arranged in a particular manner within the photophore. Type A contains big granules and look like granular cells. In the hatchet fish they are arranged side by side, but in the fish *Cyclothone* they are grouped in acini or tubes which empty into a common collecting canal. This raises the possibility that, whilst most intra-photo*phore* luminescence may also be intra-photo*cyte*, some may result from a secretion, retained within the photophore.

To resolve this problem we need to identify definitively which cells contain the various components of the chemiluminescence reaction, and where within the cell these are stored. Since most photophores and light organs contain accessory cells or structures to assist with processing the light (Table 3.14) it is essential to distinguish these from the cells containing the chemiluminescent components. Well defined accessory structures, such as cups and filters, contain melanin, xanthin, carotenoid proteins, free fluorescent porphyrins, or protein-bound tetrapyrroles. Light guides contain a large amount of collagen. Reflectors are made up of multi-layers of material with high and low refractive index. The former consists of chitin or protein

Table 3.14 — Optical accessories in photophores and luminous organs

Structure	Function	Example
A. Directional apparatus		
Pigment cup	Restricts emission to a particular direction	Smaller photophores of many fish and squid
Reflector	Directional emission more efficient than pigment cup	Most photophores in fish, squid, and crustaceans
Iridophore	Reflector cells intermingled with or surrounding photocytes	The squid *Bathothauma*
Lens ± lamellar ring	Collimated light	Many fish, squid, and euphausiid shrimp
Light guide	Directed emission	Enoplo- and lyco-teuthid squid
Light pipe	Directed emission from internal organ	Pony fish and many ceratoid angler fish
B. Alteration of colour of light emission		
Absorbing filter	Colour to suit function	Hatchet fish
Interference filter	Colour to suit function	Enoploteuthis and other squid
Plug on light guide	Colour to suit finction	The squid *Histioteuthis*
C. Switching light on or off		
Surface chromatophore	Controlling intensity	Several fish
Surface shutter	Blinking	Flashlight fish
Organ rotation	Flashing	Squids

This is intended as an illustration; many other examples exist, see Herring (1978, 1985).

in invertebrates and usually, but by no means always, guanine in fish. These accessory structures, including lenses, are easily distinguishable morphologically from the photocytes. However, other cells may intermingle with the luminous cells. For example, photophores in euphausiid shrimp and some luminous squid contain iridophores with internal platelet-like iridosomes helping the light to be reflected outwards. How can a photocyte generating intracellular chemiluminescence, or a cell secreting one or more of the components of the light-generating reaction, be identified definitively? There are five possible ways of achieving this; only three have been used to date:

(1) Association of light emission with particular cells
   Observation of light emission from within a cell in a photophore, or close to the membrane of a cell if it is a secretory system. This will require a microscope, and probably an image intensifier to provide a permanent record.

(2) The presence of fluors within the proposed photocyte, or source of luminous material
   The fluor may be the luciferin, the product of the chemiluminescent reaction (i.e. the oxyluciferin), or an energy transfer acceptor. Comparison of the fluorescence spectrum with that of the chemiluminescence will establish whether the fluor is the actual light emitter. The cellular origin of the fluorescence can be established by using a fluorescence microscope with an image intensifier if necessary. If the fluor is the luciferin then its fluorescence *in vivo* can be matched with that of the extracted luciferin, and it should decrease as the reaction proceeds. In contrast, if it is a product of the chemiluminescent reaction then its fluorescence will increase, and its fluorescence spectrum match that of the reaction product *in vitro*. However, neither the luciferin nor the final reaction product may have significant fluorescence or be present in sufficient concentration to be visible by fluorescence. Furthermore, much of the luciferin may be present in an unreactive storage form, e.g. a sulphate.

(3) Isolation of cells
   Enzymatic destruction of the extracellular matrix using collagenase and/or hyaluronidase will release the cells. Different cell types can be isolated by using Ficoll or Percoll gradients, or by flow cytometry using a fluorescence-activated cell sorter. The cell type can then be identified morphologically by using light and electron microscopy, and characterised biochemically for chemiluminescence, fluorescence, and luciferin and luciferase content.

(4) Protein cytochemistry
   Antibodies to luciferase or luciferin, labelled with a fluor, peroxidase, or gold, added to sections of the photophore, will enable the cellular origin of these components to be identified. Alternatively the luciferin or luciferase could be added to a tissue section, observed under the microscope, to generate light, and thereby identify the cell containing the other component(s).

(5) Nucleic acid cytochemistry (*in situ* hybridisation)
   Addition of labelled gene probes to microscope sections, followed by hybridisation with RNA in the section, will enable cells expressing mRNA coding for protein components of the chemiluminescent reaction to be identified.

So far, only the first two of these criteria have been widely used. Fluorescence has been found associated with the luminous cells from a wide range of organisms (Table 3.15). In sea combs, brittlestars, earthworms, marine scale and syllid worms, and the searsiid fish, there is either an increase in fluorescence intensity or a change in the colour of the fluorescence, following bioluminescence. This suggests that the fluorescence is monitoring luciferin being oxidised to oxyluciferin. In the cellular clusters secreted by searsiid fish each cell changes from green to blue. In dinoflagellates, beetles and glow-worms, and several fish, the fluorescence emission is either reduced or unchanged following bioluminescence, and has an emission spectrum quite different from that of the bioluminescence. This occurs when the luciferin is itself fluorescent. The oxyluciferin may not be reproduced in sufficient quantities to be seen by fluorescence, or may react very quickly to form non-fluorescent products, as in the dinoflagellates. In contrast, in some bacteria and cnidarians the fluorescence is from an energy transfer acceptor, not directly required for the primary chemiluminescent reaction. In secretory photocytes the fluorescence can 'disappear' on stimulation. Yet fluorescence is not necessarily a definitive method of identifying luminous cells. In the copepod *Gaussia*, antennae cells are fluorescent but not

Table 3.15 — Fluorescence associated with luminous cells

Organism group (phylum)	Example	Colour of flourescence
Bacteria	*Vibrio harveyi* (*Y*1 strain)	Yellow
Fungi	*Omphalia flavida*	Green (phosphorescence)
Dinoflagellates	*Noctiluca*	Blue
Hydroids and jelly fish (Cnidaria)	*Obelia*	Green
Sea pens and sea pansies (Cnidaria)	*Renilla*	Green
Sea combs (Ctenophora)	*Mnemiopsis*	Blue
Brittle-stars (Echinodermata)	*Acronida*	Green
Earth worms (Annelida)	*Eisenia*	Yellow–green to blue
Scale worms	*Acholoë*	Green
Syllid worms (Annelida)	*Pionosyllis*	Green
Land snail (Mollusca)	*Dyakia* (= *Quantulus*)	Yellow–green
Squid (Mollusca)	*Ommastrephes*	Yellow
Copepods (Arthropoda)	*Euangaptilus*	Blue to green
Amphipods (Arthropoda)	*Scina*	Blue
Decapod shrimp (Arthropoda)	*Sergestes*	Blue
Beetles and glow-worms (Arthropoda)	*Photinus*	Yellow–green
Fish (Chordata)	*Chauliodus*	Blue–green
	Searsia	Blue–green → blue
	Porichthys	Green
	Aristostomias (sub-orbital)	Orange
	Malacosteus (sub-orbital)	red

Note: The fluorescence is by no means universal within any group. It has usually been detected initially by using a hand UV lamp (366 nm), but in some cases more precise micro-fluorimetry has been carried out. The fluorescence may be due to the luciferin, the oxy-luciferin, or an energy transfer acceptor.

apparently luminous, whereas the blue fluorescent product of coelenterate bioluminescence is difficult to see in the cells of all but a few species.

Many photocytes and secretory luminous cells contain organelles involved in
light production (Table 3.16; Fig. 3.13). These organelles can be just a few, or may
number several thousand, as in the luminous cells of the glow-worm *Lampyris*.
Identification of the role of these organelles is based on observing fluorescence, and
sometimes chemiluminescence, from within living cells, or by biochemical analysis
following sub-cellular fractionation. In an early, elegant study by Eckert and
Reynolds, luminous microsources some 0.2–1.5 μm in diameter, were observed in
individual dinoflagellate cells of *Noctiluca* by image intensification. Hastings' group
has confirmed that these are vesicles containing luciferin and luciferase which flash
when subjected to a decrease in pH. Granules in luminous cells sometimes contain
paracrystalline structures, as in the fireflies and glow-worms, the decapod shrimp
Sergestes, and the barbel of the angler fish *Linophryre*. In other examples of
intracellular bioluminescence the light emission can be associated with intricate
membrane structures, for example the polynoid scale worms studied extensively by
Bassot. The luminous systems from searsiid fish, and the squid *Ommastrephes* with
its huge light organ, are difficult to solubilise, and are thus possible candidates for
particulate or membrane-bound bioluminescence. As one would expect, secretory
granules are found within cells in the light organs of animals releasing luminescence
extracellularly (Table 3.13b). Some luminous cells contain unusual organelles, such
as giant mitochondria, not directly responsible for luminescence. Yet others, for
example the luminous cells in *Obelia* (Fig. 3.13), appear devoid of any intracellular
structure.

Table 3.16 — Examples of intracellular structures associated with bioluminescence.

Structure or organelle	Organism
Secretory granules	All class B in Table 3.13, e.g. squid, bivalve, worms, copepods and ostracods
Secretory acini (i.e. group of cells)	Searsiid fish
Intracellular vesicles (single membrane)	Coleoptera (e.g. firefly = peroxisome) Dinoflagellates (scintillons)
Complex intracellular membrane structure (photosome)	Polynoid worms
Double membrane organelle (? intracellular bacteria)	*Pyrosoma*
Paracrystalline body or crystalloid, usually within a membrane bound vesicle	Firefly Some squid (e.g. *Bathothauma*), *Sergestes* and other decopods, barbel of the angler fish *Linophyryne*

NB Some luminous cells, e.g. in bacteria and coelenterates, appear to be virtually devoid of any
intracellular structure.

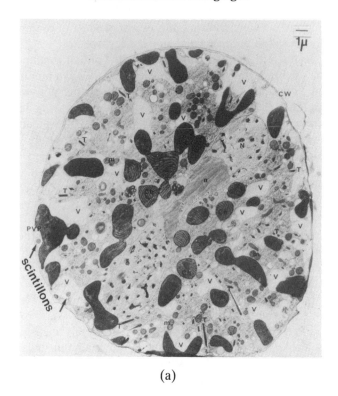

(a)

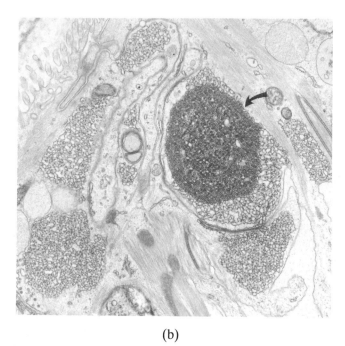

(b)

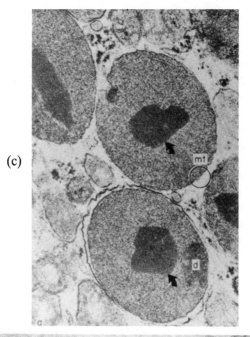

(c)

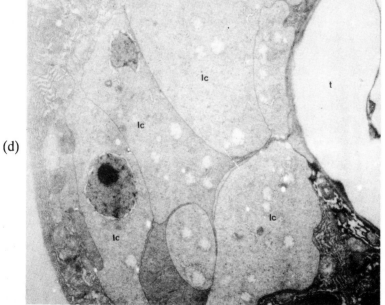

(d)

Fig. 3.13 — Intracellular structure in photocytes. (a) Scintillon in a dinoflagellate (arrows), from Nicolas *et al.* (1987) with permission of the Rockefeller University Press. Ch = chloroplasts, m = mitochondria, T = trichocysts. V = vacuole, PVB = polyvescular bodies, CW = cell wall; (b) photosome of a polynoid worm (↑) showing change in membrane convolution. Kind gift from Dr Jean-Marie Bassot. × 26000,(c) paracrystalline body (↑) in firefly (Photinus greeni). Case (1978) in Herring (1978) with permission of Academic Press. mt = microtubule, d = photocyte vesicle densification approx × 60,000,(d) amorphous cells in *Obelia* unpublished by Pulsford and Bone. lc = postulated luminous cells with the green fluorescence, t = base of tentacle.

Now that we have a method for identifying the cellular origin of living light, and know what to look for within the cells, let us return to whole organisms to see where the photophores and light organs are located. There is an amazing variety of sites (Table 3.17). Most are on the surface, but in a few, such as the decapod shrimp and pony fish, the cells responsible for luminescence are found in internal organs, usually associated with the gastrointestinal tract or the hepatopancreas. The light reaches the outside in the shrimp *Sergestes* because it is transparent. Other shrimp squirt a luminous cloud from the hepatopancreas through their mouths. Pony fish use light guides to enable the light to get out. In the bivalve *Pholas* there is a triangular light organ on each main muscle surface and also light cords. The shell of *Pholas* is, of course, opaque. The animal taken from its rock burrow can be seen to glow when the shells are not completely closed. It then squirts a luminous cloud from its siphon.

Surface light organs and photocytes occur in association with almost every part of the body (Table 3.17), the eyes, the jaws, the mouth, the body, fish tails and fins, and tentacles in squid. Perhaps the most extraordinary are some of the light organs of many oegopsid squid (Fig. 3.3a), such as *Watasenia* and *Bathothauma*, which are found on the eyeball. The structure, control, and even sometimes the biochemistry, of photophores, and the photocytes on different parts of an organism may be different. For example, the intricate lanterns on the ventral surface (i.e. underneath) of euphausiid shrimp are different from the two on the eyestalks.

Table 3.17 — Animal photophores and light organs associated with organs of the body

Main structure	Particular site	Example
Head	Jaw or mouth	*Cleidopus* (fish)
	Eyestalk	Euphausiid shrimp
	Eyeball	Some squid (e.g. *Bathothauma*)
	Orbital (sub- or post-)	Several fish, (e.g. flashlight fish and Malacosteus)
Body	Scales	Polynoid scale worms
	Epidermal surface	Many squid and fish
Tail	Near anus	Some fish (e.g. *Chauliodus*)
Appendage	Tentacles	Several squid, and starfish
	Base of tentacles	Most jelly fish
	Fin	Lure and barbel of angler fish
	Legs	Amphipod shrimp, copepods
Internal	Head	Searsiid fish
	Hepatopancreas	Decapod shrimp
	Organ of Pesta	*Sergestes* (a shrimp)
	Oesophageal or intestinal	Bacteria in pony fish
	Ink sac	Some squid
	Muscle surface and siphon	The bivalve piddock *Pholas*

The number of photophores or light organs can vary from just one to many hundred. This diversity is well illustrated by luminous fish and squid (Figs. 3.3 and 3.8a). The fish *Chauliodus* has a single, simple photophore at the tip of an elongated dorsal fin which probably acts as a lure, as in the angler fish. The deep sea fish *Malacosteus* has two pairs of light organs, one suborbital (red) and the other postorbital (blue), and multiple photophores dotted over its body surface. The firefly squid *Watasenia* has three black spots, the photophores, on the tips of each ventral (i.e. lower) pair of tentacles, five light organs on each eyeball, and several hundred photophores scattered over its body. Many other fish and squid also have multiple photophores on the surface, which can number several hundred. For example the midshipman fish *Porichthys* has over 800. In some luminous organisms the size or siting of the light organs or photophores is different between male and female (sexual dimorphism).

The sites in some luminous organisms can clearly be related to function (see section 3.6). For example, the lure of the angler fish; the ventrally sited photophores of lantern fish, euphausiid shrimp, and squid, for camouflage; the organs of fireflies and glow-worms in the last segments of the body for attracting a mate. Yet in other species the reason for the sites of photophores is not clear. Why are the photocytes of hydroids such as *Obelia* apparently randomly scattered along its stem, in the endodermal cell layer, yet rarely if ever found in the hydranths (the mouth and tentacles)? Why do luminous jelly fish such as *Obelia* and *Aequorea* have a ring of photophores at the base of their tentacles? Dorsal distribution (i.e. uppermost) of photocytes is rare, yet it is found in luminous holothurians, copepods, and amphipods. Why? What is the reason for the female octopus Japatella having luminous cells around its mouth (Fig. 3.3a)? Photophores and light organs are often found associated with the eye and orbital region. When used as a torch, as in the flashlight fish *Photoblepharon*, the eye is protected from the brightness of the light organs. However, in other organisms, the reason for close proximity to the eye is unknown. The ovaries of some echinoderms and medusae are luminous. The eggs and larvae and pupa of fireflies and glow-worms are luminous, as well as the adults. Does the luminescence have a function or not in these developmental stages? Why is the intracellular chemiluminescence in some fungi associated with the gills of the fruiting body (a group of spores), in others with part of the body of the fungus, the mycelium, and in some, occurs in both?

Now that we know what to look for and where to find it we are in a position to discover what is the chemistry and enzymology of living light.

3.4 THE BIOCHEMISTRY OF LIVING LIGHT
3.4.1 How to begin

Imagine that whilst walking, or perhaps swimming, at night you are lucky enough to discover a hitherto unknown luminous organism, be it plant or animal. How does one begin to find out what causes the light emission? Perhaps there are biological examples of tribo- or electro-luminescence. As yet we know of none. So we assume, until proven otherwise, that all living light is caused by a chemical reaction within, or released by, cells in the organism.

Our first problem is to preserve the activity of the chemiluminescent reaction until we can get it back to the laboratory. Terrestrial animals may be transported back live, but this is more difficult with marine species, particularly if they come from the deep oceans. Many systems survive freezing, freeze-drying, or precipitation as an ammonium sulphate suspension. These include most luminous fish, coelenterates, shrimp, insects, worms, and molluscs. An additional extraction in acid-methanol can also be useful for small organic moeities involved in the light generating reaction. However, several do not survive any of these procedures. For example, virtually all light is lost by freeze-thawing the tunicate *Pyrosoma*. Another approach is to prepare an acetone powder. This works well with the bivalve *Pholas*, but it is hopeless for jelly fish and hydroids (coelenterates). Whilst at the outset we may not have the ultimate storage conditions, it is essential that we begin with a preparation producing visible light. The sensitivity of our laboratory-based chemiluminometers is such that artefactual reactions can be obtained, producing apparently large recorded signals, and yet be not truly responsible for the natural luminescence of the organism. Assuming we have active luminous material, the trick then is to extract the stored material in such a way that the individual components can be separated, using chromatographic techniques and organic extractions, and then added back together to generate visible light of equivalent intensity and colour to that in the living organism. Before doing this, however, it is worthwhile carrying out one of the oldest experiments in bioluminescence, the requirement for oxygen.

3.4.2 The components

As we have already seen (Chapter 1, section 1.3.4) all bioluminescent reactions involve the oxidation of a small organic moiety catalysed by a protein. The enthalpy of the reaction must be at least some 38–71 kcal.mol^{-1} to provide sufficient energy for visible light (Eqn 1.6). ATP hydrolysis generates less than 10 kcal.mol^{-1}, and is thus not the source of energy when required in bioluminescence. A minimum of three (Eqn. (3.1)), and up to six, components are thus required to generate living light (Table 3.18). Oxygen, chemiluminescent substrate = luciferin, protein catalyst = luciferase, cofactor(s), cation(s), fluor.

Oxygen

In 1668 Robert Boyle showed that removal of air stopped the luminescence of decaying flesh and rotting wood, known to be caused by luminous bacteria and fungi, respectively. In 1810 McCartney investigated the requirement for oxygen of extracts of several luminous organisms. The problem was that it has only recently been realised how sensitive bioluminescent systems are for oxygen. In enzymological terms the Km's for O_2 are often in the μM range. Since air-saturated water contains 200–250 μM, depending on temperature, it is necessary to go to some lengths to remove enough O_2 to prove definitely that it is essential. Simply bubbling argon through a solution may not be sufficient. A good vacuum pump will do this. Harvey used Pt/H$_2$. The result is that the chemiluminescent reactions in all luminous bacteria, sea pens, molluscs, crustacea (including ostracods and shrimps), worms, insects (including fireflies and glow-worms), and fish, so far investigated require oxygen in some form in the solution. Most use molecular O_2 (Table 3.18A), though several, including fireflies, scaleworms, *Pholas* and imidalozopyrazine systems may use O_2^- as an intermediate. A few, such as the acorn worm *Balanoglossus*, earthworms, and starfish, use hydrogen peroxide. Addition of H_2O_2 to the luminous

Table 3.18 — The principal components required to generate living light

Component group	Specific component	Examples
A. Oxidant	Molecular O_2	The majority of living light, including bacteria, fungi, anthozoans, dinoflagellates, molluscs, arthropods, and fish
	O_2 with O_2^- or ROO as intermediate	*Pholas* (piddock)
	O_2 forming hydroperoxide intermediate	All coelenterates and radiolarians
	H_2O_2	Earthworms, acorn worms, brittle stars (?)
B. Protein responsible for triggering chemiluminescence	Catalytic luciferase	The majority of living light, including bacteria, fungi, anthozoans, dinoflagellates, molluscs, most arthropods, and fish
	Oxidase/oxygenase	*Pholas* (piddock)
	Peroxidase	Earthworms and acorn worms
	Photoprotein (type I)	All coelenterates (except anthozoans), radiolarians
	Catalase like protein	? brittle stars
C. Chromophore = Luciferin (oxidation produces light)	Free luciferin	Many, including bacteria, fungi, dinoflagellates, arthropods, anthozoans, earthworms, squid, fish
	Photoprotein type I[†]	All coelenterates (except anthozoans), radiolarians, euphausiid shrimp, millipede, polynoid worms
	Photoprotein type II[†]	Piddock, *Chaetopterus*, brittle stars
D. Cofactor requirement	NAD(P)H	Bacteria and fungi
	FMN	Bacteria
	Long chain aldehyde	Bacteria
	ATP	Beetles, New Zealand glowworm, some squid, millipede
	PAP or PAPS for luciferin sulphate	Anthozoans, some squid
	small fluor	*Chaetopterus*, euphausiid shrimp
	CN^-	*Odontosyllis* (worm)
E. Cation	Mg^{2+}	All ATP requiring systems (ATP Mg^{2-})
	Ca^{2+}	All coelenterates, radiolarians
	Fe^{2+} or Fe^{3+}	*Chaetopterus* (worm), brittle star
	Cu^+/Cu^{2+} (in luciferase)	Ostracods, piddock, earthworms
F. Ancillary proteins	Oxidoreductase	Bacteria, fungi
	Acid reductase	Bacteria
	Luciferin-binding protein	Anthozoans and several others
	Energy transfer protein	Several bacteria and coelenterates, some fish

[†]Photoprotein type I = luciferin tightly bound to catalytic protein.
  Photoprotein type II = luciferin tightly bound to non-catalytic protein, another protein (the luciferase) is required to trigger chemiluminescence.

glands of many fish and invertebrates generates light. This is not surprising since all bioluminescent reactions are oxidations. However, this does not necessarily mean that H_2O_2 is the actual oxidant *in vivo*.

Harvey found two exceptions to this, otherwise universal, requirement of luminous organisms for oxygen: all luminous hydroids, jelly fish, and sea combs (i.e. cnidarians and ctenophores other than anthozoa), and luminous radiolarians (Phylum: Protozoa). Thanks to the work of Shimomura and Johnson we now know that oxygen is also required by the chemiluminescent system for these. The difference between these later organisms and the others is that when extracted from jellyfish or radiolarians, the chemiluminescent components can be isolated as a complete unit, defined by Shimomura as a *photoprotein*. These particular photoproteins retain oxygen bound within them, either as O_2 or possibly in the form of a hydroperoxide.

Thus, all bioluminescent systems so far investigated use oxygen in some form. What next?

Luciferin–Luciferase

All bioluminescence requires a protein or proteins (Table 3.18B). Whilst we expect to find such proteins involved in any biological reaction, one would not expect their oxidation by O_2 or H_2O_2 to generate visible light. A small organic molecule is therefore required, capable of generating an excited state when oxidised (Table 3.18C).

The history of bioluminescence is dotted with key experiments which in in the light of current knowledge, were very lucky to have worked. One such experiment was that of Dubois (Fig. 1.3b) published in 1887. As we saw in chapter 1 he added a heated extract of the luminous organ from the bivalve *Pholas dactylus* (the common piddock Fig. 3.1) to a cold extract which was no longer luminous. This regenerated light. Some two years earlier, he had carried out a similar experiment with the West Indian click beetle *Pyrophorus*. But now he called the heat stable factor *luciférine* and the heat labile factor *luciférase*. This general nomenclature has remained ever since. Yet we know that not only is *Pholas* luciferase a protein, but also the luciferin (N.B. now without the last 'e'). Both are in fact heat labile. Heat is therefore not a good way of obtaining luciferase-free *Pholas* luciferin, but sufficient active luciferin remains to demonstrate the luciferin–luciferase reaction. Newton Harvey (Fig. 1.3c) spent many hours trying to repeat the Dubois experiment with other luminous species. Harvey was successful in generating light from luciferin and luciferase from ostracods such as *Cypridina* (now called Vargula), polychaete worms such as *Odontosyllis*, decapod shrimp such as *Systellaspsis*, and beetles such as the firefly *Photinus*. Other workers have since used organic extractions such as acid-methanol or chromatographic separations to separate the luciferin from the luciferase, followed by light regeneration on mixing the two. These experiments have identified luciferin–luciferase reactions in dinoflagellates (e.g. Gonyaulax), many fish (e.g. the midshipman *Porichthys*), lamellibranch shell-fish (e.g. *Rocellaria*), fungi (e.g. *Armillaria*), earthworms (e.g. *Diplocardia*), squid (e.g. *Watasenia*), and the sea pens and pansies (e.g. *Renilla*). However, this simple approach failed with luminous bacteria and the fish and squid harbouring them. It has also failed with searsiid fish, polynoid worms, the worm *Chaetopterus*, and hydroids, jelly fish, comb jellies, and sea jellies (i.e. coelenterates apart from the *Anthozoa*), in spite of the fact that extracts of these organisms produce plenty of light. Proteins play one or more of five main roles in the chemiluminescent reactions of living light (Table 3.18B, F): oxidant processing, catalytic, luciferin-binding, energy transfer and co-factor generation.

Cofactor(s)

The luciferin–luciferase approach has also failed with the tunicates such as the *Pyrosoma*, luminous starfish and other echinoderms, stomiatoid fish such as *Malacosteus*, and until very recently copepods. The main problem here has been the difficulty either in reconstituting the luminous reaction after freezing the organism, or in producing much light following homogenisation because of consumption of the luciferin. A similar problem was encountered by McElroy in 1947 when studying firefly bioluminescence. Addition of ATP to homogenates produced visible light. Lucky, since apart from the beetles, only two other luminous systems have shown requirements for ATP, the rare millipede *Luminodesmus* found only on mountain slopes in Sierra Nevada, USA, and the squid *Watasenia*. In none is ATP the energy source for chemiluminescence. Similarly the observation, by Strehler and Cormier in 1951, that a kidney extract evoked light emission from extracts of luminous bacteria, led to the discovery of the cofactors necessary for this luminous system: a long chain aldehyde (C_nH_{2n+1} CHO, $n > 8$), FMN and NAD(P)H.

Cofactors are required as an additional component of many bioluminescent reactions (Table 3.18D).

Cation

The fifth component required in some systems is a cation (Table 3.18E), either an alkaline earth, Mg^{2+} or Ca^{2+}, or a transition metal such as Fe^{2+}, Fe^{3+} or Cu^+ Cu^{2+}. Transition metals play a role in oxidant processing and thus can change valency states during the chemiluminescence. For example the earthworm and piddock luciferase may change between Cu^+ and Cu^{2+} states.

Fluorescent proteins

In a few systems (Table 3.18f and Fig. 3.14) a fluor bound to a protein interacts directly or indirectly with the chemiluminescent reaction so that the fluor becomes the light emitter (3.6), i.e. type II or energy transfer chemiluminescence (Chapter 1 section 1.3.1 and 1.3.6 and Chapter 9).

The structures of two of these fluors, the green fluor in certain coelenterates and some luminous bacteria, have been published (Fig. 3.14). However, there is some disagreement over the former.

Thus a minimum of three components are required for any potentially luminous organism to produce light:

$$L_{red} + (0) + Pr \rightarrow products + L_{ox}^* \rightarrow h\nu \tag{3.5}$$

where

$\quad L_{red}$ = reduced chromophore, (small) organic molecule
$\quad L_{ox}^*$ = electronically excited oxidised chromophore
$\quad (0)$ = oxygen in some form
$\quad Pr$ = protein

When type II occurs:

$$L_{ox}^* + fluor \rightarrow L_{ox} + fluor^* \rightarrow h\nu \tag{3.6}$$

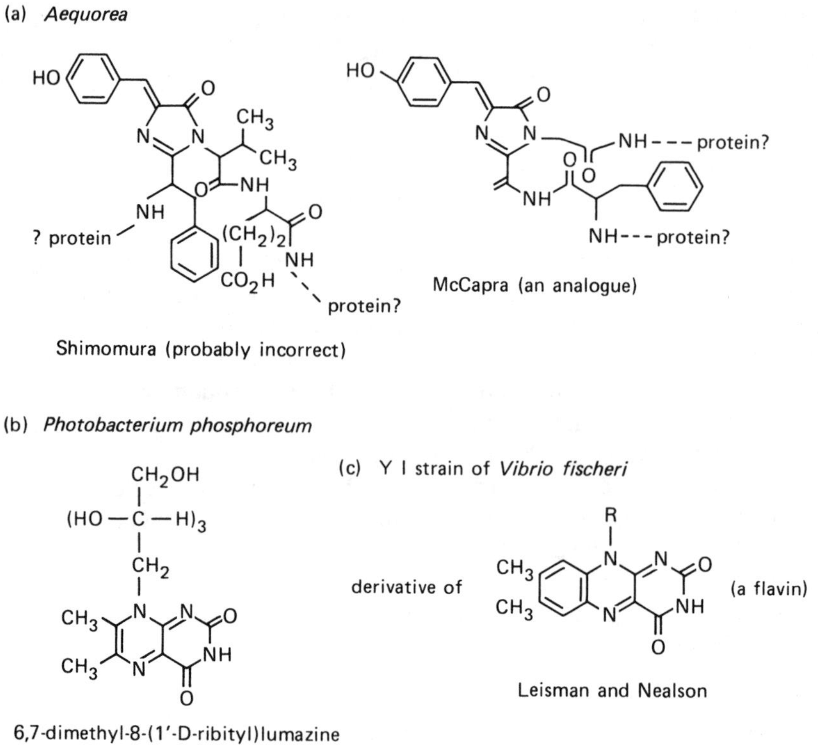

(a) *Aequorea*

Shimomura (probably incorrect)

McCapra (an analogue)

(b) *Photobacterium phosphoreum*

(c) Y I strain of *Vibrio fischeri*

6,7-dimethyl-8-(1'-D-ribityl)lumazine

derivative of ... (a flavin)

Leisman and Nealson

Fig. 3.14 — Proposed energy transfer chromophores. A, *Aequorea* and other coelenterates. B, Bacterium. *Photobacterium phosphoreum*; C, Yellow strain Yl of *Vibrio fischeri*.

In addition other proteins, cofactors, and ions may be required for maximum reaction rates and/or maximum light yields. How, then, do these components, when brought together appropriately, produce visible light?

3.4.3   How light is generated

Ultimately, we want to know how the organism itself produces light. To do this we first need the answers to five questions:

(1) What are the chemical structures of the reduced chromophore, the chemiluminescent substrate, and its excited state oxidation product?
(2) What is the chemical mechanism of this oxidation?
(3) What are the structures and roles of the other components?
(4) What are the optimal conditions for light emission *in vitro*, and how do these relate to conditions *in vivo*?
(5) What is the role and mechanism of the proteins involved in bioluminescence?

Before attempting to answer these questions, it is as well to define a few terms (see Appendix I). Not all will agree entirely with these definitions, but at least they will make the following discussion clearer, I hope.

Table 3.19 — A chemical taxonomy of luciferins

Key component for chemiluminescence (luciferin)[†]	Major group[†]	Specific example of genus	Evidence[§]
1. Aldehyde	Bacteria	*Vibrio*	1, 2
	Earthworms (phylum: Annelida)	*Diplocardia*	1, 2
	Freshwater limpet (phylum: Mollusca)	*Latia*	1
	Euphausiid shrimp (phylum: Arthropoda)	*Meganyctiphanes*	1, 2
2. Imidazolo-pyrazine[¶]	Hydroids, jelly fish (phylum: Cnidaria)	*Aequorea*	1, 2
	Sea pansies (phylum: Cnidaria)	*Renilla*	1
	Comb jellies (phylum: Ctenophora)	*Mnemiopsis*	1
	Radiolarians (phylum: Protozoa)	*Thalassicolla*	1,2
	Ostracods (phylum: Arthropoda)	*Vargula* (*Cypridina*)	1
	Copepods (phylum: Arthropoda)	*Euaugaptilus*	1, 2
	Some mysid shrimp (phylum: Arthropoda)	*Gnathophausia*	2
	Some decapod shrimp (phylum: Arthropoda)	*Oplophorus*	1,2
	Some fish (phylum: Chordata)	*Porichthys* (*Vargula*)	2
	Some fish (phylum: Chordata)	*Diaphus* (*myctophid*)	2
	Some squid (phylum: Mollusca)	*Watasenia*	1, 2
3. Benzothiazole	Fireflies (phylum : Arthropoda)	*Photinus*	1, 2
	Glow-worm (phylum: Arthropoda)	*Lampyris*	2
	Elatarid beetles (phylum: Arthropoda)	*Pyrophorus*	1,2,3
	Phengodid beetles	*Phrixothrix*	2
4. Linear tetrapyr role	dinoflagellate (phylum: Dinophyta)	*Noctiluca*	1
	Euphausiid shrimp (phylum: Arthropoda	*Meganyctiphanes*	1, 2
	Millipede (phylum: Arthropoda)	*Luminodesmus*	1
	Stomiatoid fish (phylum: Chordata)	*Malacosteus*	1
5. Flavin	Fungi	*Lampteromyces*	2
	Bacteria	*Vibrio*	

†See Fig. 3.11 for structures and Hastings (1983) and Shimomura (1985) for references.
‡Only selected organisms studied.
§Evidence: 1. Isolation and chemical identification of luciferin.
　　　　2. Cross-reactivity of luciferin with luciferase from another known species.
　　　　3. Antigenic cross reactivity of luciferase with known luciferase.
　　　　4. Spectral similarity of isolated fluor with bioluminescence.
¶Called imidazolopyrazine rather than imidazopyrazine because of the presence of a $C = O$ group.

Imidazole =   Pyrazine =

Definitions

Luciferin (Tables 3.18C and 3.19): I shall use this term for the reduced form of the small organic moiety, usually of molecular weight < 1000, which when oxidised forms the light emitting moiety. It may be isolated free, or complexed to another molecule, or bound covalently or non-covalently to a protein, or in a storage, and thus inactive, form such as sulphate. Many luciferins are highly unstable and oxidise spontaneously in a weak chemiluminescent reaction. For example if a solution of coelenterazine, the luciferin in coelenterates, is left on the bench in sunlight, more than 80% can be destroyed within an hour. Even the photoproteins isolated from ctenophores are susceptible to photo-inactivation. In some coelenterates, e.g. *Renilla*, an inactive storage form has been found, where the carbonyl group of the imidazo-ring has a sulphate linked to it (Fig. 3.11). The sulphate is transferred from 3'phospho-adenosine 5'phosphosulphate (PAPS) via a sulphokinase (3.12). In contrast, in the firefly squid *Watasenia* the *active* form of the luciferin is sulphated. However, in this case, coelenterazine is the preluciferin, being transferred from the liver to the photophores, and bisulphated using the SO_4 of PAPS, but on the two phenolic side chains (Fig. 3.11), instead of the carbonyl group. Several other squid e.g. *Pterygioteuthis* and *Eucleoteuthis* appear to use the unsulphated coelenterazine as luciferin.

Oxyluciferin: This is the final product of the reaction, but may not necessarily be identical with the actual light emitter. As we shall see, this sometimes exists only transiently during the complete reaction. Shimomura has introduced other terms such as *photogen* (initial substrate) and *photogogikon* (oxidation product) in an attempt to rationalise the nomenclature (see Appendix I). However, I shall try to retain the simpler definitions, even though further information may be necessary to identify the precise structure interacting with the catalyst of the reaction, or to identify the structure of the excited state light emitter.

Luciferase (Tables 3.18B and 3.20A): This is the protein responsible for catalysing the oxidation of the luciferin. There has been some debate as to whether all proteins interacting with luciferins during the reaction are really enzymes, i.e. truly catalytic. Data are still sparse. If a protein turns out not to be a catalyst, it will cease to be called luciferase. Luciferases act as mono-oxygenases (mixed function oxidases) or peroxidases depending on the nature of the oxidant.

Photoprotein (Tables 3.18B and 3.20B) If the luciferin is so tightly bound to a protein that it remains attached throughout the isolation procedure, *and* does not exchange with free luciferin unless structurally perturbed, then the complete unit becomes a photoprotein. Total light output is then directly proportional to the *amount* of photoprotein present. In contrast, with a conventional luciferin–luciferase reaction *total* light output is not proportional to the amount of luciferase protein but only to the amount of luciferin. In principle, the protein moiety of photoprotein could act as a luciferase, as it appears to do when 'charging up' the coelenterate system. Alternatively, the photoprotein may simply be a protein–luciferin complex which has to interact with another protein, the luciferase, to generate light. This is what happens in the bivalve *Pholas*.

Table 3.20 — Bioluminescent catalytic proteins and spectral emission

Luminous organism group (phylum)	Example	Name of protein	Approx. mol. wt. (subunits)	$^\dagger\lambda_{max}$ (in vitro)
A. Luciferases				
Bacteria	Photobacterium	Luciferase	80 000 (2 subunits)	495–500
Dinoflagellates	Gonyaulax	Luciferase	135000 (35–40000)	475
(Protozoa)				
Anthozoans (Cnidaria)	Renilla	Luciferase	35 000	480
Coleoptera (Arthropoda)	Photinus	Luciferase	100 000 (2 subunits) 62 000 = one	540–580
Cructacea (Arthropoda)	Vargula	Luciferase	68 000 (monomer ca 12000)	465
Oligochaetes (Annilida)	Diplocardia	Luciferase	300 000 (3 non-identical subunits)	500–570
Gastropod (Mollusca)	Latia	Luciferase	170 000	500
Bivalve (Mollusca)	Pholas	Luciferase	310000 (2 subunits)	490
B. Photoproteins‡ Type I				
Radiolaria (Protozoa	Thalassicolea	Thalassicollin	20 000	450
Cnidaria other than Anthozoa, and ctenophores		Aequorin, obelin, phialidin, mnemiopsin	20 000–28 000	460–490
Polychaete worm	Chaetopterus	Chaetopterin	120 000 or184 000	460
Scale worm (Annelida)	Harmothoë	Polynoidin	500 000	510
Bivalve (Mollusca)	Pholas	Pholasin	34 000	490
Euphausiid shrimp	Meganyctiphanes	Meganyctiphorin	360 000 or 940 000	476
(Arthropoda)				
Millipede (Arthropoda)	Luminodesmus	Luminodesmin	104 000	496
Type II				
Brittle star (Echinodermata)	Ophiopsila	Ophiopsilin	45 000	500

†Where a range is given this indicates the range of maximal emissions of luciferases from different species. One value means does not mean no range.

‡The classification into type I or II is not certain in several cases, e.g. worms, millipede and euphausiids. See Table 3.18 for definition.

§Two mol. wts. indicate different forms of the protein.

Luciferin-binding protein: In some organisms, for example coelenterates and dinoflagellates, the photocytes contain proteins which bind the luciferin, or a modified form of it. These proteins are distinct from photoproteins since they readily release the luciferin, take no part in the chemiluminescent reaction, and do not interact directly with the luciferase. They act as a short or long term store of the luciferin.

In order to identity the chemical structures of the luciferin and its oxidation product, the luciferin is first extracted from the organism and purified. If it is possible to extract it in a free form this usually involves a series of organic solvent extractions (e.g. methanol + $CHCl_3$), followed by chromatography, including thin layer and, now, high performance liquid chromatographies. If the luciferin is protein bound, e.g. as a photoprotein, the protein–luciferin complex is isolated, the luciferin detached and purified. The chemical conditions required for this detachment can be somewhat drastic, leading to chemical modification. Furthermore, many isolated luciferins are susceptible to autoxidation, and are light sensitive. The ultimate test is to add the pure luciferin to the other purified components and thus generate visible light. The oxyluciferin, the product of this reaction, is then isolated. Definitive structural identification of the luciferin and oxyluciferin utilises six experimental procedures:

(1) UV and IR absorbance spectra, together with the influence of pH and organic solvents on these spectra, to identify key groups on the two molecules.
(2) Fluorescence excitation and emission spectra, releating these to *in vivo* fluorescence and bioluminescence spectra.
(3) ^1H and ^{13}C NMR spectra to provide further clues about molecular structure.
(4) Mass spectrum, enabling the molecular weight to be determined as well as providing further clues to the structure based on e/m ratios of fragments.
(5) Chemical degradation, using for example CrO_3, $NaBH_4$, or Br_2, followed by identification of products by tlc or hplc.
(6) Laboratory synthesis of the component thought to be the luciferin, followed by an *in vitro* demonstration of chemiluminescence with the other components to produce light of the same physical characteristics (i.e. kinetics, intensity, quantum yield, and colour) as the native luciferin.

Using these criteria, five 'luciferin' families have so far been identified. These are based on aldehydes coupled to a chromophore, imadazopyrazines (or strictly imidazolopyrazines), benzothiazole, linear tetrapyrroles, and flavins (Table 3.20; Fig. 3.15). Of these, only the benzothiazole of luminous beetles, the aldehyde luciferin of the earthworm such as *Diplocardia*, and the two imidazolopyrazines of coelenterates and ostracod crustaceans have been synthesised artificially. Only for benzothiazole and imidazolopyrazine luciferins have the oxyluciferin products been clearly identified.

Studies carried out some 20 years ago suggested that the luciferin in the acorn worm *Balanoglossus* might be an N substituted indole. Riboflavin or flavin analogues have been proposed as luciferins or energy transfer acceptors in some polychaetes, fish and fungi. These reports remain to be confirmed by using the full criteria already outlined.

Evidence for imidazolopyrazines as the luciferins in certain decapod and mysid shrimp, as well as some fish such as the myctophids (lantern fish) and squid, is based on cross-reactivity studies. In other words, addition of coelenterazine (Figs. 3.11 and 3.15), the imidazolopyrazine from coelenterates, to extracts of the luminous organs or photophores from these other animals generates light. In contrast, a teleost, the midshipman fish *Porychthys notatus*, on cross-reactivity criteria, uses the ostracod imidazolopyrazine (Fig. 3.11). Further support for this comes from specimens collected from Puget Sound on the west coast of the United States. Most *Porichthys* from Alaska to California have more than 700 tiny photophores arranged in rows over the body capable of

1 Bacteria (no role for Ca^{2+}) — blue light

$$NADH + FMN \xrightarrow{\text{reductase}} NAD^+ + FMNH_2$$

$$FMNH_2 + RCHO + O_2 \xrightarrow{\text{luciferase}} FMN + RCO_2H + H_2O + h\nu$$

2 Photinus (the firefly) — yellow light

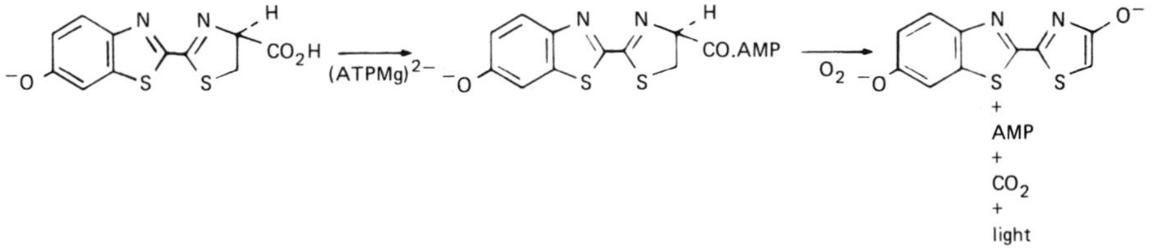

3 Aequorea (a hydrozoan jellyfish): blue light (animal: blue-green light)

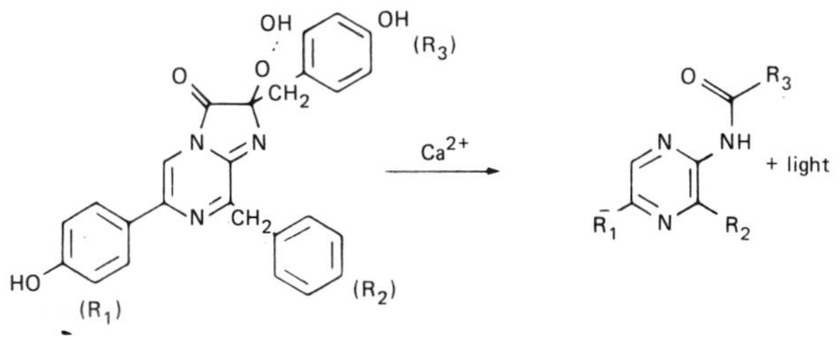

4 Latia (a freshwater limpet): pale green light

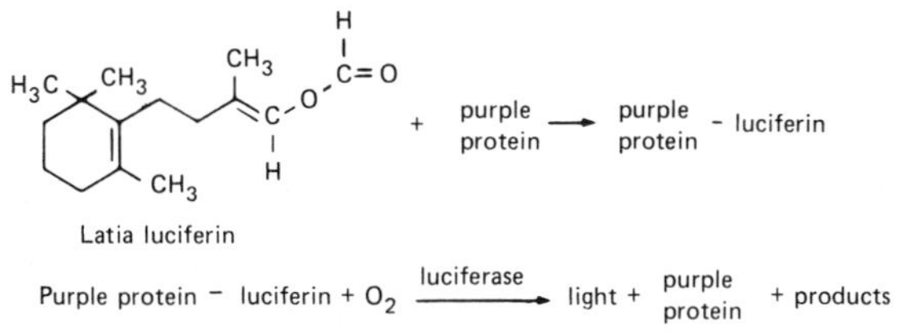

Latia luciferin

$$\text{Purple protein}^- \text{ luciferin} + O_2 \xrightarrow{\text{luciferase}} \text{light} + \text{purple protein} + \text{products}$$

5 Diplocardia (an earthworm) — blue-green light

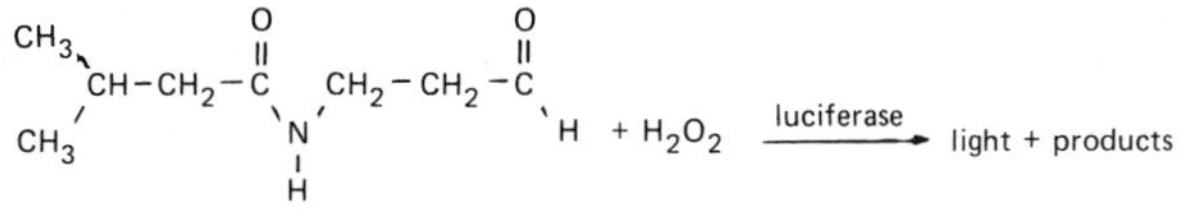

Diplocardia luciferin

(a)

Fig. 3.15 — Bioluminescent reactions (a) luciferin oxidation, (b) requirement for O_2, *in vitro* extracts.

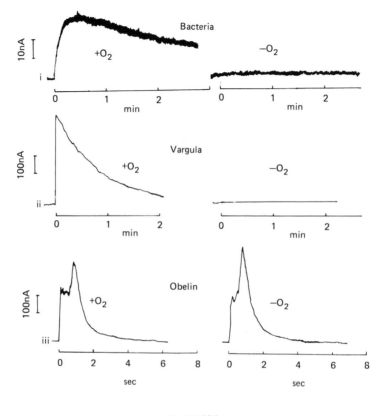

Fig. 3.15(b)

emitting blue light. These photophores fluoresce green under UV. However, specimens from Puget Sound, as well as those from a few other sites, neither chemiluminesce nor flouresce, unless injected with, or fed, ostracod luciferin.

There are still many groups of luminous organisms which remain a challenge for the chemist (Table 3.12), where little or nothing is known of the luciferin structure. These include fungi, bivalves like *Pholas*, starfish, and other echniderms, as well as many fish and squid. Flavin fluorescence has been observed in some organisms. Goto and colleagues have shown that the bioluminescent spectrum of the mushroom *Lampteromyces japonicus* is identical to the green fluorescence of riboflavin extracted from it. But it is not yet established whether this plays a primary role in the chemiluminescence, or has an energy transfer role. In luminous starfish Shimomura has recent evidence for a peroxyhaem catalase involved in the bioluminescence. It has been suggested that the luminous cells of the tunicate Pyrosoma (Fig. 3.4) contain intracellular bacteria. Bacterial luciferase has been identified in extracts, but it is very difficult to rule out bacterial contamination.

Table 3.21 — Known versus unknown*

Known	Unknown (luciferin not definitely identified)
Bacteria (1. aldehyde)	**Fungi (5. ?flavin)
Dinoflagellates (4. tetrapyrrole)	Many squid
Radiolarians (2. Imidazolopyrazine) Coelenterates (hydroids, jelly fish, sea pansies, comb jellies, 2. Imidazolopyrazine)	**Bivalve molluscs, Gastropod molluscs apart from *Latia* Sponge Sea spider **Flies
Some squid (e.g. firefly squid *Watasenia;* 2. imidazolopyrazine)	**Nemertine (ribbon) worms
Freshwater limpet (1. aldehyde)	**Polychaete worms (5. ? flavin)
Earthworms (1. aldehyde)	**Amphipod shrimp Starfish and other echinoderms
Euphausiids (4. tetrapyrrole)	Springtails
Decapod and mysid shrimp (2. imidazolopyrazine	Centipedes, **Hemichordates,
Copepods (2. imidazolopyrazine)	Sea mat
Ostracods (2. imidazolopyrazine)	Tunicates (? intracellular bacteria) **Many fish, e.g. (searsiid, stomiatoids; 4. tetrapyrrole)
Beetles, (fireflies and glow-worms; 3. benzothiazole)	
Millipede (4. tetrapyrrole)	
Some fish (e.g. *Porichthys*, myctophids; 2. imidazopyrazine)	

*Only where good evidence for true luciferin or it has been chemically identified is a group classified as 'known', though evidence is sometimes not very good (See Table 3.19). Numbers in brackets refer to structures in Fig. 3.11. (See Table 10.4).
** = something known about chemiluminescence, possible luciferin in brackets.
No asterisk = nothing definite known.

Chemical reactions

In principle, it should be possible to find a chemical intermediate, or product, of the chemiluminescent reaction which has an identical *fluorescent* spectrum to that of bioluminescence. This then provides definitive evidence for the structure of the molecule responsible for light emission. However, there are three particular reasons why this has proven difficult in several examples of bioluminescence. Firstly, the excited state intermediate (Fig. 3.16) can be chemically very unstable. This makes it impossible to isolate under normal conditions, preventing structural identification.

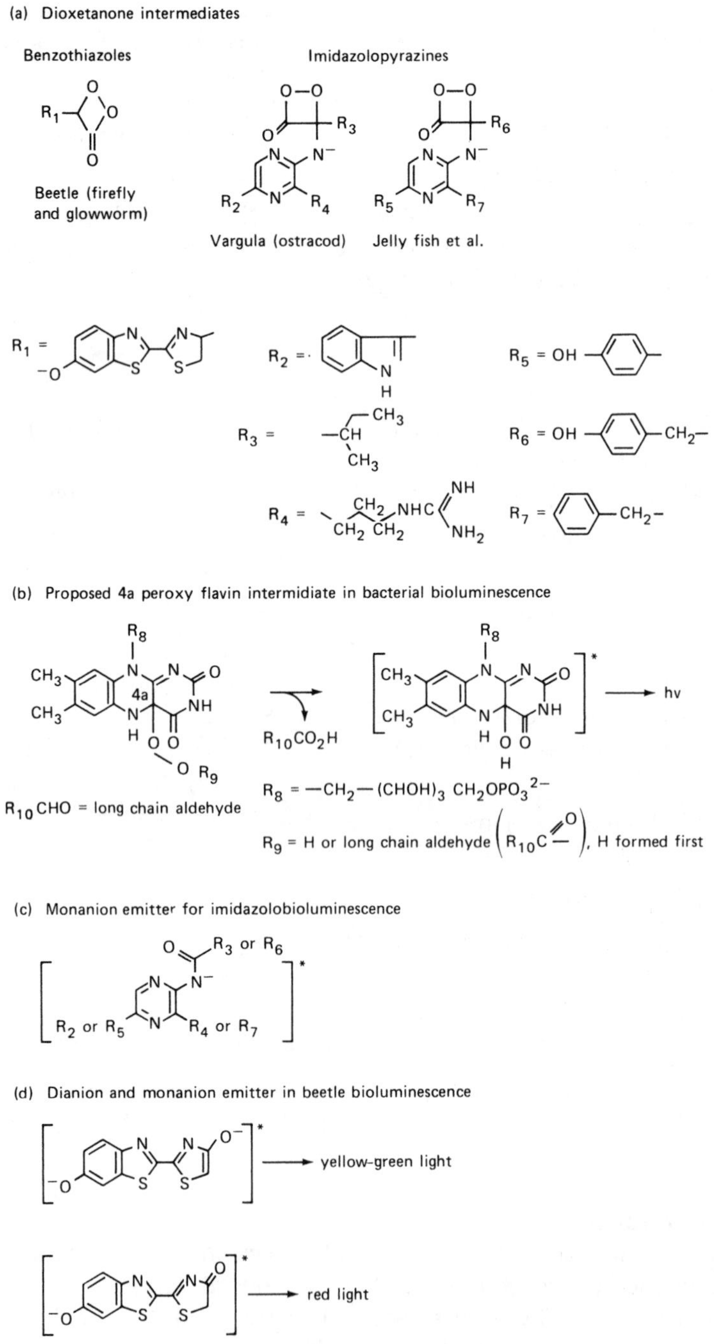

(a) Dioxetanone intermediates

Benzothiazoles

Imidazolopyrazines

Beetle (firefly
and glowworm)

Vargula (ostracod) Jelly fish et al.

(b) Proposed 4a peroxy flavin intermidiate in bacterial bioluminescence

$R_{10}CHO$ = long chain aldehyde

$R_8 = -CH_2-(CHOH)_3 CH_2OPO_3{}^{2-}$

R_9 = H or long chain aldehyde $\left(R_{10}C \diagdown {}^{O} \right)$, H formed first

(c) Monanion emitter for imidazolobioluminescence

(d) Dianion and monanion emitter in beetle bioluminescence

yellow-green light

red light

Fig. 3.16 — Some key intermediates and light emitting moieties.

This problem can sometimes be circumvented by rapid extraction at 0 to $-20°C$ or very low temperatures (liquid $N_2 = 77 °K$). Otherwise, the result is that the product isolated either has a fluorescent spectrum very different from the bioluminescent spectrum, or is hardly fluorescent at all. Secondly, the chemistry of excited state molecule is different from those in the ground state. For example, the pK_a of protonated groups is different in the two states, a factor known to influence the quantum yield and colour of some bioluminescent chromophores (Fig. 3.16c and d). Thirdly, the environment of the active centre of the protein to which the luciferin binds, and where it reacts, may be difficult to mimic when using the isolated product.

Nevertheless, the fluorescent product of imidazolopyrazine oxidation (Fig. 3.16c) has an emission spectrum quite close to that of the bioluminescence, provided that these are measured with the luciferin or oxyluciferin bound to the luciferase, or protein moiety of the photoprotein (i.e. the apoprotein). The excited state emitter is thought to be a mono-anion. The protein moiety binds O_2 and coelenterazine (Figs. 3.16c and 3.17). When Ca^{2+} binds, coelenterazine hydroperoxide rearranges to form the cyclic dioxetanone. Following oxidative decarboxylation, the anion product of coelenteramide is formed in the excited state. Protonation forms the highly 'blue fluorescent protein', in contrast to the native photoprotein which is barely fluorescent at all. Removal of the Ca^{2+} by addition of a chelator, e.g. EDTA or EGTA, causes the coelenteramide to dissociate from the protein, and the photoprotein can then be reformed by adding coelenterazine (Fig. 3.17) which becomes covalently linked to the protein. The conditions for this activation cycle are critical (see Chapters 5 and 6), but can provide a very sensitive assay for coelenterazine and a means of forming active photoprotein inside bacteria or eukaryotic cells from its mRNA or cDNA. Benzothiazole oxidation on the other hand provides a good example of how fluorescence spectra can be misleading. The singlet excited state of the oxyluciferin is more acidic than the ground state. Firefly luciferin is initially in the monoanion form in the luciferase. Reaction with the ATP leads to the formation of the AMP-luciferin dianion (Fig. 3.16d). Oxidation by O_2 forms the cyclic dioxetanone intermediate which then decarboxylates and looses AMP to form the biradical monoanion of the oxyluciferin. Intramolecular energy transfer results in the formation of excited state dianion which then emits a photon. Thus, light emission occurs naturally from the dianion (Fig. 3.16d), emitting yellow or green light. Protonation or binding by heavy metals result in red light. Perhaps this occurs in the red organs of the railroad worm, *Phrixothrix*? In the case of aldehyde systems of *Latia* and *Diplocardia* the oxidation products are non-fluorescent, and thus some sort of energy transfer process to a fluor or special protein environment is likely (see Chapter 9). In *Latia*, addition of a purple protein increases emission several-fold and recycles. A similar situation occurs with a compound, designated F, in euphausiid shrimp. Similarly, the products of tetrapyrrole luciferin oxidation are non-fluorescent, suggesting either a transient excited state intermediate, or the need for energy transfer, to produce visible light.

At one time the bacterial system presented a similar problem. Two of the substrates $FMNH_2$ and the long chain aldehyde are oxidised (Fig. 3.15). However, the excitation and emission maxima of FMN are approximately 460 and 540 nm respectively, nowhere near the peak for bacterial light emission, which is 470–500 nm. Also the acid product of the aliphatic aldehyde is non-fluorescent. Two distinct

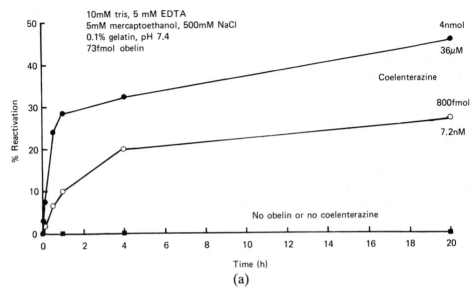

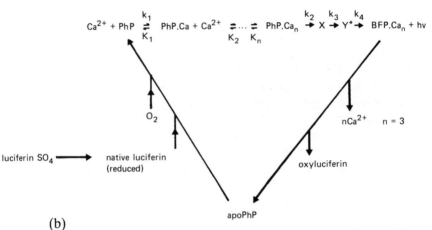

Fig. 3.17 — Reactivation of coelenterate photoproteins and other luciferins. (a) reaction scheme $n = 3$; (b) photoprotein formation; (c) sulphation of coelenterazine.

hypotheses have been proposed to resolve this problem. Hastings' group have isolated, at low temperature (-20 °C) an unstable luciferase-hydroperoxy flavin intermediate ($t_{\frac{1}{2}}$ 10 s at 23°C) (Fig. 3.15b), which reacts to form a 4a hydroxide on the luciferase with the spectral properties appropriate for the actual light emitter. $FMNH_2$ is first oxidised to 4a flavin hydroperoxide on the luciferase (Fig. 3.16), This then reacts with the long chain aldehyde to form the flavin peroxyhemiacetal. Cleavage of this molecule generates the acid plus flavin 4a hydroxide in an excited state. After light emission this looses H_2O to reform FMN. However, McCapra and co-workers believe that an intramolecular, or possibly intermolecular, energy transfer process at the active centre of the luciferase is responsible for generating the excited state emitter. This hypothesis is supported by the existence of energy transfer acceptors in some bacteria, with fluorescence excitation maxima distinct from the bioluminescence emission peak. Intramolecular energy transfer may also be involved in the benzothiazole and imidazolopyrazine systems (see Chapter 9). A similar situation may exist in euphausiid shrimp where the tetrapyrrole fluor (Fig. 3.11) is not consumed in the bioluminescent reaction. This discussion leads to the ultimate chemical question — how is the excited state intermediate or oxyluciferin actually generated?

Two mechanisms have been proposed, one involving electron transfer:

$$A^-. + B^+. \rightarrow A + B^* \rightarrow B + h\nu \tag{3.7}$$

The other involves the cyclic peroxide intermediate dioxetanone (Fig. 3.15a)

$$\text{luciferin} + O_2 \rightarrow \begin{array}{c} \text{dioxetanone} \\ \text{intermediate} \end{array} \rightarrow \begin{array}{c} \text{excited} \\ \text{carbonyl} \end{array} \rightarrow h\nu \tag{3.8}$$

A mechanism, combining the two processes may be involved in the firefly benzothiazole chemiluminescent reaction.

If these are the chemiluminescent reactions, what then are the roles of the various cofactors and cations required for the organism to produce light?

All the known cofactors in bioluminescence in some way help to process the luciferin for oxidation. There are three mechanisms by which this can occur:

(1) Reduction of one or more of the substrates by NAD(P)H.
(2) Covalent modification of the luciferin to enable it to bind to the luciferase, or to provide a good leaving group necessary to generate efficiently the excited state product (i.e. high ϕ_{ex} eq. (1.9)).
(3) Release of active luciferin from an inactive, covalently modified, storage form.

Reduced pyridine nucleotides are required in luminous bacteria to reduce acids to aldehydes and to form $FMNH_2$ (3.14), (3.15) and (3.16). The latter is very susceptible to autoxidation, a reaction which produces no light. Pyridine nucleotides also appear to produce the reduced 'luciferin' in fungi. The enzyme catalysts are oxido-reductases. $ATPMg^{2+}$ has one function in luminous beetles, and another in luminous squid *Watasenia* and the millipede *Luminodesmus*. ATP may also be involved in the luminescence of the Australasian glow-worm *Arachnocampa*, which does not seem to use firefly luciferin. In beetles, such as fireflies and glow-worms, $ATPMg^{2-}$ is necessary for proper interaction off the benzothiazole with the luciferase. Furthermore, AMP is a good leaving group in the generation of the excited oxyluciferin (Fig. 3.16d). In *Watasenia*, ATP is also involved in luciferin processing but indirectly, apparently through the following reactions:

$$\text{ATP} + \text{adenosine 5'phosphosulphate} \xrightarrow{\text{APS kinase}} \text{3'phosphoadenine 5'phosphosulphate} + \text{ADP} \quad (3.9)$$
$$\qquad\text{(APS)} \qquad\qquad\qquad\qquad\qquad\qquad\qquad \text{(PAPS)}$$

$$2\text{PAPS} + \text{luciferin} \xrightarrow[\text{sulphokinase}]{\text{luciferin}} \text{luciferin}(SO_4)_2 + \text{PAP} \qquad (3.10)$$

$$\text{luciferin } (SO_4)_2 + O_2 \xrightarrow{\text{luciferase}} \text{oxyluciferin } (SO_4)_2 + h\nu \ . \qquad (3.11)$$

The sulphates are on both of the phenolic hydroxyls of the imidazolopyrazine (Fig. 3.11). In contrast, in anthozoans the same imidazolopyrazine luciferin is used, but in this case the luciferin sulphate is inactive with the luciferase, because the sulphate is coupled with the carbonyl group of the imidazolo-ring. Here the reactions are:

$$\text{luciferin } SO_4 + \text{3'phosphoadenosine 5'phosphate (PAP)} \xrightarrow{\text{sulpho-kinase}} \text{PAPS} + \text{luciferin} \qquad (3.12)$$

$$\text{luciferin } + O_2 \rightarrow \text{oxyluciferin} + h\nu \qquad (3.13)$$

How then are cations involved (Table 3.18e)?
Three functions have so far been identified:

(1) Direct interaction with a cofactor.
(2) Oxidant formation.
(3) Triggering the protein catalyst.

$ATPMg^{2-}$ is the natural substrate for all enzymes requiring ATP, and not ATP^{4-}. Transition metals such as $Fe^{2+}Fe^{3+}$ and $Cu^+ Cu^{2+}$ are required at the active centre of some luciferases to provide the oxidant necesssary for efficient chemiluminescence. The best example of a cation trigger is Ca^{2+} in coelenterazine photoproteins. It triggers the photoproteins from hydrozoa, sycophozoa, ctenophores and radiolarians. These proteins have three Ca^{2+} binding sites, away from the luciferin-active centre. Ca^{2+}, however, does not activate anthozoan luciferases. Rather, it causes release of the luciferin from its binding protein, so that it can then interact with the luciferase.

What, then, are the precise functions and mechanisms of the proteins which interact with the luciferins, cofactors, and cations, necessary for the organism to produce light emission with the appropriate physical characteristics?

Role of proteins in bioluminescence
Four major functions have been identified:

(1) Oxidative catalysis: a luciferase or apophotoprotein is responsible for oxygen processing and for catalysing luciferin oxidation.
(2) Luciferin binding proteins: two types have been found; non-catalytic and passive (i.e. simply store of the luciferin), non-catalytic and active (i.e. involved in the light generation).

(3) Enzymes responsible for producing the cofactors, such as the NAD(P)H oxidoreductases and acid reductase producing FMNH$_2$ and long chain aldehyde respectively in luminous bacteria.

(4) Energy transfer acceptors.

Why are these proteins necessary? How do the luciferases, in particular, work?

In the absence of a protein catalyst, the luciferase, oxidation of the luciferin is usually slow, has a poor overall chemiluminescence quantum yield (i.e. one or more ϕ_c, ϕ_{ex}, and ϕ_F are low), and produces light of the wrong colour. Only a few bioluminescence quantum yields have been measured accurately but are high, in the range $\phi_{cL} = 0.1$–1 (pholasin 0.09; bacteria 0.1; aequorin 0.15; firefly luciferase 0.8–1). Furthermore, a catalyst is required *in vivo* so that it can be modified by intracellular regulators for control of light emission. Luciferins are, however, highly susceptible to autoxidation, and photoinactivation. Covalent modification and protein-binding increases their stability in the cell, enabling them to be stored for long periods and released by physiologically controlled mechanisms. Cofactors are to all intents and purposes kinetically inert without enzyme catalysts.

Oxidoreductases for producing FMNH$_2$ are found in both luminous and non-luminous bacteria. This is why transfection of the cDNA for a bacterial luciferase gene into *E. coli* will cause the latter to glow when a long chain aldehyde is added. In *Vibrio harveyi* there are three polypeptides (Mol. wts 32, 42, and 52 K) responsible for reduction of long chain acids (FA) to the respective aldehyde, resulting in the reaction sequence:

$$\text{FA precursor} \xrightarrow{32K} \text{FA} \xrightarrow[52K]{42K} \text{aldehyde} \tag{3.14}$$

e.g. tetradecanoylcoA   tetradecanoic acid   NADPH NADP   tetradecanal (TDA)

$$\text{NAD(P)H} + \text{FMN} + \text{H}^+ \xrightarrow[\text{reductase}]{\text{oxido}} \text{NADP}^+ + \text{FMNH}_2 \tag{3.15}$$

$$\text{FMNH}_2 + \text{TDA} \xrightarrow{\text{luciferase}} \text{FMN} + \text{TDacid} + h\nu \ , \tag{3.16}$$

whereas in *Photobacterium phosphoreum* a complex of three proteins replaces reaction 3.14:

$$\begin{array}{c}\text{FA} + \text{ATP} + \text{CoA} \\ \text{or} \\ \text{FACoA}\end{array} \longrightarrow \text{protein-FA} \longrightarrow \begin{array}{c}\text{aldehyde and pro-} \\ \text{tein}\end{array}$$

AMP + PPi and   NADPH   NADP
        CoA   (3.17)

Some organisms also require fluors (Table 3.19d and f). These fluors can interact directly with the chemiluminescent reaction to enhance its rate and quantum yield. Alternatively, a fluor or fluorescent protein acts as a passive acceptor of the energy from the excited state reaction product, via a non-radiative process. The fluor then becomes the actual light emitter, thereby changing the spectrum of light and quantum yield from that of the primary chemiluminescent reaction.

There has been much argument about whether the oxidation of luciferins, in the presence of a binding protein, are enzymatic or non-enzymatic. Certainly, the lack of a major effect of temperature over the range of 0–37°C on the decay constant of coelenterazine photoprotein chemiluminescence, when triggered by saturating Ca^{2+}, indicates that the final stage of the reaction may be spontaneous, rather than truly enzymatically catalysed. In any case, the division is a fine one. Formation of the photoprotein from luciferin, apoprotein, and O_2 (Fig. 3.17) behaves like a normal enzyme catalysed reaction. Thus it must be assumed, until proven otherwise, that all luciferases and apophotoproteins act as an enzyme during at least one step of the complete chemiluminescent reaction. For this dogma to be valid, not only must the protein be a true catalyst but also the molecular mechanism of catalysts must involve a direct interaction between the protein and the luciferin moiety as opposed to a 'solvent' effect of the active centre in the protein. This remains to be proven in many cases.

Enzymes using molecular oxygen can be divided into two classes: oxidases and oxygenases. The former catalyse the reaction

$$XH_2 + O_2 \rightarrow X + H_2O_2 \text{ (or } O_2^- \text{)} , \tag{3.18}$$

whereas oxygenases incorporate oxygen atoms into the other substrate. These latter enzymes are themselves divided into two groups. Di-oxygenases incorporate both atoms from O_2 into the substrate molecule, whereas mono-oxygenases (or mixed function oxidases as they are sometimes known) incorporate only one O; the other ends up as H_2O. Since there are no usable radio-isotopes of oxygen, the oxygen chemistry of bioluminescence has been examined by using ^{18}O assayed by epr or mass spectrometry.

All luciferases and apophotoproteins which use molecular oxygen, and where the chemiluminescent reaction has been defined, are mono-oxygenases, the excited state emitter within the oxidised luciferin being a carbonyl group. Luciferases catalysing the oxidation of benzothiazoles, imidazolopyrazines and aldehydes (Figs. 3.11 and 3.14) incorporate one oxygen atom into those molecules and hence are designated as mono-oxygenases. *Vargula* (formerly *Cypridina*) and earthworm luciferase contain a copper atom, whereas the others have no known transition metal. This means that oxygen must interact directly with the polypeptide chain. This has been confirmed by Shimomura, for apoaequorin using ^{13}C nmr. Furthermore in the presence of O_2, Ca^{2+} and coelenterazine coelenterate apophotoproteins act as an oxygenase, generating chemiluminescence.

The molecular weights of the mono-oxygenase luciferases and apophotoproteins range from about 20 K to 900 K (Table 3.20). Some are single polypeptides, others are oligomers, e.g. the bacterial and firefly luciferases are heterodimers. *Vargula* luciferase is a hexamer with 'identical' subunits (total M. wt. *ca* 70 K). These various luciferases and coelenterate apophotoproteins have a free SH group, which if covalently modified, can reduce or destroy enzymatic activity, but these may be required for enzyme stability rather than catalysis. However, in the firefly luciferase at least, the inactivation of the luciferase through SH modification seems to be the result of steric hindrance preventing luciferin binding, rather than inhibiting a catalytically active group *per se*. Three proteins have been cloned, bacterial and firefly luciferase, and aequorin together with part of a dinoflagellate luciferase, using either a bacterial plasmid such as pBR 322, or phage vectors such as λgt11. Crude RNA is first isolated from the organism, or light organ, and mRNA purified by using an oligo dT cellulose column. cDNA is prepared by a phage reverse transcriptase. This is then incorporated into a plasmid or phage vector using a recipe of restriction enzymes, DNAase and DNA ligase. The cDNA library is then transfected into a bacterium, e.g. *E. coli*. After conventional selection of clones containing vectors, using substrate and antibiotic selection or resistance, clones producing active luciferase or apoproteins are selected by one of three methods:

(1) An oligonucleotide probe if some of the amino acid sequence is known.
(2) An antibody, using a Western blot.
(3) Expression of chemiluminescence in the presence of luciferin.

The latter is particularly suitable if an expression vector is used as the cloning vehicle. These can express as much as 30% of the bacterial protein as the one you want. *E. coli* containing the bacterial lux genes light up when long chain aldehyde is added, since all the cofactors and other enzymes are endogenous to *E. coli*. The operon of *Vibrio fischeri* responsible for its chemiluminescence has six so-called lux genes arranged as I, C, D, A, B, E, regulated by a gene R. I and R are the sites for control of the transcription of the genes responsible for producing the α and β subunits of the luciferase (lux A and B) and the activities of the long-chain acid reductase (lux C, D and E) producing the long chain aldehyde (eqns. 3.14–3.16). I protein produces the autoinducer, R protein responds to it.

As a result of this cloning we know the amino acid sequence of the luciferases from the bacterium *Vibrio harveyi*, the firefly *Photinus pyralis* and the apophotoprotein from Aequorea (Fig. 3.18) and some of the dinoflagellate *Gonyaulax* luciferase. From these, one can begin to identify hydrophobic centres as luciferin and O_2 binding sites.

Within various eukaryotic cells are found a family of Ca^{2+} binding proteins responsible for regulating the behaviour of these cells. These proteins include troponin C and parvalbumin in striated muscle, and the ubiquitous Ca^{2+} regulatory protein calmodulin, and have close sequence homology with aequorin. The Ca^{2+} sites are high affinity (Kd Ca^{2+} 1–10 μM), and are highly selective over Mg^{2+}, present in cells at mM concentrations. Kretsinger and others in the 1970s, using X-ray crystallography, showed that the Ca^{2+} binding domains in this protein family consist of about 33 amino acid residues; 12 of, these are in a loop, where residues 1, 3, 5, 7, and 12 coordinate the Ca^{2+}, and position 6 contains a highly conserved gly required to allow the residue at position 5 and the $- CO_2^-$ at position 7 to coordinate Ca^{2+}. The residues consist of two α helices separated by a β pleated sheet, the Ca^{2+} site. This was originally designated as helix $E - Ca^{2+}$ loop-helix F, and is called the EF hand region. For a binding site to be selective for Ca^{2+} over Mg^{2+} it should be 7 or 8 coordinate, rather than 6, all of which should be oxygens, i.e. from asp, glu, thr, ser, or tyr. Nitrogen ligands reduce selectivity against Mg^{2+}. The amino acid sequence of aequorin (Fig. 3.18), obtained from conventional peptide sequence analysis and the cloned cDNA, shows that the protein can be divided into four sections. Three domains, A = residues 13–43, C = residues 106–136, and D = residues 142–172, fulfil the criteria for a Ca^{2+} site within an EF hand region, and have close sequence homology with cardiac calmodulin. Both proteins have 3 Ca^{2+}

sites. All three domains in aequorin have an invariant gly and glu at positions 6 and 12, a hydrophobic residue at position 8, and 5–6 oxygen ligands (asp, 2 asn, ser, glu, and lys, ala, or gln) in the EF hand region. Note that this gives octahedral coordination since two carbonyls donate both oxygen atoms. Domain B = residues 44–105, is the likely luciferin + O_2 binding site. One *Aequorea* medusa can yield as many as 8 isoaequorins. Thus microheterogeneity can confuse kinetic and sequence analysis. Cloning produces a pure polypeptide. The ease of obtaining amino acid sequences using cDNA, and the possibility of using site directed mutagenesis to

(a)(i)

Fig. 3.18—Amino acid sequence of two bioluminescent proteins. (a) bacterial α and β chains of luciferase, from Cohn *et al.* (1985) (with permission of the American Society of Biological Chemists) and Johnston *et al.* (1986) (with permission of the American Society of Biological Chemists). (i) lux A=α subunit; (ii) and (iii) lux B=β subunit; α helical region as indicated broad arrow=β sheet, parrallel lines=random coil, white=on surface, black=within protein, as predicted. (b) aequorin, from Tsuji *et al.* (1986) (The American Chemical Society) with permission, from cDNA; (c) aequorin, from Charbonneau, *et al.* (1985) (The National Academy of Science) with permission from conventional sequencing, compared with calmodulin (CaM). *=Ca^{2+} EF hand sites, homologies in boxes.

```
GTG ATG CCA TAT CTC AAA GAA AAA CAG TAA TTAATATTTT CTAAAAGGAA AGAGAC ATG AAA TTT GGA TTA TTC TTC CTC AAT TTT ATG AAT TCA AAG CGT TCT TCT GAT CAA GTC ATC GAA
Val Met Pro Tyr Leu Lys Glu Lys Gln  .                                MET Lys Phe Gly Leu Phe Phe Leu Asn Phe Met Asn Ser Lys Arg Ser Ser Asp Gln Val Ile Glu
                                                                       1                    10           ( )   ( )             20 (Leu)

GAA ATG TTA GAT ACC GCA CAT TAC GTA GAT CAG TTG AAG TTT GAC ACG TTG GCT GTT TAC GGA AAC CAT TTC TCG AAC AAT GGT GTG GTT GGT GCC CCA CTA ACA GTG GCT GGT TTT
Glu Met Leu Asp Thr Ala His Tyr Val Asp Gln Leu Lys Phe Asp Thr Leu Ala Val Tyr Gly Asn His Phe Ser Asn Asn Gly Val Val Gly Ala Pro Leu Thr Val Ala Gly Phe
       (Asn)  30                          39 (Tyr)        (Val)                            50                                60

TTA CTT GGT ATG ACA AAG AAC GCC AAA GTG GCT TCG TTG AAT CAC GTC ATT ACC ACG CAT CCA GTA CGT GTG GAA GAA GGT TGT CTA CTT GAC CAA ATG AGT GAA GGC CGT
Leu Leu Gly Met Thr Lys Asn Ala Lys Val Ala Ser Leu Asn His Val Ile Thr Thr His His Pro Val Arg Val Ala Glu Gln Ala Cys Leu Leu Asp Gln Met Ser Glu Gly Arg
          200*   70                             80                      90 (Ala)        (Ser)         300* 100

TTT GCC TTT GGC TTT AGT GAT TGT GAA AAG AGT GCA GAT ATG CGC TTC TTT AAT CGA CCA GAT TCT CAG TTT CAG TTG TTC AGT GAG TGT CAC AAG ATC AAT GAT GCA TTC
Phe Ala Phe Gly Phe Ser Asp Cys Glu Lys Ser Ala Asp Met Arg Phe Phe Asn Arg Pro Asp Ser Gln Phe Gln Leu Phe Ser Glu Cys His Lys Ile Asn Asp Ala Phe
         ( )    110                               (His)120         Asp Ser Gln (Phe) ( )130( Arg)   400*

ACT ACT GGG TAC TGC CAT CAC AAC AAT GAT TTT TAT AGT TTT CCT AAA ATC TCC GTT AAC CAC GCG TTC ACT GAA GGC GGT CCT GCG CAA TTT GTG AAT ACG AGC AAA GAA
Thr Thr Gly Tyr Cys His Pro Asn Asn Asp Phe Tyr Ser Phe Pro Lys Ile Ser Val Asn Pro His Ala Phe Thr Glu Gly Gly Pro Ala Gln Phe Val Asn Ala Thr Ser Lys Glu
140                 150   (Tyr Ser)              Ile Ser Val Asn Pro His Ala   160(Lys)                 170

GTG GTT GAA TGG GCG GCT AAG TTA GGG CTT CCA GTG TTT AGA TGG GAC GAC TCA AAC GCT CAA AAA GAA TAC GCC GGT TTG TAC CAC GAA GTT GCT GCA GCA CAT GGT GTC
Val Val Glu Trp Ala Ala Lys Leu Gly Leu Pro Val Phe Arg Trp Asp Asp Ser Asn Ala Gln Arg Lys Glu Tyr Ala Gly Leu Tyr His Glu Val Ala Val Ala His Gly Val
180   (Glu)            190                    600*              200         (Thr)              210

GAT GTT AGT CAG GTT CGA CAC AAG CTG ACG CTG CTG AAC CAA AAT GTA GAT GGT GAA GCA AAG GCA GAT GCT CGC GTG TAT TTG GAA GAG TTT GTC GAA TCT TAC TCA
Asp Val Ser Gln Val Arg His Lys Leu Thr Leu Leu Asn Gln Asn Val Asp Gly Glu Ala Ala Arg Ala Glu Ala Arg Tyr Leu Glu Glu Phe Val Glu Ser Tyr Ser
220                      230            700*                  240                        250

AAT ACC GAC TTT GAG CAA AAA ATG GGA GAG CTG TTG TCA GAA AAT GCC ATC GGT ACT TAT GAA GAA GGC AGT ACT CAG GCA GCA GTT GCG ATT GAG TGT TGT GGT GCC GAC CTA
Asn Thr Asp Phe Glu Gln Lys Met Gly Glu Leu Leu Ser Glu Asn Ala Ile Gly Thr Tyr Glu Glu Gly Ser Thr Gln Ala Ala Arg Val Ala Ile Glu Cys Cys Gly Ala Asp Leu
        260           800*              270                    280               290

TTG ATG TCT TTT GAG TCG ATG GAA GAT AAA GCG CAG CAA AGA GCG GTT ATC GAT GTG GAA AAC GCC AAC ATC GTC AAT TAC CAC TCG TAA CGTTTAACTG ATGCTGAAGG GGCAGCG ATG
Leu Met Ser Phe Glu Ser Met Glu Asp Lys Ala Gln Gln Arg Ala Val Ile Asp Val Val Asn Ala Asn Ile Val Tyr His Ser  .                                    MET
   900*              Ser.Met Glu Asp Lys Ala Gln Gln Arg     310             320

CCC CTT ATA TCA CCA TTC TTT TCG CCG ATA GCG CTA ACT AAT AGA GGC ATT TAT ATG GCG GTA CTT TCA GGA GTT AAG GAG AAC ATC GCA GCG AGC ACA GAA ATC GAT GAC TTG
Pro Leu Ile Ser Pro Phe Phe Ser Pro Ile Ala Leu Thr Asn Arg Gly Ile Tyr MET Ala Val Leu Ser Ala Lys Gln Glu Asn Ile Ala Ala Ser Thr Glu Ile Asp Asp Leu
                              1100*

ATT TTC ATG GGA ACT CCT CAG CAA TGG TCA TTG CAG CAA CAA AAA CAG CTG ACA TCT CGC CTT GTT AAA GGG GCA TAT CAA TAC CAT TAC CAC AAT AAT GAT TAT CGT CAG TTC
Ile Phe Met Gly Thr Pro Gln Gln Trp Ser Leu Gln Gln Gln Lys Gln Leu Thr Ser Arg Leu Val Lys Gly Ala Tyr Gln Tyr His Tyr His Asn Asn Asp Tyr Arg Gln Phe
                                   1200*

TGC GAG AGG CTG GGA GTC GGA GAG GTG GTG GAA GAT CTC AAC GAT ATC CCG GTT TTC CCT ACT TCT ATT TTT AAG TTG AAG ACC CTA TTA ACA CTT GAC GAT
Cys Glu Arg Leu Gly Val Gly Glu Val Val Glu Asp Leu Asn Asp Ile Pro Val Phe Pro Thr Ser Ile Phe Lys Leu Lys Thr Leu Leu Thr Leu Asp Asp
                 1300*
```

(a)(ii)

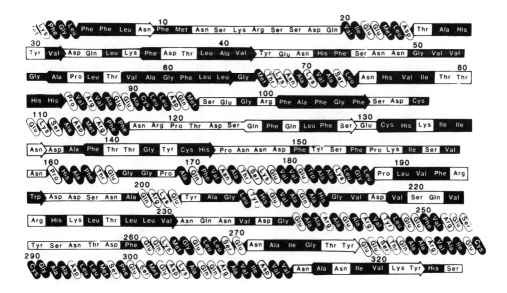

(a)(iii)

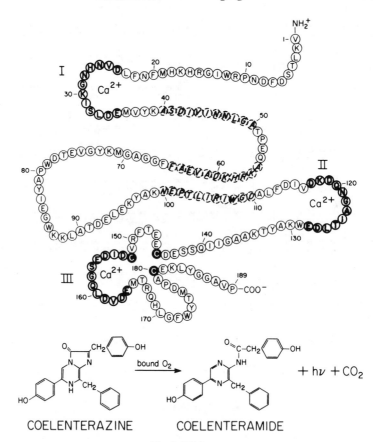

Fig. 3.18(b)

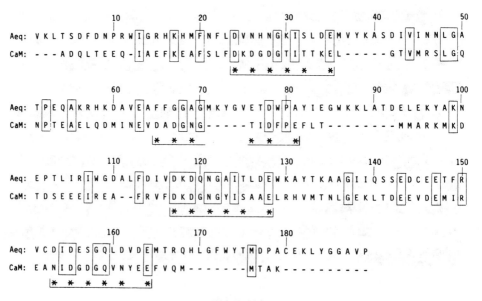

Fig. 3.18(c)

modify binding properties and catalytic activity, offers great opportunities for finding out precisely how the luciferases and photoproteins work, and from which protein families, or exons, they evolved. Such an approach has already provided key evidence for the role of particular amino acids in the Ca^{2+} sites, O_2 binding and chemiluminescence catalysis in aequorin. Substitution of the important gly by arg at position 6 (i.e. residues 29, 122 and 158) in each EF hand showed that sites A and C where necessary for Ca^{2+}-activated chemiluminescence but gly at position 158 apparently not. This study by Tsuji, Goto and colleagues assessed Ca^{2+}-activatable aequorin following reactivation of the photoprotein from apoprotein formed from an expression vector in *E. coli*, coelenterazine and O_2 (Fig. 3.16). Since coelenterazine produces a bright blue chemiluminescence in dimethyl formamide likely O_2 reaction sites were his 58 and tryp 108 both in hydrophobic regions, ala 55–phe 95 and asp NH_2 100–asp 110 respectively (Fig. 3.18b). Mutation of his 58 to phe resulted in no Ca^{2+}-activatable aequorin being formed. Finally it is known that reactivation of photoproteins is best in the presence of an $-SH$ protectant, eg. mercaptheanol or dithiothreitol, though quite reasonable photoprotein formation can be achieved in their absence. Unlike calmodulin which has only one $-SH$ (cys), aequorin has three. Replacement of cys at positions 145, 152 and 180 by ser only resulted in a large loss of aequorin formation when there was no cys 180, but complete loss was not observed. It was concluded that cys is not the sole O_2 binding site, and also an intact C terminus is required for optimum activity.

Conventionally, enzymes are characterised by substrate K_m's, turnover numbers, temperature coefficients, and pH optima. As one might expect, most mono-oxygenase luciferases have optima in the range pH 7–8. Those secreted into sea water would be expected to have pH optima near 8 *Km*s for luciferins are mainly in the 0.1–10 μM range, though *Renilla* luciferase has a Km as low as 30 nM for its luciferin, and coelenterazine apophotoproteins an apparent Km < 0.1 μM. The Km of O_2 has been poorly defined, apart from some luminous bacteria. It appears to be in the range 0.1–10 μM. The turnover number of mono-oxygenase luciferases is not particularly fast. In fact, some, e.g. bacterial luciferase, are quite slow. These have a turnover time of some 10–20 s (turnover number $= 0.1$ s^{-1}). Coelenterazine luciferases and photoproteins are a little faster with turnover numbers some 10–50 times this. These compare with a typical enzyme like lactate dehydrogenase having a turnover number of 1000 s^{-1}. However, the mono-oxygenases are good for one thing. They provide an environment for a high chemiluminescence quantum yield (ϕ_{CL}). ϕ_{CL} ranges from some 10–100% (Table 3.20).

There is one other class of luciferases. These are, or behave like, peroxidases, and contain transition metals such as $Cu^{+/2+}$ and $Fe^{2+/3+}$ at their active centres. Earthworms, the acorn worm, and brittle stars, use H_2O_2 as the oxidant, and the luciferase acts as a peroxidase. The luciferase forms a metal — OOH intermediate which increases the chemiluminescence quantum yield from $< 1\%$ in its absence to $> 50\%$ in its presence. These peroxy intermediates also form with O_2 + peroxidases. Hence peroxidases can also act as oxidases. This activity is usually weak, but in the case of the luciferase of the bivalve mollusc *Pholas*, containing two copper atoms, it appears to be the mechanism responsible for chemiluminescence. This is why *Pholas* luciferin, which is protein bound, also reacts with O_2^- or H_2O_2 + horseradish peroxidase. It is, however, also possible that O_2^- is an intermediate of the natural reaction. O_2^-, on the luciferase, may be an intermediate in other cases such as benzothiazole and some imidazolopyrazine bioluminescence. This may explain why H_2O_2 is good stimulus when added to many luminous organs or photophores. Other moieties in luciferases include carbohydrates, e.g. *Pholas* luciferin and luciferase are glycoproteins; phospholipid is required in membrane bound systems such as scaleworms and may be required for firefly luciferase. The function of these in other groups is not yet known. The luciferase in brittle stars may be a protein similar to catalase, which contains haem and normally catalyses the degradation of H_2O_2 and H_2O.

Luciferase can thus be classified into three groups:

(1) Monooxygenase (e.g. *Vargula* — an ostracod), O_2^- may or may not be an intermediate on the luciferase

$$LH_2 + O_2 \rightarrow L - O + H_2O + h\nu \tag{3.19}$$

(2) Peroxidase (e.g. Diplocardia — an earthworm)

$$LH_2 + H_2O_2 \rightarrow L - O + 2H_2O + h\nu \tag{3.20}$$

(3) Oxidase/peroxidase (e.g. *Pholas* — a bivalve)

$$LH_2 + O_2 \xrightarrow{\text{oxidase}} LH + HO_2(H^+ + O_2^-) \tag{3.21}$$

$$\downarrow O_2^-/H_2O_2$$

$$LH + HO_2 \xrightarrow{\text{peroxidase}} L - O + H_2O + h\nu \tag{3.22}$$

The other important group of proteins required for bioluminescence are luciferin-binding proteins. These are required to maintain a large store, easily accessible, of luciferin, to maintain it in a soluble form particularly when it is susceptible to autooxidation, and to prevent access to the base when both luciferin and luciferase are in the same intracellular compartment. The latter provides a convenient point for control of light emission. Thus the scintillons of dinoflagellates contain luciferin bound to a protein and luciferase. Acidification releases the luciferin and activates the luciferase. Similarly *Renilla* the luciferin is stable on the cytoplasmic binding protein until released by Ca^{2+} when it is accessible to the luciferase.

3.4.4 Sources of components

Several of the components required for light emission are synthesized within the photocyte itself. Genes are transcribed to form mRNA, which is translated to produce the proteins. Transcription may require an inducer. In the case of luminous bacteria these produce their own (Fig. 3.5). Cofactors such as ATP, NAD(P)H, FMN, aliphatic aldehydes, and PAP(S) are part of the intermediary metabolism of most cells. Cations will always be available, as will O_2 from the blood, haemolymph or surrounding fluid. H_2O_2 can be generated by oxidases, and this is what occurs in the case of luminous acorn worm *Balanoglossus*, and earthworms. The real question is, where does the luciferin come from? Surprisingly, little hard evidence is available to answer this. There are three possibilities:

(1) Synthesis from common metabolites within the photocyte, or the secretory luminous cell, itself.

(2) Synthesis in another organ, such as the liver or kidney, followed by transport to the photophores, or light organ, via the blood or haemolymph.

(3) The diet, with or without some chemical processing.

Not surprisingly, acid-methanol extraction of luminous coelenterates, teuthoidean squid such as *Watasenia* and some other squid, decapod and mysid shrimp, copepods, and myctophid fish yields the imidazolopyrazine coelenterazine detected using an appropriate luciferase or coelenterate apoprotein (see Chapter 5). However, closer examination of the various organs, particularly the non-luminous ones, often reveals luciferin in these as well as in the luminous photophores. For example, in several teuthoidean squid high enough concentrations of luciferin are found in the kidney, digestive gland, and blood for a weak chemiluminescence to be detected in homogenates of these tissues. These studies suggest that the luciferin might be synthesised in one tissue and then transported to the photophores, or ink sac for secretion, in some squid. In the firefly squid, *Watasenia*, coelenterazine, the preluciferin (Eqns. 3.9–3.11) is synthesised in the liver, and is then sulphated to the active luciferin in the photocytes. Shimomura has also detected coelenterazine in several non-luminous organisms, though there is no clear pattern recognisable as yet. Thus, coelenterazine is found in the herring *Clupea harengus*, the shrimps *Pundalus*, *Piborealis*, *P. Damae* and *P. platyuros*, *Heterocarpus* and *Sicyonia ingentis*, and the sponge *Microcioma prolifera*, all non-luminous, but was not found in the flounder *Pseudopheurorecteus americanus*, the shrimps *Alpheus heterochaelis*, *Paneus duorarum*, *Palaemonetes pugio*, *Crangon*, the lobster *Homarus*, the barnacle *Balanus nubilus*, the squid *Loligo pealei*, the abalone *Haliotis kamtschatkana*, the clam *Mya arenaria*, the bryozoan *Schizoporella unicornis*, the sea fan *Lophogorgia hebes*, and the sponge *Terpios zeteki*. No luciferase was detected in any non-luminous species. The presence of large quantities of coelenterazine in the guts of several specimens suggests that at least some of the non-luminous organisms may accumulate coelenterazine from the diet. However, this does not appear to be the whole answer. Coelenterazine may have another, as yet identified, function.

There is evidence that at least two luminous organisms, the midshipman fish *Porichthys* and the mysid shrimp *Gnathophausia*, get their luciferin via the diet. As we have already seen non-luminous *Porichthys* from Puget Sound, on the west coast of the USA, can be made luminous by feeding them the luminous ostracod *Vargula* or by injecting its luciferin. However, luminous *Porichthys* retain their luminescence for months, even when fed no luminous prey. In contrast, the mysid *Gnathophausia* (Fig. 3.3b) becomes non-luminous within two months unless fed on luminous organisms, bits of photophores from species containing coelentrazine such as jelly fish, copepods, decapod shrimp, and myctiphid fish. Feeding *Porichthys* on euphausiids does not make these mysid shrimp luminous. But whether the photocytes get their luciferin from another tissue within the same animal, or through the diet, the central question remains. What is the biosynthetic pathway producing the luciferin? There is little discrete evidence for this in any luminous organism, but we can make some guesses (Fig. 3.19a).

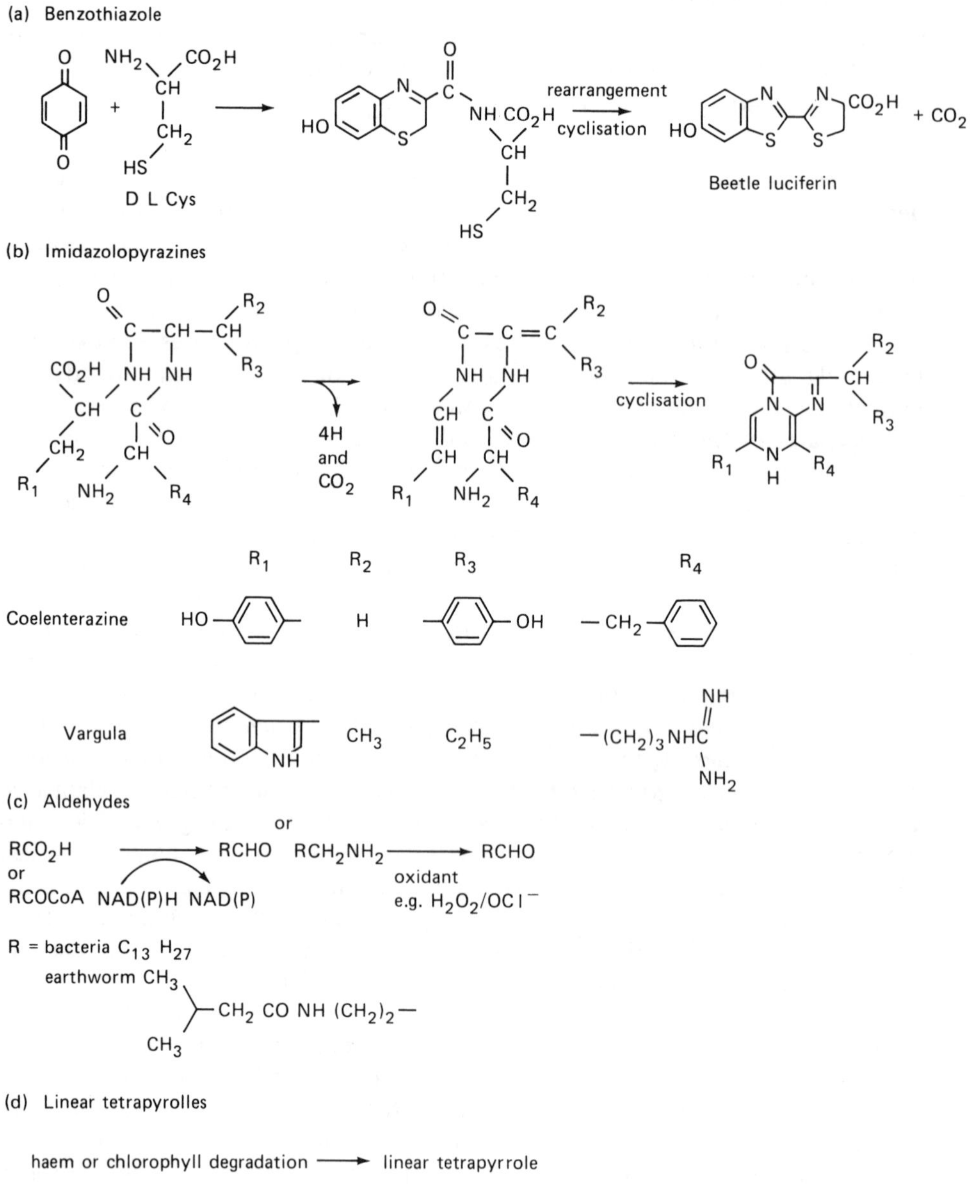

(a) Benzothiazole

D L Cys

rearrangement
cyclisation

Beetle luciferin

$+ CO_2$

(b) Imidazolopyrazines

4H
and
CO_2

cyclisation

	R_1	R_2	R_3	R_4
Coelenterazine	HO—⟨O⟩—	H	—⟨O⟩—OH	—CH_2—⟨O⟩
Vargula	(indole)	CH_3	C_2H_5	—$(CH_2)_3$ NHC (=NH, NH_2)

(c) Aldehydes

RCO_2H or $RCOCoA$ → RCHO   (or) RCH_2NH_2 → RCHO

NAD(P)H   NAD(P)

oxidant
e.g. H_2O_2/OCl⁻

R = bacteria $C_{13} H_{27}$

earthworm CH_3 (>) $—CH_2 CO NH (CH_2)_2—$ CH_3

(d) Linear tetrapyrolles

haem or chlorophyll degradation → linear tetrapyrrole

Fig. 3.19 — Possible biosynthetic route for benzothiazole (beetle) and imidazolopyrazine (coelenterates, radiolarians and ostracods) luciferins, proposed by McCapra.

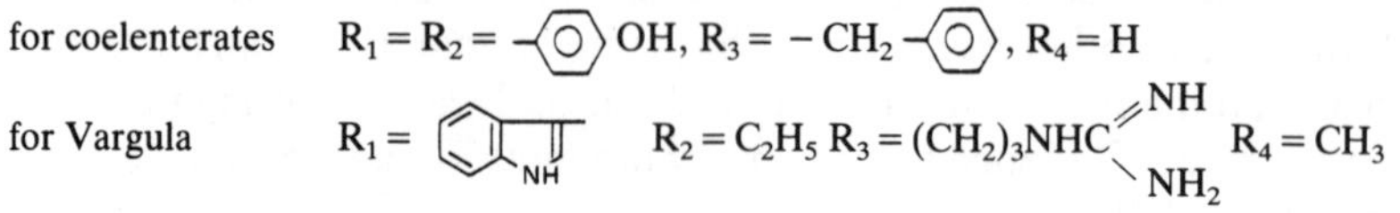

for coelenterates $R_1 = R_2 =$ —⟨O⟩OH, $R_3 = -CH_2$—⟨O⟩, $R_4 = H$

for Vargula $R_1 =$ (indole) $R_2 = C_2H_5$ $R_3 = (CH_2)_3$NHC(=NH, NH_2) $R_4 = CH_3$

The precursor for the aldehyde required for bacterial, earthworm and freshwater limpet chemiluminescence will be the corresponding acid (Fig. 3.19a). Fatty acid, and FA CoA, reductases have been isolated and characterised in luminous bacteria (section 3.4.2, (3.14)–(3.16)). The aldehyde for earthworm bioluminescence is likely to be synthesised via reaction of isovaleric acid with aminopropanoic acid, followed by reduction to form N-isovaleryl-3-aminopropanal, the luciferin. The freshwater limpet *Latia* has a luciferin containing a β-ionone ring, structurally similar to vitamin A required for vision in both vertebrates and invertebrates.

The most likely, though as yet not established mechanism for forming the imidazolopyrazine ring structure is via cyclisation of a modified tripeptide (Fig. 3.19b), from phe.tyr.dehydrotyr in the case of radiolarians, coelenterates, decapod shrimp, squid, copepods, and fish utilising coelenterazine, or arg.ileu dehydrotyr in the case of ostracods and the midshipman fish *Porichthys*. Interestingly I have found that the ostracod genus *Concoechia* may use coelenterazine and not *Vargula* luciferin. As illustrated in Fig. 3.19b, a $- CO_2H$ must be lost during the cyclisation, i.e. the reaction involves a decarboxylation before or during cyclisation. The question arises, why only two types of imidazolopyrazine, or are there others?

McCapra has provided evidence from the click beetle *Pyrophorus lucens* that the benzothiazole luciferin of beetles may be made from benzoquinone plus DL cysteine (Fig. 3.19c). Since the luciferin is optically D-, it is important to realise that the unusual amino acid D-cys is required at one stage, at least.

The structural similarity between the tetrapyrrole luciferins of euphausiids and dinoflagellates with breakdown products of chlorophyll suggests that this may be the precursor. Dinoflagellates are surface organisms containing chloroplasts. However, euphausiids are deep sea creatures, living where there is no plant life. Similarly, the stomiatoid fish, with phycobiliprotein-like pigments (Fig. 3.11), would have no direct access to blue–green or red algae as food except through plant debris sinking downwards. An alternative precursor would be a bile pigment formed from the degradation of haem.

The problem of luciferin biosynthesis provides a rich field for exploration by the biochemist.

3.4.5   *In vivo* versus *in vitro*
The ultimate objective of characterising the biochemistry of the chemiluminescent reaction extracted from a luminous organism is to understand how it works, and how it is controlled, *in vivo*. There are many examples where the physical characteristics of a reaction between purified components *in vitro* do not match those found *in vivo*. At least five reasons can be cited to explain such discrepancies:

(1) The concentrations and proportion of reagents *in vitro* are very different from those *in vivo*, resulting in a less intense, slower reaction. The photocyte has a tiny volume, perhaps only a few pl. The intracellular luciferin and luciferase concentrations may approach mM. With the preciousness of the isolated material such concentrations are rarely achievable in the test tube.

(2) It is difficult to mimic precisely *in vitro* the regulatory mechanism within the cell. For example, a hydroid flash may be the result of a pulse of calcium entering a photocyte, over a period of just a few milliseconds, through transiently open Ca^{2+} channels.

(3) The physical conditions *in vitro* may dramatically alter the kinetics and colour of the chemiluminescence *in vitro*. Most laboratory experiments are carried out at 20°C, one atmospheric pressure, and in air-saturated buffer at pH 7–7.4 containing O_2 at a concentration of some 220 μM. This O_2 concentration is probably at least ten times higher than the cells are exposed to *in vivo*. The pH of sea water is about eight; the intracellular pH of photocyte organelles is unknown. The temperature of the deep oceans is about $+4°C$, and at >100 atmospheres pressure. Luminous invertebrates are cold blooded and used to living in the dark.

(4) There may be a component missing *in vitro*, e.g. a fluor to shift the colour and produce the same quantum yield of that *in vivo*.

(5) The *in vitro* reaction is incorrect because of structural modifications during isolation of the components.

It is easy to generate dim chemiluminescence from the oxidation of organic molecules. It is much more difficult to realise *in vitro* the brightness and precise colours of living light!

3.5 SWITCHING THE LANTERN ON AND OFF

Current understanding of the control of living light is, unfortunately, a muddle. One has only to look at Harvey's books on bioluminescence, or more recent similar publications, to find a bewildering array of electrophysiological, pharmacological, and ionic experiments on luminous animals, or isolated organs and cells, which are essentially uninterpretable. There are three main problems in using these studies in an attempt to provide an overview of the mechanisms controlling bioluminescence *in situ*. Firstly, the cell biology of many systems is poorly defined. In particular, we rarely know precisely where in the cell, or even sometimes in which cell, the various components of the chemiluminescent reactions are stored. Secondly, a conceptual framework for understanding cell activation in molecular terms has, until recently, been lacking. Thirdly, few experiments have laid down the necessary criteria for designing definitive experiments required to formulate or test a hypothesis. In spite of these problems, a number of interesting observations have been made which indicate what needs to be done in the future, if we are to unravel this intriguing aspect of living light.

3.5.1 Levels of control

In principle, any one of the five physical characteristics of bioluminescence, namely direction, timing, intensity, and polarisation, could be under the control of the organism, or its environment. Yet, unless the light is switched on in the first place, mechanisms for the fine control of these parameters become redundant.

Most luciferins alone, in the presence of oxygen, oxidise spontaneously. This may produce a weak chemiluminescence, detectable by a sensitive photomultiplier tube, but invisible to the eye. However, in response to a stimulus from outside, or as a result of internal programming, the organism allows all the components of the chemiluminescent reaction to mix, thereby causing a transition threshold from 'invisible' to 'visible' light. Conventionally, the regulators of such biological pheno-mena are divided into acute or long term effectors, depending on whether the cells respond over a timescale of milliseconds to seconds, or over many minutes, hours, or even days. This temporal distinction may appear somewhat arbitrary. Yet examination of the mechanistic basis of acute and long term regulation provides a molecular rationalisation separating them. Acute regulation of cells is mediated through rapid changes in membrane potential and intracellular chemical messengers such as Ca^{2+}, H^+, cyclic nucleotides, diacyl glycerol and inositol metabolites; whereas long term regulation involves both major structural changes in the cell, as well as requiring synthesis and/or degradation of particular components.

These two levels of control, acute and long term, are imposed on the response pathway in the luminous organism. The organism, be it multi- or uni-cellular, first generates an internal signal in response to an external stimulus. This external stimulus may be touch, something the organism has seen, smelt, or heard, or it may be part of an internal programme, for example, in order to mate. The internal signal is transmitted, via an electrical conducting pathway, nerves or the circulation, to the light organ, the photocytes or the cells containing the chemiluminescent compounds. As a result the components mix and produce light.

Thus, to understand how the light is switched on, we need to define first what switches the organism on, how this is transmitted to the light generating region, and how the individual cells there respond to consummate the components of the chemiluminescent reaction.

3.5.2   The stimuli

The stimuli of bioluminescence can be considered in two distinct classes (Table 3.22): those which turn on the organism, and those which actually switch on the cells responsible for light. The former are intimately related to function (see section 3.6); the latter are related to the molecular basis of cell activation in the particular group of organisms in question. Organisms may light up spontaneously when put in the dark. They may also become luminous as the result of internal programming, for example, when hungry or during the mating season. However, more usually, luminescence occurs as the result of an external stimulus, the three commonest natural stimuli being touch (or a vibration), sound, and sight, evoking, for example, an alarm or predatory response. Many luminous organisms, including dinoflagellates, radiolarians, hydroids, jelly fish, sea combs, sea pansies, starfish, tunicates, shrimp, squid, ostracods, copepods, Pholas, and fish, light up when touched, or when mechanically disturbed. Others, such as polynoid scale worms, are more difficult to provoke. Many luminous animals are sensitive to light. A brief illumination can switch on many luminous insects, crustaceans, and tunicates. In others, for example hydroids, sea pansies, sea combs, and dinoflagellates, exposure to light inhibits the ability of the organism to become luminous.

On the beach at night, a KCl gun is a useful weapon since K^+ depolarises cells and activates the neuronal pathways, or the photocytes themselves, in most luminous organisms. A similar response can be evoked by using electric shocks or suction electrodes. In hydroids and sea pens a wave of luminescence can be observed passing through the colonies following the action potential. Other artificial stimuli include manipulation of external Na^+, distilled water, and hydrogen peroxide. However, these may simply lead to mixing of the components through cell damage, as can

Table 3.22 — The natural stimuli of living light

Organism group (example)	Stimulus
A. *Whole organism*	
Bacteria (*Photobacterium*)	Continuous once induced (i.e. genes switched on)
Glow-worms and fireflies (*Lampyris*)	Internal programming
Dinoflagellates (*Noctiluca*)	Diurnal rhythm
Coelenterates (*Obelia*)	Touch
Ostracods (*Vargula* formerly *Cypridina*)	Fear
Dipterans (*NZ glow-worm*)	Hunger
B. *Individual cells or photocytes*[†]	
Coelenterates (*Obelia*)	Electrical
Polynoid worm (*Acholoë*)	Electrical
Ophiuroid starfish (*Ophiura*)	Acetyl choline
Amphipods (*Scinia*)	Acetyl choline
Euphausiid shrimp (*Meganyctiphanes*)	5-Hydroxy-tryptamine (serotonin)
Fireflies (*Photinus*)	Octopamine (receptor 2)
Midshipman fish (*Porichthys*)	Noradrenaline and adrenaline (α)
Hatchet fish (*Argyropelecus*)	Adrenaline(β)
Some other fish (*Chauliodus* and *Stomias*)	Adrenaline(α)

[†]Evidence for many of these is not complete.

severe mechanical handling, being unrelated to any natural mechanism. In only a few organisms is the actual stimulus of the photocytes known (Table 3.22B).

Free-floating bacteria in the sea are usually non-luminous, because they require a critical concentration of an inducer (Fig. 3.5) to switch on the lux genes ultimately responsible for producing the luciferase and other proteins required for luminescence. The bacteria produce their own auto-inducer, and thus do not luminesce until they themselves reach a sufficient density to enable the inducer to reach the threshold concentration. This seems sensible in terms of energy conservation, since no organism, however tiny, would be expected to respond to just a few bacteria, each only a micrometre or so long. Once induced, the bacteria glow continuously. Starvation via cyclic AMP, and heat shock may increase sensitivity to the auto-inducer. However, fish and squid with bacterial light organs are often able to turn the lantern on and off. There are three ways in which they can do this, firstly by flaps of skin in front of the organ, secondly by rotating the organ, and thirdly by altering the degree of pigmentation of filters through which the light has to pass. All other eukaryotic cells involved in bioluminescence have to be 'switched on'. This may involve independent or synchronous triggering of many individual photocytes in one photophore or light organ, or it may involve simultaneous or sequential activation of different photophores or light organs. For example, the light organ of an adult male firefly can contain as many as 15 000 cells, the larva some 2000. The majority of these will light up when the organism flashes. Where individual components of the chemiluminescent reaction are stored in different cells, these may be needed to be switched on simultaneously or sequentially for the animal to produce a luminous secretion. For example, in the ostracod *Vargula* the luciferin is probably stored in one glandular cell, the luciferase in another. It has been suggested that the components of these cells, and those from some other organisms such as squid, may be 'squeezed' out mechanically by muscles surrounding them. Whilst these, and other, muscles may be necessary for ejection of secretions in ostracods, squid bivalves, and decapods, it seems more likely that release of the chemiluminescent components from cells is a typical secretory process from intracellular granules fusing with the cell membrane. The mechanism would then be similar, in principle, to glucose stimulating insulin secretion in mammals, or to action potentials conducted along nerve axons provoking transmitter release at a nerve terminal.

The primary stimulus of the photocyte itself can be mechanical, electrical or chemical. In unicellular organisms, such as dinoflagellates, a mechanical stimulus evokes an electrical change in the cell membrane, which then induces an intracellular chemical signal to trigger luminescence. However, in multicellular organisms the response of the animal itself has to be transmitted to the photocytes or luminous secretory cells. In most organisms this is done via neuronal pathways, or a nerve net as in the case of coelenterates and tunicates. Coelenterates have an epithelial conducting pathway connected cell to cell via gap junctions, and then to the photocyte itself. There is some evidence that there may be similar electrical synapses between neurones and the luminous scales (elytra) of polynoid worms. However, evidence is lacking for such connections between a nerve and a photocyte in any animal. Instead, the nerve terminal releases a neurotransmitter which activates the photocyte. Alternatively, the luminous cell could be activated via a hormone from the circulation.

Four criteria must be satisfied if a chemical substance is to be established as the primary stimulus of a photocyte or luminous secretory cell:

(1) The substance must provoke luminescence when injected into the intact animal, and when added to the isolated organ or a suspension of isolated cells.
(2) The substance should be extractable from the gland, or whole animal if a hormone, and localised histochemically to a nerve terminal — photocyte synapse if a neurotransmitter.
(3) Release of the substance in the whole organism should precede the onset of bioluminescence.
(4) Pharmacological antagonists and agonists should establish the nature of the receptor involved.

Of the majority of luminous organisms which have been well studied, only the first and last of these criteria have been applied (Table 3.22B). Five compounds have been most commonly investigated:

(1) Acetyl choline, enhanced by acetyl choline esterase inhibitors and inhibited by eserine or nicotinic acid, depending on whether it is a muscarinic or nicotinic receptor.

(2) Adrenergic agonists (adrenaline and non-adrenaline), blocked or enhanced by phentolamine or proparanolol depending on whether an α or β receptor is involved. It can then be classified pharmacologically as α_1, α_1, β_1, or β_2.
(3) 5-Hydroxytryptamine (serotonin).
(4) Dopamine.
(5) Octopamine.

One of the best established is octopamine as the stimulus of the firefly light organ, via octopamine-2 receptors. These are potentiated by substituted phenyliminoazolidines. The danger is that such receptors can also respond experimentally to adrenergic stimuli, albeit less well. Octopamine has not been found in vertebrates. There remain many luminous systems uncharacterised. Myctophid and many other fish do not respond to any of the above stimuli; nor have the stimuli for echinoderm, worm, crustacean, and molluscan luminous cells been identified. Neuropeptides and purinergic neurotransmitters remain unexplored. In other non-bioluminescent examples of cell activation, secondary regulators act to modify, either to enhance or reduce, the efficacy of the primary stimulus. Previous exposure to light inhibits several luminous organisms. However, other possible secondary regulators remain unknown.

3.5.3 Consummation of components

Two questions need to be answered before we can understand fully how a luminous flash or glow is initiated. Firstly, what is the act of consummation, and secondly, what are the molecular mechanisms responsible for activating the cell(s) prior to consummation? Once these are answered, a further question arises, how is the light turned off?

Because a minimum of three (luciferin, luciferase, oxidant) and a maximum of six components are required to produce visible light (oxidant, luciferin, luciferase or apophotoprotein, cofactor, cations, and a fluor), the release of any one of these to a mixture of the others could, in principle, initiate the act of consummation. It has been proposed that the availability of O_2 might be the key factor in initiating firefly luminescence. Access to O_2 has also been suggested to provoke *Pyrosoma* luminescence, because of gill movements associated with the onset of light production, and to initiate earthworm bioluminescence via H_2O_2 production. Certainly artificial removal of O_2 using argon will reduce or stop light emission. However, emission from resting organisms in a normal oxygen atmosphere is not only visible to the naked eye but very low even when measured with a sensitive chemiluminometer. O_2 diffuses readily through tissues, and O_2 in air saturated water is some 220 μM. Yet the Km's for O_2 are 1–10 μM or less. Therefore if O_2 is to be the initiator of luminescence, there has to be an extremely efficient means of keeping it in the nM range within the resting photocyte. In photocytes requiring mitochondrial respiration for ATP synthesis, this would be lethal. The photocytes would thus have to depend on glycolysis for their ATP. Although O_2 may be an unlikely trigger for bioluminescence using molecular oxygen, release of H_2O_2 in the guts a luminous earthworms like *Diplocardia*, or the hemichordate acorn worm *Balanoglossus*, appears to be a crucial regulatory factor initiating their luminescence. Luciferin and luciferase can be found in the coelomic fluid of earthworms prior to luminescence.

Release of an oxidase enables H_2O_2 to be generated, thus provoking visible light production. In organisms such as squid, the bivalve mollusc *Pholas*, ostracods, and decapod shrimp which secrete a luminous cloud, mixing of luciferin and luciferase externally is sufficient, since O_2 is always present in sea water.

There are four ways in which the luciferin can gain access to the luciferase to initiate luminescence:

(1) Secretion into the external fluid, from separate cells or from different granules in the same cell.
(2) Intracellular fusion of granules containing either luciferin of luciferase.
(3) Release of luciferin from a binding protein.
(4) Covalent modification of the stored form of the luciferin, which is inactive, to produce luciferin which can now interact with luciferase.

The first of these occurs only in secretory luminous organs of photophores. The second is a well recognised phenomenon in the turnover of membrane receptors and in phagocytosis, but has yet to be demonstrated in a photocyte. Two different mechanisms for luciferin released from binding proteins have been found. The evidence for this is based on image intensification, fluorescence, and electron microscopy to identify the source of light, luciferin and luciferase in the cell using gold labelled antibodies to *Gonyaulax* luciferase.

Nicolas and coworkers have identified cytoplasmic dense bodies, and surprisingly the trichocyst, as intracellular sites of luciferase. The dense bodies are the scintillons, the micro-light sources, which fluoresce as they also contain luciferin. Hastings and coworkers have previously shown that isolated scintillons flash when the surrounding medium is dropped from pH 8 to 5.7. In the cell these are close to the vacuole membrane which has an internal pH 4.5. An action potential generated at the plasma membrane induces an action potential in the vacuole membrane. H^+ moves rapidly into the scintillon releasing luciferin from its binding protein thereby allowing access to the luciferase and a flash of some 10^4 photons per granule occurs. A MgATPase may then restore the intragranule pH. In dinoflagellates a pulse of H^+ into vesicles may also activate the luciferase which is in the same granule. In the sea pansy *Renilla* a rise in cytoplasmic Ca^{2+} stimulates release of luciferin from its binding protein, enabling the luciferin to be oxidised by the luciferase. *Renilla* and other coelenterates also store the imidazolopyrazine luciferin as an inactive sulphate. It is not clear whether desulphation (3.12) and (3.13) is a spontaneous or controlled reaction. New luciferin would only be required in hydroids, jelly fish, and ctenophores for photoprotein regeneration (Fig. 3.16), or in anthozoans when the luciferin on the binding protein was exhausted.

In squid where imidazolopyrazine sulphate or bisulphate is the active luciferin, an obvious potential control mechanism would be access of the imidazolopyrazine to the key cofactors PAPS or ATP (3.9)–(3.11). Similarly, access to ATP is a mechanism requiring investigation in insect bioluminescence, since *in vitro* addition of μM ATP to firefly or glow-worm luciferin plus luciferase causes a fold increase in light output of as much as 10^4–10^6.

The final option is to release a cation required by the bioluminescent reaction.

The photocytes of *Obelia* and other luminous coelenterates, contain voltage sensitive Ca^{2+} channels which open when the cell is depolarised. These have been identified by using patch clamp electrodes. The isolated cells can be stimulated by replacing Na^+ in the sea water with K^+, or Li^+, but only when Ca^{2+} is also present. Ca^{2+} binds to the photoprotein and triggers it. But the kinetics of each flash are over a millisecond timescale, very much faster than when Ca^{2+} is added to a photoprotein solution in a test tube. There are two possible reasons for this. Firstly, the Ca^{2+} channels only open for a few milliseconds, and the pulse of Ca^{2+} is stoichiometrically bound by the photoprotein molecules, present at very high (μM—mM) concentrations in the cell cytosol. Secondly, there could be an equally rapid Ca^{2+} buffering or pumping system removing the Ca^{2+}. The former mechanism is the more likely.

What then is the molecular basis of photocyte activation enabling consummation of components to occur?

Bioluminescence belongs to a class of phenomena designated as 'threshold'. Activation of photophore or light organ causes a dramatic change of state, from being invisible to being visible. The question arises whether such thresholds also occur at the level of individual cells. In other words, in a population of individual cells responding to a dose of stimulus capable of provoking a half maximal response, have all of the cells been switched on to 50% of their maximum, or have only 50% of the cells been switched on? Certainly, in organisms such as dinoflagellates there are mechanical and electrical thresholds at which an individual cell switches on its lantern. This is a problem fundamental to the relationship between the cell biology and molecular biology of cell activation (see Chapter 10, section 10.3.2). Whilst we cannot, as yet, solve it for luminous cells, it is clear that in virtually all examples of living light, apart from bacteria, a primary stimulus acting at the cell membrane, be it mechanical, electrical, or chemical, switches on or activates a chemical process within the cell which causes it either to emit light from the interior of the cell, or to secrete a component(s) from intracellular granules. There has, therefore, to be an intracellular messenger transmitting the initial signal from the cell membrane into the interior of the cell, where the chemiluminescent components are to be found. Three classes of intracellular signals are known from other phenomena:

(1) Cations: particularly H^+ and Ca^{2+}, but possibly also Na^+, K^+ and transition metals.
(2) Nucleotides: cyclic nucleotides (cyclic AMP and cyclic GMP), AMP, GTP.
(3) Membrane phospholipid derivatives, including inositol tris- and tetra- phosphates, inositol sugars, and diacyl glycerol.

Other, at present speculative, possibilities include intracellular peptides, regulatory metabolites such as eicosanoids, and reactive oxygen metabolites. To understand how such second messenger systems work, we need first to identify definitively the time sequence of the intracellular signals, then discover the mechanism of their prduction or release into the cytoplasm, and find out how they act.

Ca^{2+} is the oldest intracellular messenger known. There is convincing evidence that it is the intracellular trigger for coelenterate bioluminescence, and that the primary cell stimulus increases the permeability of the cell membrane to Ca^{2+} by opening voltage sensitive Ca^{2+} channels. Indirect evidence based on inhibition of bioluminescence by a Ca^{2+} chelator EGTA, or stimulation by a Ca^{2+} ionophore

A23187, has suggested that intracellular Ca^{2+} might be the trigger for dinoflagellates, radiolarian, scale-worm, brittle star, firefly, and toad fish (midshipman) luminescence. However, the definite proof that Ca^{2+} is the trigger requires direct measurement of intracellular free Ca^{2+} and correlation of a rise with the onset of bioluminescence. This has not been done. Nothing is known of how the Ca^{2+} might act, whether by calmodulin or other Ca^{2+} binding proteins, or activation of kinases. Nor do we know whether the Ca^{2+} comes from outside the cell or is released from an internal store by messengers such as inositol 1,4,5 trisphosphate (Ip_3). Even less is known about the possible role of cyclic nucleotides and their kinases, or the possible role of proteins in coupling receptor occupancy to membrane enzyme activation. Some evidence exists for a role of cyclic AMP in firefly bioluminescence, but proof requires satisfaction of three criteria:

(1) Demonstration of an elevation in cyclic nucleotide, prior to onset of bi-oluminescence.
(2) Enhancement of the cell response by addition of an inhibitor of cyclic nucleotide phosphodiesterase, e.g. theophylline or isobutyl methyl xanthine.
(3) Stimulation of the cells by addition, extracellularly, of the cyclic nucleotide.

There is much still to be learned about the molecular basis of photocyte activation. These cells should provide a valuable model for cell biologists since the response of individual cells is easily detectable.

Some luminous fish and squid can 'switch off' their lantern by using a shutter, or by rotating their light organ. But there are three ways in which the light can be switched off, or become visible, in other species via an effect on component consummation.

(1) Reduction of the luciferin, through consumption or dispersion of components.
(2) Removal or inactivation of one component of the chemiluminescent reaction.
(3) Removal of the intracellular signals responsible for initiating and maintaining cell activation.

The first of these occurs in secretory systems such as ostracods, copepods and decapod shrimp, the latter in coelenterates, and probably dinoflagellates. In other luminous organisms, the mechanism for switching off the light is unknown.

3.5.4   Long term regulation
Long term regulation, over hours, days, or even moths, is required to enable the organism to develop from egg to adult, to enable it to adapt to environmental conditions such as light and weather, and to enable it to be ready to switch on its light at the appropriate time of day and season.

Not only do marine luminous organisms migrate upwards at night but also several of them, particularly surface species, have a circadian rhythm. Dinoflagellates can be provoked to flash more easily if they are collected at night, rather than during the day. This is reflected in an experiment where sea water from the ocean surface is passed continuously for several days in front of a photomultiplier tube. Mechanical forces in the detection chamber provoke bioluminescence. Just before dusk a large

increase in flashing occurs. This remains until just before dawn when the bioluminescence recording returns to quiescence. Diurnal rhythms have been observed in
dinoflagellates, fungi, and beetles. There are four possible mechanisms:

(1) Synthesis and regeneration of components, e.g. luciferin or luciferase, following
    consumption or secretion.
(2) Separation of components.
(3) Inhibition of cell activation.
(4) Inactivation of luciferin by light.

In one dinoflagellate, *Pyrocystis*, there appears to be a circadian movement of
vesicles to a part of the cell where they cannot be activated during the day. In the
hydroid *Obelia* light inhibits the generation of action potentials which stimulate the
photocytes, whereas in ctenophores, with apparently the same chemiluminescent
reaction, the luciferin within the photoprotein is photo-inactivated. Light also
inhibits luminescence responses in some crustaceans, hemichordates (acorn worms),
insects, and tunicates. Daylight can either switch the luminescence off, or make it
more difficult to switch on when the animal is returned to the dark. Syllid worm
sexual luminescence can show lunar or annual periodicity. The molecular basis of
these long term regulations is unknown.

3.6 WHAT IS LIVING LIGHT FOR?
3.6.1 The search for evidence
Many people would regard the science of animal behaviour as having its origins in the
letters and writings of the eighteenth century Hampshire clerygman Gilbert White.
This collection, *The natural history and antiquities of Selborne*, first published in
1789, began to put a systematic scientific approach to observations made by
naturalists in the field. Many of the letters were written to the leading zoologist of the
day, Thomas Pennant. In one poem 'A naturalist's evening walk', he wrote
perceptively of '. . . the glow-worm with her amorous fire'. Glow-worms and fireflies
can be, and have been, studied easily at night. The elegant studies of Buck, Lloyd,
and others in this century have not only amply confirmed White's belief that their
glows and flashes are there to attract the opposite sex, but also careful measurements
of spectra and flashing frequencies have been related to species and sexual
specificity.

The problem with most other examples of living light is that ideas about function
have had to rely on the ingenuity of the observer coupled with casual observation,
anecdotal evidence, common sense, and inspired guesswork. One cannot watch
organisms several thousand metres below the surface in the same way that one can
observe terrestrial luminous species. Cameras with image intensifiers, and now
human submersibles, are of some help. But one still has to rely heavily on
observations made in the laboratory, on specimens often damaged or in a quite alien
environment from their natural one. Thus, some four approaches have been used to
try to formulate hypotheses regarding the function of the light emission in a
particular species. Firstly from a knowledge of the external stimuli which switch the
light on and off, and whether the luminescence is within the photophore or secreted,
together with the siting within the animal of the photophores and light organs, it is
possible to propose a function. For example a predator may initiate a fear response if
the luminescence is for defence. Secondly, measurements of the physical properties

of the light emitted, particularly its intensity and its spectrum, followed by correlation with the characteristics of the ambient light in the organism's natural habitat are matched with the photon and spectral sensitivity of the eye of the proposed responder. The deep sea fish *Malacosteusa* has red emitting organs and is the only organism likely to be able to see the light. It thus has a torch invisible to other organisms. Thirdly, laboratory experiments can correlate luminescence with a behavioural response, either within the organism itself, or from neighbours of its own species or another species, together with comparison in behaviour between closely related luminous and non-luminous species, both in the laboratory and the wild. The role of firefly flashing in mating has been elucidated in this way. Finally experimental manipulations in the wild (and in the laboratory) using model light sources to mimic the emitter, may provoke a potential responder, for example a mate, a predator, or prey.

A combination of these approaches has provided firm evidence for luminescence having a mating function in fireflies, aiding escape reactions in squid, ostracods, shrimp, and copepods, inhibiting copepods feeding on dinoflagellates, supporting schooling and feeding in flashlight fish, and providing a camouflage for hatchet and lantern fish (Table 3.23). However, in other luminous organisms the scientific evidence is weak, in spite of the ingenuity and attraction of the ideas. Nevertheless more than 16 different functions have been ascribed to bioluminescence in living organisms (Table 3.23). These fit into the three necessities of life: obtaining food, defence, and reproduction. The light is produced to attract, to detract, to conceal, or to illuminate. Thus, the light either has to be seen by the responder, who may be close by or may be many metres away, or it has to merge with background light. So before examining in more detail the precise functions of living light, we should first relate its physical characteristics to the ambient light, at various times during the day or night.

3.6.2 Living versus ambient light

On a clear day sunlight at sea level can be as bright as 5×10^{-2} watts cm^{-2}. At 500 nm this is equivalent to about 10^{17} photons per second per sqr cm (1.6). The spectrum ranges from the near ultraviolet (290 nm) well into the infrared (3000 nm) half the energy being in the infra red, the peak 480 nm. The UV and IR are invisible to the eyes and photoreceptors of animals and plants. These only respond to a narrow spectral range, 350–750 nm. In a dense forest the daylight intensity may be 10–100 times less than in the open air. But on a moonlit night, the light is some 100 000 times dimmer than that of daylight, and it is redder. Without the moon the ambient light intensity is 1000 times less again, about 3×10^{-10} watts cm^{-2}. Comparison of these ambient intensities with those of luminous organisms (Table 3.10) reveals that only at night will the luminescence of terrestrial animals be easily visible against the background light. Thus, fireflies and glow-worms are nocturnal and mate at night. In contrast, their non-luminous beetle relatives are active during the day. However, the situation is different in the sea.

Water absorbs and scatters light. Furthermore, it absorbs and scatters UV and red light more efficiently than blue light. All the UV from the sun is absorbed within a few centimetres of the ocean surface. Virtually all the infrared, and much of the visible red, is lost within the first metre or so. Thus, any light penetrating beyond a few metres below the surface will be blue–green. Even this light is absorbed and

Table 3.23 — The function of living light

Proposed function	Example
A. *Obtaining food*	
Illumination to see prey	Flashlight fish (*Photoblepharon*) †*Malacosteus*, pinecone fish (*Monocentris*)
Luring prey — lantern	Angler fish, New Zealand glow-worm
— mimicry	†Midshipman fish (*Porichthys*), firefly (*Photuris*)
Schooling	Flashlight fish, euphausiid shrimp
B. *Defence and security*	
Camouflage (counter illumination)	Hatchet fish, lantern fish, euphausiids, some squid
Schooling (communication within species)	†Lantern fish, flashlight fish, euphausiids
Decoy	Ostracods
Frighten or 'smoke screen'	†Squid, †*Pholas* (bivalve), †hydroids, dinoflagellates
Burglar alarm	Dinoflagellates
Warning	†*Porichthys* (fish), †fireflies, amphipods
Blinding	Dinoflagellates
Sacrifice lure	†Elytra of scale worms
Vision	Flashlight fish
C. *Propagation of species*	
Schooling prior to mating	Euphausiid shrimp, syllid worms
Mating ritual	Fireflies, *Odontosyllis* (worm)
Attraction of propagating species	Bacteria, fungi, nematode
Symbiosis	Bacteria

†Based on inspired guesswork rather than good scientific evidence.

scattered eventually. The intensity (I_d) of this 'blue' light at a particular depth (d) is expressed as:

$$I_d = I_0 \exp(-Kd) \tag{3.23}$$

where I_0 = light intensity at the surface
       K = diffuse attentuation coefficient

K depends on the inherent property of water to absorb and scatter light. It also depends on local parameters such as the latitude of the sun and cloud cover. K also varies with the source of water. In clear ocean water, K is 0.02 to 0.03, resulting in a transmission of 97% per metre. Water in the sea is not, however, homogeneous. It can form layers of different density and chemical composition. As a result, some oceanic waters, particularly those near the coast, can have K's in the range 0.15 to 0.30, reducing light transmission to 74–86% per metre.

A further complication is the difference in transmission of coloured light through seawater. Pure water has a transmission for blue light ($\lambda = 475$ nm) of 98.2% per metre, but only 60.7% for red light ($\lambda = 700$ nm). This is mainly caused by absorbance, but a little from light scattering. Oceanic water is slightly worse. As a result the light reflected off an object will be decreased in intensity by 90% at 30 m if the bioluminescent torch is blue ($\lambda = 475$ nm) or at only 2 m if it is red ($\lambda = 700$ nm). There are now two problems in detecting an animal emitting light at some distance away, decreased transmission as a result of light absorption and scatter by the water,and decreased luminosity as a result of the inverse square law. For blue light losses due to scatter and absorption are usually small, between 1 and 50 metres. For example at 10 m the light intensity detected will be decreased 100 fold because of the inverse square law and only 30% because of absorption and scatter. However as the distance increases the latter become major factors limiting the light intensity observed. Remember also, phyto-plankton, near the surface, absorb light.

The result is that at depths below 500 metres the light intensity is some 10^{-5}–10^{-7}% of that at the surface. It is blue, with a peak around 475 nm. Not surprisingly, the eyes of most deep sea fish can only see blue light (478–492 nm), and those nearer the surface or in estuaries blue to blue–green light (504–512 nm). Deep sea stomiatoid fish, *Malacosteus* (Fig. 3.10b), *Aristostomias*, and *Pachystomias* can see their own red light as a result of having red-sensitive eye pigments. The red luminescence would be invisible to other deep sea creatures, but will light up organisms such as decapod shrimp with red pigments.

Between 500 and 1000 metres there is a 'twilight zone'. Many of the fish here have large eyes, with wide pupils. Their eyes may be 10–100 times more sensitive than the human eye. At this depth living light plays an important role as a camouflage, concealing the otherwise dark silhouette of a hatchet or lantern fish from a potential predator looking upwards. You can try an experiment to verify this yourselves. Take a dive, in safe waters, at night from a boat with a series of lights on its bottom. See what you notice when the lights are switched on and off. The shape of the boat will disappear. Below 1000 metres virtually all the light is living light. The eyes of fish here are much smaller than those from the twilight zone, enabling them to detect bioluminescence only over relatively short distances.

A similar relationship between eye sensitivity and the function of bioluminescence is found in terrestrial luminous animals. For example, the male glow-worm has some 2000–3000 syncittida making up its compound eye, enabling it to detect the glowing female from perhaps 100 metres away. The eye of the wingless female, on the other hand, has just a few hundred syncittida, sufficient to see the fainter luminescence of a male only when close by.

Thus in order to discover whether a luminous organism is visible or not, three pieces of information are required:

(1) Spectral sensitivity of the eye pigments of the observer, related to that of the luminescence.
(2) The transmissivity of the medium (i.e. water for marine organisms), enabling the amount of light reaching the observer, at a given distance away, to be estimated.
(3) The difference between the intensity of the luminescence and that of the ambient light.

An elegant example of this approach has been carried out by Denton and colleagues providing strong supporting evidence for the hypothesis that ventral (undersurface) photophores in mesopelagic (i.e. down to 1000 m) animals such as squid and shrimp, are used as a camouflage, the light emitted from the lower surface replacing that lost by absorption at the upper surface. Careful spectral and intensity measurements of three fish *Argyropelecus* (hatchet fish with its own photophores), *Opisthoproctus grimaldi* (bacterial light organ) and *Diaphus peleus* (own photophores) showed that the colour of the light generated inside the photophore was a poor match with the ambient submarine daylight. However *Argyropelecus* and *Opisthoproctus* pass the light through an absorbing filter pigment, whereas *Diaphus* reflects the light off a blue 'mirror'. The result is light emerging very well colour matched to the ambient light. Such 'colour matching' pigments vary in structure and colour themselves, lilac and cytochrome c like in *Argyopelecus*, red and a porphyrin in the fish *Valencienellus* and the squid *Histioteuthis*, blue in the shrimp *Sergestes*. Nevertheless careful measurement of the absorbance spectrum of the isolated pigments showed that this is just right for reducing the necessary parts of the indigenous chemiluminescence spectrum.

A second, different, example of the value of special measurements comes from studies on the deep sea fish *Malacosteus*. Its red light is made even redder by passing it through a porphyrin pigment layer. Since *Malacosteus*, and its relatives *Aristostomias* and *Pachystomias* have red visual pigments and their food is red shrimp, this is consistent with the red suborbital light organs being used as a torch. The function of the blue, post-orbital organs and those scattered over the body may be different, however.

How then does the sighting of living light, or its merger into the background, relate to its function?

3.6.3 Functions

Chemiluminescence in a live organism could simply be an incidental property of a biochemical system evolved to fulfil some other function. Haemoglobin is red, haemocyanin is blue, but it is the oxygen-carrying ability of these proteins, rather than their colour, which is relevant, and has been selected through evolutionary forces. Yet the complexity and subtlety of the chemistry as well as the structures and the control mechanisms of bioluminescence, suggest strongly that it always plays some function in the survival or propagation of the species concerned. This, at least, is the current dogma.

The diverse functions of living light (Table 3.23) could be classified in one of three ways on the basis of whether the responder is itself, one of its own species, or another species, i.e. intra- or inter-specific; whether the luminescence has evolved to attract, to detract, to conceal or to illuminate; and what behavioural response occurs in conjunction with light emission.

It is important to analyse the value of the behavioural response. For example, luminescence in many fish, euphausiid shrimp, and syllid worms appears to play a role in schooling. The animals communicate with each other by means of their living light. But the reasons for schooling may vary. In euphausiids and syllids its function is to bring them together for mating. Spontaneous luminescence in the surface krill (euphausiids) between November and June correlates with the mating season. In contrast schooling in lantern fish enables them to move up and down together in the water by night or day. This gives security to individuals within the shoal. Flashlight coral fish, on the other hand, school to help each other find food.

To relate, ultimately, the chemical and physical properties of bioluminescence to the selective forces responsible for its evolution in one species but not in another, it is perhaps most revealing to group the functions of living light into three broad categories (Table 3.23): *the obtaining of food, defence and security, or species propagation*. In some organisms, e.g. the flashlight fish or euphausiid shrimp, the light fulfils more than one of these functions. In others, different light organs or photophores have separate functions. For example, in ceratoid angler fish, its luminous esca at the end of its lure dangled in front of its mouth attracts unsuspecting shrimp as food, whereas the luminous esca on another modified fin, the caruncle on its barbel, in the female attracts the male, or can be used to detect predators.

The search for food can involve illuminating the prey, luring the prey, or helping other individuals through schooling. Only certain fish, and a few squid, have light organs large enough to act as a torch. However, clouds of light from large clusters of organisms such as dinoflagellates can provide sufficient illumination to help other organisms find food. This occurs, for example, when herring and mackerel disturb dinoflagellate blooms. Fishermen, in some parts of the world, who use bits of luminous squid or fish on their lines, know its value. They are mimicking the luminous lure of the angler fish, and the luminous sticky threads (light guides) of the New Zealand glow-worm *Arachnocampa* which ensnare insects. Other examples of mimicry are the luminescence of the midshipman fish which can fool euphausiid shrimp, and the female 'femme fatale' firefly *Photuris*, luring males of *Photinus* to their death by flashing like the female *Photinus*.

At least ten functions for bioluminescence in the defence and security of individuals, or populations, of luminous organisms have been identified (Table 3.23B). Many animals, including ourselves, jump when suddenly exposed to a flashing light at night. The ability to frighten with a flash may be particularly important in organisms which are attached to rock, or are essentially immobile like the rock-boring bivalve *Pholas*, the worm *Chaetopterus* in its chitinous tube in the mud, and hydroids like *Obelia*, growing on brown seaweeds. The brilliant flash of copepods secreted from cells in their legs and occuring close to the animal may scare or blind a potential predator. Other organisms that can move have evolved subtler defence mechanisms, including the use of luminous clouds as decoys and 'smoke screens'. Escape movements have been correlated with light emission in the squid *Heteroteuthis* emitting a luminous squirt from its siphon, either as a cloud of bacteria or which breaks up into many parts, as well as in other shrimp, copepods, the larva of the fungus gnat *Arachnocampa*, the burrowing hemichordate *Ptychodera*, and the 'blink and run' response of the flashlight fish *Photoblephoron*. The luminous camouflage of hatchet or lantern fish, on the other hand, switches on when the organism sees light around it and switches off when it goes dark again.

Other, more speculative, ideas include the possible use of detachable elytra in scale worms, living under rocks or in holes with other polychaetes. This could act as a sacrifice for the sake of survival of the organism. A 'burglar alarm' hypothesis has been proposed to explain the function of chemiluminescence in dinoflagellates. These light up when disturbed by a potential predator, which is thus illuminated and is itself then eaten by a larger organism. This seems to have been verified for copepods eaten by fish. Copepods also eat less when the dinoflagellates are shining. They appear to be 'blinded' by the luminescence. A further intriguing idea is that luminescence could serve as a warning to would-be predators that the luminous species is not very nice to eat at all. For example, *Porichthys* has a poisonous spine, and fireflies are not very popular with lizards, frogs, or birds.

Chemiluminescence can be involved in the propagation of species not only by bringing males and females together and in the actual mating ritual, but also it may help to disperse organisms such as bacteria, or fungal spores, over a wide area. A curious example of this appears to occur with the nematode *Heterohabditis*, red and conspicuous by day, luminous with bacteria, and equally conspicuous to birds or other organisms, by night.

The precise function of the luminescence in many creatures, particularly those in the deep oceans, is still far from clear. Nevertheless, what is clear is that living light plays a major role in the ecology and evolution of marine life and in some parts of the world in nocturnal terrestrial life. Bioluminescence is responsible for interactions between individuals at three levels. It can influence behavioural patterns both within and between species. It can play a role in symbiosis between a luminous and non-luminous species, though this has so far only been really established with bacteria. It is a major, or the only, source of light, in the deep oceans, and at night at the ocean surface. Living light thus enables other animals to see, and allows evolutionary forces to select animals with eyes which live in an otherwise dark environment.

3.7 PERSPECTIVES

Living light is a very diverse and unique natural phenomenon. Several quite distinct chemistries have evolved independently to generate it, and there are still great numbers of luminous invertebrates and fish in which we have little or no idea about the chemiluminescent reaction. Several different intracellular and secretory mechanisms have evolved in different luminous groups to consummate the components, and generate light. Living light has evolved to serve a wide range of functions, many of which have yet to be understood.

How did living light evolve? Did the reactions evolve originally from an ultraweak chemiluminescent reaction, which eventually became visible? In which case, how was it selected for during the 'invisible' stages? Do or did the luciferins have another function? Did the luciferases evolve from particular protein family precursors? The EF hand Ca^{2+} binding sites in coelenterate photoproteins suggest the obvious possibility that the protein was formed from a marriage of Ca^{2+} binding site exons and an O_2-luciferin binding site/catalytic exon. From the known luciferin families (Table 3.19 and Fig. 3.11) living light must have evolved independently at least ten times (3 aldehydes + 2 imidazolopyrazines + 1 benzothiazole + 3 tetrapyr-roles + riboflavin). From the number of unknown systems (Table 3.21) which do not cross-react with any of these, there may be perhaps as many as six others which had independent evolutionary origin. Bioluminescence presents the evolutionary biolo-gist with a considerable challenge (see Chapter 10, section 10.3.3 for further discussion).

Man has, since the industrial revolution began, caused major disturbances in the ecology of this planet. How stable are the chemistry and biology of the deep oceans? If we are disturbing this unique ecosystem, will it have disastrous consequences on other parts of the planet? Living light plays a major role in the ecology of the sea. This in itself makes it worthwhile considering bioluminescence as one phenomenon, in spite of the diversity in its chemistry and biology. Furthermore, it has the potential for giving us a unique perspective of some of the central problems of biology, as well as providing the chemist with excited state systems often far more efficient than he or she could conceive without the help of evolution. Five particular problems stand out:

(1) The role of chemically-generated excited states in cells.
(2) The mechanisms underlying cellular and tissues differentiation and develop-ment, as well as factors determining tissue shape.
(3) The difficulty of rationalising macro-evolutionary principles, based on Darwi-nian–Mendelian principles, with ideas about biochemical evolution based on the principles of modern molecular genetics.
(4) The role of 'threshold phenomena' in cell activation and injury, together with its rationalisation by the chemisymbiotic hypothesis.
(5) The dichotomy between the life-supporting and potentially lethal effects of oxygen.

I shall explore the value of bioluminescence in looking at these problems in the last chapter of this book.

There is no doubt about the enthusiasm exhibited by those of us who are fortunate to study living light. It is an aesthetically satisfying phenomenon to watch. It is great fun searching for luminous animals at night, on the beach, in the mountains, or on a research ship. Yet the people attracted to it are not restricted to scientists, nor have they been the only people to exploit it. Mexican women use fireflies as luminous headdressses. Japanese soldiers during the Second World War were issues with dried ostracods, *Vargula* (formerly *Cypridina*). These glowed when moist and were used for reading messages or map reading at night. Fishermen from Portugal and the Indian Ocean use bits of luminous fish or squid as bait to attract unsuspecting fish. Scientific applications (see Chapters 4–9) include its use as an ultrasensitive analytical tool, as a probe for illuminating the chemistry of living cells, as a replacement for radio-active ^{125}I in immunoassay, or ^{32}P in recombinant DNA technology, and as a tool for the genetic engineer. Yet the study of bioluminescence remains a Cinderella science. It is rarely given much, or any, exposure in student biology or biochemistry texts.

Examination of how the enthusiasts study living light, makes an important philosophical contribution to the evolution of Natural Science. It shows us how an understanding of the physics and chemistry of a biological phenomenon can enhance our understanding of Nature. It takes us on a voyage of discovery from the ecology of the deep oceans, through cell biology, biochemistry, and chemistry, to the nature of light itself. Perhaps we could listen to a remark made by one of the founders of modern biochemistry, Frederick Gowland Hopkins, when he was elected president of the London Natural History Society in 1936 — 'All true biologists deserve the coveted name of naturalist. The touchstone of the naturalist is his abiding interest in living Nature in *all* its aspects'.

3.8 FURTHER READING
Books
DeLuca, M. & McElroy, W. D. (ed.) (1987) *Methods in Enzymology* **133** Biolumi-
nescence and chemiluminescence Part B. Several chapters.
Dubois, R. (1892) *Anatomie et physologie compares de la Pholade dactyle.* Annales
de l'université de Lyon, Paris.
Dubois, R. (1914) *La Vie et la Lumière,* Paris.
Harvey, E. N. (1952) *Bioluminescence.* Academic Press, New York.
Harvey, E. N. (1956) *A history of luminescence from ancient times to* 1900. American
Philosophical Society.
Herring, P. J. (Ed.) (1978) *Bioluminescence in action.* Academic Press, London and
New York.
Johnson, F. H. & Haneda, Y. (Eds.) (1966) *Bioluminescence in progress.* Princeton
University Press, Princeton.
Wiley, E. O. (1981) *Phylogenetics: The theory and practice of phylogenetic systema-
tics.* John Wiley, New York.
Parker, S. P. (Ed) (1982) Synopsis and Classification of Living Organisms. vols I and
II. McGraw-Hill, New York.

Reviews
Buck, J. & Buck, E. (1976) *Scient. Am.* **234** 74–85. Synchronous fireflies.
Clarke, W. D. (1963) *Nature* **190** 1244–1246. Function of bioluminescence in
mesopelagic organisms.
Hastings, J. W. (1983) *J. Mol. Evol.* **19** 309–321. Biological diversity, chemical
mechanisms, and the evolutionary origins of bioluminescent systems.
Herring, P. J. (1985) *Soc. Exptl. Biol. Symp.* **38** 323–350. How to survive in the dark:
bioluminescence in the deep sea.
Herring, P. J. (1987) *J. Biochem. Chemilum.* **3** 147–163. Systematic distribution of
bioluminescence in living organisms.
Lloyd, J. E. (1971) *Ann. Rev. Ent.* **16** 97–122. Bioluminescent communication in
insects.
McCapra, F. (1982) *Proc. Roy. Soc. B.* **215** 247–272. The chemistry of
bioluminescence.
Morin, J. G. (1974) *Coelenterate Biology: reviews and new perspectives,* pp. 397–438.
Muscatine, L. & Lenkoff, H. M. (Eds.). Academic Press, New York. Coelenter-
ate Bioluminescence.
Nealson, K. H. & Hastings, J. W. (1979) *Microbiol. Rev.* **43** 490–518. Bacterial
bioluminescence: its control and ecological significance.
Nicol, J. A. C. (1960) *The biology of marine animals,* Chapter 13, pp. 536–581.
Academic Press, London and New York. Luminescence.

Seliger, H. H. (1987) *Photochem. Photobiol.* **45** 291–297. The evolution of bioluminescence in bacteria.

Shimomura, O. (1985) *Soc. Expt. Biol. Symp.* **38** 351–372. Bioluminescence in the sea: photoprotein systems.

Sweeney, B. M. (1980) *Int. Rev. Cytol.* **68** 173–195. Intracellular source of bioluminescence.

Papers
Historical
Boyle, R. (1668) *Phil. Soc. Trans. Roy. Soc.* **2** 581–600. Experiments concerning the relation between light and air in shining wood and fish.

Dubois, R. (1885) *C. R. Seanc. Soc. Biol. Paris* (ser 8) **2** 559–562. Note sur la physiologie des *Pyrophores*.

Dubois, R. (1887) *C.R. Seanc. Soc. Biol. Paris* **39** 564–566. Note sur la fonction photogenique chez le *Pholas dactylus*.

Harvey, E. N. (1916) *Science* **44** 208–209. The mechanism of light production in animals.

Macaire, J. (1827) *Anal. Chimie Physique* **17** 251–267. Memoire sur la phosphorescence des lampyres.

McCartney, J. (1810) *Phil Trans. Roy. Soc.* **100** 258–293. Observations upon luminous animals.

McElroy, W. D. (1947) *Proc. Natl. Acad. Sci.* **33** 342–345. The energy source for bioluminescence in an isolated system.

Newport, G. (1857) *J. Proc. Linn. Soc.* **1** 40–70. On the natural history of the glow-worm (*Lampyris noctiluca*).

Ultrastructure and biology
Bassot, J. M. (1966) *J. Cell Biol.* **31** 135–158. Une forme microtubulaire et paracristalline de reticulum endoplasmique dans les photocytes des Annélides Polynoinae.

Bassot, J. M. (1966) *Z. Zellforsch* **74** 474–504. Données histologiques ultrastructurales sur les organes lumineux du siphon de la Pholade.

Bassot, J. M. & Bilbaut, A. (1977) *Biol. Cell* **28** 163–160. Bioluminescence des élytoes d'Acholoë IV luminescence fluorescence des photosomes.

Herring, P. J. (1981) *J. Mar. Biol. Assoc.* **61** 161–176. Studies in bioluminescence and amphipods.

Neuwirth, M. (1981) *Tissue and Cell* **13** 599–609. Ultrastructure of granules and immunocytochemical localization of luciferase in photocytes fireflies.

Nicolas, M.-T., Nicolas, G., Johnson, C. H., Bassot, J.-M. & Hastings, J. W. (1987) *J. Cell Biol.* **105** 723–735. Characterization of the bioluminescent organelles in *Gonyaulax polyedra* (Dinoflagellate) after fast-freeze fixation and anti-luciferase immunogold staining.

Chemistry and biochemistry
Campbell, A. K. & Herring, P. J. (1987) *Comp. Biochem. Physiol.* **86B** 411–417. A novel red fluorescent protein from the deep sea luminous fish *Malacosteus niger*.

Campbell, A. K., Hallett, M. B., Daw, R. A., Ryall, M. E. T., Hart, R. C. & Herring, P. J. (1981). In: *Bioluminescence and Chemiluminescence* pp. 601–607.

de Luca, M. & McElroy, W. D. (Eds.). Academic Press, New York. Application of the photoprotein obelin to the measurement of free Ca^{2+} in cells (contains evidence for Ca^{2+}-activated photoproteins in radiolarians).

Charbonneau, H., Walsh, K. A., McCann, R. O., Prendergast, F. G., Cormier, M. J. & Vanaman, T. C . (1985) *Biochem.* **24** 6762–6771. Amino acid sequence of the calcium-dependent photoprotein aequorin.

Dunlap, J. C., Hastings, J. W. & Shimomura, O. (1981) *FEBS Lett.* **135** 273–276. Dinoflagellate luciferin is structurally related to chlorophyll.

Frank, T. M., Widder, E. A., Latz, M. I. & Case, J. F. (1984) *J. Exp. Biol.* **109** 385–389. Dietary maintenance of bioluminescence in a deep sea mysid.

Girsch, S. J., Herring, P. J. & McCapra, F. (1977) *J. Mar. Biol. Assoc.* **56** 707–722. Structure and preliminary biochemical characterization of the bioluminescent system in *Ommastrephes pteropus* (Steenstrup) (Mollusca Cephalopoda).

Isobe, M., Uyakul, D. & Goto, J. (1987) *J. Biochem. Chemilum.* **1** 181–188. *Lampteromyces* Bioluminescence-1. Identification of riboflavin as the light emitter in the mushroom *L. japonicus*.

Inouye, S., Kakoi, H. & Gogo, T. (1976) *Tetrahedron Lett.* **34** 2971–2974. Squid bioluminescence III. Isolation and structure of *Watasenia* luciferin.

Inouye, S., Kakoi, H., Murata, M. & Gogo, T. (1977) *Tetrahedron Lett.* **31** 2685–2688. Complete structure of *Renilla* luciferin and Luciferyl sulfate.

Kurfürst, M., Ghisla, S. & Hastings, J. W. (1984) *Proc. Natl. Acad. Sci. US* **81** 2990–2994. Characterization and postulated structure of the primary emitter in the bacterial luciferase reaction.

Lee, J. (1976) *Photochem. Photobiol.* **24** 274–285. Bioluminescence of the Australian glow-worm. *Arachnocampa richardsae*.

Lee, J., O'Kane, D. J. & Visser, A. J. W. (1985) *Biochem.* **24** 1476–1483. Spectral properties and function of two lumazine proteins from *Photobacterium*.

Leisman, G. & Nealson, K. H. (1982). In: *Flavins and Flavoproteins*, pp. 383–386. Massey, V. & Williams C. H. (Eds.). Elsevier North Holland, Amsterdam. Characterization of a yellow fluorescent protein from *Vibrio* (*Photobacterium*) *fischeri*.

McCapra, F. & Manning, M. J. (1973) *J.C.S. Chem. Commun.* 467–468. Bioluminescence of coelenterates: chemiluminescent model compounds.

McCapra, F. & Roth, M. (1972) *J.C.S. Chem. Commun.* 894–895. Cyclisation of a dehydropeptide derivative: a model for *Cypridina* luciferin biosynthesis.

Martin, J. P. & Fridovich, J. (1981) *J. Biol. Chem.* **256** 6086–6089. Evidence for a natural gene transfer from the ponyfish to its bioluminescent bacterial symbiont *Photobacterium leignathi*: the close relationship between bacteriocuprein and the copper-zinc superoxide dismutase of teleost fishes.

Michelson, A. M. (1978) *Meth. Enz.* **57** 385–406. Purification and properties of *Pholas dactylus* luciferin and luciferase (synopsis of several papers by Henry, Michelson, and co-workers).

Ohtsuka, H., Rudie, N. G. & Wampler, J. E. (1976) *Biochem.* **15** 1001–1004. Structural identification and synthesis of luciferin from the bioluminescent earthworm, *Diplocardia longa*.

Shimomura, O. (1979) *FEBS Lett* **104** 220–222. Structure of the chromophore of *Aequorea* green fluorescent protein.

Shimomura, O. (1980) *FEBS Lett* **116** 203–206. Chlorophyll-derived bile pigment in bioluminescent euphausiids.

Shimomura, O. (1981) *FEBS Lett* **128** 242–244. A new type of ATP-activated bioluminescent system in the millipede *Luminodesmus sequoiae*.

Shimomura, O. (1984) *Comp. Biochem. Physiol.* **79B** 565–567. Porphyrin chromophore in *Luminodesmus* photoprotein.

Shimomura, O. (1986) *Photochem. Photobiol.* **44** 671–674. Bioluminescence of the brittle star *Ophiopsila californica*.

Shimomura, O. (1987) *Comp. Biochem. Physiol.* **86B** 361–363. Presence of coelenterazine in non-bioluminescent marine organisms.

Shimomura, O. & Johnson, F. H. (1972) *Biochem.* **11** 1602–1608. Structure of the light-emitting moiety of aequorin.

Shimomura, O., Inoue, S., Johnson, F. H. & Haneda, Y. (1980) *Comp. Biochem. Physiol.* **65B** 435–437. Widespread occurrence of coelenterazine in marine bioluminescence.

Tiemann, D. L. (1970) *Natl. Geogr. Mag* **138** 56–67. Nature's toy train, the railroad worm.

Thompson, E. M., Nakpaktitis, B. G. & Tsuji, F. I. (1987) *Photochem. Photobiol.* **45** 529–533. Induction of bioluminescence in the marine fish *Porichthys* by *Vargula* (crustacean) luciferin. Evidence of de novo synthesis or recycling of luciferin.

Tsuji, F. I. (1985) *Proc. Natl. Acad. Sci.* **82** 4629–4632. ATP-dependent bioluminescence in the firefly squid *Watasenia scintillans*.

Young, R. E., Roper, C. F. E., Mangold, K., Leisman, G. & Hochberg, F. G. (1974) *Marine Biology* **53** 69–77. Luminescence from non-bioluminescent tissues in oceanic cephalopods.

Cloning of bioluminescent proteins

Cohn, D. H., Mileham, A. J., Simon, M. I., Nealson, K. H., Rausch, S. K., Bonam, D. & Baldwin, T. O. (1985) *J. Biol. Chem.* **260** 6139–6146. Nucleotide sequence of the lux A gene of *Vibrio harveyi* and the complete amino acid sequence of the α subunit of bacterial luciferase.

Cohn, D. H., Ogden, R. C., Abelson, J. N., Baldwin, T. O., Nelson, K. H., Simon, M. I. & Mileham, A. J. (1983) *Proc. Natl. Acad. Sci.* **80** 120–123. Cloning of the *Vibrio harveyi* luciferase genes: use of a synthetic oligonucleotide probe.

de Wet, J. R., Wood, K. V. Helsinki, D. R. & de Luca, M. (1985) *Proc. Natl. Acad. Sci. US* **82** 7870–7873. Cloning of firefly luciferase cDNA and the expression of active luciferase in *Escherichia coli*.

Inouye, S., Noguchi, M., Sakaki, Y., Takagi, I., Miyata, T., Iwanaga, S. & Tsuji, F. I. (1985) *Proc. Natl. Acad. Sci. US* **82** 3154–3158. Cloning and sequence analysis of cDNA for the luminescent protein aequorin.

Inouye, S., Sakaki, Y., Goto, J. & Tsuji, F. I. (1986) *Biochem.* **25** 8425–8429. Expression of apo aequorin complementary DNA in *Escherichia coli*.

Johnson, T. C., Thompson, R. B. & Baldwin, T. O. (1986) *J. Biol. Chem.* **261** 4805–4811. Nucleotide sequence of the lux-B gene of *Vibrio harveyi* and the complete amino acid sequence of the β subunit of bacterial luciferase.

Prasher, D., McCann, R. O. & Cormier, M. J. (1985) *Biochem. Biophys. Res. Commun.* **126** 1259–1268. Cloning and expression of the cDNA coding for aequorin, and bioluminescent calcium-binding protein.

Tsuji, F. I., Inouye, S., Goto, J. & Sakaki, Y. (1986) *Proc. Natl. Acad. Sci. US* **83** 8107–8111. Site-specific mutagenesis of the calcium binding protein aequorin.

Control

Anctil, M., Brunel, S. & Descarries, L. (1981) *Cell Tissue Res.* **219** 557–566. Catecholamines and 5-hydroxytryptamine in photophores of *Porichthys notatus*.

Case, J. F., Reynolds, G. T., Buck, J., Burns, J. & Halverson, R. (1970) *Biol. Bull.* **139** 417 Comparative pharmacology of neurally controlled luminescence.

Eberhard, A., Burlingame, A. L., Eberhard, C., Kenyon, G., Nealson, K. H. and Oppenheimer, N. J. (1981). *Biochemistry*, **20**, 2444–2449. Structural identification of antoinducer of *Photobacterium fischen*.

Engebrecht, J. and Silverman, M. (1984) *Proc. Natl. Acad. Sci.* US **81**, 4154–4158. Identification of genes and gene products necessary for bacterial bioluminescence.

Johnson, C. H., Roeber, J. F. & Hastings, J. W. (1984) *Science* **223** 1428–1430. Circadian changes in enzyme concentration account for rhythm of enzyme activity in *Gonyaulax*.

Nathanson, J. A. (1985) *Proc. Natl. Acad. Sci. US* **82** 599–603. Characterization of octopamine-sensitive adenylate cyclase: elucidation of a class of potent and selective octopamine-2 receptor agonists with toxic effects in insects.

Ulitzur, S. and Kuhn, J. (1988) *J. Biolum. Chemilum.* **2**, 81–93. The transcription of bacterial luminescence is regulated by Sigma 32.

Properties and function

Anctil, M. (1981) *Comp. Biochem. Physiol.* **68C** 187–194. Luminescence control in isolated notopods of the tube-worm *Chaetopterus variopedatus*: effects of cholinergic and gabaergic drugs.

Eckert, R. & Reynolds, G. T. (1967) *J. Gen. Physiol.* **50** 1429–1458. The sub-cellular origin of bioluminescence.

Denton, E. J., Herring, P. J., Widder, E. D., Latz, M. F. & Case, J. F. (1985) *Proc. R. Soc. Lond. B.* **225** 63–97. The roles of filters in the photophores of oceanic animals and their relation to vision in the oceanic environment.

Hastings, J. W. (1971) *Science* **173** 1016–1017. Light to hide by: ventral luminescence to camouflage and silhouette.

Herring, P. J. (1983) *Proc. R. Soc. Lond. B.* **220** 183–217. The spectral characteristics of luminous marine organisms.

Morin, J. G., Harrington, A., Nealson, K., Krieger, N., Baldwin, T. O. & Hastings, J. W. (1975) *Science* **190** 74–76. Light for all reasons: versatility in the behavioural repertoire of the flashlight fish.

Pavans de Ceccatty, M., Bassot, J.-M., Bibaut, A. & Nicolas, M.-T. (1977) *Biologie Cellulaire* **28** 57–64. Bioluminescence des elytres d'Acholoë I. Morphologie des supports structuraux.

Reichelt, J. L., Nealson, K. & Hastings, J. W. (1977) *Arch. Microbiol.* **112** 157–161. The specificity of symbiosis: pony fish and luminescent bacteria.

Ward, W. W. & Cormier, M. J. (1978) *Photochem. Photobiol.* **27** 389–396. Energy transfer via protein–protein interaction in *Renilla* bioluminescence.

Widder, E. A., Latz, M. I., Herring, P. J. & Case, J. F. (1984) *Science* **225** 512–514. Far red bioluminescence from two deep-sea fishes.

Widder, E. A., Latz, M. I. & Herring, P. J. (1986) *Photochem. Photobiol.* **44** 97–101. Temporal shifts in bioluminescence emission spectra from the deep-sea fish *Searsia koefoedi*.

4

The need for chemiluminescence analysis in biology, medicine, and biotechnology

4.1 FROM THE OCEAN DEPTHS TO THE HOSPITAL BED

4.1.1 The chain

Man has exploited chemiluminescence in some remarkable ways (Table 4.1). Many examples over the past centuries are known where luminous bacteria, fungi, ostracods or beetles have been used either to illuminate or to pin-point something in

Table 4.1 — Exploitation of chemical light by man

Application	Example
1. Torch	'Lamps' containing luminous bacteria, fungi, glow-worms, or firefly tails. Wartime use by Japanese soldiers of luminous ostracods for message or map reading.
2. Fishing	Bits of luminous squid as a lure on Portuguese fishing lines. Live flashlight fish in a bamboo as a lure on nets in the Indian Ocean. Detection from the air of fish and shrimp shoals.
3. Military	Detection of nuclear submarines. Detection of harmful germs or chemicals.
4. Detection and ritual	Fireflies and cucujus in head-dresses of Mexican women. Luminous fungi in rituals of natives in Ecuador. Light sticks on night club dancers.
5. Fun	Toys and objets d'art. Luminous golf balls.
6. Scientific	Model for the chemistry of the excited state. Model for anaesthetic and drug action. Model for biochemical evolution. Probe in genetic engineering. Microbe assay in industry and clinical microbiology.

the dark. Luminous fungi, either the fruiting bodies or rotting wood, have had many such applications. One of the earlier accounts of these comes from Bishop Olaus Magnus in 1652. He reported that 'when it is expedient or there is an urgent need, people of the far north use an ingenious method for lighting their way through forests in the dark and when continuous night prevails before and after the winter solstice: they place pieces of rotten oak bark at certain intervals on the proposed route in order that by their glow they may complete their journey'. The fungus in this case was almost certainly the honey-tuft fungus, *Armillaria mellea*. Magnus also refers to the use of luminous wood to illuminate 'places full of combustible material, such as winter barns, full of harvested crops or hay'. During World War I soldiers in the trenches used small pieces of luminous wood in their helmets and bayonets to prevent collisions on very dark nights, whilst in World War II paratroopers at Arnhem had to hide luminous timber with tarpaulins as the light was so conspicuous. A more recent use of chemical light, is the use of synthetic oxalate ester chemiluminescence, to pin-point objects in the dark, the use of light sticks (see Fig. 4.1) and safe lights, replacing radioisotopic illumination on ships and luminous golf balls. Bioluminescent adornments can be found in the headdresses of Mexican women using fireflies or cucujus, in the rituals of natives from Ecuador, and in Indonesia are used by girls to adorn their bodies and thereby attract and guide their lovers by night.

There is perhaps a fine line between scientific investigation and application of a phenomenon. As we have already seen this can be traced back to the experiments of Boyle and others at the end of the seventeenth century. However widespread use of chemi- and bio-luminescence to study phenomena and problems other than the luminescence itself really began towards the end of the nineteenth century, and has only exploded since the second world war. Dr Allen Macfadyen in 1900 at the Royal Institution under the supervision of Professor James Dewar showed that cooling luminous bacteria in liquid air (− 190°C) for 20 h to 7 days caused them to become non-luminescent but on raising the temperature again the 'phosphorescence recommenced with full vigour'. This, they claimed showed that 'mere cooling did not destroy putrefying organisms or bacteria'. A more recent application of the quenching of luminous bacteria as an indicator of cell death is in the quantification of phagocytosis and consequential microbial killing by blood neutrophils and macrophages.

Living light, and the sources of their luminescence, have been regarded through the centuries by many communities as having 'magic powers'. Reports of luminous wounds and cadavers on the battle field are found as far back as the earliest Chinese literature. An intriguing anecdote related to this is the ability of such wounds infected with luminous bacteria to recover better than 'dark wounds', and are less susceptible to gangrene. This story has been accredited to Florence Nightingale in the Crimean War, but I have been unable to confirm this from her letters. The anecdote has received further credibility recently as a result of the discovery that luminous bacteria from the soil, *Xenorhabdus*, can produce a potent antibiotic.

The widespread use of phosphorus in the treatment of various human ailments, including 'epilepsies, insanity and malignant fevers', during the seventeenth and eighteenth centuries may have been related, at least in part, to its ability to produce light. One patient of a French doctor, Alphonse Leroi, received such heavy doses that on autopsy the internal organs were found to be luminous.

A

B

Fig. 4.1 — Light stick containing oxalate ester chemiluminescence. A: No fluor. B: + fluor.

Fishermen in different parts of the world have learnt to mimic the luminous lure of the deep sea angler fish, whilst the triggering of luminous organisms at the surface can be used to detect fish or shrimp at night by using image intensification from the air. Submerged nuclear submarines also provoke a luminous display as they bump into the diverse organisms of the deep ocean.

Chemiluminescence offers exciting opportunities for the scientist (Table 4.1). It provides models for studying the chemistry of the excited state, for the interaction of drugs such as local and general anaesthetics with proteins, and for marrying the micro- and macro-approaches to biochemical evolution. But chemiluminescence also provides the chemist and the biologist with a unique analytical tool and probe in genetic engineering, which has aroused the interest not only of the research scientist but also scientists in clinical laboratories, as well as in biotechnology and industry. The protein-bound luciferin from the common piddock *Pholas* Fig. 3.1 provides us with a means of examining the workings of single human cells. These cells occur in large numbers in the joints of patients with rheumatoid arthritis, one of the major causes of invalidism in the Western world. The living light from these same cells has also provided new clues about heart attacks, the most common cause of death in the Western world, Wales, Scotland, and Finland being particular black spots. Another luminous organism, this time of worldwide distribution, is the tiny jelly fish *Obelia*. The use of its photoprotein has enabled a link to be established between the role of calcium inside cells and the mechanism underlying another crippling disease, multiple sclerosis. Such links between living light and medicine are not restricted to marine creatures. The glow-worm can be found displaying her 'amorous fire' in a quiet Welsh valley during the summer. This beetle, and its American relative, the firefly, provide the laboratory scientist with an exquisitely senstive means of assessing whether any material on earth, or extra-terrestrial for that matter, contains life.

4.1.2 The links

The key to the unique potential of chemiluminescence as an analytical tool, and as a probe in genetic engineering, lies in the remarkable sensitivity with which chemiluminescent reactions can be detected. The naked eye can perceive an image if the photons hitting a square centimetre of the retina number about 30 000 per second. The superiority of chemiluminescence over its rivals, absorbance, fluorescence, and phosphorescence, only really becomes evident when the light emission is invisible to the naked eye, and to photographic film, though some commercially-available assays manage to exploit successfully brighter chemiluminescence. The photomultiplier tube (see Chapter 2) can detect just a few photons per second. The location of these very low light emissions from live cells can be recorded by means of an image intensifier. Thus, just a few hundred molecules reacting per second can be measured and visualised.

Many of the researchers into chemiluminescence over the past one hundred years have realised that by coupling one of the components of their particular chemiluminescent reaction to a substance of chemical or biological interest, that substance can be measured. Beijerinck, at the end of the nineteenth century, was one of the first to realise that luminous bacteria could be used to measure the concentration of oxygen in solution. Harvey, in 1916, whilst studying the catalysis of pyrogallol and H_2O_2

chemiluminescence by plant peroxidases, or 'oxidases' as he called them, noted the exquisite sensitivity of this reaction for measuring these enzymes. McElroy, after discovering the ATP requirement in the firefly chemiluminescent reaction in 1947, immediately realised its potential as an analytical method for this central cellular metabolite. And, some thirty years ago, Erdy, in Budapest, was exploring the analytical potential of the lophine, luminol and lucigenin reactions (Fig. 1.5).

The last twenty years have seen many hundreds of papers containing assays utilising chemiluminescent rections. These applications are based on coupling any one of the six possible components of the chemiluminescent reaction to the substance or material of interest (Table 4.1):

(1) The chemiluminescent compound giving rise to the excited state molecule responsible for light emission (e.g. chemiluminescent label in immunoassay or recombinant DNA technology).
(2) The oxidants (e.g. O_2 produced by photosynthesis or a reactive metabolite of oxygen from phagocytic cells).
(3) A catalyst (e.g. a luciferase or peroxidase) as a label in immunoassay or recombinant DNA technology).
(4) Other substrates or cofactors necessary for the chemiluminescent reaction to take place (e.g. ATP in the firefly reaction).
(5) A cation which may interact with the substrate, cofactor(s), or catalyst (e.g. Ca^{2+} provokes coelenterate photoprotein chemiluminescence for measurement of free Ca^{2+} in living cells).
(6) A fluorescent energy acceptor in the case of sensitised (energy transfer) chemiluminescence (e.g. homogeneous immunoassay, see Chapter 9).

In Chapter 2 (section 2.1.3) we saw that the equation describing a first order chemiluminescent reaction was:

$$\text{Intensity}(I) = dhv/dt = \phi_{CL}(LH_2)_0 \cdot k \exp(-kt) \tag{4.1}$$

where ϕ_{CL} = overall quantum yield; $(LH_2)_0$ = amount (N.B. *not* concentration) of chemiluminescent substrate at time zero; k = exponential decay rate constant.

The analyte or material to be measured may alter ϕ_{CL}, $(LH_2)_0$, k, colour or polarisation, or a combination of these parameters. In practice analyte assays either alter the rate of a chemiluminescent reaction or are end-point assays involving measurements of absolute photon yield. An enzyme or metabolite which changes the rate of a chemiluminescent reaction can then be quantified by monitoring intensity (I), the rate constant or the peak height ($\phi_{CL}(LH_2)_0 k$). Sometimes, to a first approximation, a mini-integral e.g. of 10 seconds, is sufficiently precise, particularly if the decay rate is very slow. This is often used in the firefly luciferase assay for ATP. The concentrations of the reagents can be adjusted to ensure that light intensity is directly proportional to ATP concentration, so that an enzyme decreasing or increasing ATP will produce a linear fall or rise in intensity proportional to enzyme activity. In contrast when using chemiluminescent compounds as a label in immu-noassay or recombinant DNA technology an end-point assay is needed, usually requiring measurement of total light yield ($\phi_{CL} \cdot (LH_2)_0$). Experimental conditions are designed to make the reaction fast so that an integration time of just a few seconds is necessary. On the other hand if the label is an enzyme, such as peroxidase, then a rate parameter is once again used, total light yield simply being proportional to the number of chemiluminescent substrate molecules present. Since the peak height ($\phi_{CL} \cdot (LH_2)_0 k$) is proportional to both the amount of chemiluminescent substrate and the rate of the reaction this can be used as a rapid parameter in either rate or end-point assays. But, substances in biological material may alter the rate constant and thus change the peak height in the absence of a change in the amount of $(LH_2)_0$, the chemiluminescent substrate.

Homogeneous ligand — ligand interactions may be quantified by monitoring the ratio of intensities or rates at different wavelengths or degrees of polarisation.

In some assays:

$$\text{cofactor} + \text{chemiluminescent substrate} \xrightarrow[\text{cation}]{\text{catalyst}} \text{light} \qquad (4.2)$$

If the unknown substance increases cofactor concentration there will be an increase in light intensity, but no effect on total light yield. Similarly, increasing or decreasing catalytic activity will alter light intensity and the rate of its decay without affecting total light yield. However, altering the amount of chemiluminescent substrate alters both light intensity and total light yield. An important feature of chemiluminescence is that the total light yield is dependent on the absolute amount of chemiluminescent substrate present and not its concentration. One attomol (10^{-18} mol) of a chemiluminescent substance with an overall quantum yield of 0.1 will produce 6022 photons whether it is in $10 \, \mu l$ or 1 ml. However, the *rate* of reaction, and the binding to the active centre of an enzyme, *will* depend on concentration and thus affect light intensity, or the rate of decay of light emission.

Thus ATP can be measured by its effect on the *rate* of firefly luciferin oxidation by firefly luciferase, and Ca^{2+} can be measured by its effect on the fractional utilisation (k) of aequorin or obelin (see Chapter 3 for reactions), whereas the amount of an acridinium labelled antibody will be assayed from the total light yield. In the case of assays using chemiluminescence energy transfer (see Chapter 9), a change in colour, measured as the intensity at particular wavelengths, of the light emission may provide the parameter relating chemiluminescence to the concentration or amount of substance being analysed. The use of other physical parameters of the light, such as polarisation, have yet to be realised with chemiluminescence.

Equation (4.1) not only identifies the parameters to measure but also helps us to select the appropriate chemiluminescent reaction. To use such a reaction to measure a substance of biological interest, four criteria must be satisfied:

(1) The substance to be measured must itself, or after coupling it to one of the components of the chemiluminescent reaction, alter one or more of the physical parameters: the rate of light emission, the total light yield, the colour, or the state of polarisation of the light emitted.

(2) The physical conditions (e.g. pH, temperature, ionic strength), must be compatible with the stability of the substance to be measured or that of the biological system to be studied.

(3) The quantum yield (ϕ_{CL}) of the chemiluminescent reaction should be high enough to generate sufficient light. Bioluminescent systems with $\phi_{CL} = 0.1$–1 or synthetic compounds with CL = 0.01–0.1 are usually adequate. However, if ϕ_{CL}

< 0.001, then the chemiluminescence assays may not have any advantages over one based on absorbance of fluorescence. One important exception is cellular ultraweak chemiluminescence (see Chapter 6).

(4) The kinetics of the chemiluminescence should be appropriate for the system being studied. For example, a slow reaction taking many hours to accumulate the integral of total light yield would be of little use when quantifying chemiluminescent labels. In contrast, the complete consumption of a chemiluminescence compound within a few seconds in a cell system, might be too fast to allow sufficient time for the biological phenomenon to be observed. As we saw in Chapter 2 (section 2.1.3):

$$k \text{ for an exponential reaction } = 0.693/t_{\frac{1}{2}} \text{ ,}$$

thus if $t_{\frac{1}{2}} = 2$ seconds, $k = 0.35 \text{ s}^{-1}$, and the fractional utilisation of the chemiluminescent substrate in five half times $= (\frac{1}{2})^5 = \frac{1}{32}$ i.e. about 97%– consumption, 10 seconds (5 half times) would be sufficient in most assays where the substance to be measured is related to total light yield (see Chapter 8). In other assays, e.g. coelenterate photoproteins (aequorin and obelin) for measurement of free Ca^{2+} (see Chapter 7), k describes the fractional saturation of the photoprotein by Ca^{2+}. This fractional saturation is dependent on the binding constant (K_d^{Ca}) for the Ca^{2+} sites on the protein, and fortunately is very small (about 10^{-6}) at the concentration of free Ca^{2+} in the resting cell (i.e. at about $0.1 \mu M$). These photoproteins can therefore be used to measure intracellular free Ca^2 in the range 0.1–$10 \mu M$ without losing all of the photoprotein chemiluminescence too quickly. However, when the protein is saturated by Ca^{2+} ($Ca^{2+} > 100 \mu M$) the rate constant for aequorin is 1.4 s^{-1} ($t_{\frac{1}{2}} = 0.5 \text{ s}$), and for obelin 4 s^{-1} ($t_{\frac{1}{2}} = 0.18 \text{ s}$). Thus, more than 97% of the photoprotein prosthetic group will be consumed within a few seconds. Furthermore, the rate constant is very insensitive to changes in free Ca^{2+} close to saturation of the photoprotein. Photoproteins are therefore quite hopeless for measuring changes in free Ca^{2+} in the mM range, such as occur in extracellular fluids.

For these reasons, it is essential that before using a chemiluminescent reaction in a new biological assay, the quantum yield (ϕ_{CL}) and maximum rate constant (k_{sat}) are determined.

The most important feature of chemiluminescence analysis is sensitivity, giving it its unique potential over absorbance, fluorescence or radioactivity in certain circumstances. But remember the noise. A chemiluminescent compound may generate 10^{22} detected luminescent counts in a few seconds. With an apparatus background of < 10 counts per second as little as 10^{-21} mol will be detectable (see section 4.3). But if the chemical noise, generated by the reagents necessary to provoke the chemiluminescent reaction itself produces several thousand photon counts, as can often be the case, the detection limit is worsened by several orders of magnitude. We shall return to this problem during the following chapters. Even the water used to make up reagents has to be of the highest purity. One fungal cell may be sufficient to ruin precision at this exquisite level of sensitivity.

This, then, is the basis behind the major biomedical applications of chemiluminescence (Table 4.2). But are these chemiluminescent assays of any real use? Have they led to any real insights into a biological or medical problem? Are clinical

Table 4.2 — Some analytical applications of chemiluminescence

Chemiluminescent system	Application
A. *Synthetic compounds*	
1. Phthalazine diones (luminol, isoluminol, and derivatives)	Reactive oxygen metabolites Phagocyte activation Haem compounds Transition metals Label in immunoassay Peroxidase as label NO, NO_2
2. Acridinium salts (lucigenin and acridinium esters)	Reactive oxygen metabolites Phagocyte activation Label in immunoassay
3. O_3	NO, NO_2, NO_3^-
B. *Bioluminescent reactions*	
1. Firefly (*Photinus*, luciferin–luciferase)	ATP and related metabolites Enzymes producing or degrading ATP Immunoassay
2. Bacterial (*Photobacterium* or *Vibrio*) oxidoreductase–luciferase	NAD(P)H, FMN and related metabolites Enzymes coupled to NAD(P)H or FMN Immunoassay O_2 concentration
3. Coelenterate (*Aequorea* or *Obelia*) photoprotein	Intracellular free Ca^{2+}
4. Sea pansy (*Renilla*) luciferin–luciferase	Phosphoadenosine phosphate (SO_4)
5. Earthworm (*Diplocardia*) or acorn worm (*Balanoglossus*)	H_2O_2
6. Piddock (*Pholas*) luciferin	O_2^-, peroxidase, neutrophil activation
7. Polychaete worm (*Chaetopterus*)	Fe^{2+}

See Chapters 6–10 for details and other examples.

decisions based on their use? Why have industries such as brewing, food manufacture, and textiles, as well as new biotechnology companies, and even the military and space industries, taken such an avid interest recently in chemiluminescence? Anyway, why not use absorbance or fluorimetric analysis?

Three particular features of chemiluminescence provide the explanation for this awakening:

(1) The unique sensitivity of chemiluminescence analysis.
(2) The ability to detect, and visualise, chemiluminescence signals from living cells.
(3) Chemiluminescent compounds are non-radioactive.

To see why these features are so attractive, it is necesary first to remind ourselves what we might want to measure.

4.2 THE ANALYTICAL PROBLEMS

4.2.1 What do we want to measure?

The study of cell activation and cell injury in both the research and clinical laboratory requires the identification and quantification of chemical changes, both within the cell and in the surrounding fluid, bound and free. Substances requiring analysis include:

(1) Substrates, e.g. oxygen, glucose, fatty acid, amino acids.
(2) Metabolites, e.g. ATP, AMP, cyclic AMP, cyclic GMP, NAD(P)H, inositol phosphates, diacyl glycerol.
(3) Enzymes, e.g. peroxidase, lactate dehydrogenase, creatine kinase.
(4) Non-enzymatic proteins, e.g. albumin, haemoglobin, calmodulin, embryonic proteins in cancer diagnosis.
(5) Hormones, e.g. thyroxine (T_4), thyroid stimulating hormone, insulin, steroids.
(6) Vitamins, e.g. B_{12}, biotin.
(7) Ions, e.g. Na^+, K^+, Ca^{2+}, Mg^{2+}, Fe^{2+}, Cl^-, HPO_4^{2-}, HCO_3^-.
(8) Pharmacological substances and drugs of abuse, e.g. digoxin, barbiturates, heroin.
(9) Pathogens and microbiological contaminants, e.g. bacteria, viruses, protozoa, yeasts, moulds, and toxins.

Quantification of cell viability, cell number, and biomass also plays a vital part in interpreting effects of inhibitors, or long-term stimulators of cell activity, e.g. drugs. This can be assessed by measuring ATP, present as ATP Mg^{2-} at about 1–10 mM in all living cells. ATP also provides a quantitative assessment of bacteria in urine and other biological fluids, or bacterial contamination of foods. It can even be used to look for extraterrestrial life! However, ATP cannot be used, directly, to detect viruses, since these, so far as we know, do not contain ATP. Dead cells or cell fragments also contain little or no ATP.

Assays based on absorbance spectrophotometry, fluorimetry, atomic absorption spectrometry, and flame emission spectroscopy, and radioimmunoassay, are all apparently available for a wide range of analyses. Why then do we need chemiluminescence? We can begin to see the answer to this by examining first how small is the concentration or amount of substance, known as the analyte, we need to measure.

4.2.2 How little do we need to measure?

The concentration of an analyte in a biological sample can vary over some eleven orders of magnitude (Table 4.3), from 140 mM (1.4×10^{-1} M) for Na^+ to just a few pM (10^{-12} M) for some free hormones in the plasma. These substances can be divided conveniently into four main groups (Table 4.3), based on the concentration range into which they fall (mM, μM, nM, and pM where M = moles per litre). Whether there will be a need in the future for measuring analytes down to fM (10^{-15} M) is not yet clear. If substances are to be measured in the pM–fM range, then the assay must be capable of detecting 0.1 fmol (10^{-16} mol) to 0.1 amol (10^{-19} mol) in a sample volume of 0.1 ml. Similarly, to measure enzymes, ATP and other metabolites in small tissue samples weighing μg–mg, e.g. biopsies or isolated cell clusters such as pancreatic acini or individual islets of Langerhans, the assays

Table 4.3 — The range of analytes in human plasma

Analyte	Approx. plasma concentration	Analyte	Approx. plasma concentration
A. M–mM (10^{-4}–10^{-7} mol in 0.1 ml)		B. mM–μM (10^{-7}–10^{-10} mol in 0.1 ml)	
Na$^+$	0.14 M	Albumin	0.65 mM
Cl$^-$	0.1 M	Venous-	
HCO$_3$	30 mM	arterial O$_2$	40–100 μM
K$^+$	5 mM	Phenylalanine	125 μM
Glucose	5 mM	Anticonvulsants	20–500 μM
Urea	4 mM	Barbiturates	10–100 μM
Cholesterol	4 mM	IgG	85 μM
Ca^{2+}	2.5 mM	α, antitrypsin	50 μM
Triglyceride	1 mM	Ammonia	30 μM
		IgM	1.5 μM
		Antibiotics	1–20 μM
		Total Bilirubin	10 μM
C. μM–nM (10^{-10}–10^{-13} mol in 0.1 ml)		D. nM–pM (10^{-13}–10^{-16} mol in 0.1 ml)	
Complement proteins	0.1–1 μM	Vitamin B$_{12}$	0.4 nM
Oestriol	0.6 μM	Ferritin	0.2 nM
Cortisol (total)	0.4 μM	Insulin	0.15 nM
T4 (total)	0.1 μM	Aldosterone	0.12 nM
Thyroid binding		Parathyroid	
globulin	0.2 μM	hormone	0.1 nM
Creatine kinase	1 nM	Creatine kinase[†]	
Corticosterone		MB	75 pM
(total)	20 mM	Vitamin D	60 pM
T3 (total]	2 nM	Free T4	50 pM
Digoxin	1.5 nM	Free T3	10 pM
Prolactin	1 nM	ACTH	10 pM
Progesterone	1 nM	Thyroid	
Adrenaline	1 nM	stimulating	
		hormone	5 pM
		Oxytocin	1 pM

These figures are only approximate, and are intended to indicate the order of magnitude of each analyte. millimol = 10^{-3}, micromol = 10^{-6}, nanomol = 10^{-9}, picomol = 10^{-12}, femtomol = 10^{-15}, attomol = 10^{-18}, tipomol = 10^{-21}, impossomol = 10^{-24}.
M = moles per litre.
[†]Following myocardial infarction CKMB increases 10–100 $\times$. Ultimately it may be necessary to assay as many as 100 analytes in one blood sample (e.g. allergens).

must be sensitive at least down to the pmol (10^{-12} mol) to fmol (10^{-15} mol) range, and sometimes down to fmol to amol. Metabolites such as cyclic nucleotides are present at concentrations some 10^3 to 10^4 times lower than ATP. To detect live bacteria we need an ATP assay capable of measuring 10–100 attomol ATP (10^{-17}–10^{-16} mol), equivalent to 10–100 bacteria per millilitre. Ultimately we need to detect 1 eukaryotic cell (*ca.* 10^{-15} mol ATP) or 1 bacterium (*ca.* 10^{-18} mol) in a volume of some 1–10 ml.

Over the last ten years, there has been an increasing awareness of the need for the chemical analysis of single cells. The need arises not only because of the detection of small numbers of viable pathogens, but also because of the heterogeneity of cells within any given population. Conventional biochemical analyses of tissue extracts measure a mean of thousands, if not millions, of cells. Yet many cell preparations contain sub-populations which respond differently, a good example being found in populations of lymphocytes. The biochemistry of both eu- and pro-karyotes varies greatly during the growth/division cycle. Furthermore, there is increasing evidence that individual cells respond differently to stimuli, drugs and pathogens, in terms of the magnitude, the time course and dose–response relationship.

In its extreme form, the heterogeneity of response within a population of cells occurs via electrical and chemical 'thresholds' within individual cells (see Chapter 10 section 10.3.2 for an explanation of the chemisymbiotic hypothesis). Many examples of such threshold phenomena can be found in cell activation and cell injury (Table 4.4). The question then arises, does a particular primary stimulus act by 'switching on' different numbers of cells and at different times? At 50% of the maximal response of a population of cells, are all of the cells switched on to 50% of the individual maximum (a graded response), or have only 50% of the cells been switched on (a threshold response)? Does a secondary regulator or drug, which alters the magnitude or time course of the response in a cell population induced by the primary stimulus, act by altering the threshold level or time to threshold within any one cell? Alternatively, does it act by enhancing or inhibiting the cell's response once the threshold point has been crossed? Such thresholds can be provoked by an increase in intracellular Ca^{2+} and modified by cyclic nucleotides, or by diglyceride activation of protein kinase C. The characterisation of such thresholds holds the key to understanding the molecular basis of many examples of cell activation, including platelet aggregation and secretion, neutrophil activation, insulin action and secretion, as well as certain types of cell injury, such as complement-mediated cell lysis and non-lytic effects of the membrane attack complex of complement (a complex of the five components C5b6789n). The chemisymbiotic hypothesis provides a new conceptual framework for understanding immune and inflammatory based disease (see Chapter 10).

The identification and characterisation of threshold responses requires the chemical analysis of individual cells. Giant cells such as the nerve axons of the squid *Loligo forbesi* and the muscle fibres of the barnacle *Balanus nubilus* have long been a good target for the electrodes and microsyringes of the electrophysiologist. They can be up to 10–100 μl in volume (Table 4.5), providing ample material for microanalysis. Mammalian cells, however, can be some six to seven orders of magnitude smaller than this (Table 4.5). Whereas a giant cell can provide as much as 0.1–1 μmol of ATP, one platelet may contain only 100 amol. The cyclic AMP content of a platelet

Table 4.4 — Some examples of threshold phenomena

Phenomenon	Example	Primary stimulus	Secondary regulator
A. *Cell activation*			
Cell movement	Skeletal muscle	Action potential	?
	Heart muscle	Action potential	ACh (i), adrenaline (a)
	Smooth muscle	Neurotransmitter	Eicosanoids (a or i)
	Paramecium (cilia)	Collision	?
	Neutrophil chemotaxis	C5a	Eicosanoids (a or i)
Vesicular secretion	Insulin from β cell	Glucose	Adrenaline (i)
	Neurotransmitter from nerve terminal	Action potential	Adenosine (i)
Vision	Retinal rod	Photon	Dark adaptation (a)
Fertilisation	Egg	Sperm	Eicosanoid (a)
Cell transformation	Lymphocyte	Antigen	?
Reversal of intermediate metabolism	Hepatocyte	Insulin/glucagon	Corticosteroids (a)
Bioluminescence	Firefly	Neurotransmitter	?
Cell fusion	Viral induced	Sendai	?
B. *Cell injury*			
Cell death	Complement lysis	Membrane attack complex	Adenosine (a)
	Toxins	Phalloidin	?
Viral infection	Neutrophil attack	$(O_2^-, OCl^-, {}^1O_2)$ Viral protein	Adenosine (i)
Cell transformation	Cancer	Virus	?
Inherited disorders	Phenylketonuria	—	?
Autoimmune disease	Grave's	TSH receptor antibody	?
Cell shape change	Sickle cell anaemia	Abnormal haemoglobin	?

i — inhibitor, a — activator.

Table 4.5 — The volume of some invertebrate and vertebrate cells

Organism (systematic name)	Cell type	Dimensions	Cell volume
Squid (*Loligo forbesi*)	Giant nerve axon	> 1 cm long, 1 mm diam.	60 μl
Barnacle (*Balanus nubilus*)	Giant muscle	2 cm long, 1 mm diam.	12 μl
Sea hare (*Aplysia californica*)	Giant neuron	> 1 cm long, 0.4 mm diam.	1.2 μl
Sea urchin (*Arbacia punctulata*)	Egg	100 μm diam.	0.5 μl
Frog (*Rana pipens*)	Skeletal muscle	7 mm long, 100 μm diam.	70 μl
Starfish (*Marthasterias glacialis*)	Oocyte	160 μm diam.	1.8 ml
Frog (*Rana pipens*)	Cardiac muscle	250 μm long, 10 μm diam.	5–10 pl
Rat (*Rattus norvegicus*)	Hepatocyte	20 μm diam.	4 pl
Human (*Homo sapiens*)	Neutrophil	13 μm diam.	0.45 pl
Human (*Homo sapiens*)	Lymphocyte	10 μm diam.	0.23 pl
Human (*Homo sapiens*)	Erythrocyte	7 μm radius	86 fl
Human (*Homo sapiens*)	Platelet	1 μm radius	10 fl

Volume = $r^2 \times$ length or $\frac{4}{3} \pi r^3$ except for human cells where figures represent mean measured values (i.e. not perfect spheres). micro = 10^{-6}, n = nano = 10^{-9}, p = pico = 10^{-12}, f = femto = 10^{-15}.

would be some three to four orders of magnitude less than this! Since no unit appeared to exist to represent 10^{-21} mol (or 6022 molecules) I suggested, some years ago, affectionately in my laboratory that we use a Welsh word, tipyn = little to derive a tipomol, and that a good name for 10^{-24} mol might be an impossomol!!

Returning to the more serious vein, these tiny values for enzymes and metabolites in the femto–tipo mol range for single mammalian cells (Table 4.6) clearly require a very special method of analysis since this is far beyond the capacity of conventional spectrophotometric assays.

4.3 SENSITIVITY AND SPEED
Most people would be prepared to accept an absorbance change of 0.005 as being near to the detection limit of the average spectrophotometric assay, assuming no light scattering and an instrument in good order. Since

$$\text{absorbance (A)} = \varepsilon . c . 1 \tag{4.3}$$

we can calculate from this approximately the detection limit of a spectrophotometric assay based on NADH.

$$\varepsilon_{340} \text{ for NADH} = 6000 \, M^{-1} \, cm^{-1}$$

Table 4.6 — Some approximate values for analytes in single mammalian cells

Substance	Cell	Assumed intracellular concentration	Approx. content per cell (mol)
Total K$^+$	Hepatocyte	130 mM	520 fmol (5.2×10^{-13})
Total Ca^{2+}	Hepatocyte	2 mM	8 fmol (8×10^{-15})
Free Ca^{2+}	Hepatocyte	0.1 μM	0.4 amol (4×10^{-19})
ATP	Hepatocyte	10 mM	40 fmol (4×10^{-14})
ATP	Neutrophil	2 mM	1 fmol (10^{-15})
ATP	Platelet	10 mM	100 amol (10^{-19})
AMP	Hepatocyte	0.2 mM	0.8 fmol (8×10^{-16})
Cyclic AMP	Hepatocyte	1–10 μM	4–40 amol ($4\text{–}40 \times 10^{-18}$)
Cyclic AMP	Neutrophil	0.1–1 μM	45–450 tmol ($45\text{–}450 \times 10^{-21}$)
Cyclic AMP	Hepatocyte	0.1–1 μM	0.4–4 amol ($0.4\text{–}4 \times 10^{-18}$)
Free SH groups	Hepatocyte	25–250 μM	0.1–1 fmol ($10^{-16}\text{–}10^{-15}$)
Lactate dehydrogenase	Hepatocyte	10 μM	0.4 fmol (4×10^{-15})
Calmodulin	Hepatocyte	10 μM	40 amol (4×10^{-17})
Phosphorylase	Hepatocyte	1 mM	4 amol (4×10^{-18})
Cyclic AMP protein–kinase	Heart muscle	0.7 μM	3 amol (3×10^{-18})
Haemoglobin	Erythrocyte	5.3 mM	0.45 fmol (4.5×10^{-16})
Haem	Fibroblast	80 μM	7 amol (7×10^{-18})
Oxygen consumption	Neutrophil (resting)	?	0.1 fmol . min^{-1}
Oxygen consumption	Neutrophil (activated)	—	1 fmol . min^{-1}
H$_2$O$_2$	Hepatocyte	10 nM	40 tmol (4×10^{-20})
O$_2^-$	Hepatocyte	10 pM	40 imol (4×10^{-23})!!
O$_2$ consumption	Hepatocyte	?	0.1 fmol . min^{-1}
DNA	Hepatocyte	—	30 pg (ca. 6×10^{10} nucleotides)
RNA	Fibroblast	—	30 pg (rRNA 71%; tRNA 15%; mRNA 3%)
Total protein	Hepatocyte	—	1.7 ng

These figures are approximate and are only intended to give an an indication to the values in a single cell. For volume of cell see Table 4.5.

f = femto = 10^{-15}, a = atto = 10^{-18}, t = tipo = 10^{-1}, i = imposso = 10^{-24}.

Therefore the detection limit $\doteq$ 1 μM (or 1 nmol in 1 ml) for 1 cm cuvette. Extinction coefficients of useful compounds for biological assays are mainly in the range of 1000–100000. Thus, spectrophotometry may be adequate for analyses in group A of Table 4.3 and for some in group B, but it falls short of the sensitivity required for most hormones, drugs, and vitamins, and for single cell or tissue biopsy analysis.

Fluorimetry can increase the sensitivity some 10–100 times compared with absorbance spectrophotometry. Thus, NADH can be easily detected down to 0.1 μM. Fluorescein, with its superior extinction coefficient of $ca.$ 70000 M^{-1} cm^{-1} and high fluorescence quantum yield ($ca.$ 50%) can be detected down to 1–10 nM. This is only sensitive enough for single cell analysis if the volume of sample is very small ($ca.$ nl–pl). This can be achieved in the fluorescence activated cell sorter, with the use of a laser light source.

As with any assay technique, fluorescence is limited by the signal from the 'blank', i.e. the complete reaction mixture without added analytes. The blank signal in a fluorescence assay can arise from any one of at least five sources:

(1) Native fluorescence of the 'blank' solution, including the solvent
(2) Fluorescent impurities in the reagents and solvent.
(3) Fluorescence of the cell housing.
(4) Rayleigh light scattering, without change in wavelength, a problem if emission and extinction spectra are close (i.e small Stokes shift).
(5) Raman light scattering (with change in wavelength).

With care, many of these can be minimised, but several, particularly the endogenous fluorescence of the solvent, even H$_2$O, and Raman scattering, cannot. Thus, it is rarely possible in most fluorimetric assays to achieve greater sensitivity than 1–10 nM (i.e. 1–1 pmol in 0.1 ml).

The fluorescence lifetime of fluorescein is about 4 ns, close to that of the 'blank' signal lifetime. Phosphorescent, and certain fluorescent, compounds with longer lifetimes (e.g. μs-ms or even seconds) enable the fluorescent signal from the solvent and contaminants to be ignored by detecting the phosphorescence after switching off the incident light, or by using time-resolved fluorescence. This has become more feasible with the advent of lasers. However, apart from one or two special cases (see Chapter 8), phosphorescent or slow-decaying fluorescent compounds do not have the necessary chemistry to couple them to analytes of biological interest.

Lowry ingeniously attempted to improve the sensitivity of fluorescence assays by introducing an enzyme cycle as an amplifier. The success of the method depends on the acid lability of NADH and alkali lability of NAD. For example:

A.   Initial step

$$\text{Analyte} + \text{NADH} \rightarrow \text{NAD} + \text{reduced analyte} \qquad (4.4)$$

Excess NADH is removed by strong acid, leaving the NAD.

B.   Cycle

$$\text{NAD}^+ + \text{lactate} \rightarrow \text{NADH} + \text{pyruvate} + \text{H}^+ \qquad (4.5)$$

$$\text{NADH} + \alpha \text{ oxoglutarate} + \text{NH}_4^+ \rightarrow \text{NAD}^+ + \text{glutamate} \qquad (4.6)$$

Tens of thousands of pyruvate and glutamate molecules accumulate for every NAD molecule present per hour.

After stopping the reaction, pyruvate is measured by adding excess NADH plus lactate dehydrogenase. The resulting NAD^+ is measured by converting it to a highly fluorescent compound in strong alkali.

C. Final assay

$$\text{pyruvate} + \text{NADH} \rightarrow \text{lactate} + \text{NAD}^+ \tag{4.7}$$

$$\text{NAD}^+ + \text{OH}^- \rightarrow \text{highly fluorescent product} \tag{4.8}$$

Although some of these cycling assays enable metabolites to be measured in the range of 0.1–10 pmol, serious problems arise in others because of contaminants in the reagents or the sample. This is exacerbated by the large amplification factor essential for the method, which may be as much as 15 000 times. The result is an unacceptably high blank reading, reducing the sensitivity of the assay well below theoretical expectations. Furthermore, these assays are tedious to set up, requiring experienced and skilled hands because of the tiny initial volumes (0.1–1 μl). This requires that the first reaction be carried out under oil, and great care is needed to minimise contamination.

In contrast, a chemiluminescent compound with a quantum yield (ϕ_{CL}) of 10% has the capacity to generate 6×10^{22} photons per mol. If the chemiluminometer has an overall efficiency of 10% (photomultiplier + electronics + geometry, see Chapter 2, section 2.7) and a machine background of 10 counts per second, then 1 cps above background is detectable, and is equivalent to *ca.* 0.2 tipomol (2×10^{-22} mol), provided that the reaction is fast enough to be complete in about 10 s — an incredible sensitivity relative to conventional spectrophotometry of fluorimetry.

The sensitivity of detection of a chemiluminescent indicator may, however, not quite achieve this, for a number of reasons:

(1) $\phi_{CL} < 10\%$.
(2) Rate of reaction slow ($t_{\frac{1}{2}} > 1$ s).
(3) Chemiluminometer efficiency $< 10\%$, usually because of poor geometry.
(4) Luminescence blank, for which there are three main sources:

    (a) phosphorescence of tubes or media,
    (b) an ultraweak chemiluminescence source (see Chapter 6),
    (c) chemical blank.

Nevertheless, several synthetic chemiluminescent compounds, for example phthalazine diones and acridinium salts (Fig. 4.2) can be detected in the 1–100 amol range, and potentially down to 0.1 amol, whereas some bioluminescent substances, for example the coelenterate photoproteins aequorin and obelin, can be detected down to 1 tipomol (10^{-21} mol).

This exquisite sensitivity may not always be realised in an analyte assay because of other limitations in the reagents or increased luminescent blank. For example, using the firefly luciferin–luciferase reaction, it is difficult to detect < 10 amol ATP. However, by careful selection of the chemiluminescent reaction, with a high quantum yield and fast kinetics, together with the use of highly pure reagents, assays

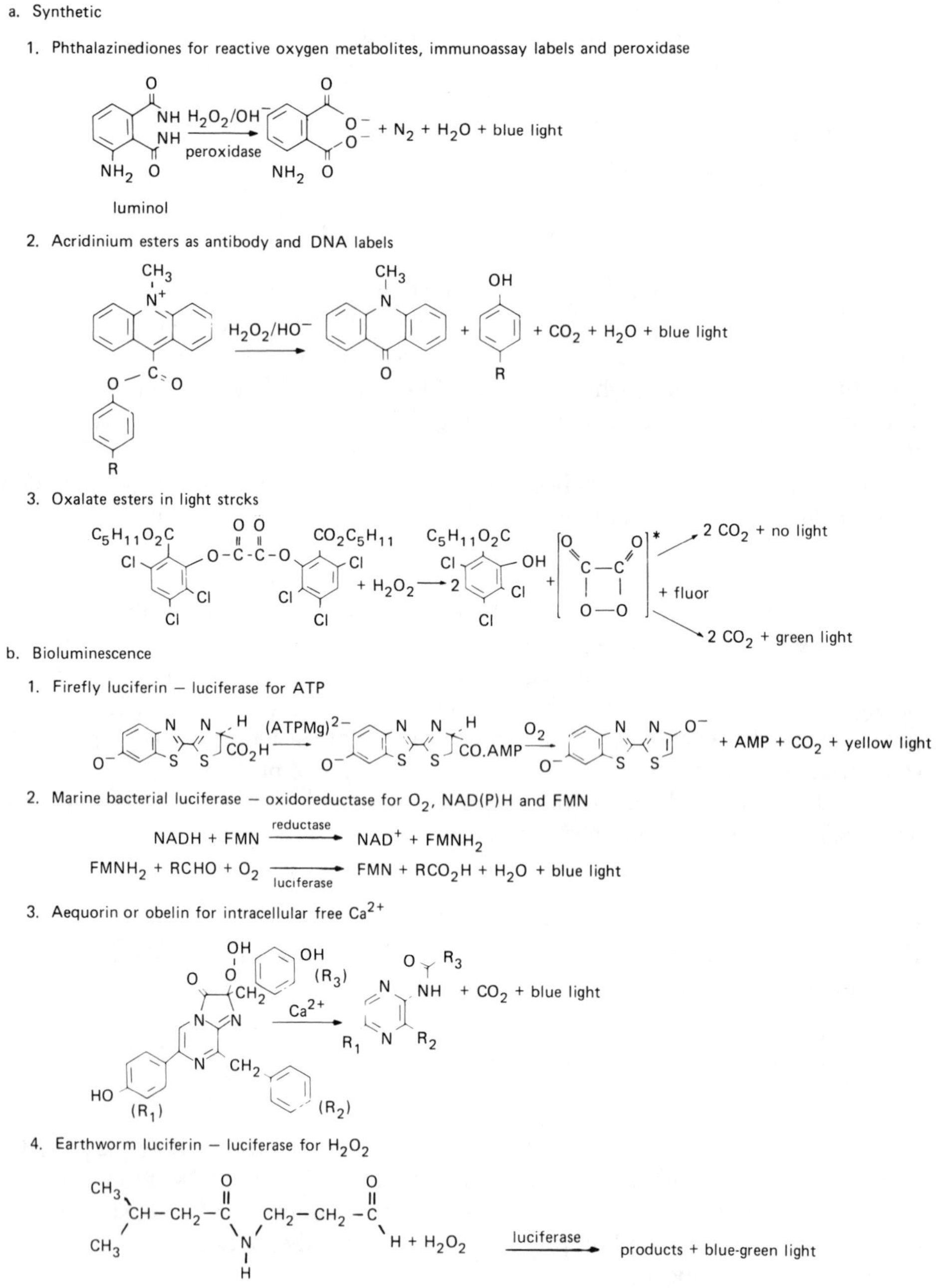

a. Synthetic

1. Phthalazinediones for reactive oxygen metabolites, immunoassay labels and peroxidase

$$luminol \xrightarrow[\text{peroxidase}]{H_2O_2/OH^-} \cdots + N_2 + H_2O + blue\ light$$

luminol

2. Acridinium esters as antibody and DNA labels

$$\xrightarrow{H_2O_2/HO^-} \cdots + CO_2 + H_2O + blue\ light$$

3. Oxalate esters in light strcks

$$C_5H_{11}O_2C \cdots + H_2O_2 \longrightarrow 2 \cdots + \left[\begin{array}{c} O \quad O \\ C-C \\ O-O \end{array}\right]^* \begin{array}{l} \nearrow 2\ CO_2 + no\ light \\ + fluor \\ \searrow 2\ CO_2 + green\ light \end{array}$$

b. Bioluminescence

1. Firefly luciferin – luciferase for ATP

$$\xrightarrow{(ATPMg)^{2-}} \cdots CO.AMP \xrightarrow{O_2} \cdots + AMP + CO_2 + yellow\ light$$

2. Marine bacterial luciferase – oxidoreductase for O_2, NAD(P)H and FMN

$$NADH + FMN \xrightarrow{reductase} NAD^+ + FMNH_2$$

$$FMNH_2 + RCHO + O_2 \xrightarrow{luciferase} FMN + RCO_2H + H_2O + blue\ light$$

3. Aequorin or obelin for intracellular free Ca^{2+}

$$\xrightarrow{Ca^{2+}} \cdots + CO_2 + blue\ light$$

4. Earthworm luciferin – luciferase for H_2O_2

$$\cdots H + H_2O_2 \xrightarrow{luciferase} products + blue\text{-}green\ light$$

Fig. 4.2 — Chemi- and bio-luminescent reactions in biology and medicine.

can be established with sufficient sensitivity 'for measuring many enzymes and metabolites in single cells. Incorporation into cells of the genes or mRNA coding for bioluminescent proteins enables each cell to synthesise its own chemiluminescent indicator. Light yields of several thousand photons per second can be generated in this way, sufficient to detect one cell. A further consequence of this high sensitivity is the ability to measure the analyte quickly. The peak of chemiluminescent emission usually occurs with 1–250 ms, where the signal : noise is greatest, the reaction being complete within 1–10 s. This provides the potential for measuring very large numbers of samples quickly (e.g. 1000's per hour), given the appropriate mechanical engineering.

4.4   CHEMICAL EVENTS IN LIVING CELLS

Physiologists have long recognised the need to study electrical phenomena whilst the cell, or the cell membrane at least, remains essentially intact. Breaking the cell membrane destroys its selective ionic permeability and the ion gradients across it. The membrane potential and excitability are then lost. Yet the biochemist has been happy to use the 'grind and find' approach for nearly one hundred years, to measure enzymes and metabolites in tissue extracts. Gowland Hopkins, one of the founders of modern biochemistry, argued during the first quarter of this century that this approach was justified, provided that the event which occurred in the intact cell also occurred in the broken cell preparation, or that the extract reflected directly an event which had previously taken place in the living cell. Much has been learned by using this approach, enabling metabolic pathways to be defined and the effects of cell stimuli, pathogens, and drugs on them to be characterised. However, it has become clear, over the last twenty years or so, that some chemical events are difficult, or even impossible, to reproduce in broken cells. Hopkins's criterion is therefore not satisfied.

The ability of a primary stimulus to activate a cell through an increase in intracellular Ca^{2+}, released from an internal store or through the cell membrane, can only be properly characterised whilst the electrochemical gradient of Ca^{2+} across the cell membrane remains intact (see Chapter 7). The 'oxidase' responsible for generating reactive oxygen metabolites in phagocytes is extremely unstable, and attempts to activate it, or even to maintain its activated state after cell stimulation, have met with great difficulties or have been altogether unsuccessful (see Chapter 6). Three possible reasons for this are first — the need for intracellular messengers produced by the living cell, second — the need for membrane fusion and recruitment of 'oxidase' components for full activation, and third — conditions within the cell necessary for stabilisation of the enzyme complex are not yet understood.

A further argument for studying chemical events in the intact cell is the need to maintain the polarity of components, or the movement or organelles and molecules within the cell. All cells exhibit some polarity. Organelles or pathways may be unevenly distributed throughout the cell; a primary stimulus may act at one surface and provoke an event at another. Examples of polarisation in cell activation can be found in the exocrine secretion of the pancreas, or the exocytotic burst following fertilisation of some eggs. Cells such as lymphocytes and neutrophils have the capacity to undergo capping, where proteins in the cell membrane congregate at one end of the cell. If the mechanisms underlying these polarisable events are to be fully elucidated, it is essential that distributional changes in intracellular signals, as well as cellular polarisation of metabolic pathways and individual enzymes, are identified and quantified. They must then be related directly to where the physiological or pathological event has occurred in the cell. This cannot be done once the cell has been broken.

Intracellular signals provide the link between an event at the plasma membrane and a chemical or morphological event within the cell. Three groups have been discovered:

(1) Cations — e.g. Ca^{2+}, H^+, Na^+ and possibly K^+ and transition metals.
(2) Nucleotides — e.g. AMP, GTP, cyclic nucleotides.
(3) Lipid derivatives — e.g. phosphatidyl inositol producing inositol trisphosphate + diacyl glycerol and other inositol derivatives.

Eicosanoids and related compounds have been proposed as intracellular signals, but the evidence is not convincing. A cornerstone of the experimental investigation of the chemisymbiotic hypothesis (Chapter 10, section 10.3.2) is the ability to measure intracellular signals, energy balance and threshold end-responses in the same individual cells.

This argument is not intended to encourage a return to the organo- or chemo-vitalist philosophies of the nineteenth century. Rather, it is intended to highlight the long-term objective of the biochemist — to understand fully the chemistry of the living cell and whole tissue. This philosophy is embodied in a well known nursery rhyme:

> All the King's horses and all the King's men
> Couldn't put Humpty Dumpty together again!

Chemiluminescence provides a technique for detecting, quantifying, and visualising chemical events in living cells. Chemiluminescent indicators have the necessary sensitivity and can produce a detectable signal from within, without the need for exposure to a high energy light source which may heat up the cell or result in damaging photochemical reactions. Chemiluminescence offers a more sensitive alternative to fluorescence or nuclear magnetic resonance indicators for studies in intact cells. But methods are required to incorporate the chemiluminescent indicators into living cells. Six techniques are available (Fig. 4.4; Chapters 7 and 10):

(1) Diffusion of extracellular indicator through the cell membrane, only relevant for lipophilic molecules.
(2) Microinjection.
(3) Permeabilisation of the cell membrane, for example by ghosting, swelling, or electric shock.
(4) Vesicle–cell fusion, e.g. vesicle = erythrocyte 'ghost' or liposome.
(5) Release from intracellular micropinocytotic vesicles filled with indicator from the extracellular fluid.
(6) Transformation, transduction, or conjugation of genetic material coding for bioluminescent proteins into prokaryotic cells, or transfection into eukaryotic cells.

(N.B. Transformation of bacteria is the acquisition of new genetic material e.g. in a cDNA; transduction is the transfer of a gene to a bacterium in a bacteriophage; conjugation is the transfer of chromosomal material from one bacterium to another; transfection of eukaryotic cells is the acquisition of new genetic material via DNA.)

4.5   THE REPLACEMENT OF RADIOACTIVE ISOTOPES

Since the late 1940s, radioactive isotopes have provided the biochemist with unique, highly sensitive probes for studying biological process, both within and outside cells. ^{14}C and ^{3}H labelled substances such as glucose, amino acids, and fatty acids have enabled metabolic pathways and the mechanism of protein synthesis to be defined. ^{32}P has enabled phospholipid and nucleic acid metabolism to be studied in cells, and has been esssential for the development of recombinant DNA technology. ^{125}I and ^{131}I have proved valuable labels for tagging ligands and for studying antibody–antigen reactions, for example, in immunoassay (see Chapter 8).

The detection limit for a radioactive isotope can be estimated from the half time of its decay (Table 4.7). If A_0 = amount of pure isotope at time O and A_t = amount of isotope at time t, then the familiar equation for decay is:

$$A_t = A_0 \exp(-kt) \tag{4.9}$$

$$k = \text{decay rate constant} = 0.693/t_{\frac{1}{2}} . \tag{4.10}$$

At time t the number of disintegrations per second (dps) = dA_t/dt

$$\tag{4.11}$$

From (4.9)

$$dA_t/dt = -kA_0 \exp(-kt) , \tag{4.12}$$

and at time 0

$$dA_t/dt = -kA_0 . \tag{4.13}$$

Table 4.7 — Detection limits for radioisotopes and non-isotopic labels

Label	Particle	$t_{\frac{1}{2}}$	Approx. detection limit (mol)†
Isotopic			
^{14}C	β-	5568 years	4×10^{-13}
^{3}H	β-	12·26 years	10^{-15}
^{125}I	γ	60 days	10^{-17}
^{32}P	β	14.4 days	3×10^{-18}
^{131}I	βi, γ	8.09 days	2×10^{-18}
Non-isotopic‡			
Enzyme	β galactosidase		10^{-13}
Fluor	Fluorescein		10^{-14}
Chemiluminescent	acridinium ester		10^{-17}

† Assuming an ability to detect 10 dps over background.
‡ Assuming a volume of 0.10^{-1} ml.

If the detection limit is equivalent to 1 dps, then

$$A_0 = \text{detection limit} = 1/k = t_{\frac{1}{2}}/0.693 \text{ molecules} \qquad (4.14)$$

where $t_{\frac{1}{2}}$ is in seconds.

However, $t_{\frac{1}{2}}$ is usually at least days or years for a usable isotope, and the

$$\text{Avogadro constant} = 6.023 \times 10^{23},$$

then the approximate detection limit for any isotope $= (2.07 \times 10^{-19}/t_{\frac{1}{2}})$ and where $t_{\frac{1}{2}}$ is in days.

For ^{125}I, $t_{\frac{1}{2}} = 60$ days,

Therefore the approximate detection limit for ^{125}I $= 12$ amol (1.2×10^{-17} mol). Thus, when using iodine 125 its detection limit is not a limiting factor for the majority of analytes shown in Table 4.3.

In the absence of a chemical amplifier, such as an enzyme, an assay can never be more sensitive than the detection limit of the indicator or tracer. In practice the assay is usually considerably less sensitive, because of limitations imposed by the other reagents or problems of reagent blank. Nevertheless, the detection limit of the isotope gives us a guide to the ultimate sensitivity attainable. For iodine, it is within the limitations of most of the analytes required from plasma or single cells. However, ^{14}C or ^{3}H are much less sensitively detectable than ^{125}I (Table 4.7). As a result, many assays utilising them do not have sufficient sensitivities for measuring key analytes in plasma, small tissue samples, or single cells. Thus, in spite of the undoubted success of radioisotopes, they have a number of disadvantages:

(1) Hazard — All radioisotopes are potentially hazardous. Iodine readily accumulates in the thyroid gland. Legislation in many countries is now limiting their use to specialised laboratories with the necessary safety procedures, contamination screening, and disposal facilities.

(2) Production of isotope — This is expensive and requires highly specialised equipment such as a cyclotron. Furthermore, the isotopes with the best sensitivity are the ones with the shortest half lives, and have to be made in batches at regular intervals since the decay leads inevitably to loss of activity.

(3) Limited shelf life of labelled derivatives — Production is rarely done in large batches since not only is the isotope continuously decaying, but the decay often leads to radiolytic damage of the reagent molecules. ^{125}I-labelled polyclonal antibodies survive only a few weeks, whereas ^{125}I monoclonal antibodies often do not survive freeze-thawing.

(4) Sensitivity and speed — Even when using the high sensitivity isotopes ^{32}P or ^{125}I, an increase in sensitivity would be desirable. This should extend the analyte range down to tipomols (10^{-21} mol). It also should reduce assay times to seconds, from the many minutes often required to produce statistically acceptable results.

(5) Homogeneous ligand: ligand interactions — To quantify antibody–analyte binding or receptor–agonist interacting using radioisotopes, a step is required to separate bound from free ligand. Not only does this limit investigations to slowly dissociating ligand–ligand interaction with $K_D < 10^{-9}$ M, but also the separation step is laborious and introduces imprecision. A homogeneous method,

not requiring a separation step, would be very convenient and would enable rapidly equilibrating systems to be quantified, e.g. enzyme–substrate or inhibitor interactions, or rapid protein–protein interactions, or RNA–DNA and DNA–DNA hybridisation in free solution.

Chemiluminescent substrates are apparently safe, can be synthesised in any laboratory in gram quantities, and are stable for years on a laboratory shelf in the absence of the triggering conditions. Yet they are equivalent to a very fast decaying isotope under optimal conditions (Table 4.7; $t_\frac{1}{2}$ 0.1–1 s). All of the label can thus be quantified in a few seconds, and potentially in a few milliseconds (the time to peak height). A detection limit in the range 10^{-15}–10^{-18} mol is easily achievable, and in some cases down to 10^{-20}–10^{-21} mol. The 'specific activity' of a chemiluminescent labelled reagent can be increased without the problem of radiolytic damage, which usually limits ^{125}I to 1 mol per mol. Chemiluminescent compounds can be readily coupled to biological molecules by standard mixed anhydride, carbodiimide, isothiocyanate, N-hydroxy succinimide, or imidate coupling methods. Furthermore, ligand–ligand interactions can be quantified 'homogeneously' by chemiluminescent quenching, enhancement or energy transfer (see Chapters 8 and 9).

There are, however, two potential problems with chemiluminescent indicators. Firstly, the assay is transient. Each chemiluminescent molecule can produce a photon only once. After the reaction has been triggered it cannot be repeated with that tube, whereas an isotope label can be recounted, or a fluorescent substance re-excited. Secondly, the triggering conditions often produce a 'chemical blank' which reduces the detection limit. For example, luminol (Fig. 4.2) generates *ca.* 6×10^{21} photons per mol. Using micro- peroxidase and H_2O_2 at pH 13, one attomol (10^{-18} mol) will generate 6000 photons within 10 seconds. However, the chemical blank may be as much as 500 000 photons in the same time. If we accept a detection limit of 2 standard deviations above this blank, this means that the practical detection limit for luminol is *ca.* 10 amol under these conditions. Acridinium esters (Fig. 4.2) produce approximately the same number of photons per mol as luminol, but the conditions for triggering their chemiluminescence are simpler ($H_2O_2 + OH^-$) and the reaction faster, resulting in a 10 to 100-fold reduction in chemical blank and thus an improved detection limit.

4.6   THE MAJOR BIOMEDICAL APPLICATIONS OF CHEMILUMINESCENCE

Five main areas of chemiluminescence analysis can be identified, not only where chemiluminescent assays have been established (Table 4.2) but also where these have been used either to investigate a biological problem or to make a clinical decision. They are:

(1) Measurement of enzymes and metabolites in small tissue extracts using firefly luciferin–liciferase (ATP), or NAD(P)H using the oxidoreductase–luciferase from marine bacteria (see Chapter 5)
(2) Measurement of reactive oxygen metabolites produced by cells, tissue extracts, and enzymes using endogenous ultraweak chemiluminescence or the addition of luminol, lucigenin or pholasin (see Chapter 6).

(3) Measurement of intracellular free Ca^{2+} using the coelenterate photoproteins aequorin or obelin (see Chapter 7).

(4) Non-isotopic labels in immunoassay and recombinant DNA technology (see Chapters 8 and 9).

(5) Genetic engineering using bioluminescence genes; bacterial lux and firefly luciferase genes (see Chapter 5), for detection of microbes and gene expression following transformation, transduction, conjugation or transfection.

Before examining how these applications can be realised, it is important to remind ourselves of two vital differences between assays based on absorbance or fluorescence on the one hand, compare with those based on chemiluminescence on the other.

4.7 TWO IMPORTANT FEATURES OF CHEMILUMINESCENCE

In spectrophotometric or fluorimetric analysis the light source is continuous and the analyte *concentration* is directly proportional to absorbance (Beer–Lamberts law) or to the fluorescence signal respectively. For an increase in product formation a gradual increase in signal will be recorded on the chart recorder (Fig. 4.3A), or conversely a gradual decrease if a reduction in substrate concentration is being detected. As the number of excitable molecules change, so the number of excited molecules increases or decreases accordingly.

In contrast, the signal from a chemiluminescence assay, the photon, is transient. Like the other two analytical procedures, light intensity (dhv/dt) from the reaction cuvette is the parameter being detected. However, unlike the other two, where the molecules are being excited by continuous irradiation, excitation of a particular molecule formed by a chemiluminescent reaction occurs only once during the course of the assay or experiment. The signal recorded by the chart recorder in a chemiluminescent is thus a differential (Fig. 4.3B) of that recorded when using absorbance of fluorescence. A steady state enzyme rate, under conditions where there is insignificant consumption of the chemiluminescent substrate, will generate a constant light intensity (trace B in Fig. 4.3b). In contrast, direct coupling of a metabolite to the chemiluminescent substrate will result in a trace similar to trace *A* (Fig. 4.3b), where a peak in light intensity is reached, followed by a decay. This decay will only be exponential if either the analyte or chemiluminescent substrate is consumed significantly during the reaction, but not if both are consumed concomitantly. A continuously increasing trace can be generated by plotting the integral of light emission versus time. Sometimes conditions can be arranged so that the chemiluminescence emission is directly proportional to metabolite concentration. Thus firefly luciferin–luciferase can generate continually increasing or decreasing luminescence traces as ATP is produced or degraded respectively.

The other distinctive feature of chemiluminescence analysis is that light intensity has no volume term incorporated in it, unlike molarity. The chemiluminescent measures light intensity and total light yield. The latter is related directly to the number of chemiluminescent *molecules* consumed, and not to the molarity (strictly molality) of this compound. In fluorescence, 1 mol in 0.1 ml will generate a signal F, whereas the same sample in 1 ml will generate a signal $F/10$. In contrast, if this compound were chemiluminescent, assuming the geometry of the apparatus is unaffected by sample volume, the total light yield will be independent of concentration. However, light emission in a chemiluminescence assay can be made to be related to analyte concentration. This can be achieved through a change in reaction kinetics, for example because of the *Km* of an enzyme. Two important features of chemiluminescent analysis therefore are:

(1) The signal is transient.
(2) The signal is proportional to the absolute number of chemiluminescent molecules reacting, not their molar concentration.

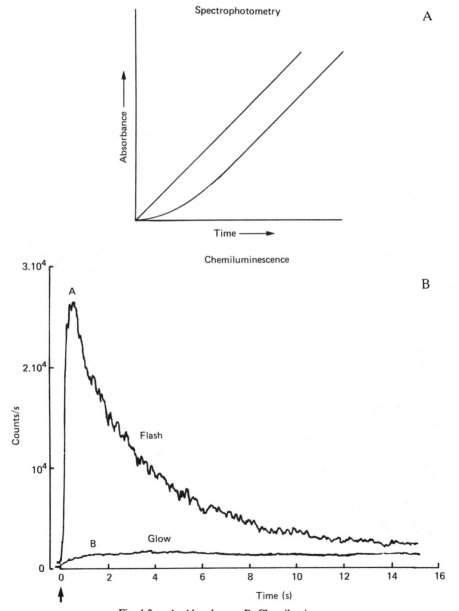

Fig. 4.3 — A. Absorbance. B. Chemiluminescence.

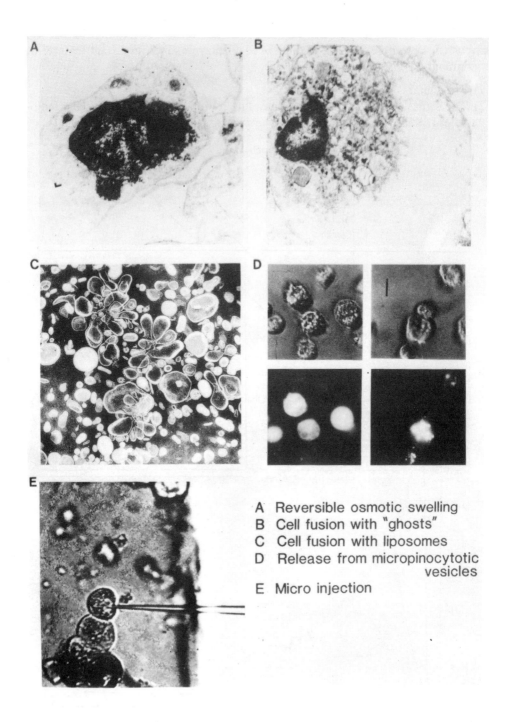

Fig. 4.4 — Incorporation of chemi- and bio-luminescent indicators into living cells. A–D = methods used by the author. E = microinjection method of Cobbold *et al.*

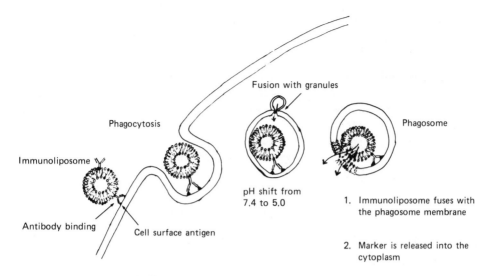

Fig. 4.5 — Efficient delivery of chemi- and bio-luminescent indicators by pH-sensitive or temperature sensitive immunoliposomes reprinted by permission from Campbell *et al.* (1988) *Biochem. J.* 252, 143–150, copyright ©1988 The Biochemical Society London.

4.8 APPLICATIONS IN MICROBIOLOGY, INDUSTRY, AND BIOTECHNOLOGY

Detection of small numbers of microorganisms is vital both in medicine and in industry. These include bacteria, viruses, protozoa, algae, moulds and yeasts. In medical microbiology particular current problems exist in the rapid detection, i.e. within 1 h, of *E.coli* in urinary tract infections causing bactiuria, of *Shigella* causing dysentery and of *Salmonella* in food poisoning. Concomitant assessment of the sensitivity of these bacteria to particular antibiotics is also required for successful therapy. Specific assays are required for each bacterium down to 10–100 per mol of sample. Culturing using selective media is not very sensitive nor specific, and takes time, often a day or more. Likewise in the food industry active *Salmonella* at a level of 1 organism per 25 g is significant. Present assays requiring culturing take at least 24–48 hours and sometimes 4–7 days, by which time the food may be on the shelf! Sensitive detection of bacterial contaminants is equally important in the dairy industry, whilst in brewing the major problem is persisting live yeast. Microorganisms cause costly havoc as 'spoilers' of many man made products. Moulds such as *Penicillium* and fungi such as *Aspergillus* can ruin textiles, or even painting masterpieces and buildings. Detection and quantification of microorganisms plays a crucial role in waste disposal, and in checking water purity. Would that the Third World had a single, specific test for pathogenic protozoa and bacteria in water supplies.

Let us not forget about viruses. Current concern with HIV (AIDS) is obvious. But how many realise that the test for AIDS relies mainly on detecting antibodies once the infection has taken hold. A sensitive test for antigen i.e. the virus itself, prior to clinical manifestations of the disease would enable intervention to be made at the earliest possible time.

Growth of microorganisms or viruses using selective media or cells are insensitive, lack specificity and take too long. Three current procedures attempt to circumvent these problems, all couplable to chemiluminescence analysis:

(1) A specific antibody used to extract a particular bacterium or eukaryotic microorganism from a 'soup', and then assay live cells using ATP (immunoanalysis), or develop an immunoassay. The latter will also detect fragments, and may be useful for toxin detection also.
(2) The detection of specific bacterial genes or RNA using DNA probes.
(3) The detection of bacteria using a labelled phage or a phage containing bioluminescent genes within its DNA.

The latter approach developed by Ulitzur and colleagues in Israel is a particularly exciting development. It is very sensitive (down to 1–10 bacteria per 1–10 ml), highly specific for the bacteria, homogeneous, produces no false positives, and is relatively quick. Antibiotic sensitivity can be assessed simultaneously (see Chapter 5).

The immunoassay market worldwide is already at the level of billions of dollars per year. Recombinant DNA technology and genetic engineering are on their way to equally vast new markets. The strict budgeting in medicine and industry requires that assay methodology not only works but is cheap. An immunoassay in clinical biochemistry or medical microbiology which costs more than $2 per tube is too expensive, whilst one costing only 20 cents per tube will be highly competitive. The speed, sensitivity and potential cheapness of chemiluminescence analysis is providing a spur to the ingenuity of the biotechnologist not only keen to make a fast buck but also concerned with the quality of his product. The commercial exploitation of chemi- and bio-luminescence requires original science, coupling equipment to the appropriate biochemistry and chemistry, protection of inventions with patents, identification of the markets, cost effectiveness and good marketing. Quite a combination if your biotechnology company is to be a success!

We now have the framework to examine the detail of chemiluminescence as a unique analytical tool in cell biology, medicine and biotechnology.

4.9 RECOMMENDED READING

Campbell, A. K. (1987) *The University of Wales Science and Technology Review* **1** 38–45. Living light: from the ocean depths to the hospital bed.
Campbell, A. K., Hallett, M. B., & Weeks, I. (1985) *Methods Biochem. Anal.* **31** 317–416. Chemiluminescence in cell biology and medicine.
De Luca, M. A. & McElroy, W. D. (eds) (1978, 1987) *Methods in Enzymology* **57** and **133**. Bioluminescence and chemiluminescence.
Kricka, L. J. & Carter, T. J. N. (eds) (1982) *Clinical and biochemical luminescence*. Marcel Dekker, New York and Basel.
Stannard, C. J. & Gibbs, P. A. (1986) *J. Biolum. Chemilum.* **1** 3–70. Rapid microbiology. Applications of bioluminescence in the food industry — a review.

5

Measurement of enzymes and metabolites

5.1 MEANS AND REQUIREMENTS

Many enzymes and metabolites can be coupled to the production or utilisation of ATP, NAD(P)H or H_2O_2. Firefly luciferin–luciferase can detect ATP and bacterial oxidoreductase–luciferase NAD(P)H, both down to 0.01–0.1 fmol (10^{-17}–10^{-16} mol). This is some million times more sensitive than conventional spectrophotometric assays. Several synthetic chemiluminescent compounds such as luminol, lucigenin, and oxalate esters, and some bioluminescent systems such as the earthworm *Diplocardia* and the marine acorn worm *Balanoglossus*, can be used to measure H_2O_2 at least down to 1 pmol (10^{-12} mol). This also is several orders of magnitude more sensitive than spectrophotometric assays using, for example, peroxidase plus a chromophore.

How do these chemiluminescent assays work? Have they any real biological or clinical application?

Once an appropriate chemiluminescent reaction has been identified, the establishment of a successful assay for a particular analyte involves three stages:

(1) Preparation of reagents

The source of the chemiluminescent compounds may be organic synthesis, extraction, and purification from a luminous organism, or, in the case of a bioluminescent protein, it may be possible to clone it in a bacterium. The latter requires at least seven steps (Fig. 5.1):

(a) extraction of RNA (or DNA in the case of luminous bacteria) using, for example, guanidine isothiocyanate;

(b) purification of mRNA, with its poly A tail, on oligo dT cellulose. mRNA can be assayed using translation to luciferase in rabbit reticulocyte lysate. With the vector we use this is done on the vector, which has an oligo dT tail at the 3′ end. Thus crude RNA can be used;

(c) cDNA is then synthesised using reverse transcriptase;

(d) incorporation of the cDNA into a cloning vector, usually a plasmid, or a bacteriophage such as λgt 11;

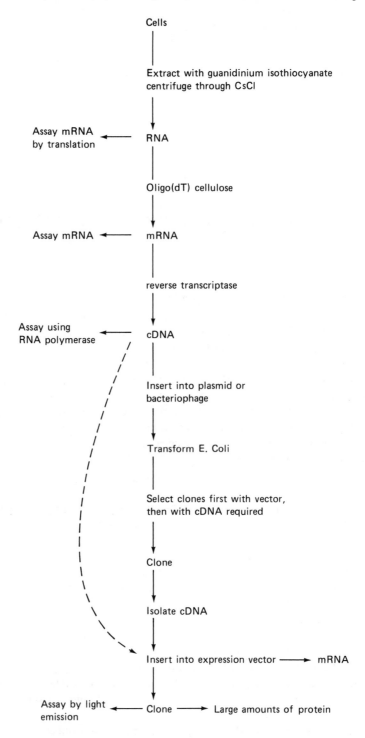

Fig. 5.1 — From RNA to DNA to protein. A typical cloning sequence.

(e) transformation, growth, and selection of bacteria containing the vector producing the protein we want. A second, expression, vector may be used to amplify production;

(f) extraction and purification of the protein, plus regeneration of bioluminescence using synthetic, or natural, luciferin;

(g) Nucleotide sequencing of the gene to give the amino acid sequence.

A key step in cloning is selection of appropriate clones. This is usually done either with an oligonucleotide probe, if some of the amino acid sequence is known, or by using an antibody to the protein (Western blotting). In view of the high sensitivity available for detecting chemiluminescence, addition of luciferin to produce light should be a good method. Both this and the antibody method work best if an expression vector is used. These vectors incorporate a promoter region enabling the RNA polymerase in the bacterium to produce a large amount of mRNA for the protein being studied. The result is that as much as 30–50% of the protein extracted from the bacteria will be the one required. For our purposes, it is important that the complete, active protein, and not just a small fragment, is isolated from the clone, even if attached to another peptide in a 'fusion protein'. Plasmids work well up to molecular weight 50k but higher molecular weight luciferases will require phage or cosmid vectors. For example the phage λgt 11 accepts a length of DNA up to 8.3 kilobases (kb), equivalent to a protein of molecular weight around 300k.

Whether cloned or not, the luciferase, or photoprotein, has eventually to be extracted and purified. Purification usually involves a combination of ion exchange chromatography, salt precipitation (e.g. $(NH_4)_2SO_4$), and gel filtration. Further purification may be attainable by affinity chromatography and hplc. Luciferins, like all small organic molecules, require organic solvent separation techniques, e.g. Sephadex LH20, followed by tlc and hplc. Conditions for drying and storing both the luciferase and luciferin need careful examination for maintenance of high activity.

(2)   Establishment of the assay using pure analyte
This involves coupling the analyte to a component of the chemiluminescent reaction, using enzymes and metabolites, followed by measurement of the rate or total yield of chemiluminescence. Optimal conditions of reagent concentrations, pH, ionic strength, –SH protectors, protein carriers, and temperature are then defined to produce the best signal for a given analyte concentration, with the lowest 'noise'. Any assay should be sensitive, precise, and reproducible, and ideally linear over a wide range of concentrations. A chemiluminescent assay should be able to detect analytes at least down to the pmol–fmol (10^{-12}–10^{-15}) range, if it is to be superior to fluorescence, and ideally down to 1 attomol (10^{-18}). At this level of sensitivity great care is required regarding purity and sterility of reagents and stock solutions. Bacterial and fungal contamination are removed, prior to assay, by filtration. Remember that one fungal or algal cell may contain >10 fmol ATP, equivalent to some 1000 bacterial cells! A second problem is reagent stability. Several luciferases require –SH group protectors and a protein carrier, e.g. albumin or gelatin, whereas luciferins are often destroyed by light and autoxidation.

The key to an ultra-sensitive assay is a low chemical blank, in the absence of added analyte. Many reagents are often contaminated with the analyte itself. For example, commercially-available firefly luciferase contains ATP, and microorganisms may release ATP into stock solutions, even when stored in the fridge or freezer. This ATP can be removed by repurification, or by using an ATPase enzyme, either free or on solid phase e.g. potato apyrase. Luciferases, and even albumin, have an inherent chemiluminescence in the presence of luciferin and O_2, with no added analyte. This, and not the instrument background noise, usually determines the ultimate detection limit of your assay. As a rule of thumb, the detection limit is the analyte concentration providing a signal two standard deviations above the 'blank'. The sensitivity of an assay (detection limit) should be expressed both as molarity and as absolute moles of analyte in the assay. Thus 1 fmol in 1 ml gives a final concentration of 1 pM, but in 10 μl gives 0.1 nM.

(3)   Use of analyte in a real assay
The analyte must be extracted from a biological sample in high yield and be stable on storage. The extract should contain no interfering substances. This can be checked by addition of an internal standard. The complete procedure is illustrated in Fig. 5.2.

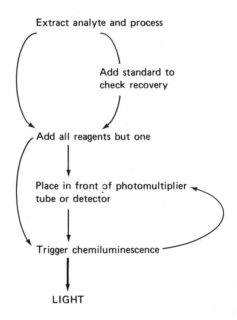

Fig. 5.2 — The principle steps in chemiluminescence analyte assay.

5.2   ATP AND THE FIREFLY SYSTEM
Harvey (Fig. 1.3c) in 1916 showed that fireflies contained an O_2 requiring luciferin--luciferase reaction, in line with Dubois' concept. One of the observations which puzzled Harvey was that not only would heated extracts of the lanterns produce light when added to spent cold water extracts, but also extracts from non-luminous parts of the firefly, and even extracts from non-luminous insects. This puzzle was solved by

McElroy in 1947 when he showed that addition of ATP to extracts of lanterns produced light. This experiment is one of many throughout science which produced an important result and yet was carried out for the wrong reason. McElroy, realising the importance of ATP in the energy metabolism of cells, expected it also to be the energy source of the light emission from fireflies. The direct effect of ATP, or strictly $MgATP^{2-}$, on the firefly luciferin–luciferase is not because it is the energy source, for this comes from oxidation of the luciferin, as with all other bioluminescent reactions. Rather, ATP converts the luciferin into a form capable of being catalytically oxidised by the luciferase in a high quantum yield chemiluminescent reaction. This is because AMP is a good leaving group.

The measurement of metabolites or enzymes using firefly luciferin–luciferase uses a combination of the equations (5.1) or (5.2) plus (5.3):

$$\text{metabolite} + \text{ATP} \xrightarrow{\text{enzyme}} \text{loss of ATP} \tag{5.1}$$

$$\text{metabolite} + \text{ADP} \xrightarrow{\text{enzyme}} \text{increase in ATP} \tag{5.2}$$

$$\text{MgATP} + \text{luciferin} \longrightarrow \text{PP}_i + \text{AMP luciferin} \xrightarrow{O_2} \text{oxyluciferin} + \text{AMP} + \text{light.} \tag{5.3}$$

Under optimal conditions, and at low ATP concentrations, light intensity is linearly related to ATP concentration. A constant ATP concentration thus produces a constant light intensity for many seconds, whereas an enzyme produces a rise, or fall, linearly related to enzyme concentration (Fig. 5.3). This is because both the rates of ATP and luciferin degradation are relatively slow, and because the ATP concentration is within the 'linear' portion of the Michaelis curve (i.e. [ATP] versus reaction rate is linear). Oxyluciferin is a potent inhibitor. Thus, at low luciferin concentrations the kinetics of light emission show a rise to a peak followed by a decay, as oxyluciferin accumulates. However, this inhibition can, to all intents and purposes, be wiped out by using excess luciferin.

A further possibility is to use interference with the kinetics of the luciferase to measure an analyte. General anaesthetics inhibit, competitively with the luciferin. This may provide some useful information regarding the mechanism of anaesthetic action, but has yet to find an analytical application.

5.2.1 Source of reagents

Fireflies and glow-worms are beetles, from the order Coleoptera, as we saw in Chapter 3. The family, Lampyridae, to which they belong, contains more than 2000 species worldwide. More than 120 genera contain luminous species (Appendix II). Luminous species can be found in North and South America, Europe, Africa, the far East including Japan, Indonesia and Malaya, and Australasia, but not in the Arctic or Antarctica. In Europe the glow-worm *Lampyris noctiluca* (Table 5.1) can be found easily in particular localities on a warm summer's night, via its green luminescence, though it appears to be less common than it used to be. Three North American genera have received particular scientific attention, *Photinus*, *Photuris*, and *Pyractomena*. *Photinus pyralis* (Table 5.1) emits yellow light (λ_{max} 565 nm), but in many other fireflies and glow-worms the emission is green.

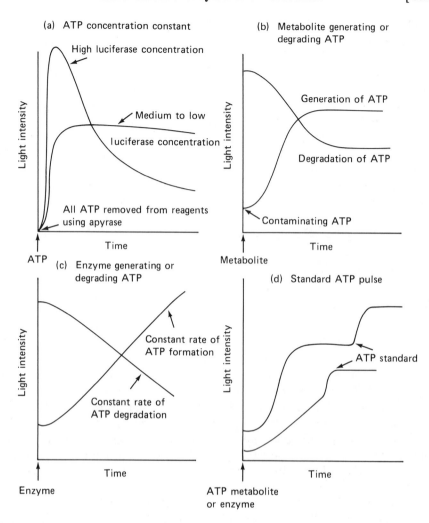

Fig. 5.3 — The different responses from a firefly luciferin-luciferase assay. A: ATP constant. B: metabolite generating or degrading ATP. C: enzyme generating or degrading ATP. D: ATP standard. N.B. B–D = medium to low luciferase concentration, high luciferase results in a 'spike'.

The life cycle of lampyrids consists of four states: egg, larva, pupa, and adult. In some species only one stage is luminous; in others all four stages may produce visible luminescence. Sometimes both sexes can fly, whereas in certain species, like the European glow-worm *Lampyris*, the female is wingless. However, both male and female of each species are usually luminous, though in some cases, exemplified by the European glow-worm *Lampyris noctiluca*, the female is brighter than either male or larva. Adult specimens are thought to be the best source of luciferase, where light emission plays a role in sexual communication (see Chapter 3). They need to be collected in hundreds for an adequately sized preparation, though some experiments

Table 5.1 — The taxonomy of the firefly and the European glow-worm

Kingdom:	Animalia
Phylum:	Arthropoda
Class:	Insecta
Subclass:	Pterygota
Order:	Coleoptera
Super family:	Cantharoidea
Family:	Lampyridae
†Tribe:	Lampyrini
Genus and species:	*Lampyris noctiluca* (European glow-worm)
Tribe:	Photinini
Genus and species:	*Photinus pyralis* (North American firefly)

† Just a few representatives. See Appendix II for complete list.

can be done with a single organism. The luminous organs are usually on the last few segments of the undersurface of the tail (Fig. 3.3A). Thus the female *Lampyris noctiluca* has two large light organs on the ventral surface of the fifth and sixth abdominal segments and two smaller ones on the seventh segment (a further two have been reported on the fourth segment though I have never been able to see them). The male has two small light organs on the last abdominal segment. In the North American firefly *Photinus* the light organs are on the ventral (under-belly) surface of the sixth and seventh abdominal segments. These organs are easy to recognise in both live and dried specimens, even though they may not be visibly luminous after collection.

Our understanding of the biochemistry of the luminescence of lampyrids is almost exclusively based on the pioneering work of McElroy and colleagues on one of the North American fireflies *Photinus pyralis*. This has been the sole source, to date, of luciferase for ATP analysis. Identical reactions utilizing the same luciferin and ATP Mg, catalysed by similar, but structurally distinct, luciferases are thought to occur in all luminous beetles. However, this has been confirmed directly in only a few species. The differences in colour, between species, from green to yellow, are caused by the environment created by the active centre of each luciferase to which the excited AMP-oxyluciferin product is bound (see Chapter 3).

The biochemistry of related coleopterans has yet to be fully established. The only other fully documented bioluminescent systems known to have a direct requirement for ATP are in the chemiluminescent reactions in the millipede *Luminodesmus sequoiae*, belonging to a different class of arthropod, *Diplopoda*, and the squid *Watasenia*. The luciferins in these organisms are quite different from the benzothiazole of the lampyrids (Fig. 3.11)

5.2.2 Preparation of reagents

Fireflies can be desiccated in the field by using $CaCl_2$. This, however, takes many hours and results in loss of both luciferin and luciferase. It is better to remove the lanterns from the ventral surface and freeze-dry (lyophilise) them. Better still, prepare an acetone powder. This not only removes H_2O, thereby preventing rapid inactivation of the luciferase, but also removes the large amount of unwanted lipid in the lanterns. Dried extracts are stable for years in a deep freeze and contain both luciferin and luciferase.

Crude extracts in arsenate buffer have been available commercially for many years (Sigma FLE50). The soluble luciferin–luciferase in this preparation can be reconstituted simply by redissolving in H_2O and centrifuging off particulate material. A few mg of dried extract enables several hundred ATP assays to be carried out within a sensitivity range 0.1–100 pmol in a 10–100 μl sample. Though this type of assay has been satisfactorily used by many workers it suffers from two major disadvantages:

(1) Poor sensitivity, due to: (a) contamination of luciferase with ATP and other nucleotides; (b) low luciferin concentration; (c) inhibition by arsenate.
(2) Poor specificity, due to ATP utilising enzymes in the luciferase preparation, including nucleoside kinases, pyrophosphatase, apyrase, and adenylate kinase.

Attempts to remove contaminating nucleotides from the firefly extract include pre-incubation with apyrase (an ATPase from potato) to destroy ATP, or gel filtration. Addition of synthetic luciferin increases light yield. However, for maximum sensitivity and specificity the assay for ATP requires purification of luciferase, together with optimisation of assay conditions including type of buffer, pH, and luciferin concentration.

Crude luciferase can be prepared by extraction in 10% ammonium sulphate + 1 mM EDTA, pH 8. Several other buffers can be used, but the pH must be maintained above 7.5 to prevent acid inactivation of the enzyme. The frozen solubilised luciferase is stable for months. Further purification can be achieved by using ammonium sulphate fractionation, phenyl sepharose, hydroxyapatite chromatography, and electrophoresis. Euglobulins are proteins which are insoluble in water. The luciferase, a euglobulin, crystallises on dialysis against water or low ionic strength buffer (e.g. 1 mM EDTA, 10 mM NaCl, 2 mM Na_2HPO_4 pH 7.8). 60 g of acetone powder of the lanterns can yield 70 mg of pure (single band on polyacrylamide electrophoresis) luciferase with a yield of 40–50%. It is now commercially available (see Appendix V). Literally tens of thousands of fireflies are processed each year, a worry for the environmentalist.

Firefly (*Photinus pyralis*) luciferase has been cloned by De Wet and colleagues. Total RNA was extracted from isolated lanterns by homogenising in guanidinium thiocyanate, and then sedimenting the RNA through a 3 M CsCl cushion. RNA was further purified by phenol extraction and stored as a precipitate in 70% ethanol. mRNA was purified by using an oligo (dT) cellulose column, since mRNA has a poly(A) tail. Great care is necessary to prevent RNAase contamination at all stages. Even bare fingers contain enough RNAase to degrade the isolated mRNA. A λgt.11 cDNA library was then constructed. cDNA was prepared by using reverse transcriptase, with RNAaseH and DNA polymerase I to produce the second strand of the cDNA. The cDNA was incorporated into bacteriophage λgt11. The cDNA library was amplified in *E. coli* strain Y1088 or 1092. Clones expressing luciferase were detected by using chromogen-labelled antiluciferase antibody. One clone, λLuc1, contained 1.5 kb of cDNA which hybridised to a band of RNA, separated by electrophoresis from firefly lantern RNA. The largest clones λLuc23 = 1.8 kb, though luciferase mRNA was *ca* 2 kb. *In vitro* synthesis of luciferase could be demonstrated from mRNA by using a translation assay kit based on rabbit reticulo-cyte lysate. When the cloned cDNA from λLuc23 was incorporated into a plasmid expression vector the fusion protein expressed in *E. coli* produced light when luciferin was added. Expressing clones can be detected using photographic film. The organism itself should therefore now be safe from over-collecting.

Luciferin can be extracted from lanterns in organic solvents such as ethyl acetate, as was done by Bitler and McElroy in 1957. However, the synthesis was published by White and co-workers in 1961 and modified by Seto and colleagues in 1963 (Fig. 5.4). The key intermediate is 2-cyano-6-methoxybenzothiazole (Fig. 5.4 I), which has been synthesised in three different ways, route C being the most convenient. The methyl group from compound I is then removed and the product condensed with D-cysteine to form the luciferin. The pure luciferin is a pale-yellow solid and should be stored in the dark, desiccated under N_2, resulting in negligible oxidation, or photodecomposition. Aqueous solutions, pH 7, can be stored at 4°C for a month or so, preferably in the dark under N_2 or Ar. Purity can be assessed by:

(1) The *in vitro* chemiluminescence has a peak emission at 565 nm using firefly luciferase.
(2) ϕ_{CL}, should be >0.7.
(3) Absorbance at neutral pH A_{327}/A_{268} > 2.3
     at alkaline pH A_{385}/A_{285} = 2.5.
(4) Fluorescence emission spectrum λ_{max} 537 nm; λ_{max} dehydroluciferin = 555 nm.
(5) Silica gel tlc R_F = 0.31 on ethyl acetate:methanol (10:1)
(6) Cellulose tlc R_F = 0.55 in 95% ethanol: 1 M ammonium acetate pH 7.3 (7:3) (visualise at 366 nm by fluorescence).

Many analogues have also been synthesised. Synthetic luciferin is commercially available in 1–10 mg (3.6–36 μmol) batches, sufficient for several hundred experi-ments or assays.

5.2.3 Properties

Some properties of the firefly luciferin–luciferase system are listed in Table 5.2.

Firefly luciferase is a euglobulin, protein dimer, molecular weight approx. 100 000 with 1 ATPMg (in the form $ATPMg^{2-}$ i.e. $ATP^{4-} + Mg^{2+}$) binding site. However recent cloning experiments show only one polypeptide (62K) is necessary. Euglobulins are virtually insoluble in distilled water; they require salt to dissolve them. On SDS polyacrylamide gel electrophoresis a single band, 62 000 molecular weight, is found. Reaction of the luciferin (LH_2) occurs in two stages (Eqn (5.3), Fig. 3.14), first the formation of AMP-luciferin, and then its oxidation by O_2 to the excited state oxyluciferin. Under normal assay conditions, i.e. neutral-alkaline pH, the colour of light emission is yellow to green (λ_{max} 565 nm), from the excited phenolate. However, at acid pH, or in the presence of heavy metal cations, the emission is red (λ_{max} 615 nm). The peak in light intensity is reached within 250–300 ms, and is followed by a decay, which at saturating ATP Mg^{2-} (1 mM) and saturating luciferin (*ca.* 100 μM) has a half time of about 1 s. This decay is not due to the consumption of either luciferin or ATP, but to the potent competitive inhibition

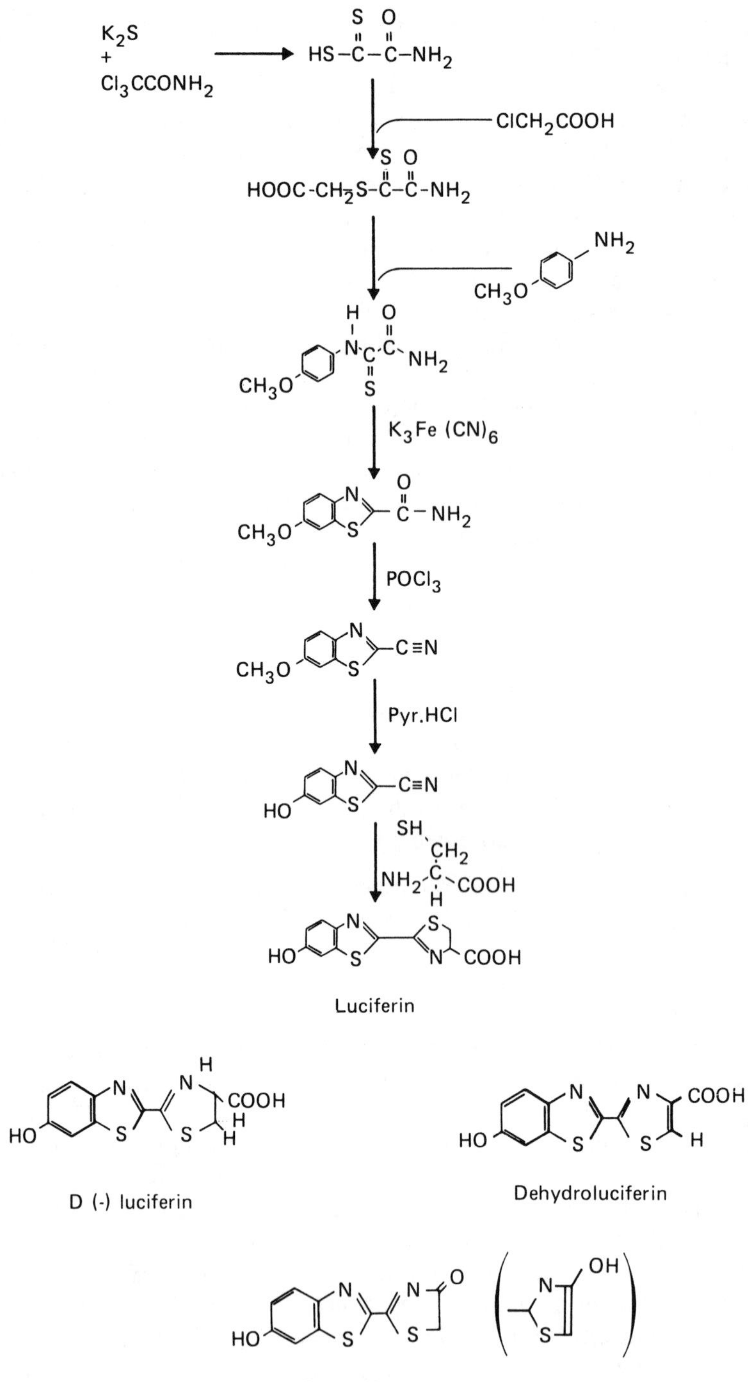

Fig. 5.4 — Synthesis of firefly luciferin and some other structures, from Seto *et al. Bull. Chem. Soc. Japan* **36** 332 (1963) and White *et al.* (1965) *J. Org. Chem.* **30** 2244.

Table 5.2 — Properties of the firefly luciferin–luciferase system

1.	Components	luciferin (LH_2), O_2, ATP+Mg^{2+} ($ATPMg^{2-}$), luciferase
2.	Reaction products	oxyluciferin, AMP, pyrophosphate, Mg^{2+}, H_2O, CO_2
3.	Colour of chemiluminescence	
	in vivo	yellow–green λ_{max} 565 nm
	in vitro	yellow-green at high pH λ_{max} 565 nm
		red at low pH or +Pb^{2+}, Hg^{2+}, Zn^{2+}, Cd^{2+}, λ_{max} 615 nm
4.	Luciferin	D-2-(6'hydroxy-2'benzothiazolyl)-Δ^2-thiazoline-4-carboxylic acid
	composition	$C_{11} H_8 O_3 N_2 S_2$
	mol. wt.	276
	absorbance	phenol in acid; λ_{max} 327 nm = 1.82×10^4 l.mol^{-1}, cm^{-1}
		phenolate in base; λ_{max} = 385 nm
	fluorescence (phenolate)	λ_{max} emission 537 nm; ϕ_F at pH 11 = 0.62
	ionisable groups	phenol OH and –CO_2H
	solubility	H_2O, pyridine
5.	Luciferase	*Photinus* luciferin: oxygen 4-oxidoreductase EC 1.13.12.7
	structure	*dimer, mol. wt. *ca.* 100 000 = 2 × 62 000
	K_m for substrates	luciferin = 1–10 μM, O_2?, ATPMg 0.25 mM (1 site per dimer), AMP K_i = 0.24 mM
	pH optimum	7.75
	inhibitors	oxyluciferin (K_i 0.23 μM), dehydroluciferin, pyrophosphate, ATP (i.e. free not ATPMg) K_i 1 mM, anions ($SCN^- > I^- \sim NO_3^- > ClO_4^- > Br^- > Cl^- >$ acetate)
	kinetics + ATPMg	25 ms lag, 250–300 ms to reach peak, $t_\frac{1}{2}$ decay for saturating ATPMg (1 mM) $\doteqdot$ 1 s
	chemiluminescence quantum yield	ϕ_{CL} = 0.8−1.0
	A_{280} 1 mg ml^{-1}	0.75

*SDS polyacrylamide gel electrophoresis gives only one band=61–62K.

by the product, oxyluciferin. At the very low ATP concentrations used in biological assays the chemiluminescence decay is much slower since the build-up of oxyluciferin is slower.

The kinetics of the reaction are in fact complicated by a number of factors:

(1) The two stage reaction results in a 25 ms lag at room temperature before any light is detected, peak light intensity not being reached until 250–300 ms after injection
(2) Inhibition by products, pyrophosphate and oxyluciferin (Fig. 5.4)
(3) Inhibition by contaminating dehydroluciferin (Fig. 5.4) and its potentiation by contaminating pyrophosphatase

Crude luciferin–luciferase preparations, as well as old purified luciferin, can contain substantial concentrations of dehydro-luciferin. The luciferase is able to catalyse the adenylation of the dehydroluciferin, just as it does the luciferin itself:

$$\text{dehydroluciferin} + \text{ATP Mg}^{2-} \rightarrow \text{dehydroluciferin-AMP} + \text{PP}_i \qquad (5.4)$$

But here the reaction stops. The enzyme is unable to catalyse oxidative decarboxylation of the dehydroluciferin-AMP to oxyluciferin. Not only is no light emitted, but also the dehydroluciferin-AMP remains tightly bound to the luciferase, inhibiting the chemiluminescent reaction from the real luciferin. Addition of pyrophosphate reverses the adenylation reaction and can generate a light signal from inhibited luciferase containing excess ATP + luciferin. Pyrophosphatase, contaminating crude luciferase preparations, hydrolyses the pyrophosphate, and the inhibition returns. Paradoxically pyrophosphate *inhibits* the normal light reaction since it tends to reduce the formation of luciferin-AMP because of the reversibility of the first step in the reaction (Eqn (5.3)).

Contamination of crude preparations with dehydroluciferin and pyrophosphatase therefore result in a rapid and biphasic decay in light emission, even though there may be no significant consumption of ATP or luciferin. A fast decay lasting about a second is followed by a much slower decay phase lasting many seconds to minutes. Up to the end of the 1960s many workers used arsenate buffer to prevent this 'flash' effect. Under these conditions a steady light emission, detectable in a scintillation counter, without the need for injection in front of the photomultiplier, is produced on addition of ATP. Under these conditions, little consumption of ATP or luciferin occurs over the time of measurement, e.g. 10 s. However, arsenate inhibits the chemiluminescence, greatly lowering the potential sensitivity of the assay for ATP.

ATP Mg^{2-} is the best substrate for the luciferase, the enzyme reacting poorly with dATP Mg^{2-}. Uncomplexed ATP can act as a competitive inhibitor at high concentrations ($K_i = 1$ mM). Other nucleotides such as GTP Mg^{2-} do react but are much less potent than ATP Mg^{2-}. Contamination with ATP of other nucleotides always causes problems in interpretation. Pure nucleoside di- and mono-phosphates produce no light with pure luciferase, though some such as AMP can be competitive inhibitors of the ATP Mg^{2-} provoked reaction. The pH optimum of the normal ATP reaction of firefly luciferin–luciferase is 7.5–8 (mean 7.75).

Firefly luciferin is water soluble and highly fluorescent. In alkali it rapidly oxidises in the presence of O_2 to dehydroluciferin, which is structurally different from oxyluciferin (Fig. 5.4). In some solvents, for example pyridine, racemisation may also take place. The luciferin undergoes ground state and exicted state ionisation. The 6'OH results in an absorbance maximum of 327 nm, and 6'O⁻ a maximum at 385 nm. However, in aqueous media only the phenolate results in fluorescence (λ_{max} emission 537 nm), whereas in non-polar solvents, where O⁻ is not formed, λ_{max} for fluorescence emission is 420 nm. These fluorescence properties are useful in studying the micro-environment of the luciferase active centre from different organisms and in assessing the purity and stability of luciferin preparations.

5.2.4 Measurement of ATP
This knowledge of the properties of the firefly luciferin–luciferase reaction (Table 5.2) enable the optimal conditions for assay of ATP to be defined. These are:

(1) Pure reagents
The luciferase should be free of contaminating enzymes which could lead to the production of inhibitors, degrade or produce ATP significantly, or reduce the specificity of the reaction for ATP. The enzyme preparation should also be free of nucleotides, luciferin, dehydro-, and oxy-luciferin. The luciferin should be pure, and

free of dehydroluciferin. Even commercial reagents such as ADP may be too severely contaminated with ATP for high sensitivity assays. A 0.01% contamination of 1 mM ADP by ATP is equivalent to 10 pmol ATP in 100 μl! Treatment of the assay cocktail (i.e. buffer + luciferin + luciferase) with apyrase reduces the chemical blank some 100-fold. The result is a reduction in the detection limit of the assay from about 10–100 fmol without apyrase treatment to 0.01–0.1 fmol, after treatment (Fig. 5.5). Although it is better to remove the apyrase, this is not always absolutely essential. ATP is everywhere, in the water, in the buffers, in the dust in the tubes, and in the luciferase itself. Destruction of this ATP, particularly from luciferase, is essential if assay sensitivity is to be optimised by increasing the luciferase concentration.

(2) Storage of reagents
The luciferase has two SH groups which, if covalently modified, cause loss of activity. The luciferase should therefore be stored in solution with an SH group protection such as dithiothreitol, and albumin. Luciferin should be stored crystalline or in aqueous solution under N_2 to Ar to minimise production of dehydroluciferin. In assay cocktails luciferase may lose as much as 10% of its activity per hour because of accumulation of oxy- or dehydroluciferin. The whole assay should therefore be completed within 1–2 hours. However, aliquots of frozen or freeze-dried cocktail can be stored for weeks at −20°C with little or no loss in activity, though occasionally >95% loss of activity occurs inexpliciably.

(3) Buffer
A non-inhibitory buffer such as tris, and not arsenate, should be used. Freeze dried luciferase is usually first dissolved in 0.5 M tris-acetate pH 7.8. Anion inhibition, even Cl^-, should be minimised (Table 5.2). Thus the best Mg^{2+} concentration is usually about 5–10 mM. The pH optimum is in the range 7.5–8.0 (mean 7.75), although a range of 6–8 is acceptable. Red emission may be significant below the optimal pH, thereby resulting in apparent inhibition of chemiluminescence if a blue sensitive, red insensitive bialkali photomultiplier tube is used (see Chapter 2).

(4) Temperature
Room temperature to 25°C (optimal) is usual, and should be maintained constant. The rate of reaction, monitored by light intensity, increases with temperature up to 25°C and then decreases. Extremes of temperature which inactivate the luciferase, or other enzymes being assayed, must obviously be avoided.

(5) Concentrations of reagents
The concentrations of luciferin and luciferase are usually set to prevent significant accumulation of product inhibitors (pyrophosphate, oxyluciferin, and AMP) and to keep ATP consumption <1% during the experiment. O_2 consumption is negligible. Typical concentrations in the assay are: luciferin, 0.1–0.3 mM (i.e. >10 × K_m); luciferase ca. 10 nM (i.e. 1 μg.ml^{-1}). However, for ultrasensitive assays, at very high luciferase concentration, significant ATP consumption may occur.

(6) Sterility
All stock reagents and the final cocktail, plus tubes, syringes, and plastic tips should

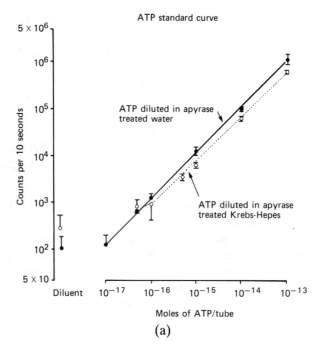

(a)

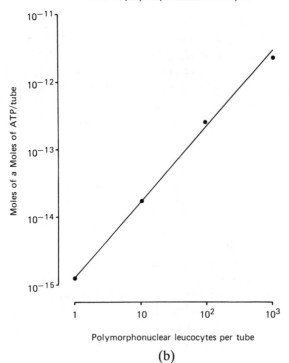

(b)

Fig. 5.5 — Measurement of ATP: (a) A typical ATP standard curve; (b) Single cells. From Sala-Newby, Campbell *et al.*, unpublished.

be autoclaved, filtered, or acid washed for assays <10 fmol ATP (N.B. ATP is acid labile. Concentrations of stock ATP e.g. 1 mM have pH<4 and are thus unstable on storage, even when frozen).

(7) Linearity
Since the K_m for ATP is 0.25 mM, from the Michaelis–Menten equation the upper limit at which the light emission ceases to be linearly related to ATP concentration can be estimated:

$$\text{light intensity} = \text{rate of reaction} = V_{max}\, \text{ATP}/(K_m+\text{ATP}), \qquad (5.5)$$

where ATP = molarity of ATP in the assay.

If ATP $\ll K_m$ then the assay standard curve (Fig. 5.5) will be linear. This means, to a first approximation, the ATP concentration in the assay should be less than 1–2 μM (0.1–0.2 nmol in a final volume of 0.1 ml).

When these conditions are satisfied, addition of ATP to the other reagents results in a rapid rise in light emission to a constant light intensity, lasting many seconds or more (Fig. 5.3). The light emission can therefore be measured either with a chemiluminometer or a scintillation counter (out of coincidence) since the ATP can be added before the assay tube is placed in front of the photomultiplier. A plot of light intensity, or an integral over a short time interval (e.g. <10 s), against ATP added is linear over the range 1 pM–1 μM in the assay tube, some six orders of magnitude, equivalent to 0.1 fmol–0.1 nmol (10^{-16}–10^{-10} mol) in 0.1 ml final volume. Reduction of the assay volume may improve the detection limit still further to approximately 1–10 attomol, sufficient for analysis of individual eukaryotic or just a few prokaryotic cells (Fig. 5.5b). Increasing luciferase (ATP free) concentration produces 'spike' kinetics, i.e. light intensity reaches a peak and then decays rapidly. However this increases sensitivity, down to 10 amol ATP (10^{-13} M). Since the *apparent* K_m for ATP may vary from 0.1 mM–1 mM depending on the analytical conditions the linear range should always be determined for a defined set of conditions.

The method of extracting ATP from a cell or tissue sample should be critically examined, not only for optimal yield but also for interference with the firefly assay. Some cells are surprisingly difficult to break. It is essential that >95% ATP is extracted if meaningful results are to be obtained. Extraction methods have included boiling buffer (e.g. tris-EDTA), trichloroacetic acid (TCA), perchloric acid (PCA), alkaline ethanol and detergent lysis. Neutralisation of PCA can result in coprecipitation of some ATP, whilst residual TCA and PCA can inhibit the luciferase, and thus need to be stringently removed. Lack of inactivation of ATPases and interfering enzymes, as well as interference by other nucleotides, are usually the major problems. Internal ATP standards should always be added to extracts before relying on results from a new tissue. ATP is stable between pH 7 and 9, but is very rapidly degraded at acid pH. Na_2 ATP made up in H_2O will be acid (pH<4) and must be buffered if it is to be stored, or added to an enzyme preparation which will be denatured at acid pH.

Cellular ATP can be measured by direct lysis of the cells with a detergent, in the presence of the luciferin–luciferase cocktail, in front of the photomultiplier tube. Mammalian cells can be lysed rapidly (<0.1 s) using a non-ionic detergent such as triton X-100 (e.g. Nonidet P40) at a concentration of about 1%. Bacteria or yeast cells, with a cell wall, are more difficult to lyse efficiently. A popular detergent is a quaternary ammonium salt such as tetra decyltrimethyl methyl ammonium bromide. Others include mixed alkyltrimethyl ammonium bromide (mainly $C_{14}H_{29}$ $N(CH_3)_3Br$) and benzalkonium chloride (alkyl dimethyl benzyl ammonium chloride). Unfortunately these detergents result in a rapid inhibition of the luciferase, up to 90% activity sometimes being lost within 1 minute. It is therefore not possible to 'spike' the reaction with a standard amount of ATP at the end of the cell ATP assay. Attempts to alleviate this potent inhibition include the solution of glycerol or phospholipids such as lecithin. Indeed the luciferase *in vivo* may even be associated with phospholipid. In spite of this problem, direct ATP analysis following detergent lysis still allows the detection of single eukaryotic cells (ATP = 0.1–10 fmol per cell) and about 10–50 bacteria (Fig. 5.5), even though the ATP standard curve has reduced sensitivity compared with that in pure buffer.

The three major practical problems to be aware of in the firefly luciferin–luciferase assay for ATP are therefore:

(1) ATP contamination of reagents and cell culture broths.
(2) Anion inhibition.
(3) Detergent inhibition.

5.2.5 Coupling to other enzymes and metabolites

Any enzyme or metabolite capable of being coupled to ATP formation or degradation can be measured by firefly assay. A large number of enzymes and metabolites can be measured in the range 10 fmol–20 nmol in this way (Table 5.3).

(1) Formation of ATP

(a) $\quad$ ADP + Pi $\xrightarrow{\text{E}}$ ATP $\hfill$ (5.6)

e.g. ATP formation by mitochondria, chloroplasts, and reversal of ion pumps, or measurement of ADP or Pi

(b) $\quad$ ADP + X – P $\xrightarrow{\text{E}}$ ATP + X $\hfill$ (5.7)

measures X – P, ADP or E
e.g. creatine phosphate and creatine kinase

$\qquad$ ADP + creatine phosphate $\rightarrow$ ATP + creatine $\hfill$ (5.8)

this enzyme is reversible
or phosphoenol pyruvate and pyruvate kinase

$\qquad$ ADP + PEP $\rightarrow$ ATP + pyruvate $\hfill$ (5.9)

(c) $\quad$ X + Y $\xrightarrow{\text{E}_1}$ X – P $\xrightarrow[\text{+ ADP}]{\text{E}_2}$ ATP + X $\hfill$ (5.10)

measures X, Y, or E_1 or E_2
e.g. myokinase and AMP

Table 5.3 — Applications of the firefly system

.1. *Direct measurement of ATP*
    (a) Cellularity   — biomass (sewage, marine and fresh water, marine and freshwater
                                 sediments, soil, brewing, milk, food, textiles, urine (bacteriuria),
                                 cell growth in culture, extraterrestrial life, germ warfare)
                             — cell viability *in vitro* (erythrocytes, platelets, white cells, cultured cells,
                                 and organs)
    (b) Cell activation   — release from nerve endings, platelets, and muscle; plasma
                             — change on activation (lymphocytes, neutrophils)
                             — change during life cycle and cell growth
    (c) Analyte bioassay   — vitamins necessary for cell growth (folic acid, B_{12})
                             — antibiotics reducing cell numbers (gentamycin, tobramycin,
                                 doxycycline, ampicillin, rifampicin, cephaloridine)
    (d) Direct ATP generation from ADP and Pi
                             — oxidative phosphorylation and photosynthesis (mitochondria,
                                 chloroplasts, bacterial chromatophores)
                             — reversal of ion pumps (Na^+ pump, Ca^{2+} pumps)
    (e) Nonisotopic label in immunoassay

2. *Metabolites coupled to ATP formation or loss*
    (a) Generation of ATP   — ADP, AMP, cyclic AMP, adenosine tetraphosphate, adenosine
                                 phosphosulphate, GTP, GDP, GMP, CTP, UTP, PEP, creatine
                                 phosphate 1,3 diphosphoglycerate, PPi
    (b) Degradation of
          ATP   — glycerol, triglyceride, creatine, cyclic GMP, glucose
    (c) Effect on product   — coenzyme A (relieves)
          inhibition of
          luciferase

3. *Enzymes coupled to ATP formation of loss*
    (a) Generation of ATP   — creatine kinase, pyruvate kinase, ATP sulphurylase, ATP
                                 phosphoribosyl transferase, adenylate kinase, nucleotide
                                 phosphokinases, cyclic AMP phosphodiesterase
    (b) Degradation of
          ATP   — hexokinase, ATPases, apyrase, cyclic GMP phosphodiesterase,
                                 histidine biosynthesis

4. *Solid phase luciferase*

5. *Non-analytical applications of luciferase*
    (a) Model for general anaesthetic action
    (b) Probe in genetic engineering

See section 5.7 and in particular De Luca (1978) and De Luca and McElroy (1987), Campbell and Simpson (1979), Lundin (1982), Campbell *et al.* (1985) for references and further details.

$$\text{AMP} + \text{ATP} \longrightarrow 2\,\text{ADP} \xrightarrow[+\,\text{PEP}]{} 2\,\text{ATP} \tag{5.11}$$

e.g. triose phosphate isomerase (TPI)

$$\text{glyceraldehyde 3 phosphate (G3P)} + \text{Pi} + 2\text{NAD} \xrightarrow{\text{TPI}} 1{,}3\ \text{diphosphoglycerate} \tag{5.12}$$

$$\begin{aligned}\text{1,3 diphosphoglycerate (1,3 DPG)} &\xrightarrow[\text{kinase (PGK)}]{\text{phosphoglycerate}} \quad \text{ATP} + \\ + \text{ADP} &\qquad\qquad\qquad 3\ \text{phosphoglycerate}\end{aligned} \tag{5.13}$$

these reactions assay TPI, PGK, G3P, or 1,3 DPG

(2)   Utilization of ATP

(a) $\text{ATP Mg}^{2-} \rightarrow \text{ADP} + \text{Pi}$ (5.14)

e.g. ion pumps and ATPases

(b) $\text{ATP Mg}^{2-} + \text{X} \xrightarrow{\text{E}} \text{ADP} + \text{Mg}^{2+} + \text{X} - \text{P}$ (5.15)

measures X or E
e.g. glucose or hexokinase

$$\text{ATP Mg}^{2-} + \text{glucose} \rightarrow \text{ADP} + \text{Mg}^{2+} + \text{G6P}$$ (5.16)

F6P or phosphofructokinase

$$\text{ATP Mg}^{2-} + \text{F6P} \rightarrow \text{ADP} + \text{Mg}^{2+} + \text{FDP}$$ (5.17)

(c) $\text{X} + \text{Y} \xrightarrow{\text{E}_1} \text{Z} \underset{+ \text{ATP Mg}^{2-}}{\xrightarrow{\text{E}_2}} \text{ADP} + \text{Mg}^{2+} + \text{Z} - \text{P}$ (5.18)

measures X, Y, or E_1 or E_2.

Optimisation of the firefly assay means that these reactions can be carried out in a single tube since steady state light intensity is directly proportional to the ATP concentration. Three main types of chemiluminescence traces result (Fig. 5.3). Firstly (Fig. 5.3A) addition of ATP without significant destruction results in a rapid increase in light intensity (I) reaching a plateau within 1 s, which is then measured as I (dhv/dt) or an integral of I over a short time period (e.g. 1–10 s). Secondly (Fig. 5.3B) addition of a metabolite through a coupled reaction produces an endpoint determination. The metabolite reacts rapidly and quantitatively, converting ADP to ATP (Eqns (5.6)–(5.11)) or vice versa (Eqns (5.12)–(5.18)). Within a few seconds or minutes a new plateau in light intensity is reached. The difference in light intensity between this plateau and that before metabolite addition is directly proportional to the ATP formed or utilised, and thus the amount of metabolite added. Again I, or a small integral of I, is plotted against amount of metabolite added. Thirdly (Fig. 5.3C) addition of an enzyme or organelle producing or degrading ATP results in a linear increase or decrease in light intensity respectively. The slope is directly proportional to enzyme activity.

In all of these assays the linear relationship between light intensity (photons per second) and the amount or concentration of ATP in the assay tube is calibrated by addition of a minimum of two ATP standards. This can be done at the beginning or end of the experiment (Fig. 5.3D).

5.2.6 Solid phase reagents
Solid phase enzymes have the advantage, particularly in the clinical laboratory, that they can be washed and re-used after each assay. A method based on glutaraldehyde coupling to arylamine glass beads attached to glass rods has been published for attaching firefly luciferase to a solid phase. Unfortunately, although as much as 20% of the enzyme protein could be attached, more than 99% of the catalytic activity was lost. The spectrum of light emission also shifted from 565 nm to 615 nm. Nevertheless, the K_m for ATPMg^{2-} and the pH optimum appeared unaltered, and the reagent

could be stored for several weeks at 4°C in 0.1 M phosphate +1 mM mercaptoetha-nol. Each rod could be used for at least 30 consecutive ATP measurements. A linear range for ATP from 10 nM to 10 μM was obtained. An improved yield of up to 20% has been achieved by using Sepharose 4B beads and cellophane film. The luciferase is relatively unstable in this form ($t_{\frac{1}{2}}$ for loss = 1 week at 4°C). The recovery and stability can be improved further using CNBr coupling to Sepharose 4B and 6B.

5.2.7 Real applications of the firefly system

In spite of the plethora of assays for ATP, and for systems generating or degrading it (Table 5.3), do we really need to measure ATP at all?

All cells, prokaryote and eukaryote alike, require ATP to remain alive and to carry out their specialised functions. Whilst some ATP is found in extracellular fluids, most ATP is inside living cells. Intracellular ATP is maintained by oxidative phosphorylation in respiring cells and by substrate phosphorylation in cells with few or no mitochondria, e.g. red and white blood cells and the renal medulla. ATP is also formed by chloroplasts during photosynthesis. ATP is the link between catabolic and anabolic processes in all cells. Lack of oxygen, substrates or injury results in a rapid decrease in cytoplasmic ATP within seconds or minutes, even in cells with a good creatine phosphate buffer. Measurement of ATP is therefore fundamental to the study of living processes. Cell activation and cell injury result in significant changes in ATP concentrations.

ATP per cell varies over four orders of magnitude from *ca.* 1 amol (10^{-18}) in bacteria and 100 amol (10^{-16}) in red and white blood cells, to 1–10 fmol in hepato-cytes or large cells such as the protozoan *Paramecium* and the dinoflagellate *Perinium*.

The firefly luciferin–luciferase assay is now the method of choice for measuring ATP. It is the most sensitive method available, down to 0.01 fmol (10^{-17} mol), and has the widest linear range (1 pM–1 μM equivalent to 0.1 fmol–0.1 nmol in 0.1 mol). It is very quick, the signal being recordable within 1–10 seconds after addition of ATP or analyte, enabling many hundreds of samples to be analysed in one experiment. It is precise and accurate. Generation or degradation of ATP by cellular organelles, particles, or enzymes can be monitored continuously. This assay is thus far superior to spectrophotometric or fluorimetric assays. HPLC analysis has the advantage of providing values for many nucleotides from the same sample simulta-neously. However, it is laborious and some 10^5 times less sensitive than chemiluminescence.

The firefly assay for ATP has, not surprisingly, been widely used since its introduction in the early 1950s. Purified reagents, now available, have improved its performance considerably. Dead or dying cells contain < 1–10% of the cell's normal ATP. ATP is thus being used as a measure of biomass in several industries, including water purification, brewing, textiles, milk, and food. The efficiency of waste disposal and sewage plants can be monitored simply by measuring ATP. It has even been used to search for life on Mars! Clinically, the sensitivity of the assay has attracted microbiologists and haematologists. It can be used to assay for pathogenic bacteria, e.g. *Salmonella*. Infection of the urogenital tract, including the kidneys, ureter, bladder, and urethra, leads to microorganisms in the urine. These include not only bacteria like *E. coli*, but also fungi such as *Candida* (thrush), mycobacteria, mycoplasma, and viruses. The number of bacteria in a given volume of urine which are chemically significant has been the source of much debate. 10^4–10^6 clone-forming units (CFU) per ml, or more, are certainly indicative of urinary tract infection, but as little as 10^2–10^3 CFU per ml^{-1} may be significant. To be useful in assessing bacteriuria, the method must therefore be able to detect down to 10^2 bacteria per ml. Because of the low ATP content per cell (*ca.* 1 amol) the present method can only detect 10^3–10^5 bacteria per ml. It is therefore necessary to concentrate the organisms by filtration or centrifugation. A further problem is the lack of specificity for a particular species of bacteria. This could be solved by using an antibody or a DNA probe. In some applications, for example detection of microorganisms in meat, ATP can come from other cells beside microorganisms.

The ATP assay has been used to establish clinically viable assays for antibiotics and vitamins based on their effects on bacterial cell growth and thus cell number, though a few laboratories have yet to replace existing immuno- or culture plate assays by the one based on firefly luciferin-luciferase. The firefly assay provides a standard technique for assessing, biochemically, cell viability *in vitro*. This is essential if an effect on cell death of an inhibitory agent of cell metabolism, e.g. a drug, is to be ruled out. Interpretation can, however, be complicated by changes in cellular ATP which often occur quite normally following cell activation or during the cell growth cycle.

Secretory cells containing catecholamines, and some other cells containing hormones or neurotransmitters, have ATP bound to these substances within the secretory granules. Stimulation of the cell therefore results in ATP release. The continuous monitoring of ATP concentration using the optimised firefly assay has enabled not only continuous production of ATP by mitochondria to be monitored, but also ATP release from nerve endings and platelets to be studied by using living cells. A lumiaggregometer has been developed to measure secretion via chemiluminescence, and platelet aggregation, concomitantly. ATP release from muscle *in vivo* and in plasma has been measured, though these values must be treated with caution because of the ease with which ATP can be released from damaged red cells.

The ability of the plasma membrane sodium pump and the sarcoplasmic reticulum Ca^{2+} pump to generate ATP by reversal of the natural ion gradient has provided, for the first time, direct evidence for the reversibility of these two membrane enzymes. However, the assays for other enzymes and metabolites based on their coupling to ATP generation or degradation have not been widely used. The sensitivity of the assays is not usually quite as good as for ATP itself, perhaps only 0.1–100 pmol, particularly if the reagents are contaminated with small amounts of ATP. Another reason for relatively poor application in this field is that the surge of studies on the regulation of metabolic pathways which occurred in the late 1950s and early 1960s has been redirected more towards measurement of covalent modification of proteins, e.g. by mechanisms such as phosphorylation or ribosylation.

Two clinical assays for creatine kinase have attracted some attention, one measuring the enzyme in blood spots as a screen for muscular dystrophy, the other, produced commercially, for measuring creatine kinase MB isoenzyme, released from damaged myocytes, as a test for myocardial infarction. This latter assay uses an antibody to inhibit the MM isoenzyme activity. More recent two-site monoclonal immunoassays and conventional spectrophotometric assays, following removal of

MM with a monoclonal antibody, have attracted more interest in clinical laboratories. The necessity for the sensitivity of chemiluminescence has not really been convincingly demonstrated for this enzyme, found at about 0.08 nM (6 ng ml^{-1}) in normal serum and up to 6 nM (500 ng ml^{-1}) after an infarct.

Further applications of the firefly ATP assay will depend on the imagination behind the questions being asked by the investigator. Its use inside intact cells is limited because the enzyme is saturated by the concentration of ATP Mg^{2-} found in normal living cells, around 1–10 mM. Serious cell injury may reduce ATP Mg^{2-} to sub-saturating concentrations, but is also likely to lead to luciferin–luciferase release. Now that the luciferase gene has been cloned, site-directed mutagenesis may increase the range of ATP concentrations measurable by the luciferase.

5.2.8   A model for the anaesthetist and the genetic engineer; a model for anaesthetic action

In the United Kingdom alone some three million operations are carried out every year under general anaesthesia. Yet we still do not know what the molecular mechanism underlying this vital part of medicine is. The structures of the compounds capable of inducing anaesthesia are extremely diverse (Fig. 5.6a), and include alcohols, hydrocarbons, either plain or halogenated, ethers and gases as chemically different as xenon and nitrous oxide. Two opposing hypotheses have been proposed to explain how such a variety of structures can have essentially the same effect on the central nervous systems: either they interact directly with the phospholipid bilayer of the membrane, or they bind to a hydrophobic pocket within a key membrane protein (or proteins). These would then lead to effects on ionic permeability, preventing excitability, i.e. stopping the generation of action potentials and nerve–nerve

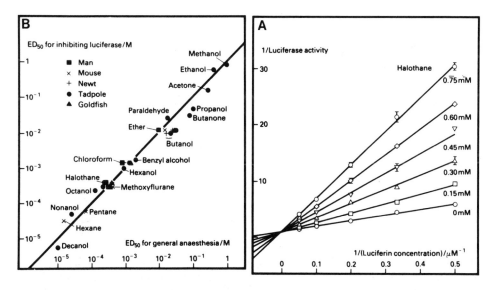

Fig. 5.6 — General anaesthetics and inhibition of the firefly reaction. A: Inhibition. B: Correlation with anaesthetic potency. From Franks & Lieb (1985) *Chem. Brit.* 919–921, with permission of The Royal Society of Chemistry.

stimulation. However, attempts to relate anaesthetic potency to the solubility and effect on lipid fluidity have failed to provide convincing evidence for the 'lipid' hypothesis.

Harvey, in 1917, made the puzzling observation that extracts of the firefly *Luciola viticollis* lost their luminescence when saturated with ether or chloroform. Other workers have also shown direct effects of anaesthetics on the firefly chemilumines-cent reaction, in addition to anaesthetising effects on the beetle's nervous system or the luminous cells. Franks and Leib have ingeniously followed this observation through to the firefly luciferase itself (Fig. 5.6b). They have shown that there is a remarkable correlation between the concentration of anaesthetic for half maximal inhibition of the luciferase and half maximal potency of anaesthesia. This suggested to Franks and Leib that the hydrophobic luciferin binding site may be very similar to the protein binding site for general anaesthetics in nerve cell membranes. By using recombinant DNA techniques it should be possible to use a cloned sequence of cDNA from the firefly luciferase gene to pick out in a human cDNA library protein(s) which have similar amino acid sequences to the luciferin binding site. Thus the target protein(s) for general anaesthetics can be identified, if this hypothesis is correct.

Genetic engineering
Transfection of foreign DNA into animal and plant cells provides a powerful new approach to the study of cell differentiation, the treatment of genetic abnormalities, and the improvement of foodstuffs. Yet a problem in developing the optimal conditions for this is how to assess, particularly in the initial experiments, whether the gene or DNA fragment being incorporated into a foreign cell is actually being expressed to give mRNA and protein, and how well. By coupling the firefly luciferase to the same RNA polymerase promoter region in the transfected DNA, both genes must be expressed together. Addition of firefly luciferin to the transfected cells thus enables an assessment of gene expression to be made, ultimately down to the single cell level. This has been achieved in the tobacco plant. The firefly gene is transfected into the young plant and the leaves then glow when sprayed with luciferin.

5.3 PYRIDINE NUCLEOTIDES AND LUMINOUS MARINE BACTERIA
The oxidoreductase (5.19)–luciferase (5.20) reactions from marine bacteria utilise $NAD(P)^+/NAD(P)H$, $FMN/FMNH_2$, long chain aliphatic aldehyde and O_2:

$$NAD(P)H + FMN \rightleftharpoons NAD(P)^+ + FMNH_2 \qquad EC1.6.8.1 \quad (5.19)$$

$$FMNH_2 + RCHO + O_2 \rightarrow FMN + RCO_2H + light \qquad EC1.14.14.3 \quad (5.20)$$

$$RCO_2^- + NADPH + 2H^+ + ATP \rightarrow RCHO + NADP^+ + H_2O + AMP + PP_i \qquad (5.21)$$

These reactions can thus be used to measure any one of these metabolites, as well as enzyme reactions producing or degrading them. Substances capable of interfering with the kinetics of these reactions, e.g. proteases degrading luciferase, can also be analysed. Because of the inherent instability of $FMNH_2$, which autoxidises sponta-neously and rapidly in the presence of O_2 in a 'dark' reaction, the kinetic relationship between analyte and light emission is different from that between ATP and the firefly luciferin–luciferase. A third set of reactions is involved in bacteria, to generate the long chain aldehyde (see Chapter 3) from fatty acid or fatty acyl coA (5.21). These can be used to assay fatty acids or coA derivatives. Light intensity from the luciferase depends on how much aldehyde is present.

5.3.1 Sources

As we discovered in Chapter 3, free-living bacteria capable of being luminous are found throughout the oceans of the world (Table 3.7). Several fish, including the celebrated flashlight and angler fishes (Fig. 3.3), and some luminous squid, contain specialised organs with cultures of luminous bacteria, controlling light emission by rotating the organ or by flaps of skin enabling the organism to 'flash'. It was thought at one time that such bacteria were restricted to the sea. However, a few examples of luminous bacteria have been found in fresh water and in the soil, including one infecting a nematode, making it luminous. The marine bacteria readily infect decaying animal matter, such as dead fish or invertebrates, causing then to glow at night. However, it is tempting to believe that the spectacular cases of 'glowing' meat observed by many people, including Boyle (Fig. 1.3a) during the 17th century were caused by terrestrial, luminous bacteria rather than by a marine species as has usually been presumed.

Luminous bacteria also occur as intestinal or parasitic symbionts, without necessarily emitting much light externally. It has even been suggested that the luminescence of the tunicate *Pyrosoma* (Fig. 3.b(i)) is caused by *intra*cellular bacteria. This is based on the identification of bacteria-like organelles within the luminous cells, and on detection of bacterial luciferase in extracts. The latter observation may be difficult to interpret because of the ubiquitous occurrence of luminous bacteria in and around marine organisms.

The taxonomy of luminous bacteria has long been a source of controversy (see Chapter 3 section 3.2.3). Three species have been used as a chemiluminescence source for biomedical applications: *Vibrio fisheri*, *Photobacterium phosphoreum*, and *Vibrio harveyi*. These were known formerly as *Photobacterium fischeri*, *Photobacterium phosphoreum*, and *Beneckea harveyi* respectively. They emit light continuously, mainly in the blue region (*ca.* 470–500 nm), though one yellow emitting strain (Y1) of *P. phosphoreum* has been found by random culturing of sea water samples (see Chapter 9 for explanation of colour).

Several species and different strains of individual luminous species are available from culture banks (see Appendix V), and can be stored for long periods freeze dried or in 50% glycerol at -70–$80°C$ or liquid N_2. They do not store well, in general, on agar slopes, particularly since the luminosity is easily lost. Many media compositions have been published, and are commercially available, for growing the luminous bacteria. A simple nutrient sea water medium is, however, usually sufficient. Sea water should be natural or defined, maximal growth and luminescence requiring 0.3–0.5 M Na^+. Nutrients include glycerol, peptone, and yeast extract. Five problems have been encountered in growing the bacteria to produce maximunm luminescence (*ca.* 10^3–10^4 photons per cell per second).

(1) Complete media may contain an inhibitor(s) which delays synthesis of the three

protein groups responsible for chemiluminescence (luciferase, oxido-reductase, long chain acid reductase). The inhibitor can be removed by first growing bacteria until induction occurs and then transferring to a minimum medium not containing the inhibitor(s).

(2) An autoinducer (Fig. 3.5) is required to switch on the operon containing the genes for the luciferase and oxidoreductase. Each species of bacterium appears to produce a different autoinducer, which has to reach a critical concentration before induction of protein synthesis will occur. This may explain why isolated bacteria present in the oceans at low concentration are not luminous, and explain the difficulty some people have in generating brightly emitting flasks containing bacteria in the laboratory.

(3) Glucose, and to a lesser degree some other sugars, cause catabolite repression of luciferase synthesis. This can be reversed by cyclic AMP.

(4) Although O_2 is the substrate for the luciferase and gassing a culture may increase luminosity, it is not necessarily beneficial for the development of the chemiluminescent system. In some species luciferase synthesis can be higher in low O_2 concentrations; in others the reverse may be true. Thus, the optimal O_2 concentration (O_2 tension) needs to be defined for each species.

(5) Growth on amino acids causes acetic acid, and on glucose pyruvic acid, to accumulate in the medium. A decrease in pH can inhibit luciferase synthesis.

Only when these problems are circumvented will maximum production of the chemiluminescent system be achieved. More recently the lux genes for the two polypeptides in the luciferase together with the rest of the operon have been cloned in *E. coli*. Addition of long chain aldehyde to these *E. coli* caused them to luminesce.

The reactions of the three proteins, luciferase, oxido-reductase, and long chain acid reductase are shown in equations (5.19)–(5.21). Only the first two are required for most biomedical applications since synthetic long chain aldehyde can be added to the reaction *in vitro*. As many as three oxidoreductases (FMN reductases) have been found in an individual species (e.g. *V. harveyi*). This can be useful as the selectivity for NADH or NADPH varies with each enzyme. One is selective for NADH (K_m NADH 23 μM, K_m for NADPH ten times K_m NADH), a second is selective for NADPH (K_m NADPH 22 μM), the third is equipotent with both (K_m NADH 71 μM; NADPH 44 μM). The selectivity is further increased by a 4–10-fold difference in V_{max} between the two substrates. In contrast *V. fischeri* appears to have only one enzyme, equally active with NADH or NADPH. Similar reductases are also present in non-luminescent bacteria, including *E. coli*.

5.3.2 Preparation of reagents

In the fully induced bacteria more than 5% of the soluble protein is luciferase. This, together with the oxidoreductase, can be released from stored bacteria (*ca*. 500 g) by freezing–thawing and osmotic lysis. The enzymes are separated and purified from the soluble fraction using DEAE cellulose batch elution (0.15 M to 0.35 M phosphate pH 7), ammonium sulphate precipitation (40 to 75% cut), and DEAE Sephadex chromatography. The remaining 5–10% protein impurities can be removed from the luciferase by rechromatography on DEAE Sephadex or by using G200 Sephadex or aminohexyl Sepharose 4B or 6B. The overall yield of pure luciferase is 30–50%. The buffers usually contain dithiothreitol and EDTA as protecting and antimicrobial agents respectively.

The oxidoreductases can be separated from the luciferase and the individual isoenzymes from each other by using DEAE cellulose, 40–75% ammonium sulphate, followed by Sephadex G100 and affinity purification on an NADP-agarose column (elution with NADP) for the NADPH-FMN reductase, or 5'AMP-Sepharose 4B (elution with NAD) for the NADH-FMN reductase or non-selective enzyme.

A FMN Sepharose 4B column has also been used for the latter. A problem with the oxidoreductases are their extreme instability, even in the presence of thiol-protecting agents (0.1 mM). More than 90% of their activity can be lost on storage at 4°C or −20°C, or through freeze–thawing. However, >50% activity can be maintained using 0.1–1 mM dithiothreitol + bovine serum albumin (0.2%) followed by storage in liquid N_2 or freeze-drying. Removal of oxygen improves stability if the enzyme is to be stored for several months. A freeze-dried preparation is available commercially (see Appendix V).

The luciferase also requires protection with dithiothreitol. Unless care is taken, activity can be lost on freeze-drying. It is rapidly inactivated by proteases if solutions become contaminated with bacteria. Luciferase is best stored frozen at *ca.* 20 mg/ml in $(NH_4)_2 SO_4$ or phosphate pH $7.0 + 0.1$–0.5 mM dithiothreitol, and 1 mM EDTA or 0.02% sodium azide to reduce microbial growth. However, satisfactory freeze-dried preparations are available commercially (see Appendix V).

From 500 g of bacteria it is possible to obtain nearly 400 mg (5 μmol) of luciferase and 1–3 mg (*ca.* 60 nmol) of the NADH or NADPH selective reductases.

The lux A and B genes coding for the α and β subunits of the luciferase from *V. harveyi* have been cloned in *E. coli* from the DNA, using plasmids and the phage construct, charon 13. A mixed oligonucleotide probe, based on a partial amino acid sequence, was used originally to identify clones. By using an expression vector it is possible to increase the enzyme content per cell up to 30–50%. However, for purification of normal luciferase it is simpler to use the original bacteria as the source of the luciferase.

5.3.3 Properties of the chemiluminescent system

Some properties of the bacterial luminous system are listed in Table 5.4.

Bacterial luciferase is a heterodimer with a mean molecular weight of about 80,000 ($\alpha = 42,000$; $\beta = 37,000$). Mutant enzyme studies, together with chemical modification, have shown that there is a single active centre primarily on the α subunit, yet both α and β are required for luciferase activity. The complete amino acid sequences of both subunits have been obtained via cloning (Fig. 3.18) using an M13 cloning vector and the dideoxy method of Sanger. They have considerable homology, particularly at the amino terminus, suggesting that the two genes, lux A and lux B, arose by tandem mutation from an ancestral gene.

In 1915 E. N. Harvey demonstrated an O_2 requirement for the luminescence of moistured dry bacteria. Yet in spite of the assumptions of many workers up to the early 1950s, he failed to isolate the presumed luciferin. The route to resolving this puzzle began with the discovery by Cormier and Strehler in 1953 that extracts of kidney cortex was an effective stimulant of bacterial chemiluminescence *in vitro*.

Table 5.4 — Properties of the bacterial luminous system

1. Components	FMN, NAD(P)H, O_2, long chain aldehyde (C_8–C_{14}) luciferase, oxido-reductase
2. Light emitter	hydroxy derivative of FMN
3. Reaction products	NAD(P), FMN, H_2O, RCO_2H
4. Colour of luminescence	(*in vivo* varies from 470–505 nm) *in vivo P. fischeri* blue (λ_{max} 492 nm) *P. phosphoreum* blue (λ_{max} 476 nm) *P. phosphoreum* Y1 yellow (λ_{max} 545 nm) *in vitro* all blue, λ_{max} 492 nm

5. Luciferase = alkanal, reduced-FMN: oxygen oxidoreductase (1-hydroxylating, luminescing) EC 1.14.14.3

(a) Structure	mol. wt. *ca.* 80 000, heterodimer (α = 42 000, β = 37 000)	
	α = 42 000	β = 37 000
(b) K_m for substrates	FMN H_2	0.3 μM
	O_2	10 μM
(c) Inhibitors	SH reagents, e.g. p Cl mercuribenzoate, lysine, and histidine reagents, Cu^{2+}, Zn^{2+}	
(d) pH optimum	6.4–7.2	
(e) Temperature optimum	28°C	
(f) 1.6×10^{11} photon units/s/mg luciferase		
k cat = 0.3 (decanal s^{-1})		

6. Oxidoreductase properties (FMN reductase or NAD(P)H:FMN oxidoreductase) EC 1.6.8.1. ·

(a) Structure	monomer 30–50 000	
(b) K_m for substrates	NADH selective	NADH 23 μM; NADPH 250 μM
in *Vibrio harveyi*	NADPH selective	NADPH 22 μM
	unselective	NADH 71 μM, NADPH 44 μM
	FMN	2–80 μM
(c) pH optimum	broad range (5–10, some optimal at 6–6.4)	

7. Other properties
Chemiluminescence quantum yield $\phi_{CL} = 0.1$

Careful analysis showed that the active ingredient was palmitaldehyde. In fact, aldehydes with six carbons and up were effective. Although it was originally thought that the aldehyde was catalytic, in 1955 McElroy and Green showed loss of aldehyde during the chemiluminescent reaction, suggesting it might be converted to its acid. Thanks particularly to the elegant and detailed work of Hastings and his colleagues the complete chemiluminescence reaction in luminous bacteria has been defined (Eqns (5.19)–(5.21), Chapter 3, section 3.4.3).

NADH or NADPH reduce FMN to $FMNH_2$, catalysed by the oxido-reductase. Some of the $FMNH_2$ binds to the α subunit of the luciferase, resulting in chemilumi-nescence after oxidation by O_2 and reaction with the long chain aldehyde. However, the turnover of the luciferase is very low, the time for a single enzyme cycle being about 10 s. This is one of the slowest known, so any $FMNH_2$ not immediately bound autoxidises spontaneously with a half time of about 0.1 s but without producing any light. Thus, the remaining unbound $FMNH_2$ contributes nothing to the chemilumi-nescence as it is autoxidised, either before it can bind to free luciferase or if other luciferase molecules become free to bind another $FMNH_2$. The enzyme therefore turns over only once, *in vitro* at least.

The enzyme bound $FMHH_2$ is oxidised to a stable peroxy intermediate (Fig. 3.15b), which has been isolated bound to the luciferase at low temperature, ($t_{\frac{1}{2}}$

at $0°C = 30$–60 min at high ionic strength). FMN cannot be the actual light emitter since not only does it fluoresce with maximum emission at 530 nm, whereas the *in vitro* chemiluminescence maximum is 490 nm, but also when bound to the luciferase FMN is vitually non-fluorescent. In the absence of long chain aldehyde the peroxy-$FMNH_2$ derivative breaks down to produce FMN and H_2O_2 in a 'dark' reaction with a ϕ_{CL} of 10^{-3}–10^{-5}. However, in the presence of the aldehyde, an excited state 4a hydroxide is formed, resulting in light emission with an overall chemiluminescence quantum yield of about 0.1 (i.e. 10%), followed by the release of acid and FMN. Aliphatic aldehydes from C_8 to C_{16} are all good substrates, though the enzyme turnover varies with each, from 2–20 s depending on the luciferase and aldehyde chain length. Both $FMNH_2$ and the aldehyde bind to the α subunit of the luciferase, though the β subunit is essential for luciferase activity. The role of the β subunit is still not fully understood. $FMNH_2$ is by far the best hydrogen source, but $FADH_2$ and other flavin derivatives appear to have some weak activity.

5.3.4   Measurement of FMN, NAD(P)$^+$, and NAD(P)H

In principle, the bacterial luciferase-oxido-reductase can be used to measure any one of the components of the reaction (Table 5.5), though measurement of $FMNH_2$ is really impracticable because of its extremely short half life ($t_{\frac{1}{2}}$ *ca.* 0.1 s in air saturated

Table 5.5 — Applications of the bacterial system

---

1. Direct reactants
    NADH, NADPH, FMN, $FMNH_2$,long chain aldehyde (including certain insect pheromones), amino ethyl NADH (AENADH) as a non-isotopic label in immunoassay, O_2

2. Metabolites or enzymes coupled to NAD(P)H formation
    NAD^+, $NADP^+$, FAD, glucose, G6P, G/P, lactate, various glycolytic inter-mediates, malate, glycerol, glycerol-3-phosphate, NH_3, 3-hydroxybutyrate, ethanol, oestrogen, glycogen, trinitrotoluene, myo-inositol, free fatty acids, various dehydrogenases (malate, lactate, G6P, 3OH butyrate, isocitrate), lipase and phospholipase, ATP-NMN adenylate transferase, hexokinase, amino transferases

3. Metabolites or enzymes coupled to NAD(P)H utilisation
    oxidised glutathione, 3 OH butyrate, oxaloacetate, pyruvate

4. Destruction of luciferase
    protease, e.g. trypsin

5. Solid phase reagents (luciferase + oxidoreductase ± coupling enzymes), consecutive metabolites and enzymes on various pathways, testosterone, progesterone using steroid dehydrogenase, free and conjugated bile acids using 7α-hydroxy-steroid dehydrogenase, ethanol, homogeneous immunoas-say (e.g. for progesterone and α fetoprotein)

See section 5.7 for references and in particular Campbell and Simpson (1979), De Luca (1978, 1987), Whitehead *et al*. (1979), Jablonski & De Luca (1982), Bergmeyer (1985), Térouanne *et al*. (1986).

media at room temperature). The requirements therefore are: (a) the two enzymes, luciferase and FMN:NAD(P)H oxido-reductase, selecting the latter for specificity to NADH or NADPH if required; (b) FMN+ reduced pyridine nucleotide; (c) long chain aldehyde, and (d) O_2. The concentration of O_2 in air saturated media is about 210–250 μM, depending on temperature. Since the K_m for O_2 is about 10 μM, oxygen is unlikely to be a limiting factor in *in vitro* assays.

Three polypeptides (34K, 50K, and 58K) are involved in aldehyde synthesis in *P. phosphoreum*; one + ATP forms an acyl-enzyme, the other two are involved in reduction of this to aldehyde. *Vibrio harveyi* uses acyl-CoA to reduce fatty acid to aldehyde. However, *in vitro* addition of *ca.* 50 μM aldehyde (a few μl of a saturated solution in methanol) provides ample substrate for the luciferase. High concentrations can be inhibitory. Tetradecanal (C_{14} = myristic aldehyde) has been popular. It is solid at room temperature so some workers prefer to use the liquid decanal (C_{10}). FMN is used at about 0.1 mM, unless it is itself being measured. Luciferase is usually at a concentration of about 100 μg ml^{-1} (*ca.* 1 μM) in phosphate pH 7 containing dithiothreitol (0.1–1 mM) and bovine serum albumin 0.1–1 mg ml^{-1}. Sometimes EDTA is present to remove inhibitory Ca^{2+} or Zn^{2+}. CN^-, N_3^-, and F^- at 1 mM are not inhibitory. Many of the published assays use the small amount of oxido-reductase contaminating most luciferase preparations. However, greater specificity and sensitivity can be achieved by using purified oxido-reductase and luciferase added in defined amounts. Addition of the reductase to commercial luciferase can increase light signals several-fold and speed up the reaction considerably. 1 pmol of NADH can be detected easily when using 4 μg (50 pmol) luciferase plus 50 ng (1 pmol) oxido-reductase.

What then are the kinetics of this unusual enzymatic reaction? Addition of $FMNH_2$ to luciferase + aldehyde + O_2 results in a peak in light intensity within less than a second followed by an exponential decay ($t_{\frac{1}{2}}$ 5–10 s, $k \doteq 0.2$ s^{-1}) as the enzyme bound $FMNH_2$ reacts. Under assay conditions, when NAD(P)H is added the light emission reaches a peak in 1–2 s and decays as the NAD(P)H is consumed (Fig. 5.7A). The intensity of light at any particular time is directly proportional to the amount of E.$FMNH_2$−OOH, and the total light emitted is proportional to the amount of $FMNH_2$ which reacts through the light emitting route. Any $FMNH_2$ which oxidises spontaneously contributes nothing to the chemiluminescence. Thus the reaction proceeds as follows: addition of NAD(P)H to the assay cocktail leads to the production of $FMNH_2$. Within a few seconds a pseudo steady-state is reached and the maximum concentration of $FMNH_2$ is achieved. Some of this $FMNH_2$ binds to the enzyme and produces light; the rest autoxidises to give no light. The peak in light emission at maximum $FMNH_2$ will be directly proportional to the concentration of NAD(P)H added, providing the initial NAD(P)H concentration is $\ll K_m$ of the oxido-reductase, i.e. on the linear part of the Michaelis curve. The lowest reported K_m is *ca.* 20 μM, thus the peak height will be linearly related to NAD(P)H concentration up to about 1 μM (i.e. 1 nmol.ml^{-1}). Since the concentration of $FMNH_2$ at any time is much less than the K_m for the luciferase, the total light produced during the reaction is also linearly related to the initial concentration of NAD(P)H. The NAD(P)H will eventually all be consumed since O_2, aldehyde and FMN are in excess.

Paradoxically the decay constant (Fig. 5.7A) is *independent* of NAD(P)H

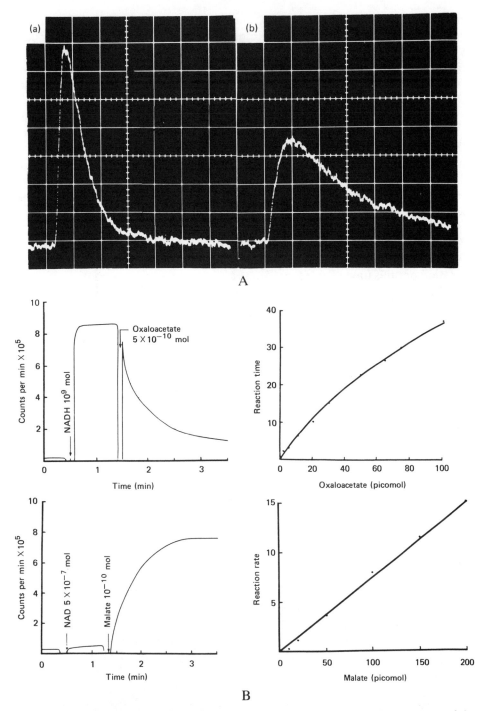

Fig. 5.7 — NAD(P)H and bacterial luciferase. A. Typical chemiluminescence trace: a, 2.4 pmol NADH; b, 2.4 pmol NADPH. From Brolin *et al.* (1972) *Anal. Biochem.* **50** 50–55, with permission of Academic Press. B. Assay of oxaloacetate or malate. From Stanley (1978) *Methods Enz.* **57** 181–188, with permission of Academic Press.

concentration under these conditions, and is therefore the same throughout the standard curve. The chemiluminescence decay rate from NAD(P)H depends both on the concentration of luciferase and the reductase, and their respective turnover numbers. Crude commercial preparations with a little reductase have $t_{\frac{1}{2}}$ as long as 30 s, thus total integrals take a long time to amass. Addition of purified oxido-reductase can decrease the half time to just a few seconds. However, in view of the competition between the light and dark reactions of $FMNH_2$ the amount of reductase and luciferase should be titrated to produce the maximal signal, with minimum blank ($-NAD(P)H$), in the shortest possible time. Peak light intensity is linear over the range 1–1000 pmol NAD(P)H. The useful range is 10 pM–10 μM (1 fmol–1 nmol in 0.1 ml), and as little as 0.1 fmol may be detected by reducing volumes and optimising signal to noise. The major limitation of the assay is the high blank intensity with no added NAD(P)H. Some of this may be due to contamination in the reagents.

5.3.5   Coupling to other metabolites and enzymes
A wide range of metabolites and enzymes have been coupled to either NAD(P)H formation or utilisation (Table 5.5):

$$AH_2 + NAD(P)^+ \overset{E}{\rightleftharpoons} A + NAD(P)H + H^+ \tag{5.22}$$

Formation

$$\text{Glucose} + \text{ATP} \xrightarrow{\text{hexokinase}} \text{glucose 6 P} + \text{ADP} \tag{5.23}$$

$$\text{glucose 6 P} + \text{NADP} + \xrightarrow[\text{dehydrogenase}]{\text{G6P}} \text{gluconate P} + \text{NADPH} \tag{5.24}$$

Utilisation

$$\text{30H butyrate} + \text{NADH} \xrightarrow[\text{dehydrogenase}]{\text{30H butyrate}} \text{acetoacetate} + \text{NAD}^+ \tag{5.25}$$

$$\text{oxidised glutathione} + 2 \text{ NADPH} \xrightarrow[\text{reductase}]{\text{glutathione}} 2\text{GSH} + \text{NADP}^+ \tag{5.26}$$
$$(\text{GS}-\text{S G})$$

Metabolites can be detected easily in the range 1 pmol–1 nmol and some enzymes down to 1 fmol. Four problems can arise, limiting specificity and sensitivity, or making the assay cumbersome.

(1)   Lack of selectivity for one pyridine nucleotide
Tissue samples will contain NAD^+, $NADP^+$, NADH, and NADPH.
   Selective destruction of the oxidised form can be achieved with dilute alkali, or of the reduced form using strong acid. Selectivity of NADH over NADPH or vice versa is best achieved by using a selective oxido-reductase.

(2)   Blank
Many biomedical reagents are contaminated with small amounts of NAD(P)H. It is not clear whether this is the sole cause of the substantial reagent blank, some hundreds of chemiluminescence counts, in many assays.

(3)   High redox potential of NAD⁺/NADH

(3)   High redox potential of NAD ⁺/NADH

The equilibrium position of many dehydrogenases favours NAD^+. This can be improved by using acetyl NAD^+.

(4)   Reaction kinetics

Because of the unusual 'dark' v 'light' reaction of $FMNH_2$ it is not so easy to establish the simple, flexible, linear relationship in *continuous* assays achievable with the firefly system. Most assays are therefore *discontinuous*. The change in redox state of the pyridine nucleotide is carried out first, the enzymes killed, selective destruction of the oxidised form carried out if required, and the remaining NAD(P)H added to the luciferase-oxido-reductase cocktail. Peak height or integral are linearly related to the original metabolite concentration or enzyme activity. Continuous assays are, however, possible, adding the metabolite or enzyme to the complete reaction cocktail, i.e. coupling system to NAD(P)H and luciferase/oxido-reductase. Light intensity is measured against standards of the metabolite or enzyme. Standard curves are not necessarily linear, though careful titration with small amounts of the oxido reductase can enable linear enzyme standard curves to be constructed. Solid phase reagents solve many of these problems (section 5.3.6).

Other useful applications of the bacterial system include the use of enzyme cycles to amplify the production of NAD(P)H from an enzyme to metabolite, the measurement of NADH derivatives as non-isotope protein labels and the sensitivity of the luciferase to proteolysis as an assay for proteases in the pmol–fmol range.

A further innovation has been the use of bacterial mutants to assay substrates added directly to the bacteria. A weakly emitting mutant of *Vibrio* (formerly *Beneckea*) *harveyi* lacks the machinery to generate high intracellular aldehyde concentrations. Addition of long chain aldehydes and acids increases light emission from the bacteria, enabling pmol quantities of these substrates to be detected. Other mutants containing enzymes with altered affinities and V_{max} for the substrates can be produced, which may also aid optimisation of assay conditions.

5.3.6   Solid phase reagents

Bacterial luciferase has been coupled to acrylamine porous glass beads or polyacrylic hydrazide by diazo coupling, as well as to cyanogen bromide activated Sepharose 4B and 6B. A major problem initially was marked loss of activity, which could be 95%. However, by carefully adjusting the amount of CNBr more than 60% of luciferase activity could be retained. Coupling reduced V_{max} and decreased the affinity of the enzyme for $FMNH_2$ as indicated by the K_m being increased some 10–15 times (N.B. K_m not necessarily equal to K_d for $EFMNH_2$), yet the catalytic turnover of the enzyme was unaltered. Other modifications on solid phase may include shifts in pH optimum and chemiluminescence spectrum. Solid phase luciferase/oxido-reductase should be left for at least 24 hours before use.

Using glass rods containing arylamine beads, luciferase and reductase (molar ratio *ca.* 10:1) have been coupled simultaneously, providing a solid phase reagent capable of 200 consecutive assays for NADH with a linear range of 1 pmol–50 nmol (10 mM–0.5mM), as well as dehydrogenases in the 1 fmol—1 pmol range. The enzymes even appear more stable once they are in the solid phase than they are in free solution. Other enzymes have been co-immobilised onto the beads to create solid phase NADH or NADPH generating pathways, e.g. G6P dehydrogenase/ oxido-reductase/luciferase for glucose, and alcohol dehydrogenase/oxido-reductase/ luciferase for ethanol.

Initial light intensity, in the presence of other reactants in excess, is proportional to the concentration of analyte. As with continuous soluble assays the linear relationship between maximum light intensity, or total light output, and the analyte concentration, either metabolite or enzyme, has to be established with standards. Unlike spectrophotometry, light output cannot easily be related directly to absolute units of enzyme activity.

A particularly ingenious application of solid phase reagents has been developed by Crastes de Poulet and co-workers. They coupled the luciferase and oxido-reductase from *V. harveyi* to Sepharose 4B, together with an antibody to an analyte (e.g. progesterone or α fetoprotein) they wished to assay. A fixed amount of pure antigen, labelled with an NAD glucose-6-phosphate dehydrogenase from the bacterium *Leuconostoc mesenteroides*, was added with varying amounts of unlabelled analytes (N.B. the mammalian G6P dehydrogenase uses NADPH). Thus some of the dehydrogenase labelled antigen bound to the antibody on the solid phase. Addition of glucose generated NADH, but the NADH generated within the solid phase produced much more light than NADH generated from unbound dehydrogenase. This effect was presumably the result of the instability of $FMNH_2$. Only $FMNH_2$ produced close to the luciferase will generate light. The selective effect was enhanced by addition of NADH scavengers such as lactate dehydrogenase + pyruvate, FMN oxido-reductase or diaphorase in the medium, reducing still further any $FMNH_2$ reaching the solid phase luciferase from free solution. Since it was unnecessary to separate bound from free antigen, this can be considered as a 'homogeneous' immunoassay (see Chapter 8).

5.3.7 Real applications of the bacterial system
The bacterial system has not yet been used as widely in the research laboratory as the firefly system. Furthermore, in spite of the potential of the solid phase reagents, clinical chemists have yet to be convinced that the sensitivity and speed of these chemiluminescence assays offer any great advantage over the already established, highly automated spectrophotometric and fluorometric assays for enzymes and metabolites in plasma and urine. Nevertheless a small number of researchers have found the sensitivity of assays utilising bacterial luciferase invaluable. These assays have enabled enzymes and metabolites associated with glucose metabolism to be measured in ng quantities of individual islets of Langerhans. An intra-cellular metabolite of glucose has been thought to be the trigger for insulin secretion in these cells. Previous studies have had to use the laborious Lowry cycling techniques, not always successfully. Enzymes such as isocitrate dehydrogenase have been measured in skin biopsies following exposure to drugs.

This disappointing paucity of results using bacterial luciferase as an analytical tool may change when a more imaginative conceptual approach to metabolic regulation in cells is adopted, requiring the sensitivity of this chemiluminescent system. The unusual properties of solid phase reagents offer interesting prospects for the future in the clinical laboratory and in biotechnology.

On a more positive note, Ulitzur and colleagues in Israel have developed some ingenious applications of bioluminescence in whole bacteria. These can be divided into four sections:

(1) Quantification of phagocytosis and bacterial killing by live phagocytes.
(2) Measurement of genotoxic (i.e. mutagenic) agents.
(3) Measurement of antibiotics.
(4) Ultrasensitive detection of particular pathogenic bacteria, or other microbes, using genetically engineered phages containing bacterial lux genes. This also enables antibiotic sensitivity to be assessed at the same time.

A particular problem in phagocytic assays in neutrophils, eosinophils and mononuclear cells is the difficulty in relating particle uptake to bacterial killing. The fresh water luminous bacterium *Vibrio cholerae* var. *albensis* is phagocytosed readily by polymorphonuclear leucocytes. Loss of bioluminescence indicates killing of the bacteria taken up into the cells.

Genotoxic agents cause mutagenesis and DNA damage by one of three mechanisms: base substitution for frame shift, direct DNA damage or inhibition of DNA synthesis, intercalation of DNA. The bioluminescence assay for these agents is based on their ability to restore luminescenc to dark mutants of strains of *Photobacterium leignathi* or *fischeri*; by blocking repressor formation, altering repressor or operator structure, inactivating the repressor of the luminescence operon, or causing a change in the configuration of the DNA or whole chromosome. Typical genotoxin assays range from $0.1–100$ ng ml^{-1} for mitomycin C and $10^2–10^4$ mg ml^{-1} for methotrexate.

The operon for bioluminescence in *Photobacterium fischeri* consists of 7 genes, designated in order RICDABE. R and I are regulatory genes, C, D, and E coding for enzymes responsible for aldehyde production, and A and B the lux genes for α and β sub-units of the luciferase. Transformation of bacteria by a plasmid such as pBR322 containing the DNA of the bioluminescence operon enables some 10^5 cells of *E. coli* to be detected within 10–15 h. However, as a means of detecting specific microbes the method lacks sufficient sensitivity, specificity and speed to be useful. However, by inserting the operon into bacterial specific phage DNA, particular bacterial species can be transduced, and will thus luminesce. A combination of the phages, combined with a selective culture medium using tetrathionate, enables 1–10 *Salmonella* in 1–10 ml to be detected specifically, without false positives from *E. coli* or false negatives. Several hundred strains or species of *Salmonella* have been tested by Ulitzur and coworkers with >95% success. If this genetically engineered procedure can be reduced to 1 hour then not only will it be the best test available for the food industry but also usable in clinical microbiology. Clearly the method is in principle applicable to any microbe providing a specific phage or virus is available or can be engineered. A further advantage of this method is that the sensitivity to particular antibiotics of the pathogen can be determined at the same time. The more potent the antibiotic the greater the inhibition in expression of bioluminescence in the transduced bacteria. The method has six particular strengths: it is very sensitive (down to 1–10 bacteria per sample), specific for a particular live bacterium, homogeneous (i.e. no separation procedure is required, unlike antibody dependent methods), produces no background (i.e. no false positives, the viral DNA will not produce light unless it gets into a live bacterium), result at least within 24 h and potentially quicker than this, easily automatable.

5.4   OXYGEN AND ITS METABOLITES
5.4.1   Measurement of oxygen concentration

Beijerinck was the first, in 1902, to realise that luminous bacteria could be used to detect oxygen. He placed some crushed clover leaves together with luminous bacteria plus 3% salt in a stoppered bottle in the dark. The bacteria ceased to glow once the oxygen had been removed by respiration. Exposure to sunlight, or even a lighted match, caused the bacteria to luminesce. He used this O_2 assay to define the region of the spectrum responsible for photosynthesis in the marine green alga *Ulva* and the red alga *Porphyra*. Several other workers at the turn of the century used a similar technique to study oxygen production by chloroplasts. It was also shown by Hill in 1928 that luminous bacteria could be used to measure the diffusion of oxygen through a variety of 'inert' media such as rubber and paraffin.

The concentration of oxygen in air-saturated water is about 250 μM at room temperature, and in the arterial blood about 120 μM at 37°C. These concentrations are easily measurable polarographically by using a Clarke electrode, which can detect O_2 in solution down to about 0.1–1 μM, being limited ultimately by the 'unstirred layer'. In contrast, the bacterial luciferase is virtually saturated by oxygen in air-saturated media. Yet Chance and Oshino have shown that luminous bacteria can measure O_2 over a useful range of 10 nM–1 μM, with a detection limit of 0.1 nM. They added the oxygen sample to a 5 l flask containing the bacteria, calibrated using pulses of air-saturated buffer (0.1 ml air-saturated buffer gives 5 nM O_2 in 5 l). A refinement of the technique is to enclose the bacteria, stored first at 4°C for two days, within a polypropylene or silicone rubber membrane (Fig. 5.8A). This device can be calibrated by exposing media, into which the probe is placed, to defined gas mixtures. Light emission can be linear with respect to O_2 over a range of 30 nM–8 μM, close to the value of 9 μM O_2 for half maximal luminescence. The detection limit is about 1 nM, and the response time very similar to that of an oxygen electrode ($t_{\frac{1}{2}}$ for bacterial O_2 probe 2–12 s exposed to the gas phase, or 20–100 s exposed to the liquid phase).

Why do we need such a sensitive oxygen probe if a Clarke electrode is adequate for air saturated media? There are five potential applications:

(1) Quantification of O_2 <1 μM, e.g. where anaerobes grow, at sites of inflammation in the body, and within living cells.
(2) Detection of small quantities of O_2 produced by organelles, e.g. chloroplasts.
(3) Measurement of diffusion of O_2 through relatively impermeable membranes or where O_2 utilisation produces large gradients across membranes.
(4) Measurement of the O_2 affinity of high affinity oxidases with K_ms for O_2 in the range 0.1 nM–0.1 μM. For example, terminal oxidases such as the mitochondrial cytochrome oxidase have an O_2 affinity of about 10 nM, outside the range of the Clarke electrode.
(5) Measurement of respiration rates at very low oxygen concentrations.

Life on this planet depends on oxygen, yet oxygen is also toxic. Eukaryotic cells

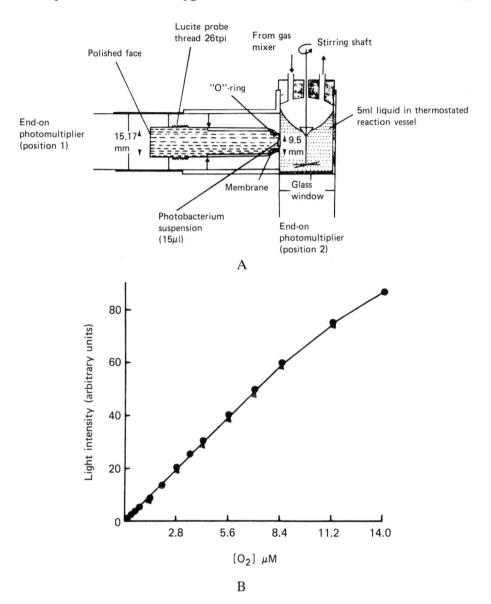

Fig. 5.8 — Measurement of O_2 using luminous bacteria. From Lloyd *et al.* (1981) *Anal. Biochem.* **116** 17–21, with permission of Academic Press. A, Apparatus. B, Calibration curve using *Photobacterium fischeri* (strain MJ-1) at 20°C.

require oxygen to provide energy, to oxidase unwanted metabolites, or to synthesise external and internal regulators. Phagocytes require O_2 to produce O_2^- and other oxygen metabolites (H_2O_2, OCl^- and $^1O_2$) as part of the killing mechanism for microorganisms and their toxins (see Chapters 1 and 6). The acute and long term relationship between the oxygen concentrations surrounding cells, the gradients of O_2 within them, and their development, together with their ability to respond to stimuli or pathogens, is not understood. This requires measurement of O_2 gradients within individual cells.

Twenty or thirty years ago, measurement of respiration and changes in oxygen concentration played a central role in biochemistry. The increasing awareness of the imprtance of O_2 gradients in living systems may lead to its revival (see Chapter 10).

5.4.2 Chemiluminescent indicators for H_2O_2 and related enzymes

For some time a number of analyte assays have been based on measurement of H_2O_2 production, using horseradish peroxidase plus an oxidisable substrate to produce a coloured product. Examples are glucose plus glucose oxidase, xanthine plus xanthine oxidase, cholesterol plus cholesterol oxidase and choline plus choline oxidase. At least six different chemiluminescent reactions have been used to establish assays for H_2O_2, together with enzymes or metabolites coupled to its formation or degradation (see Table 5.6). Several produce a linear relationship between peak height, or the integral of chemiluminescence, and H_2O_2 concentration over a range 10 nM to approximately 1 mM. One of the earliest assays to be developed used ferricyanide to catalyse luminol oxidation by H_2O_2. For maximum light emission and fast kinetics, alkaline conditions are required. Unfortunately, the reagents without H_2O_2 generated a high 'blank' light emission, and the reaction was subject to interference by small molecules and proteins present in biological samples. Nevertheless a linear relationship between peak light intensity, achieved within a few 100 ms, and H_2O_2 concentration over the range 10 nM–0.1 nM (1 pmol–10 nmol in 0.1 ml) was established.

Attempts have been made to circumvent some of the problems associated with the $K_3Fe(CN)_6$-luminol assay by using oxalate ester sensitised chemiluminescence, an energy transfer system (see Chapter 9). bis(2,4,6-trichlorophenyl) oxalate (TCPO) with perylene as the fluorescent acceptor provides an assay for H_2O_2 over a

Table 5.6 — Measurement of oxygen and its metabolites

Substance	Indicator	Approx. detection limit (useful range)
Oxygen	Intact luminous bacteria	0.1 nM (10 nM–1 uM)
O_2^-	Luminol	?
	Lucigenin	?
	Pholasin	?
H_2O_2	Luminol + peroxidase	0.1 nM (1 nM–1mM)
	Lucigenin and acridinium esters	?
	bis (trichlorophenyl) oxalate + perylene	70 nM (70 nM–1 mM)
	Pholasin + peroxidase	10 nM (10 nM–0.1 mM)
	Balanoglossus luciferin + peroxidase or its luciferase	10 nM
	Diplocardia luciferin + peroxidase or its luciferase	10 nM (10 nM–1 mM)

wide pH range, linear from 70 nM–1 mM, but some 1–2 orders of magnitude less sensitive than the best luminol assay, now using peroxidase instead of ferricyanide as the 'catalyst'. Oxalate esters also tend to require peculiar organic solvents not compatible with living systems.

Lucigenin provides a poor assay for H_2O_2 and requires alkaline conditions. An assay for blood glucose has been established by McCapra using its derivatives, the acridinium esters. However, the detection limits of these are poor because of high blanks and the problem of pseudo-base formation at pH<6 (see Chapter 8).

Three bioluminescent systems, the piddock *Pholas*, the marine acorn worm *Balanoglossus*, and the earthworm *Diplocardia*, offer promise for continuous assays where H_2O_2 is generated over several orders of magnitude in concentration, down to less than 1 nM. They can also be used as continuous assays for enzymes such as glucose oxidase, galactose oxidase, and putrescine oxidase in the pmol–fmol range. As yet, the chemiluminescent reagents for these are not readily available in sufficient quantities. The proteins have yet to be cloned. The structure of *Pholas* luciferin is unknown.

Luminol is thus still the most popular method for measuring H_2O_2 and related enzymes. Assays are carried out in two steps:

$$(1) \qquad AH + O_2 \xrightarrow{\text{oxidase}} A + H_2O_2 \qquad\qquad (5.27)$$

$$(2) \qquad H_2O_2 + \text{luminol} \xrightarrow{\text{peroxidase}} \text{aminophthalate} + N_2 + \text{light} \qquad (5.28)$$

Peroxidase is the best 'catalyst' in terms of signal-to-noise. Horesradish peroxidase can be used, but microperoxidase, a proteolytic cleavage product of cytochrome c still containing the haem, appears to be the best. It has a pH optimum of about 11, but will work at physiological pH with a reduced apparent quantum yield *and* reduced chemical blank. The assay is linear for H_2O_2 over several orders of magnitude, peak light intensity being reached within 100 ms (i.e. the mixing time without a stopped-flow apparatus). The detection limit, imposed by the chemical blank without H_2O_2, is about 0.1 μM (10 fmol in 0.1 ml).

This assay can be used in the continuous or discontinuous mode to measure acetylcholine with sufficient sensitivity to detect its release from nerve terminals of the electric organs of *Torpedo* or *Electrophorus*. Choline and acetylcholine esterase can also be measured by continuous assay by this method.

$$\text{acetyl choline} \xrightarrow{\text{esterase}} \text{acetate} + \text{choline} \qquad\qquad (5.29)$$

$$\text{choline} + O_2 \xrightarrow{\text{oxidase}} \text{betaine} + 2H_2O_2 \qquad\qquad (5.30)$$

$$H_2O_2 + \text{luminol} \xrightarrow{\text{peroxidase}} \text{aminophthalate} + \text{light} \qquad\qquad (5.31)$$

Since the luminol chemiluminescent reaction works best at alkaline pH, care must be taken to optimise the assay under conditions compatible with the biological system being studied. Using 0.5 mM luminol in the assay, choline or acetylcholine in the range 1 nM–10 μM can easily be detected.

Luminol can also be used to measure a number of other analytes containing transition metal cations (Table 5.7), including haem, catalase, peroxidase, and the cations themselves, using their ability to catalyse luminol oxidation by H_2O_2 (see

Table 5.7 — Applications of phthalazine diones

Class of analyate	Specific substance
1. Reactive oxygen metabolites	O_2^-, H_2O_2, OCl^- cellular production
2. Metabolite and enzymes producing H_2O_2	Glucose, xanthine, hypoxanthine, adenosine, inosine, cholesterol, choline, acetylcholine, diamines, polyamines, glucose oxidase, xanthine oxidase, choline oxidase
3. Metabolites and enzymes utilising H_2O_2	NAD^+ Peroxidase (horse radish, myelo-lacto-), superoxide dismutase, catalase
4. Degradative enzymes using chemiluminescent leaving group	Chymotrypsin (Boc ala-ala-phe-isoluminolamide)
5. Haem and haemoproteins	Haem Haemoglobin Cytochrome c
6. Corrin	Vitamin B_{12}
7. Trace metals	Transition metal cations (see Chapter 7) Ferritin
8. Non-isotopic labels in immunoassay and recombinant DNA technology	Diazo-or succinyl luminol Isoluminol derivatives (see Chapter 8) Peroxidase label

See section 5.7 for references.

Chapter 7). Again alkaline conditions are optimal. A haem assay, using sodium perborate as oxidant, has been suggested as a means of detecting bacteria in small numbers.

There are five main problems with luminol as a detector of H_2O_2 and related enzymes, which limit sensitivity and specificity:

(1) Luminol has a lower chemiluminescence quantum yield (ϕ_{CL} *ca.* 0.01) than most bioluminescent reactions (ϕ_{CL} 0.1–1).
(2) Reaction conditions generate high blanks, sometimes as high as 10^4 photons sec^{-1}.
(3) High quantum yield and fast kinetics require alkaline conditions (pH 11–13).
(4) The luminol reaction is not specific for H_2O_2 (O_2^- and OCl^- also react), nor for the catalyst.
(5) The reaction is subject to interference by reactive oxygen metabolite scavengers (e.g. superoxide dismutase, catalase, RNH_2, and proteins) and quenchers present in biological samples.

In spite of these problems the luminol reaction does provide a very sensitive and useful assay for H_2O_2, with many biological applications.

5.4.3 Other reactive oxygen metabolites
Addition of OCl^- to an alkaline solution of luminol results in a flash of light with a very fast decay ($t_{\frac{1}{2}} < 1\,s$). Luminol could therefore, in theory, be used to detect

OCl^-. The oxidation of xanthine or hypoxanthine, catalysed by xanthine oxidase, produces O_2^- (i.e. before H_2O_2). That O_2^- can cause luminol chemiluminescence can be shown by the inhibition of light emission by superoxide dismutase ($2O_2^- + 2H^+ \rightarrow H_2O_2 + O_2$). The ability of luminol to react with reactive oxygen metabolites has been exploited mainly in the study of phagocyte activation (see Chapter 6). Singlet oxygen can be detected by its inherent chemiluminescence or using an indicator (see Chapters 6 and 9), and ozone by inducing or enhancing the chemiluminescence of many synthetic or biological substrates. Nitric oxide might seem an unlikely candidate for the biologist to be interested in. However, it appears that a factor responsible for smooth muscle relaxation, known as endothelial derived relaxing factor, is NO. It may have importance in myocardial infarction and can be assayed by chemiluminescence. Atmospheric and soluble NO can be detected by its chemiluminescence with O_3. Reduction or photodissociation of NO_2 and NO_3^- to NO enables these to be measured (e.g. nitrate in the sea). Alternatively luminol can be used to detect NO_2 (and thus NO) down to 10 parts per 10^{12} by volume. Nitric acid and peroxyacetyl nitrate play an important part in acid rain chemistry.

5.5 PHOSPHOADENOSINE DERIVATIVES AND THE SEA PANSY
Our knowledge of the chemiluminescent system in the sea pansy (Table 5.8) is based, almost exclusively, on the beautiful work of Cormier and his colleagues, mainly using

Table 5.8 — Properties of the chemiluminescent system from the sea pansy *Renilla reniformis*

1. Components *in vitro*: luciferin, O_2, luciferase
2. Chemiluminescence quantum yield *in vitro*: $\phi_{CL} = 0.06$ (λ_{max} 480 nm)
3. Addition components *in vivo*: luciferyl sulphate, sulphokinase, luciferin binding protein, green fluorescent protein
4. Luciferin (coelenterazine = 2-(p-hydroxybenzyl)-6-(p-hydroxyphenyl)-3,7-dihydro-8 benzyl-imidazo[1,2-a)pyrazin-3-one

composition	$C_{26} H_{21} O_3 N_3$
mol. wt.	423

5. Luciferase

glycoprotein	3% carbohydrate
monomer	mol. wt. $= 35\,000$
K_m luciferin	30 nM
K_i oxyluciferin	23 nM
K_m O_2	?
pH optimum	*ca.* 7.6
turnover number	111 min^{-1}

6. Luciferin binding protein

monomer	mol. wt. $= 18\,500$
2 Ca^{2+} sites	K_d Ca $= 0.14\ \mu M$

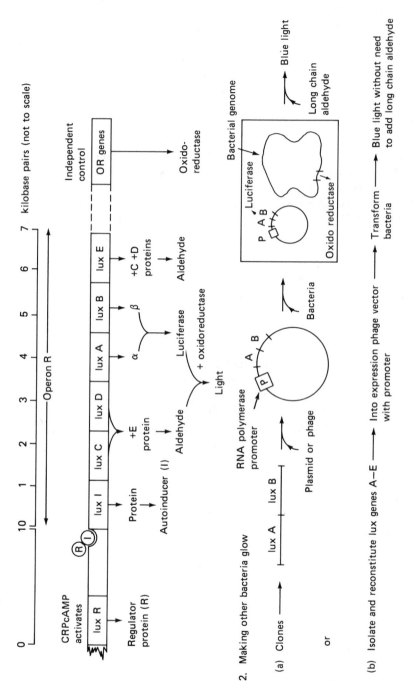

Fig. 5.9 — Applications of bacterial lux genes through genetic engineering.

Renilla reniformis. In such anthozoans, as in other luminous coelenterates (see Chapter 3), the chemiluminescent reaction is an oxidation of an imidazolopyrazine coelenterazine, (Figs 3.16 and 5.10). However, the luminous anthozoans have a conventional O_2 requiring luciferin–luciferase reaction (Fig. 5.10), in contrast to luminous hydrozoans, scyphozoans, and ctenophores which have the coelenterazine covalently linked to the luciferase as a photoprotein. In *Renilla* itself, in addition to the luciferin and oxygen, four other components are involved in the chemiluminescent reaction. Firstly there is a store of luciferin in the cell in the form of luciferyl sulphate (Fig. 3.16). This is converted to the luciferin by a sulphokinase, the sulphate being transferred from the carbonyl group of the luciferin to 3′phosphoadenosine 5′phosphate (PAP) to form 3′phosphoadenosine 5′phosphosulphate (PAPS). The luciferin binds to another protein, a luciferin binding protein, until it is made available to the luciferase by the binding of Ca^{2+} to the luciferin-binding protein. The final complication is that in the animal after the luciferin is oxidised to O_2, catalysed by the luciferase, to form the excited oxyluciferin monoanion an energy transfer takes place to the fluorophore on a 'green fluorescent protein' which becomes the actual light emitter (λ_{max} *ca.* 508 nm; blue-green).

Since the sulphokinase reaction is reversible it can be used to measure either PAP or PAPS continuously by measuring the chemiluminescence of the luciferin formed by the reaction.

5.5.1 Preparation of reagents
Only three of the components of the complete reaction *in vivo*, plus O_2, are required to measure PAP or PAPS; luciferin sulphate or luciferin, the sulphokinase, and the

Fig. 5.10A.

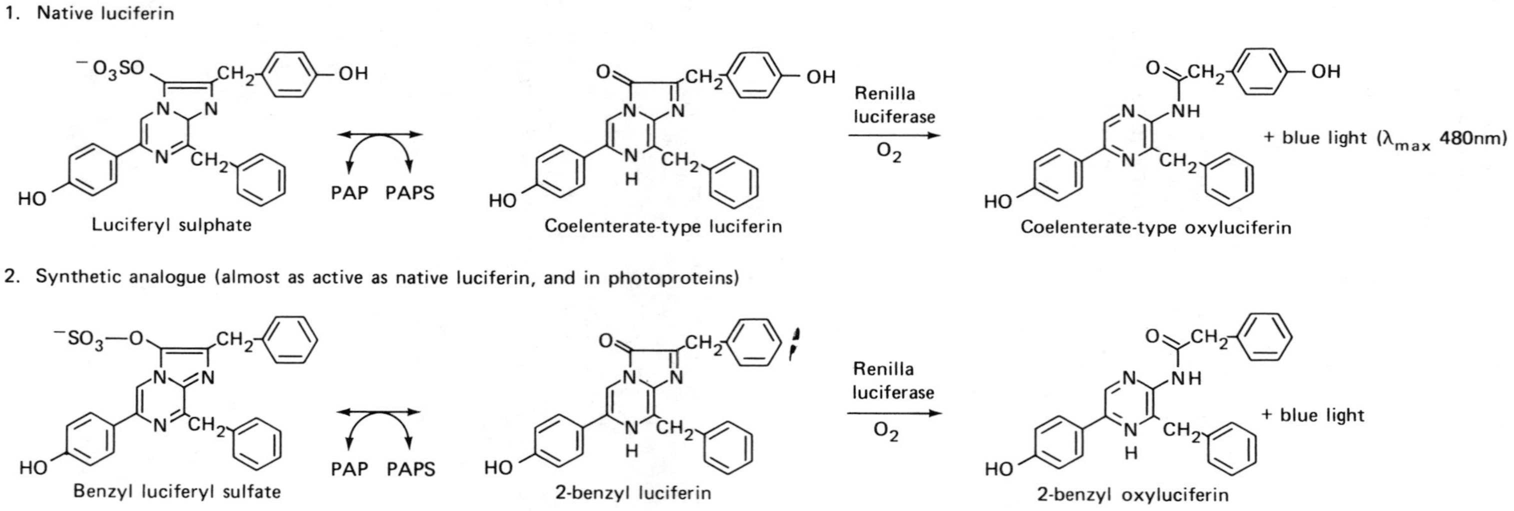

Fig. 5.10 — A: Sea pansy *Renilla reniformis* from Herring (1978), with permission of Academic Press. B: Reactions in the sea pansy *Renilla*.

luciferase. λ_{max} for light emission without the green fluorescent protein is about 480 nm (blue).

Some 40 000 *Renilla* are required to provide 5 mg (*ca.* 10 μmol) of luciferin. Coelenterazine can be synthesised, but it is easier to synthesise the benzyl derivative (Fig. 5.10), which is oxidised by the luciferase at the same rate as the native luciferin, with the same quantum yield and chemiluminescence spectrum. Benzyl luciferin sulphate is synthesised by condensing the methyl ether of 2-amino-3' benzyl-5' (*p*-methoxyphenyl) pyrazine (etiobenzyl luciferin) with the diethyl acetal of benzyl glyoxal. 1 mg of benzyl luciferin sulphate is sufficient for 16 000 assays.

The two enzymes required can be purified together, by a slightly different procedure from that required for homogeneity of the individual proteins. Several species of luminous anthozoans can be found in the shallow waters off both the Florida and California coasts (Table 5.9). They can be purchased (see Appendix V). Cormier's group have used mainly *Renilla reniformis* found in large numbers off the coast of Georgia, USA. Some modifications in the purification procedure may be necessary with each species.

Fresh animals (*ca.* 100–200 g), or specimens stored frozen at $-80°C$ in a 'relaxing' medium (tris, EGTA, Mg^{2+}, pH 7.8), are homogenised in tris + EDTA, + mercaptoethanol pH 7.5. The soluble components are purified by salt-gradient elution from DEAE cellulose, gel filtration on Sephadex G100, and then DEAE agarose, after a preliminary cleaning up and desalting of the crude suspension by centrifugation, alumina gel, and BioGel G-100. The luciferase and sulphokinase separate slightly on the last column and are mixed together to provide the final reagent. Luciferin is removed from the binding protein during the purification by adding Ca^{2+}. There should therefore be no luciferin contamination in the reagents. However, some PAP

Table 5.9 — Some luminous anthozoans

Common name	Order	Genus (species)
Sea pansies	Pennatulacea	*Renilla* (*reniformis, kolikeri, mulleri*)
		Acanthoptilum (*gracile*)
		Stylatula (*elongata*)
		Veretillum
		Cavernularia
		Funiculina
		Umbellula
		Virgularia
Sea pens	Pennatulacea	*Ptilosareus guerneyi*
		Pteroides
		Pennatula
Zoanthids	Zoanthiniaria	*Parazoanthus*
		Epizoanthus
Phylum: Cnidaria; Class: Anthozoa		

does come through the procedure, limiting the sensitivity of the eventual assay. The yield of luciferase + sulphokinase is about 30–40%, and the proteins can be stored stable at $-80°C$.

5.5.2 Assays for PAP and PAPS

The assay for PAP requires about 125 nM benzyl luciferin sulphate ($BLSO_4$), and $3–6 \times 10^{-6}$ units of the enzymes in buffer pH 7.6. The reaction can be started either with PAP (0.1–10 nM) or benzyl luciferin sulphate ($BLSO_4$):

$$PAP + BLSO_4 \xrightarrow{\text{sulphokinase}} PAPS + BL \qquad (5.32)$$

$$BL + O_2 \xrightarrow[\text{luciferase}]{} B \text{ oxy } L + CO_2 + \text{light.} \qquad (5.33)$$

Light emission rises slowly over the first 2–3 minutes, reaching a plateau at about 4–5 minutes. A plot of the plateau of light intensity versus PAP is linear over the range 1 pmol–15 pmol in the assay. A more sensitive standard curve (0.1–1 pmol) can be obtained by using twice as much enzyme. This assay for PAP is more sensitive and simpler than the spectrophotometric one based on phenol sulphokinase.

PAPS can be measured by loss of benzyl luciferin using the reverse reaction of equation (5.32), or by converting it to PAP using a bacterial sulphatase:

$$PAPS \xrightarrow{\text{sulphatase}} PAP + SO_4. \qquad (5.34)$$

Enzymes associated with PAP metabolism can also be measured by using the *Renilla* system:

$$MgATP + SO_4 \xrightarrow[\text{sulphurylase}]{\text{ATP}} APS + PPi + Mg \qquad (5.35)$$

$$APS + Mg \ ATP \xrightarrow[\text{kinase}]{\text{APS}} PAPS + ADP + Mg \qquad (5.36)$$

$$\text{(Benzyl) luciferin disulphate} \xrightarrow[\text{sulphatase}]{\text{aryl}} \text{(benzyl) luciferin sulphate} + S0_4 \qquad (5.37)$$

5.5.3 Applications

This is an area of nucleotide metabolism perhaps less familiar than most.

Some micro-organisms have the ability to reduce SO_4^{2-}, $S_2O_3^{2-}$ to S. These act as the terminal electron acceptor of the transport chain. This is in contrast to eukaryotes, where the electron transport chain in mitochondria ends in cytochrome oxidase and the reduction of O_2 to H_2O_2. Higher plants also have to utilise SO_4. The initial step in SO_4 utilisation is the formation of PAPS by a two step process using a sulphurylase and a kinase. The electron transport chain then reduces the SO_4, using reducing equivalents from NADPH. However, PAPS also acts as a general donor of SO_4 for esterification of alcohols, phenols, polysaccharides, and for the formation of cerebroside sulphate. PAP, PAPS, and their associated enzymes are therefore found in most pro- and eukaryotic cells. They are found at a concentration some 10 times lower than ATP, values varying from *ca.* 20 nmol per g wet wt in mammalian liver to *ca.* 0.5 nmol per g in higher plants.

Removal of PAP contamination in the assay reagents may enable fmol–amol amounts to be assayed, sufficient for analysis of just a few cells.

5.6 THE FUTURE

Synthetic and bio-chemiluminescent reactions provide a uniquely sensitive and rapid method for measuring many enzymes and metabolites of both research and clinical interest. Many are measurable easily in the 1 pmol–1 fmol (10^{-12}–10^{-15} mol) range. Several are potentially measurable down to 1 amol, by improving the quality of reagents and by working with minute reaction volumes. This will normally be sufficient for single cell analysis.

It is unlikely that, in the near future, these assays will find wide use in the clinical laboratory, particularly since expensive fully automated and computerised machines based on absorbance measurements have already been purchased by most laboratories. However, certain tests may benefit from the increased speed and sensitivity offered by chemiluminescence. If analyte assays in the fM region, i.e. amol per ml, become necessary, then chemiluminescence will come into its own.

A more promising avenue has been opened up by the increasing development of 'on the spot' tests at the bedside, in the clinic, in the general practitioner's surgery, and even at home. Chemiluminescence assays detected with a pocket chemiluminometer, for example based on a photodiode or ultrasensitive photographic film, are therefore worthy of exploration. These would also be of benefit in the Third World where simplicity and cheapness are crucial.

Analyte assays based on chemiluminescence are likely to become of increasing importance in the research laboratory as the need to measure enzymes and metabolites in small numbers of cells or individual cells increases. Some, such as bacterial measurement of O_2, and luminol detection of H_2O_2 and other reactive oxygen metabolites, are adaptable to studies in live cells. However, the firefly luciferase is virtually saturated by ATPMg at the concentrations of 1–10 mM found in most cells (K_m ATP $ca.$ 10 μM). Thus it is unlikely to be of any value, without modification, in intact cells. Furthermore, all cells contain relatively high concentrations of NADH and NADPH. Site directed mutagenesis may extend the working range of the bioluminescent systems.

5.7 RECOMMENDED READING

Books

Bergmeyer, H. U. (1985) *Methods of enzymatic analysis*, 3rd ed., Vols VI and VIII. VCH Federal Republic of Germany.

De Luca, M. A. (ed) (1978, 1987) *Methods Enzymol.* Vols 57 and 133. Bioluminescence and Chemiluminescence.

Kricka, L. J., & Carter, T. J. N. (eds) (1982) *Clinical and biochemical luminescence.* Marcel Dekker, New York and Basel.

Kricka, L. J., Stanley, P. E., Thorpe, G. H. G., & Whitehead, T. P. (eds) (1984) *Analytical applications of bioluminescence and chemiluminescence.* Academic Press, London and New York.

General reviews

Brolin, S. E., Wettermark, G., & Hammar, H. (1977) *Strahlentherapie* **153** 124–131. Chemiluminescence microanalysis of substrates and enzymes.

Campbell, A. K. (1986) *Trends in Biochemical Sciences* **11** 104–108. Living light: principles and biomedical applications.

Campbell, A. K. & Simpson, J. S. A. (1979) *Techniques in Life Sciences* **B213** 1–56. C. Pogson (ed). Elsevier/Amsterdam. Chemi- and bioluminescence as an analytical tool in biology.

Campbell, A. K., Hallett, M. B., & Weeks, J. (1985) *Methods Biochem. Anal.* **31**, 317–416. Chemiluminescence as an analytical tool in cell biology and medicine.

Glick, D. (1977) *Clin. Chem.* **23** 1465–1471. Microchemical analytical techniques of potential clinical interest.

Gorus, FG. & Schram, E. (1979) *Clin. Chem.* **25** 512–519. Applications of bio- and chemiluminescence in the clinical laboratory.

Whitehead, T. P., Kricka, L. J., Carter, T. J. N., & Thorpe, G. H. G. (1979) *Cin. Chem.* **25** 1531–1546. Analytical luminescence: its potential in the clinical laboratory.

Some papers of historical interest

Beijerinck, M. W. (1902) *Proc. Acad. Sci. Amst.* **4** 45–49. Photobacteria as a reactive in the investigation of the chlorophyll-function.

Cormier, M. J. & Dure, L. S. (1963) *J. Biol. Chem.* **238** 785–789. Studies on the bioluminescence of *Balanoglossus biminiensis* extracts. 1. Requirement for hydrogen peroxide and characteristics of the system.

Cormier, M. J. & Strehler, B. L. (1953) *J. Am. Chem. Soc.* **75** 4864. The identification of KCF: requirement of long-chain aldehydes for bacterial extract luminescence.

Harvey, E. N. (1916) *Science* **44** 208–209. The mechanism of light production in animals.

Harvey, E. N. (1915) *Am. J. Physiol* **37** 230–240. Studies on light production by luminous bacteria.

McElroy, W. D. (1947). *Proc. Natl. Acad. Sci. US* **33** 342–345. The energy source for bioluminescence in an isolated system.

Papers

Data sheets from LKB/Wallac and BCL (see Appendix V)

ATP

Harber, M. J. & Asscher, A. W. (1977) *J. Antimicorb. Chemother.* **3** 35–41. A new method for antibiotic assay based on measurement of bacterial adenosine triphosphate using the firefly bioluminescence system.

Kricka, L. J. & De Luca, M. A. (1982) *Arch. Biochem. Biophys.* **217** 674–681. Effects of solvents on the catalytic activity of firefly luciferase.

Lundin, A., Rickardsson, A., & Thore, A. (1976) *Anal. Biochem.* **75** 611–620. Continuous monitoring of ATP converting reactions by purified firefly luciferase.

Lundin, A. & Styrelius, I. (1978) *Clin. Chim. Acta.* **87** 199–209. Sensitive assay of creatine kinase isoenzymes in human serum using M subunit inhibiting antibody and firefly luciferase.

Stanley, P. E. (1982) *Lab. Equip. Digest.* Rapid measurements of bacteria by ATP assay.

Strehler, B. L. & Totter, J. R. (1954) *Methods Biochem. Analysis* **1** 341–356. Determination of ATP and related compounds; firefly luminescence and other methods.

Wettermark, G., Stymne, H., Brolin, S. E., & Peterson, B. (1975) *Anal. Biochem.* **63** 293–307. Substrate analysis in single cells. *I. Determination of ATP.*

NAD(P)H

Ågren, A., Berne, C., & Brolin, S. E. (1977) *Anal. Biochem.* **78** 229–234. Photokinetic assay of pyruvate in the islets of Langerhans using bacterial luciferase.

Barak, M., Ulitzur, S. & Merzbach, D. (1983) *J. Immunol. Meth.* **64** 353–363. The use of luminous bacteria for determination of phagocytosis.

Gudermann, J. W. & Cooper, T. K. (1986) *Anal. Biochem.* **158** 59–63. A sensitive bioluminescence assay for myo-inositol.

Hastings, J. W. & Ta, S.-C. (1981) Luminescence from biological and synthetic macromolecules.

Jablonski, E. & De Luca, M. (1977) *Proc. Natl. Acad. Sci. US* **73** 3848–3851. Immobilisation of bacterial luciferase and FMN reductase on glass rods.

Kather, H. & Wieland, ER. (1984) *Anal. Biochem.* **140** 349–353. Bioluminescence determination of free fatty acids.

Hastings, J. W. & Tu, S.-C. (1981) In: *Luminescence from biological and synthetic macromolecules. Morawetz, H. & Steinberg, I. Z. (eds), Annal. NY. Acad. Sci.* **366** 315–327. Bioluminescence of bacterial luciferase.

Terouanne, B., Carrie, M.-L., Nicolas, J.-C., & Crastes de Paulet, A . (1986) *Anal. Biochem.* **154** 118–125. Bioluminescent immunosorbent for rapid immunoassays.

Watanabe, H. & Hastings, J. W. (1982) *Mol. Cell. Biochem.* **44** 181–187. Specificities and properties of the reduced pyridine nucleotide-flavin mono-nucleotide reductases coupling to bacterial luciferase.

Stanley, P. E. (1971) *Anal. Biochem.* **39** 441–453. Determination of subpicomole levels of NADH and FMN using bacterial luciferase and the liquid scintillation spectrometer.

Oxygen

Lloyd, D. James, K. Williams, J., & Williams, N. (1981) *Anal. Biochem.* **116** 17–21. A membrane-covered photobacterium probe for oxygen measurement in the nanomolar range.

Oshino, R., Oshino, N., Tamura, M., Kobilinsky, L., & Chance, B. (1972) *Biochim. Biophys. Acta* **273** 5–17. A sensitive bacterial luminescence probe for O in biochemical systems.

H₂O₂

Auses, J. P., Cook, J. L., & Maloy, J. T. (1975) *Anal. Chem.* **47** 244–249. Chemiluminescence enzyme method for glucose.

Freeman, T. W. & Seitz, W. R. (1976) *Anal. Chem.* **50** 1242–1246. Chemiluminescence fibre optic probe for hydrogen peroxide based on the luminol reaction.

Bachrach, U. & Plesser, Y. M. (1986) *Anal. Biochem.* **152** 423–431. A sensitive, rapid, chemiluminescence-based method for the determination of diamines and polyamines.

Bellisario, R., Spencer, T. F., & Cormier, M. J. (1972) *Biochem.* **11** 2256–2266. Isolation and properties of luciferase, a non-heme peroxidase, from the bioluminescent earthworm *Diplocardia longa*.

Birman, S. (1985) *Biochem. J.* **225** 825–828. Determination of acetylcholinesterase activity by a new chemiluminescence assay with the natural substrate.

Kather, H., Wieland, E., & Waas, W. (1987) *Anal. Biochem.* **163** 45–51. Chemiluminescent determination of adenosine, inosine and hypoxanthine/xanthine.

NO₂

Wendel, G. J., Stedman, D. H., Cantrell, C. A. & Damrauer, L. (1983) *Anal. Chem.* **55**, 937–940. Luminol-based nitrogen dioxide detector.

PAP(5)

Burnell, J. N. & Anderson, J. W. (1973) *Biochem. J.* **134**, 565–579. Adenosine 5'-sulphatophosphate kinase activity in spinach leaf tissue.

Cormier, M. J., Hori, K., & Karkhanis, Y. D. (1970) *Biochem.* **9** 1184–1190. The conversion of luciferin to luciferyl sulphate by luciferin sulphokinase.

Anaesthetics

Franks, N. P. & Lieb, W. R. (1984) *Nature* **310** 599–601. Do general anaesthetics act by competitive binding to specific receptors?

Franks, N. P. & Lieb, W. R. (1985) *Chemistry in Britain* **21** 919–921. The firefly throws light on anaesthesia.

Cloning

Cohn, D. H., Mileham, A. J., Simon, M. I., Nealson, K. H., Rausch, S. K., Bonam, D., & Baldwin, T. O. (1985) *J. Biol. Chem.* **260** 6139–6146. Nucleotide sequence of the LuxA of *Vibrio harveyi* and the complete amino acid sequence of the subunit of bacterial luciferase.

Cohn, D. H., Ogden, R. C., Abelson, J. N., Baldwin, T. O., Nealson, K. H., Simon, M. I., & Mileham, A. J. (1983) *Proc. Natl. Acad. Sci. US* **80** 120–123. Cloning of the *Vibrio harveyi* luciferase genes: use of a synthetic oligonucleotide probe.

Johnson, T. C., Thompson, R. B., & Baldwin, T. O. (1986) *J. Biol. Chem.* **261** 4805–4811. Nucleotide sequence of the luxB gene of *Vibrio harveyi* and the complete amino acid sequence of the subunit of bacterial luciferase.

Ow, D. W., Wood, K. V., De Luca, M., de Wet, J. R., Helsinki, D. R., & Howell, S. H. (1986) *Science* **234** 856–859. Transient and stable expression of the firefly luciferase gene in plant cells and transgenic plants.

De Wet, J. R., Wood., K. V., Helsinki, D. P., & De Luca, M. (1985). *Proc. Natl. Acad. Sci. US* **82** 7870–7873. Cloning of firefly luciferase cDNA and the expression of active luciferase in *E. coli*.

6

Ultraweak chemiluminescence

Oxygen might burn the candle of life too quickly and too soon exhaust the life within.
Joseph Priestley, 1775

6.1 WHAT IS ULTRAWEAK CHEMILUMINESCENCE?

There are no luminous mammals, yet some of our cells are capable of producing a weak chemiluminescence, invisible to the naked eye but nevertheless an indicator of some important oxidative reactions inside cells.

Individual luminous bacteria are strongly chemiluminescent, emitting up to 10^3–10^4 photons per second. The luminous cells of eukaryotes, being much larger, may emit light intensities several orders of magnitude greater than this, particularly at the peak of a flash. Such examples of strong chemiluminescence are visible, being greater than the 30 000 photons per second per square centimetre of retinal surface, necessary for light perception. Furthermore, the luminescence plays a function in the natural history of the organism concerned (see Chapter 3). The chemiluminescence quantum yields (ϕ_{CL}) of the reactions responsible for the light emission in luminous organisms are high, most being in the range of 0.1–1. Similarly, the synthetic chemiluminescent reactions we have examined so far (see Chapters 1, 4, and 5) are also visible, with relatively high quantum yields (ϕ_{CL} 0.01–0.5).

In contrast, there are a number of oxidative reactions which can be provoked *in vitro*, and some which occur in living cells, which emit a much weaker light signal. Luminescence is invisible, and can be some 10^3–10^6 less intense than that from luminous animals. The emission is often equivalent to less than one photon per second per cell, which in whole organs results in less than 10 000 photons emitted per second per square centimetre of surface. The light itself is not thought to have any function, although this remains to be proved, particularly in view of the controversial work of Gurvich and his followers, begun in the 1920s and claiming it has mitogenic activity. The chemiluminescent reactions responsible usually, but not always, have extremely low quantum yields (ϕ_{CL} often 10^{-3}–10^{-8}, and sometimes 10^{-9}–10^{-15}). This, then, is *ultraweak chemiluminescence*, also sometimes called *dim*, '*dark*' or *low-level chemiluminescence*. It only became possible to find such reactions because of the sensitivity of the photomultiplier tube, capable in principle of detecting chemiluminescent reactions with quantum yields at least down to 10^{-17}. It is important to distinguish *indigenous*, or *inherent*, *ultraweak* chemiluminescence from *indicator-dependent* chemiluminescence (Fig. 6.1). The latter results from the

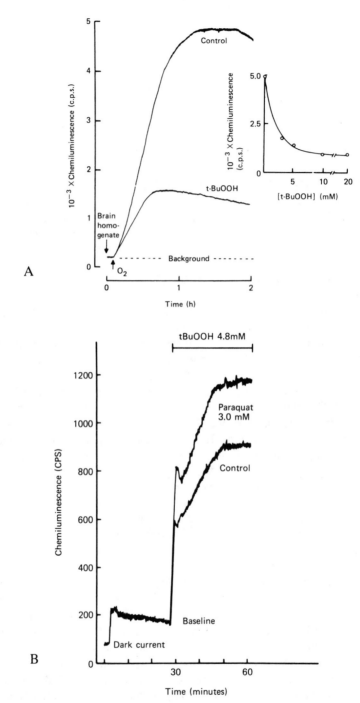

Fig. 6.1 — Two examples of ultraweak chemiluminescence. A. brain homogenate + O_2 ± butyl hydroperoxide. Reprinted by permission from Cadenas *et al.* (1981) *Biochem. J.* **198**, 645–654, copyright ©1981 The Biochemical Society of London; B. Intact rat lung ± butyl hydroperoxide- + paraquat; from Aldrich *et al.* (1983) *J. Lab. Clin. Med.* **101** 66–73, with permission of C. V. Mosby Co.

addition of chemiluminescent compounds, such as luminol, which have become popular as means of detecting reactive oxygen metabolite production by cells and acellular reactions (see sections 6.4 and 6.6). Because the amount of luminol reacting is very small, the luminescence is not visible, nor does it seem easily detectable by image intensification. Nevertheless, the chemistry of the luminol reaction is essentially no different from that producing visible light at higher reaction rates, or under more efficient conditions. This latter phenomenon of indicator-dependent chemiluminescence will, however, also be considered in this chapter, because the chemical reactions responsible for generating the oxygen metabolite(s) which reacts with luminol, and other indicators, are often the same as those responsible for the *indigenous* ultraweak chemiluminescence.

Ultraweak chemiluminescence can occur spontaneously in cells (Table 6.1). It can also be induced, or enhanced, by the addition of a variety of compounds or by cell stimuli (Table 6.2) These compounds and stimuli may themselves be chemiluminescent, or they may act on the pathway required to generate the endogenous ultraweak

Table 6.1 — Cellular sources of ultraweak chemiluminescence

A. *Whole organ*
        Liver
        Lung
        Heart
        Skeletal muscle
        Brain
        Seedling (root, stem)

B. *Isolated cells*
        Neutrophil
        Eosinophil
        Monocyte-macrophage
        Kupffer cells
        Fertilised egg (sea urchin, starfish, fish)
        Amoeba
        Hepatocyte
        Platelet
        Yeast
        Fungus

C. *Tissue homogenates and subcellular organelles*
        Liver homogenate
        Heart homogenate
        Brain homogenate
        Mitochondria — liver, adrenal
        Microsomes — liver
        Seminal vesicles
        Chloroplasts

Table 6.2 — Compounds inducing, enhancing or inhibiting ultraweak chemiluminescence

Class of compound	Specific substance	Source of chemiluminescence
A. Induction or enhancement		
Inorganic peroxide	H_2O_2	Whole heart, microsomes
Organic hydroperoxides	(ROOH)-t butyl	Whole lung and heart preparations, liver and brain homogenate
	hydroperoxide	
	Cumene hydroperoxide	
	Ethyl hydroperoxide	Lung
Hyperbaric oxygen		Isolated hepatocytes
Toxins	CCl_4	Liver
	Paraquat	Lung
Stimuli of respiratory burst	Phagocytic and chemotactic	Neutrophils
	Sperm	Egg
Stimulation of prostaglandin synthesis	N-ethylmaleimide	Platelets
	Thiomersal	Platelets
	Arachidonic acid	Platelets
	Linoleic acid	
	Eicosatetraenoic acid	Vesicular gland
Carcinogens	Benzo (a) pyrene	Liver microsomes
Source of reducing electrons	NADPH	Liver microsomes
	Fe^{2+}	Mitochondria
Depletion of oxygen metabolite scavengers	Glutathione	Whole heart
		Isolated hepatocytes
Enhancers of $^1O_2$	D_2O, 1,4-diazabicyclo[2.2.2] octane	Several tissues
dimol emission		
Chemiluminescent compounds	luminol, lucigenin, pholasin, aromatic aldehydes	Phagocytes
		Fungus
Energy transfer acceptors	Fluorescein	Fungus
B. Inhibitors		
		Liver microsomes
Removal of oxygen	N_2, $Na_2S_2O_4$	Lung, heart, liver
Oxygen metabolite scavengers	Superoxide dismutase	neutrophils, lung + paraquat
	Glutatione, dithio-threitol, di-tert-butyl quinol	Hepatocytes
	Ascorbate	Catecholamine oxidation
Singlet oxygen Scavengers	N_3^-, furans (diphenyl-2,5 dimethyl) carotene	Lung
Peroxidase inhibitors	N_3^-	Neutrophils
Cytochrome P450 inhibitors	CN^-, SKF-525A	Liver microsomes
Prostaglandin synthesis inhibitors	Indomethacin, aspirin	Platelets
Non-steroidal anti-inflammatories	Methylsalycilate, dimercaptopropanol	neutrophils

chemiluminescent reaction (see section 6.5 for classification). The substrate(s) producing the excited state molecules responsible for such ultraweak chemiluminescence must therefore also be endogenous.

Inducers or enhancers fall into four broad categories:

(1) Provision of chemiluminescent substrate.
(2) Provision of oxidant, e.g. H_2O_2 or a hydroperoxide.
(3) Activation of interacellular pathways to provide endogenous chemiluminescent substrate (e.g. arachidonic acid), or to consummate and stimulate the components of the chemiluminescent reaction (e.g. phagocyte activation).
(4) Energy transfer acceptors (see Chapter 9).

In contrast, inhibitors of ultraweak chemiluminescence act by one of two mechanisms:

(1) Scavenging the oxidant or the excited state product of the chemiluminescent reaction.
(2) Inhibition of the pathways responsible for generating one or more of the components of the chemiluminescent reaction, or the consummation.

There are a plethora of acellular and synthetic reactions which can generate ultraweak chemiluminescence (Table 6.3). The time course of ultraweak chemiluminescence can vary from a very weak glow lasting many minutes or hours, to a mono-, bi- or tri-phasic light emission, with individual phases lasting only a few seconds or minutes. The spectrum of ultraweak chemiluminescence can also be very broad, embracing almost all of the complete 'visible' range, from 400–750 nm. However, the spectrum is often not smooth, peaks being detectable in certain regions. These are important pointers to the origin of the light emission (see section 6.3).

Table 6.3 — Acellular sources of ultraweak chemiluminescence

H_2O_2
H_2O_2 + variety of organic compounds including amino acids
H_2O_2 + proteins (e.g. albumin, gelatin)
H_2O_2 + peroxidase
Any reaction generating singlet O_2, e.g. myeloperoxidase ($OCl^- + H_2O_2$)
Fe^{2+} + bleomycin ($+ H_2O_2$)
Oxidation of a range of lipids
Prostaglandin synthesis pathway (lipoxygenase route)
Photosensitivity of porphyrins

As we saw in Chapter 1 (section 1.3.5) at the end of the 1920s A. G. Gurvich reported an ultraweak luminescence in the ultraviolet from 190–325 nm or 320–350 nm, emanating from onion root, regenerating frog cornea, dividing yeast cells, and blood cells. He claimed that this radiation had 'mitogenetic' properties, being capable of stimulating mitosis in neighbouring cells. This story has long been a source of much controversy and argument. Many hundreds of papers, particularly in the German and Russian literature, contain similar claims, as well as refutations. Perhaps it warrants re-examination with the more sophisticated photon-counting equipment now available. I shall not consider it further, but the interested reader can find a path into this literature from section 6.9.

The distinction between ultraweak chemiluminescence and the other examples of chemiluminescence examined in this book may seem somewhat arbitrary, particularly since a precise definition is difficult (see section 6.5). Yet the identification of ultraweak chemiluminescence highlights a group of chemiluminescent reactions which, in living cells at least, involve a number of oxygen intermediates and play an important role in certain types of cell activation, in the body's defence system, as well as having a potential role in the pathogenesis of disesases such as rheumatoid arthritis and ischaemic heart disease. The two key features which make a chemiluminescent reaction ultraweak are firstly it is invisible to the naked eye, and secondly the light emission itself has no function.

Furthermore, ultraweak chemiluminescence does not involve an oxygen metabolite — luciferin-luciferase reaction as is the case in all bioluminescence (see Chapter 3).

Four questions now arise:

(1) Where do ultraweak chemiluminescent reactions occur in Nature?
(2) Are they *ultraweak* because of a low quantum yield, or because the kinetics and/ or the concentration of the chemiluminescent components are too low to generate a visible, functional light emission?
(3) Whilst the light emission may have no function, what about the excited state molecules responsible for the ultraweak emision?
(4) Can we use ultraweak chemiluminescence to discover anything new about the role of excited states and oxygen in cell physiology and pathology?

6.2 SOURCES

The first well documented observation of ultraweak chemiluminescence from living tissue, in the 'visible' region, seems to have been made in 1954 by Colli and co-workers. They observed a weak luminescence (a few 100 pulses per second) emanating from various seedlings, as well as from extracts of various parts of the plant. The broad spectrum had a peak emission at about 550 nm. The roots were more luminescent than the stems or the seeds. There are no mammals with luminous organs equivalent to those found in some fish or luminous invertebrates. However, in 1961 Tarusov and colleagues detected an ultraweak chemiluminescence from mouse liver *in situ*, and extended their observations to homogenates of this and other tissues. Ultraweak chemiluminescence has since been detected from a wide range of intact organs, isolated cells, and tissue homogenates from vertebrates, invertebrates, plants, and higher protists (Table 6.1), as well as from a number of reactions in the test tube.

These emissions were detected by placing the whole tissue, cells, or extract in front of a highly sensitive photomultiplier tube, in the dark. The ultraweak chemiluminescence is associated with some important cellular functions, depending on the cell type (Table 6.1):

(1) Mitochondrial respiration (a byproduct).
(2) Photosynthesis.
(3) Egg fertilisation.
(4) Cell division.
(5) Eicosanoid production.
(6) Killing of micro-organisms by phagocytes.
(7) Detoxification (e.g. the cytochrome P450 and P448 system).

6.2.1 Organ chemiluminescence

Since the original observation on liver by Tarusov, spontaneous chemiluminescence has also been detected from the surface of the brain and the lung, both perfused and *in situ*, as well as from perfused liver and heart. Rather than place the photomultiplier directly over the organ *in situ*, or the perfused organ, light is usually channelled on to the surface of the cooled photomultiplier by a lucite rod as a light guide. The intensity of such spontaneous emissions varies considerably between tissues and the area being investigated, but is usually about 10^4 photons per second per cm^2 of tissue surface. Since the overall efficiency of the apparatus will be about 0.1% (taking account of the geometry plus electronics, see Chapter 2), this corresponds to a minimum of 10 cps per cm^2 of surface. Thus, a circle of diameter about 3.6 mm is required to register 1 cps.

Can we estimate what is the mean ultraweak photon emission from an individual cell? Let us assume that only light emitted from the top 0.5 cm of tissue is able to reach the detector. A spontaneous light emission of 10^4 photons $s^{-1}cm^{-2}$ is then equivalent to 20 000 photons $s^{-1}cm^{-3}$. If the individual cells have a radius of 20 μm (e.g. a hepatocyte) there will be approximately 2.5×10^8 cells per cm^3 (volume of cell $= 4/3\pi r^3 = 4$ pl). The photon emission is therefore 8×10^{-5} per cell per second, or more than 10^8 times lower than light emission from the cells of luminous organisms (Table 3.10).

Hyperbaric oxygen can enhance the light signal some 2 to 10-fold, depending on the tissue, and hydroperoxides increase the chemiluminescence some 20 to 30-fold. Perfusion with the hydroperoxide causes a detectable increase in chemiluminescence within one minute, a plateau being reached within about 4-5 minutes. Removal of the hydroperoxide results in a decay in light emission to $<20\%$ of maximum within about 15 minutes. The concentration range of hydroperoxide required is 0.2–20 mM, equivalent to a perfusion rate of 20–200 mol min^{-1}, a half maximal concentration being around 4 mM, and maximal at about 12 mM. Prolonged concentrations > 10–20 mM tend to be toxic. The relative potency of different hydroperoxides varies from tissue to tissue, indicating the intracellular organelle mainly responsible for the chemiluminescence. If the ultraweak chemiluminescence is mainly of mitochondrial origin, as often in the liver, cumene hydroperoxide is more potent than *t*-butyl or ethyl hydroperoxides. Whereas when it is microsomal, as in the lung, ethyl hydroperoxide is more potent than *t*-butyl hydroperoxide. The signals can be enhanced further by substances such as paraquat, or by glutathione depletion (Table 6.2). Removal of O_2, using N_2 or argon, markedly reduces the overall spontaneous and induced chemiluminescence signal.

6.2.2 The respiratory burst

In 1908, Otto Warburg, a pioneer in the study of respiration and the biology of oxygen, discovered that when sea urchin eggs are fertilised they undergo a rapid burst of oxygen utilisation. Some 25 years later, Ado, and Baldridge and Gerard, observed a similar 'respiratory burst' when a white blood cell preparation containing phagocytes was exposed to particles. Baldridge and Gerard observed a 100–400% increase in respiration in cat white cells, using a Warburg monometer. Interestingly, all of these workers probably misinterpreted the cause of these 'respiratory bursts', and did the experiment for what turned out to be the wrong reason. They argued that fertilisation and phagocytosis must be energy-requiring processes, and thus an increase in respiration was to be expected. However, in the 1950s several workers showed that these respiratory bursts were not inhibited by cyanide. They were therefore not of mitochondrial origin, and must have a function or cause other than satisfying the energy demand of the activated cells. By the 1960s it had been found that a reduced pyridine nucleotide was required, reducing O_2 only partially to H_2O_2, rather than completely to H_2O as it is in the mitochondrial respiratory chain. One of the reasons that it is easy to observe this respiratory burst in phagocytic cells is that they have few mitochondria, the cells relying on glycogen and glycolysis for their supply of ATP. Most of the respiration of phagocytic cells is required for the two, non-ATP generating pathways, microbial killing and eicosanoid production. In 1972 Allen argued that such an oxidative system ought to give rise to indigenous chemiluminescence. Ultraweak chemiluminescence occurs in a variety of phagocytic cells, including neutrophils, eosinophils, monocytes, macrophages, Kupffer cells, and amoebae, when exposed to a phagocytic stimulus. Chemotactic peptides and a range of artificial and pathological stimuli can also provoke the response in neutrophils and other phagocytes (Table 6.4). As with whole organs, the light emission is very weak, although it can sometimes be as high as 10^{-2} photons per second per individual cell. Thus all phagocytes appear to be able to generate ultraweak chemiluminescence. Whether all fertilised eggs chemiluminesce, e.g. mammals, is unknown.

6.2.3 Other cellular sources

Although the so-called 'mitogenetic radiation' in the ultraviolet, from dividing cells is still controversial, ultraweak chemiluminescence, in the 'visible' region of the spectrum, has been detected from both dividing and non-dividing cells. These include a yeast and fungus during their life cycle, and non-dividing platelets, hepatocytes, or plant cells following irradiation. Ultraweak chemiluminescence can be detected from the yeast *Saccharomyces cervisiae*, both during the logarithmic growth phase and during the stationary phase of culture, though the former is more intense. Light emission from a cell population can last more than 10 hours. Light intensity during the growth phase increases some three-fold, compared with a 20 to 100-fold increase in cell number. The maximum light emission from some 10^{10} cells is about 5×10^5 photons per second, equivalent to one photon for every 20 000 cells. However, when estimated on the basis of photons per dividing cell the value is much higher, perhaps as much as 0.1–1 photon per dividing cell, suggesting a particular event leading to a molecule or molecules in an excited state at the moment of cell division.

Table 6.4 — Stimuli and modifiers of neutrophil reactive oxygen metabolite production

Class of stimulus	Example
A. Primary stimuli	
1. Phagocytic	Fc receptors (immune complexes)
	C3b receptors (opsonised particles)
	unopsonised particles (bacteria, viruses, dusts, crystals)
	Surfaces (frustrated phagocytosis, basement membranes)
2. Chemotactic	Bacterial peptides
	C5a receptors (complement activation)
3. Experimental	Unopsonised particles (latex)
	N formyl met leu phe (bacterial chemotactic receptor)
	Opsonised zymosan (C3b recptor)
	phorbol esters (phorbol myristate acetate)
	Concanavalin A
	Ca^{2+} ionophores (A23187, ionomycin)
	Oleoyl acetyl glycerol
	Platelet activating factor
4. Pathological	Unopsonised particles (asbestos and erionite dusts, urate crystals, unphagocytosed viruses)
	membrane attack complex of complement (C5b6789n)
	Bacterial toxins e.g. streptolysin O
B. Secondary stimuli	
Cellular metabolite	Leukotrienes (B_4)
	Interferon (long term control)
C. Secondary regulators	
1. Cellular metabolite	Adenosine (inhibits chemotactic stimuli only)
2. Experimental	Cytochalasin B (inhibits phagocytic, stimulates chemotactic, stimuli)
	forskolin (inhibits chemotactic stimuli only)
	dibutyryl cyclic AMP (inhibits chemotactic stimuli only)
	Pertussis toxin (inhibits)
	Chlorea toxin (inhibits)
D. Drugs	
	tBOC Nfmlp inhibits
	indomethacin
	tamoxifen (inhibits phagocytic stimuli and phorbol esters)
	Penicillamine

6.2.4 Organelle chemiluminescence

The fungus *Blastocladiella emersonii* has a life cycle involving formation from a zoospore of so-called 'round' cells I and II, the germling, and the zoosporangium which releases the zoospores again. These cells have a number of interesting features associated with their organelles: a flagellum, a nucleus with a single nucleolus, a membrane-bound nuclear cap over the nucleus, one giant mitochondria and several lipid granules, but no cell wall. Addition of phenyl acetaldhyde generates a biphasic, ultraweak, chemiluminescence. This is a particularly intriguing reaction since it appears to involve generation of triplet state benzaldehyde which is capable of transfering its excitation energy either to an endogenous flavin or to added fluors such as fluorescein, eosin or rose bengal.

Ultraweak chemiluminescence has also been detected in non-dividing cells, including yeast and cells from tissues such as the liver, known to be capable of spontaneous or inducible chemiluminescence in the intact organ. Not only can this be induced by hydroperoxides (Table 6.2) but also it can be enhanced by exposure of the animals to drugs such as barbiturates, prior to isolation of the cells. These drugs induce enzymes, including the cytochrome P450–448 detoxification system. One other blood cell, in addition to those capable of the CN^--resistant 'respiratory burst', has been shown to produce chemiluminescence, the platelet. It is rarely possible to detect any light emission from resting platelets. However, addition of arachidonic acid, or any one of several other unsaturated acids which are substrates for prostaglandin synthesis, results in ultraweak chemiluminescence. Once again, the emission is very weak, perhaps $< 10^{-4}$ photons per platelet per second, some 10^9 platelets being required to produce an easily detectable response. About 3 ml of blood are required to provide this many platelets. In contrast to organ or phagocyte chemiluminescence, the time course in platelets is very fast, being detectable within six seconds, reaching a peak within 10–40 seconds, and decaying to $<20\%$ of maximum within 1–2 minutes. Whether natural stimuli, such as thrombin, which stimulate prostaglandin synthesis from endogenous substrates, also give rise to ultraweak chemiluminescence has yet to be fully established. More than 90% of the platelet chemiluminescence can be shown to require oxygen. It disappears on blowing through N_2. The chemiluminescence is also markedly inhibited by aspirin, and other inhibitors of the prostaglandin synthesising pathway. Even platelets isolated from people who have taken such inhibitors may show a reduced chemiluminescence response to arachidonate. As with all studies on cellular ultraweak chemiluminescence, it is essential to characterise the purity of cell suspensions, particularly for contamination with granulocytes which are an active source of such chemiluminescence. This appears to have been done for platelets, but not for some other cell types. For example reports that lymphocytes and natural killer cells exhibit ultraweak chemiluminescence remain unproven because of this problem.

6.2.5 Acellular reactions

The oxidation of many organic compounds, not normally defined as chemiluminescent, can produce light detectable by a sensitive photomultipier tube (Table 6.3) but invisible to the naked eye. Many can be related to reactions in intact cells responsible for cellular or organ ultraweak chemiluminescence. Generators of reactive oxygen

metabolites, peroxides, and singlet oxygen frequently give rise to such chemiluminescence. This can be a nuisance in some analyte assays, for example using
H_2O + peroxidase, as it gives rise to high luminescence blanks, reducing the
detection limit of the assay. Organic contaminants in the air, or detergents and other
contaminants in the water may exacerbate this problem. It sometimes seems that the
addition of H_2O_2 to virtually any oxidisable organic compound will generate light!

Irradiation of porphyrins produces chemiluminescence, detectable after switching off the exciting light source. Such chemiluminescence may occur *in vivo* in
diseases such as porphyrias where large accumulations of porphyrins occur close to
the skin, or in cancer therapy utilising these compounds where they accumulate in
several tissues, including the skin. Incorporation of a metallo-cation into the
tetrapyrrole nucleus greatly decreases chemiluminescence.

6.3 ORIGINS OF ULTRAWEAK CHEMILUMINESCENCE
6.3.1 Identification of source
To identify the source of ultraweak chemiluminescence, it is necessary first to be sure
which cells are responsible, secondly it is necessary to identify from where in the cell
the light is being emitted (Table 6.5), and thirdly the pathway and ultimately the
reaction responsible for generating the excited-state molecules which give rise to the

Table 6.5 — Source of ultraweak chemiluminescence in cells

Cell type	Organelle or site	Inducer or stimuli	Pathway responsible for excited state
Lung, brain	Microsomes	Hydroperoxide	Lipid peroxidation
Liver	Microsomes	Hyperbaric oxygen	Lipid peroxidation
Liver Mitochondria			Respiratory chain — oxygen metabolites lipid peroxidation
Vesicular gland cells	Microsomes	-OOH or O_2	
Platelets		Arachidonate	Prostaglandin synthesis
Phagocytes	Cell membrane. Intracellular vesicles and granules. Extracellular	Cell stimuli	Reactive oxygen metabolites Lipid peroxidation and $^1O_2$
Fertilised eggs	Cell membrane and extracellular	Sperm	Reactive oxygen metabolites + ovoperoxidase

luminescence must be isolated and identified. Organs such as the liver contain at least
three other cell types in addition to the predominant hepatocyte, namely bile duct
cells, fibroblasts, and the cells of the reticulo–endothelial system (Kupffer cells), as
well as contaminating blood cells. Pure hepatocytes do chemiluminesce, but so do
Kupffer cells and several of the blood cells. Thus, interpretation of cell experiments
becomes very confused if the suspensions are impure. As we shall see, this has led to
much debate about natural killer cells as a source of chemiluminescence in the
presence of luminol (see section 6.4). The following criteria enable the cellular

source of chemiluminescence to be identified, and can eliminate cellular contamination as an explanation of results:

(1) Estimation of cell purity, and identification of any contaminants, however few they are. Only 1% contamination with neutrophils has explained many a puzzling observation!
(2) Effect of reducing cell contaminants, e.g. by gradient separation, flow cytometry, or killing using antibody plus complement (but beware, complement does not kill all the cells in a population).
(3) Addition of excess cells likely to be contaminating the preparation should produce a response much less than that from the cell preparation under investigation.
(4) Comparison of effects of inducer, cell stimuli, or inhibitors on contaminating cells (e.g. neutrophils) to show a different response in terms of magnitude, shape, time course, and spectrum compared with that from major cell types (e.g. platelets).
(5) Use of blood containing genetically deficient cells, e.g. neutrophils, from patients with chronic granulomatous disease or myeloperoxidase deficiency which show an absence or a great reduction of ultraweak chemiluminescence, yet the response thought to arise from the major cell type (e.g. platelets) remains unaltered.
(6) Isolation of ultraweak chemiluminescence in an organelle or subcellular fraction absent or low in any cell type likely to be contaminating the major cell type, e.g. P450/448 containing microsomes from hepatocytes. Subcellular fractionation has led to the identification of NADPH-microsomes containing cytochrome P450 as a major source of ultraweak chemiluminescence. This can be confirmed by induction of cytochrome P450 using barbiturates in the animal, which enhances the chemiluminescence of isolated microsomes. Microsomes, free of mitochondria, which may contain fragments of the endoplasmic reticulum and the cell membrane, have also been identified as sources of ultraweak chemiluminescence in sheep vesicular gland cells and platelets. In cells undergoing a respiratory burst, for example phagocytes or fertilised eggs, the chemiluminescence appears to be both intracellular, in vesicles and with phagolysosomes, as well as very close to the outside surface of the cell membrane. However, the evidence so far for this has been based not on studying indigenous *ultraweak* chemiluminescence but on *indicator-dependent* chemiluminescence (see section 6.4).

6.3.2 Reactions responsible for generating indigenous ultraweak chemiluminescence

Indigenous ultraweak chemiluminescence can be inorganic or organic in origin. Identification of the reactions responsible for the chemiluminescence requires, firstly, the identification of the excited state molecule emitting the light, and secondly, identification of the pathway generating this metabolite. The two principal metabolites thought to be responsible directly for ultraweak chemiluminescence emission in cells are:

(1) Excited carbonyls; $>C=0^*$
(2) Singlet oxygen; (dimol, i.e. $^1O_2$-$^1O_2$, in the $^1\Delta g$ state)

The formation of excited carbonyls requires an oxidant. This oxidises an endogenous organic substrate to form a dioxetan ($\begin{smallmatrix} O-O \\ | \quad | \\ -C-C- \end{smallmatrix}$) intermediate, which then breaks down spontaneously to form the excited carbonyl. Singlet oxygen, on the other hand, can be formed through both inorganic and organic reactions. A third possible source of ultraweak chemiluminescence is the transfer of energy from an excited state molecule (X^*), which is not necessarily a carbonyl, to an endogenous fluor (F) such as a flavin.

$$\text{Reaction} \rightarrow X^* \xrightarrow{+F} X + F^* \rightarrow F + \text{light}. \qquad (6.1)$$

Much of the evidence for excited carbonyls versus O_2 is circumstantial, the most convincing being the spectral data. Singlet oxygen dimol emission emits in the red (634 and 703 nm), whereas excited carbonyls emit usually in the blue (380–460 nm), but these spectral parameters are not definitive (see section 6.3.3). Carbonyls are formed from dioxetan intermediates after the reaction of C=C double bonds in lipids with reactive oxygen metabolites, including singlet oxygen, adding to the confusion in the interpretation of chemiluminescence spectra from cells. Four pathways capable of generating the two excited state groups, $>C=O$ and $^1O_2$, are mainly responsible for indigenous ultraweak chemiluminescence:

(1) Pathways generating oxygen some of which may be singlet, e.g. chloroplast + light, superoxide dismutase, $H_2O_2 + OCl^-$.
(2) Reactive oxygen metabolites oxidising endogenous substrates such as lipids or proteins. These metabolites include O_2^-, H_2O_2, $\cdot OH$, and hydroperoxides.
(3) Peroxidase catalysed reactions.
(4) Prostaglandin synthesis.

Oxygen itself is often the endogenous oxidant, since the requirement for molecular oxygen may be circumvented experimentally by adding a hydroperoxide. However, *in situ* formation of any such hydroperoxides, including H_2O_2, will have required dioxygen (O_2). A source of electrons, such as NAD(P)H or a cytochrome chain, is required to form the reactive oxygen metabolite O_2^-. H_2O_2 can then form spontaneously via dismutation (enhanced by superoxide dismutase). This may be followed by formation of $\cdot OH$ through the Haber–Weiss reaction catalysed by iron. Hypochlorite formation occurs from H_2O_2 and Cl^- via enzymes such as myeloperoxidase. Several other peroxidases with various anion specificities exist in different cells, e.g. lactoperoxidase in milk utilises thiocyanate (SCN^-), thyroid peroxidase uses I^-, whereas ovoperoxidase in eggs probably also uses Cl^-, and eosinophils can form OBr^- via a peroxidase catalysed reaction. Peroxidases are thus also potential sources of chemiluminescence in many tissues.

In 1947 Chance, using the absorbance spectrum of catalase, estimated that the intracellular concentration of H_2O_2 in a liver cell was about 10 nM. The estimated superoxide anion (O_2^-) concentration is then 10 pM, and the rates of formation of hydroxyl radical ($\cdot OH$) and endogeneous hydroperoxides (ROOH) $10^{-9}–10^{-12}$M s^{-1} and 10^{-6}M s^{-1} respectively. These oxidants, together with singlet oxygen, play an important part in lipid peroxidation. This process is a common source of indigenous ultraweak chemiluminescence. The metabolism of lipid peroxides can be considered in five stages:

(1) Formation of a radical chain reaction

initiation $RH + \cdot OH \rightarrow R\cdot + H_2O$ (6.2)
propagation $R\cdot + O_2 \rightarrow ROO.$ (6.3)
completion of the chain $ROO\cdot + RH \rightarrow R\cdot + ROOH.$ (6.4)

(2) Accumulation of metastable products from the chain reaction, e.g. ROOH
(3) Reaction of the metastable products and radicals to form excited states:

$ROO\cdot + ROO\cdot \rightarrow \rangle CHOH + O_2 + \rangle C = O^* \rightarrow$ blue light   (6.5)

or

$ROO\cdot + ROO\cdot \rightarrow ROOR + O_2 \rightarrow$ red light   (6.6)
$R = \rangle CH-$

Equation (6.5) may produce red light if $^1O_2$ is also formed. The lipid peroxy radical $(= ROO\cdot = \rangle C\underset{O-O}{\overset{H}{\diagup}})$ is where the reaction may terminate if scavenged by vitamin E.

(4) Reactions with $^1O_2$ or other inorganic oxygen metabolites

$\rangle C-C\langle$ in the lipid $+ ^1O_2 \rightarrow - \underset{O-O}{C-C} - \rightarrow \rangle C = O + \rangle C = O^* \rightarrow$ blue light   (6.7)

(dioxetan)

(5) Energy transfer

$\rangle C = O^*$ or $R\cdot^*$ or $^1O_2 +$ fluor $\rightarrow$ fluor$^* \rightarrow$ light.   (6.8)

If energy transfer is efficient the chemiluminescence spectrum will then be that of the fluor.

The four most common unsaturated fatty acids in mammalian tissues are:

C16 palmitoleic acid (*cis*-Δ^9-hexadecenoic) $CH_3(CH_2)_5CH = CH(CH_2)_7CO_2H$
C18 Oleic acid (*cis*-Δ^9-octadecenoic) $CH_3(CH_2)_7CH = CH(CH_2)_7CO_2H$
C18 linoleic acid (*cis,cis*-Δ^9,Δ^{12} octadecadienoic)
    $CH_3(CH_2)_4(CH = CHCH_2)_2(CH_2)_6CO_2H$
C20 arachidonic acid ($\Delta^5,\Delta^8,\Delta^{11},\Delta^{14}$-eisocatetraenoic)
    $CH_3(CH_2)_4CH = CHCH_2)_4(CH_2)_2CO_2H$

Oleic and palmitoleic are the most abundant in animal lipids. They all occur in membrane phospholipids as the fatty acid moieties linked to two of the $-OH$ groups in glycerol, and they occur in the plasma in lipoproteins. The occurrence of $-CH = CHCH_2CH = CH-$ in both linoleic and arachidonic acids results in the formation of a soluble, easily measurable, product released from cells after lipid peroxidation, malonyl dialdehyde ($MDA = OHCCH_2CHO$). Lipid peroxidation is occurring in aerobic cells all the time, but may be exacerbated if cells are exposed to more oxygen than usual, for example following microbleeding or reperfusion injury. Certain pathogens, drugs, and toxins may also increase lipid peroxidation, and thus lead to alternations in membrane structure and function. Plants often contain high levels of unsaturated fatty acids, for example ·soya beans and the Mediterranean salad plant purslane are rich in $\Omega 3$ unsaturated fatty acids. Any unsaturated fatty acid is, in principle, a source of indigenous ultraweak chemiluminescence. The role of saturated fatty acids, still capable of reacting with ·OH for example, in ultraweak chemiluminescence, is less clear.

A major problem in unravelling the chemistry of lipid peroxidation, and its associated ultraweak chemiluminescence, is the number of oxygen derivatives potentially capable of acting as oxidants or sources of singlet oxygen. There are at least eight:

(1) O_2^- formed by the univalent reduction of O_2. Superoxide dismutases will inhibit the chemiluminescence if O_2^- is involved directly, provided that they have access to the O_2^-.

(2) H_2O_2 formed by dismutation

$$2\,O_2^- + 2H^+ \rightarrow H_2O_2 + O_2(^1O_2) \tag{6.9}$$

Catalase inhibits chemiluminescence dependent on H_2O_2 as oxidant by destroying it:

$$2H_2O_2 \rightarrow 2H_2O + O_2 \tag{6.10}$$

(3) ·OH formed by the Haber–Weiss reaction

$$H^+ + H_2O_2 + O_2^- \rightarrow \cdot OH + O_2 + H_2O \tag{6.11}$$

This reaction is catalysed particularly well by iron salts, and some other transition metals.

(4) OCl^- formed by the myeloperoxidase reaction in neutrophils

$$H_2O_2 + Cl^- \rightarrow OCl^- + H_2O \tag{6.12}$$

Other Cl intermediates can be ClO_2^- or CCl_3O_2, occurring after CCl_4 administration to the liver.

(5) OBr^- formed by the peroxidase of eosinophils

$$H_2O_2 + Br^- \rightarrow OBr^- + H_2O. \tag{6.13}$$

(6) Iron-oxygen derivatives at the active centre of peroxidases, cytochromes, and other haem containing proteins, e.g.

$Fe^{3+}-O$ and $Fe^{3+}-O-O$ formed from O_2 or $H_2O_2 + Fe^{3+}$

(7) Organic oxyradicals, e.g. RO· and ROO.

(8) Singlet oxygen ($^1O_2$) itself.

Evidence for the precise reactions responsible for indigenous ultraweak chemiluminescence is controversial, particularly that for singlet oxygen. Nevertheless, the following experimental procedures can be used in an attempt to identify the pathway and ultimate source of cellular ultraweak chemiluminescence:

(1) Peaks in the chemiluminescence spectra; red for $^1O_2$, blue for $\rangle C = O$. Other spectral peaks may be explained by energy transfer to endogenous fluors such as flavins or tetrapyrroles.

(2) Requirement for oxygen, and oxygen concentration dependence.
(3) Addition of specific substrates, e.g. H_2O_2, arachidonate, or linoleic for prostaglandins induce or enhance chemiluminescence, whereas oleic and saturated fatty acids give little or no chemiluminescence; the potency of particular lipid oxidants, such as organic hydroperoxides.
(4) The effect of specific inhibitors, e.g. aspirin or indomethacin in platelets for prostaglandin synthesis; azide as an inhibitor of myeloperoxidase.
(5) The effect of inducers of enzyme synthesis, e.g. barbiturates for cytochrome P450/448.
(6) Comparison of genetic abnormalities, e.g. myeloperoxidase deficiency, with normal cells.
(7) Inhibition of chemiluminescence by scavengers of oxygen metabolites, e.g. azide or urate for $^1O_2$, superoxide dismutase for O_2^-, catalase for H_2O_2, mannitol or benzoate for $\cdot OH$, amines for OCl^-.
(8) Direct measurement of oxygen metabolites using specific indicators (see section 6.6.4).
(9) Identification of, and correlation with formation of, other reaction products, e.g. malonyl dialdehyde in lipid peroxidation or oxidation of endogenous scavengers, such as glutathione, vitamin E, ascorbate, and free SH.
(10) Addition of energy transfer acceptors, e.g. fluorescein, to identify excited state intermediates.

Whatever the chemical basis is for indigenous ultraweak chemiluminescence, three particular puzzles remain to be answered. Firstly why is the emission ultraweak? Is this because the endogenous reactions responsible have very low quantum yields (ϕ_{CL}), low reaction rates, or are occurring in only tiny amounts? If neutrophils are anything to go by, the last is an unlikely explanation, since oxygen uptake in stimulated cells can be as high as 0.1 fmol (10^{-16}) per second per cell. If all this oxygen reacted with a good chemiluminescent substrate it would produce a very bright light indeed, i.e. about 6×10^5 photons per second per cell if $\phi_{CL} = 0.01$, equivalent to a photocyte in bioluminescence! Secondly, what are the spin states of the excited state emitters? Thirdly is there any really convincing evidence for singlet oxygen in cellular ultraweak chemiluminescence?

6.3.3 The problem of singlet oxygen

The electron configuration of the oxygen atom is $1s^2 2s^2 2p^4$, superscripts representing the number of electrons in each orbital. Dioxygen becomes $\sigma 1s^2$, σ^*1s^2, $\sigma 2s^2$, σ^*2s^2, $\sigma 2p^2$, $\pi 2p^4$, π^*2p^2. These latter two electrons are in different orbitals but have the same, i.e. parallel, spin. Ground state dioxygen, its normal state, therefore is in the triplet state, having two unpaired electrons in separate π^* ($*$ = antibonding) orbitals and is designated as $^3\Sigma_g^-$. It is thus, in a sense, a biradical, expected to be very reactive but kinetically relatively inert because of its triplet spin state. If the two lone electrons become paired with the opposite spins, a so-called 'forbidden' transition, we have singlet oxygen. There are now two forms possible, one designated as $^1\Sigma_g^+$ again has the electrons in different orbitals, now of opposite spin. The other is designated as $^1\Delta g$ and has both electrons in the same orbital. This would be impossible if they were not of opposing spins.

Singlet oxygen, being of a higher energy level than ground state triplet oxygen, can lose its excess energy via luminescence. The luminescence emission of a relatively simple molecule like O_2 is much sharper than the broad spectra from the complex organic chemiluminescent molecules we have encountered in previous chapters. There are two types of photon emission ($h\nu$) possible:

(1) Unimolecular

$$^1\Delta g \rightarrow \Sigma^- g + h\nu \text{ at } 1268 \text{ nm} \tag{6.14}$$
$$^1\Sigma g \rightarrow \Sigma_g^- + h\nu \text{ at } 762 \text{ nm} \tag{6.15}$$

(2) Bimolecular, so-called dimol emission, by energy pooling (see Chapter 9)

$$2^1\Delta g(0,0) \rightarrow 2\Sigma_g^- + h\nu \text{ at } 633.4 \text{ and } 703.2 \text{ nm} \tag{6.16}$$
$$2\Sigma_g^+ \rightarrow 2\Sigma_g^- \tag{6.16}$$
$$^1\Delta g + \Sigma_g^+ \rightarrow 2\Sigma_g^- + h\nu \text{ at } 478 \text{ nm} \tag{6.17}$$

Here is one other possibility:

$$2^1\Delta g(1,0) \rightarrow 2\Sigma_g^- + h\nu \text{ at } 570\text{--}580 \text{ nm}$$

The numbers in brackets refer to rotational states.

Not only are both species of singlet oxygen very reactive, but they are readily quenched back to ground state, without luminescence, by collision with surrounding molecules, particularly those from the solvent. As a result, Σ_g^+ singlet oxygen has a lifetime of only a nanosecond in aqueous media, and is therefore unlikely to make any contribution to cellular ultraweak chemiluminescence. This compares with normal lifetimes of 1–10 ns for electronically excited states producing light. $^1\Delta g$ singlet oxygen is a little more stable, but even so, still only has a lifetime, τ (time to 1/e of the initial concentration) of 2–4 μs in H_2O. These very short lifetimes, the presence of other quenchers and reactants in biological systems, together with the lack of sharpness in the spectrum of cellular ultraweak chemiluminescence, have led to much scepticism about the existence or significance of $^1O_2$ in living systems. Yet there are many potential generators of $^1O_2$ in cells (Table 6.6). Under certain conditions, the spectrum of cellular ultraweak chemiluminescence is predominantly in the red, for example in hepatocytes exposed to high O_2. Furthermore, $^1\Delta g\ O_2$ is considerably more stable in organic solvents than in water. In CCl_4, the lifetime is about 1 ms, sufficiently long for it to emit significant light, or to react chemically with other molecules, e.g. ›C=C‹. Oxygen is some ten times more soluble in many organic solvents than in water. Since all of the examples of ultraweak chemiluminescence in living systems occur in the presence of phospholipid membranes, the possibility exists that the phospholipid bilayer may provide an important environment in which $^1O_2$ can participate in both chemiluminescence and oxidative reactions. In organic solvents, the generation of singlet oxygen can also be very efficient. For example, in methanol the generation of $^1\Delta g\ O_2$ from the reaction:

$$H_2O_2 + OCl^- = H_2O + Cl^- + {}^1O_2 \tag{6.19}$$

can be as high as 80%, all of the O_2 coming from H_2O_2 and none from OCl^- or H_2O, as shown by using $H_2^{18}O_2$.

Table 6.6 — Possible sources of singlet oxygen

A. In cells

1. Dismutation $2O_2^- + 2H^+ \rightarrow H_2O_2 + {}^1O_2$

2. Peroxide $H_2O_2 + OCl^- \rightarrow H_2O + Cl^- + {}^1O_2$
   decomposition $H_2O_2 + OBr^- \rightarrow H_2O + Br^- + {}^1O_2$
   + peroxidase $ROO\cdot \text{ (lipid)} + ROO. \rightarrow ROOR + {}^1O_2$

3. Electron $O_2^- + X \rightarrow X^- + {}^1O_2$
   transfer $O_2^- + \cdot OH \rightarrow OH^- + {}^1O_2$

4. Peroxidase + $(FeO)^3 + \text{oxene donor (PhI = 0)} \rightarrow$
   oxene $Fe^{3+} + PhI + {}^1O_2$

5. Aromatic
   endoperoxide

6. Energy transfer   chlorophyll, porphyrins, retinal, flavins

B. Chemical

1. Electric discharges.
2. Electron transfer reactions.
3. Thermal decomposition, e.g. O_3 and phosphite.

There are four types of reaction which could lead to the formation of ${}^1O_2$ (Table 6.6):

(1) Peroxide decomposition, e.g. superoxide dismutation or $H_2O_2 + OCl^-$.
(2) Aromatic endoperoxide decomposition.
(3) Electron transfer reaction, analogous to ${}^1O_2$ formation in the gas phase through electric discharges.
(4) Energy transfer, e.g. photosensitisation of biological pigments such as porphyrins (Eqn (6.20)):

$$X + h\nu \rightarrow X^* \xrightarrow{{}^3O_2} X + {}^1O_2 \tag{6.20}$$

But do they?

It is fairly well established that the last of these four occurs in chloroplasts, in the eye, in porphyrias, and following porphyrin therapy of cancer. When porphyrins accumulate in the skin, this tissue becomes very prone to photo-oxidation reactions caused by ${}^1O_2$. Addition of a cation into the porphyrin nucleus greatly reduces its ability to generate ${}^1O_2$ on irradiation.

Evidence for ${}^1O_2$ generation as a source of ultraweak chemiluminescence can be provided by:

(1) Infrared luminescence at 1268 nm.
(2) Red chemiluminescence with equal maxima near to 634 and 703 nm, with a trough at 668 nm.
(3) Enhancement of chemiluminescence by D_2O, which increases the lifetime of ${}^1O_2$ from 2-4 μs to 30–50 μs.
(4) Inhibition by quenchers (Q) of ${}^1O_2$ (Table 6.7),

$$ {}^1O_2 + Q \rightarrow {}^3O_2 + Q^* \rightarrow \text{loss of energy by collisions} \tag{6.21}$$

Table 6.7 — Chemiluminescent indicators or quenchers of singlet oxygen

Indicators
Endogenous unimolecular (1268 nm) or dimol (634,703 nm) emission
Polycyclic hydrocarbons (e.g. rubrene)
methoxyvinyl pyrenes, e.g.   CH$_3$O

Quenchers
N$_3^-$
β carotene
urate
$^1O_2$

---

(4) Indicators of $^1O_2$ (Table 6.7); these include chemical traps and chemiluminescent indicators.

The first and third of these have been used mainly as evidence for $^1O_2$ generation in
aqueous living systems. But one place where solvent quenching of $^1O_2$ is likely to be
much less of a problem is in the gas phase, for example in the air inhaled and exhaled
in the lung. In the atmosphere, H_2O_2 exists at around 0.1–10 nM and ozone, a $^1O_2$
generator, at 10–100 nM. Other oxygen metabolites exist at much lower concentrations, e.g. HO_2 (the protonated form of O_2^-) at 0.1–10 pM, ·OH at 1 fM–1 pM and ·O
at about 0.1 fM. Human breath contains peroxide at about 2 nM, and as many as a
hundred organic components in the nM–fM range. Chemiluminescence at a rate of
several thousand photons per second can be detected in air exhaled from the lung. It
decays with the long halftime of about 20 minutes. The peak emission is in the red,
though there is also a small UV component which may be caused by an excited
derivative of CO_2. In the gas phase other red emitters apart from the 634 and 703 nm
emission of $^1O_2$ must also be considered, e.g. ·OH (632.9 nm) and H_2O^+ (614 and
619 nm). There are, therefore, three possible, as yet not definitely identified, sources
of the chemiluminescence (Fig. 6.2) in animal breath:

(1) Direct emission, from ·OH or $^1O_2$ dimols.
(2) Reaction of ozone (O_3) with linoleic acid and other unsaturated fatty acids to
   produce trioxides which decompose to aldehydes and peroxyradicals, leading to
   excited carbonyl or $^1O_2$ chemiluminescence.
(3) Lipid peroxidation induced by H_2O_2 or ·OH leading to excited carbonyl or $^1O_2$
   emission.

When singlet oxygen is formed in living systems it is extremely reactive, and
potentially damaging. It attacks nucleic acids, particularly guanine residues. It
attacks proteins through oxidation of methionine to the sulphonate, as well as
oxidising cysteine, histidine, tryptophan, and tyrosine. It oxidises lipids, converting
for example ›C=C into ›C=O, leading to membrane damage. This is why so much
effort has been put into proving definitively that $^1O_2$ does occur in cells, as well as
into the identification of naturally-occurring scavengers, and to establishing its role

in phenomena such as the microbial killing of phagocytes. Chemiluminescence, both dimol at 634 and 703 nm and the infrared unimolecular emission at 1268 nm, is playing an increasingly important role as a criterion for the existence and importance of $^1O_2$ in biology.

The development of new, specific chemiluminescent indicators (Table 6.7) is also likely to provide crucial evidence for or against $^1O_2$. In spite of much scepticism about the existence of significant amounts of $^1O_2$ in solution because of its μs lifetime, there is now much evidence consistent with its formation during porphyrin photo-oxidation, during lipid peroxidation induced by hyperbaric oxygen or hydro-peroxide formation, and in oxyanion reactions involving halogenation of proteins and halogen or other anion peroxidases. $^1O_2$ also appears to play an important role in paraquat poisoning. The much longer duration of singlet oxygen in organic solvents, and in the gas phase, suggests that the lipid bilayer and air–tissue interfaces could be an important site to search more diligently for it. Whether you are convinced or not of its existence, the ingenious experiments which have been designed to search for it have, and will in the future, lead to new discoveries and to a better understanding of the mechanism underlying oxidative damage inside and outside cells.

6.4 ENHANCEMENT WITH CHEMILUMINESCENT COMPOUNDS

Addition of a chemiluminescent compound to cells, organelles, or acellular reactions producing reactive oxygen metabolites results in light emission often more than ten thousand times more intense than the indigenous ultraweak chemiluminescence (Fig. 6.3), though it is still usually invisible to the naked eye. For example, addition of 10 μM luminol to activated neutrophils can produce as many as 1–100 photons per second per cell. A signal can thus be detected from as little as 5–10 cells. Peroxidase-enhancement may improve this still further and single cell activation may be detectable using integration and image intensification. Three compounds have been particularly well investigated (Fig. 6.4), luminol, lucigenin, and pholasin the protein-bound luciferin from the bivalve mollusc *Pholas dactylus* (Fig. 1.1). Luminol at a concentration of 1 μM–1 mM has been particularly popular, a luminol-dependent chemiluminescent signal having been detected down to 1 nM luminol in nuetrophils, and in many other cell types (Table 6.8), as well as in several sub-cellular organelles such as mitochondria and with isolated enzymes such as xanthine oxidase and peroxidase.

Several other phthalazine diones and acridinium salts have been investigated in an attempt to either increase the light signal or improve specificity for detection of a particular oxygen metabolite. One compound, 7-dimethyl amino-naphthalene-1,2-dicarbonic-acid hydrazide produces a signal some 2–5 times that of luminol when added to neutrophils or monocytes. Pholasin results in an even bigger light signal. The pholasin light output is some 50–100 times that of luminol at the same molarity, enabling the activation of a single neutrophil to be detected (Fig. 6.6). When neutrophils are activated by chemotactic stimuli the order of sensitivity is pholasin > luminol > lucigenin. However, with other stimuli, or in other cell systems, the sensitivity can be reversed. An extreme example of this is the hepatic cytochrome P450 system induced by barbiturates which in the presence of NAD(P)H appears to generate significant chemiluminescence only from lucigenin, and not from luminol.

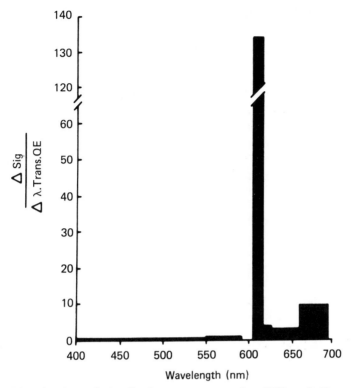

Fig. 6.2 — An ultraweak chemiluminescence spectrum from Williams & Chance (1985) with permission (see section 6.9). Spontaneous chemiluminescence in human breath.

Table 6.8 — Indicator-dependent chemiluminescence

Cell type	Indicator
Neutrophil	Luminol, lucigenin, pholasin
Eosinophil	Luminol
Macrophage	Luminol, lucigenin, DMNH†
Monocyte	Luminol, lucigenin
Kupffer cells	Luminol
Spleen cells	Luminol
Thymocytes	Luminol
NK cells	Luminol
Platelets	Luminol
Amoeba	Luminol
Fungus	Aromatic aldehyde
Mitochondria	Luminol
Microsomes	Lucigenin, benzo(a) pyrenes
Enzymes (hypoxanthine, xanthine + xanthine oxidase)	Luminol

†DMNH = 7-dimethyl amino-naphthalene-1,2-dicarbonic acid hydrazide

Aryl acridinium esters are, in general, much weaker than lucigenin. Other compounds, not normally thought of as being chemiluminescent, can nevertheless generate light from some cells. These compounds include $^1O_2$ indicators (Table 6.7), and aromatic aldehydes capable of generating an excited state in an O_2 + peroxidase reaction, followed by energy transfer to an endogenous or added fluor.

Four questions arise. Firstly which reactive oxygen metabolites are responsible for generating the indicator chemiluminescence? Secondly, where is the chemiluminescence being generated (i.e. inside or outside the cell)? Thirdly, how closely does this *indicator-dependent* chemiluminescence parallel endogenous *ultraweak* chemiluminescence? Fourthly, is it of more value as an analytical tool to use the indicator rather than the indigenous emission?

Luminol reacts well *in vitro* with O_2^-, H_2O_2 + peroxidase, OCl^- and $^1O_2$, though the reaction with singlet oxygen may be a 'dark' reaction. Lucigenin, on the other hand, may be more specific for O_2^-, involving a three-step reaction:

$$\text{lucigenin}^{2+} + O_2^- \rightarrow \text{lucigenin}^+ \text{ radical } + O_2 \tag{6.22}$$

$$\text{lucigenin radical} + O_2 \rightarrow \text{lucigenin dioxetan} \tag{6.23}$$

$$\text{lucigenin dioxetan} \rightarrow N\text{-methyl acridone}^* \rightarrow N\text{-methyl acidone} + h\nu \tag{6.24}$$

Under alkaline conditions, HO_2^-, formed from H_2O_2, is also a potent oxidant for lucigenin chemiluminescence. Like luminol, pholasin also reacts with O_2^-, H_2O_2 + peroxidase, and OCl^-, and also with peroxidase + O_2 (the natural Cu containing luciferase in *Pholas* has peroxidase activity). Pholasin also appears to react with longer lived, and in some cases, high molecular weight, oxidants released from cells such as activated neutrophils. For example, the extracellular fluid from these cells will provoke pholasin chemiluminescence, but not that of luminol. These oxidants may include metastable hydroperoxides and chloramines formed from protein or polyamine $- NH_2$ groups plus OCl^-. None of these indicators is therefore entirely specific for a particular oxygen metabolite but pholasin appears to detect mainly O_2^- from neutrophils.

Oxidases reduce O_2 univalently to O_2^-, bivalently to H_2O_2, or both. Enzymes such as xanthine oxidase produce O_2^- and provoke luminol chemiluminescence, which can be inhibited by more than 90% by superoxide dismutase. This latter enzyme converts all of the O_2^- formed rapidly to H_2O_2. In contrast, this chemiluminescence is unaffected by catalase, which degrades the H_2O_2. Other enzymes such as glucose oxidase, which produce mainly H_2O_2, result in poor luminol chemiluminescence. A better stimulus *in vitro* of luminol chemiluminescence is H_2O_2 plus any peroxidase, though differences in potency between different peroxidases are observed. This can be abolished by azide (1 μM–1 mM). Azide is acting here to inhibit the peroxidase, not as a quencher of $^1O_2$. In neutrophils, superoxide dismutase results in a variable inhibition of luminol-dependent chemiluminescence of 20–80%, but can inhibit by more than 90% lucigenin or pholasin-dependent chemiluminescence induced both by cells or by xanthine oxidation. In contrast, azide (N_3^-), which virtually abolishes ($>95\%$ inhibition) the luminol-dependent chemiluminescence of neutrophils when added with the stimulus, has no effect on luminol chemiluminescence induced by xanthine–xanthine oxidase (O_2^-).

The shapes of the chemiluminescence traces from cells vary with stimulus and indicator (Fig. 6.6), as well as with indicator concentration. They can also be markedly biphasic. The interpretation of these traces has been a source of much argument. Biphasic responses have, for example, been interpreted as being caused by two metabolites O_2^- on the one hand and H_2O_2 + peroxidase on the other. This has been supported by the differential inhibition caused by inhibitors such as azide or

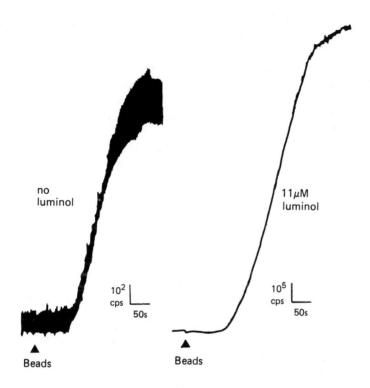

no
luminol

10^2
cps
50s

▲
Beads

11μM
luminol

10^5
cps
50s

▲
Beads

Fig. 6.3 — Ultraweak versus indicator-dependent chemiluminescence. Rat neutrophils ± 11 μM luminol, stimulated by latex beads.

superoxide dismutase added during the response, relative to the inhibition when these are added with the stimulus. However, the site of chemiluminescence can change during the course of activation of the cell, thereby confusing the interperation of such experiments.

Where then does the chemiluminescence of luminol, lucigenin, and pholasin occur when added to cells such as phagocytes? No good data exist on the permeability of these chemiluminescent indicators across lipid bilayers. It is usually assumed that they diffuse across biological membranes slowly, or not at all, particularly since lucigenin is charged, pholasin has a molecular weight of 34 000, and all three prefer hydrophilic or polar solvents rather than hydrophobic or non-polar ones. Attempts have been made to synthesise lipid-soluble luminol derivatives in order to get them across biological membranes, including those of organelles such as mitochondria.

There is good evidence from absorbing and fluorescent indicators that reactive oxygen metabolite production in phagocytes occurs both within intracellular granules, and externally at the cell surface. The relative amounts at each site vary with stimulus and time. Luminol-dependent chemiluminescence emanating from a suspension of neutrophil, activated either by phagocytic or by chemotactic stimuli, disappears from the medium when the cells are centrifuged. It is found entirely with the cells. Cytocholasin B, a disrupter of microfilaments, enhances the response to chemotactic factors and increases chemiluminescence at the cell surface, whereas it inhibits by more than 95% the response to phagocytic stimuli (Fig. 6.7). In contrast cytochalasin B enhances the pholasin response under both of these conditions, but only slightly. Intracellular chemiluminescence may occur as the result of uptake of the chemiluminescent indicator during phagocytosis, or via pinocytosis. The precise magnitude of the indicator- or indigenous-chemiluminescence at different locations has yet to be fully established.

Indigenous ultraweak chemiluminescence and indicator-dependent chemiluminescence are shown to reflect the same cellular process if:

(1) Both depend on O_2.
(2) Both are induced, enhanced, or inhibited by the substances which affect the concentration of the oxygen metabolite reacting with both endogenous substrates or the indicators.
(3) The time of onset, time to peak emission, and time of decay are the same.
(4) Both are temperature-sensitive.
(5) Both are reduced or lost in cells with the appropriate genetic abnormality, namely one preventing formation of the oxygen metabolite(s) responsible.

For example, the indigenous ultraweak and luminol-dependent chemiluminescence of neutrophils are both stimulated, with a similar time of onset, by the same phagocytic, chemotactic, and experimental stimuli (Table 6.4), both depend on O_2, and cytochalasin B enhances chemotactic stimuli. Both show similar time courses with a lag of 10–30 seconds, time to peak of 5–6 minutes, and $t_{\frac{1}{2}}$ of decay of about 10 minutes. Both responses are lost in cells from patients with chronic granulomatous disease. However, the two sources of light, indigenous and indicator-dependent, can be separated on the basis of:

(1) The chemiluminescence spectrum: in the presence of high concentrations (μM–mM) of the indicator its spectrum predominates. By adjusting the indicator concentration to about 10 nM both the blue chemiluminescence of luminol and the predominantly red emission of the endogenous ultraweak chemiluminescence can be detected simultaneously with a rapidly spinning disc spectrometer. The blue component of the indigenous emission will, however, still be obscured by the indicator.
(2) It may be possible to locate the two light emissions at different sites, related events occurring within or outside the activated cells.
(3) Cells from patients with myeloperoxidase deficiency have little or no luminol-dependent chemiluminescence, whereas the indigenous ultraweak emission

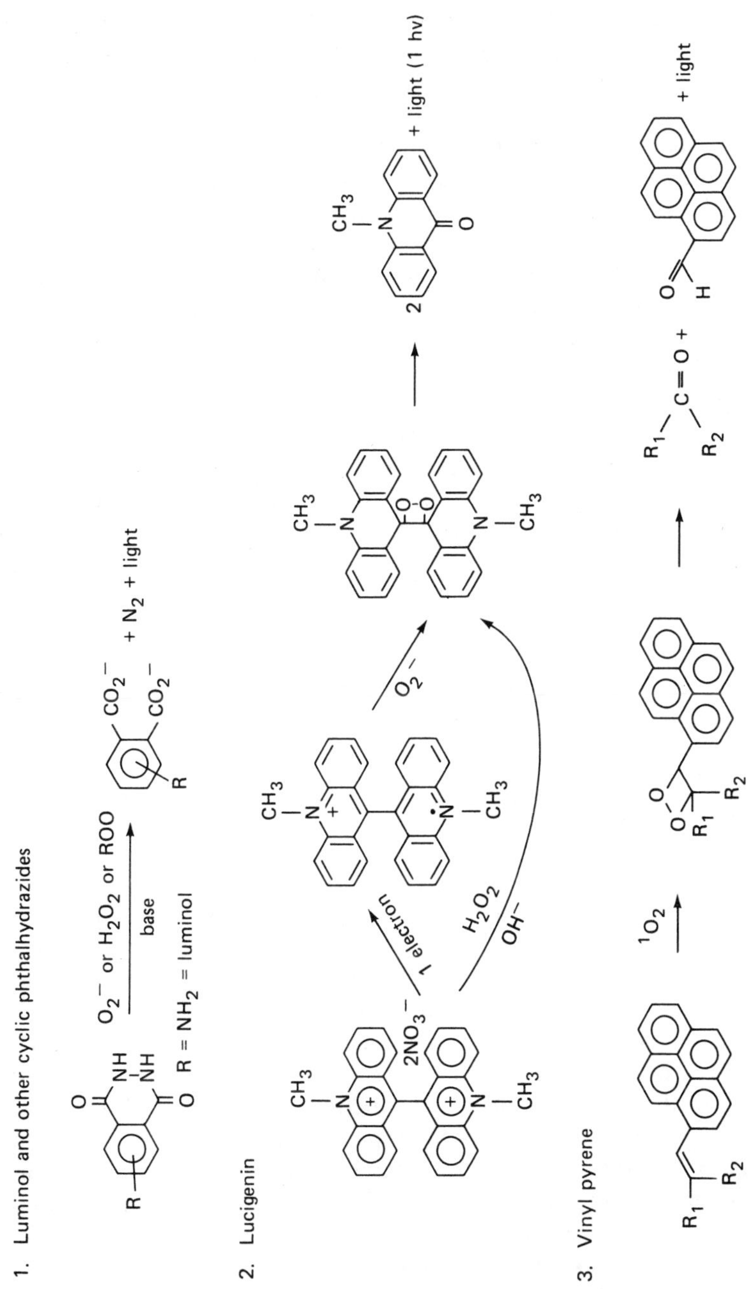
1. Luminol and other cyclic phthalhydrazides
O₂⁻ or H₂O₂ or ROO
base
R = NH₂ = luminol
+ N₂ + light
2. Lucigenin
2NO₃⁻
1 electron
H₂O₂
OH⁻
O₂⁻
+ light (1 hν)
3. Vinyl pyrene
¹O₂
+ light

Fig. 6.4 — Some reaction involved in indicator-dependent chemiluminescence.

remains and simply shows a more prolonged response. This is consistent with the importance of myeloperoxidase in the luminol reaction.

The increased light emission of cells, organelles, and acellular reactions in the presence of a chemiluminescent indicator has several consequences for its application as an analytical tool, in the study of cell activation and cell injury. Firstly, fewer cells are required to produce a detectable signal than for indigenous ultraweak chemiluminescence. Secondly, signals can ultimately be detectable in single cells, by using high quantum yield chemiluminescent indicators. Thirdly, different reactive oxygen metabolites can be monitored concomitantly.

In spite of the lack of apparent specificity for one oxygen metabolite, the use of luminol and other chemiluminescent compounds has enabled important insights into the mechanism of phagocyte activation to be obtained. But the increased sensitivity resulting from the addition of chemiluminescence indicators also has its dangers. For example, a very small contamination with neutrophils of an apparently pure cell type can produce misleading signals. Lymphocytes do not apparently produce significant quantities of reactive oxygen metabolites. However, natural killer (NK) cells, thought to be related to the lymphocyte cell line, have been reported to provoke luminol-dependent chemiluminescence. This has been related directly to their tumour cell killing activity. This suggested that they should be re-allocated, to the monocyte–macrophage cell line. Unfortunately, the original results could be explained by a neutrophil contamination of only 1–5%. NK cells, 99% pure, have been reported not to chemiluminesce. A similar problem has been encountered in identifying the cell type responsible for the luminol-dependent chemiluminescence from thymocytes. To resolve these problems, the criteria outlined in section 6.3.1 should be applied.

6.5   A CLASSIFICATION

In a book such as this the value of any classification, however controversial, is that it clarifies how an individual author, or a consensus of scientists, use particular terms. The low intensity, 'invisible' chemiluminescence now being extensively used to study oxidative reactions inside and outside cells is perhaps best separated into three distinct groups: *ultraweak*, *dim*, and *indicator-dependent* chemiluminescence. The two parameters which distinguish them from one another are the overall quantum yield of the reaction, and the nature of any added compound necessary to induce the luminescence. The quantum yield can be assessed by relating light intensity (dhv/dt) to substrate utilisation, product formation, or consumption of a scavenger responsible for removing the bulk of a metabolite capable of generating the chemiluminescence. For example, the absolute rates of O_2 consumption (the substrate), of malonyl dialdehyde (the product), and of oxidised or total glutathione released (the scavenger) during lipid peroxidation can be as much as 10^{14} greater than the light intensity, photons being related to moles through the Avogadro constant, 6.022×10^{23}.

The three types of feeble chemiluminescence can be defined as follows:

Ultraweak chemiluminescence (UWCL): the emission of light, invisible to the naked eye, from chemical reactions with very low quantum yields. ϕ_{CL} (Eqn 1.9) is usually

$< 10^{-3}$ and often as low as 10^{-9}–10^{-12}. The low quantum yield may arise from any one of four factors:

(1) More than one chemical pathway, only one of which is chemiluminescent.
(2) Low ϕ_{EX}, the fraction of molecules generated in the excited state.
(3) Low ϕ_F, the fluorescence quantum yield of the light emitter.
(4) Quenching of the excited state by solvent or solute molecules, or by other chemical reactions. This is a particular feature of low light yields from $^1O_2$ in aqueous solution.

When considering cells and living tissues, this term will be used synonymously with indigenous ultraweak chemiluminescence, i.e. the excited state molecules arise solely from endogenous molecules. If inducers are required, they can only be included within this category if they do not contribute directly to the formation of the excited state, unless this is oxygen itself (i.e. singlet oxygen). They may, however, contribute reactants which provide oxidants of native substrates. Examples are shown in Table 6.2. We therefore now exclude from this category cellular or acellular chemiluminescence detectable on addition of compounds such as luminol, lucigenin, pholasin, aromatic aldehydes, or $^1O_2$ indicators. Also excluded is any luciferin-luciferase or photoprotein chemiluminescence.

Dim chemiluminescence (DC): the emission of light, invisible to the naked eye, from chemical reactions with high quantum yields (ϕ_{CL} usually > 0.001), but which produce very low light intensities either because of a very slow reaction rate or because of the small quantity of chemiluminescent compound present. The substrate producing the excited state is again endogenous. Examples could be luminous bacteria emitting a weak emission because of lack of the auto-inducer, or firefly luciferin-luciferase in the presence of low ATP concentrations, e.g. in an assay.

But where do we put Grignard reaction chemiluminescence which has a very low quantum yield, yet, because of the high concentrations of reagents, can produce visible light? The answer is that either ultraweak or dim chemiluminescence become 'bright' if visible.

Indicator-dependent chemiluminescence (IDCL): the emission of light as a result of a chemical reaction between an added indicator (Table 6.2) and one or more substances produced by the system, e.g. cells or an enzymatic reaction. The spectrum and quantum yield are therefore characteristics of the indicator and not the system being studied. Examples are the luminol- or lucigenin-dependent chemilumines-cence of phagocytes. It is important to distinguish this from ultraweak chemilumines-cence, just defined, and not to call it indicator-*enhanced* chemiluminescence, as the indicator is not 'enhancing' the ultraweak reaction as such, but rather producing light as a result of a completely different reaction, albeit sometimes involving a reactant common to both. Indicator-dependent chemiluminescence can be ultraweak, dim, or bright (visible). The indicator can be a chemiluminescent compound or an energy transfer acceptor.

These definitions, with their inherent distinctions, are proposed not for the sake of semantic pedantry, but because they provide the basis for identifying truly endogenous excited-state molecules, which themselves play a key role in cell physiology and cell pathology even though the light itself apparently does not.

Indicator-dependent chemiluminescence has become very popular in the study of phagocyte activation, in particular neutrophils and other phagocytes. Yet workers who favour other methods for quantifying neutrophil activation, based on spectrophotometry, fluorimetry, or electron spin resonance, are sometimes sceptical about the value of such investigations. Let us therefore see whether anything has been learnt about the neutrophil by using such chemiluminescence.

6.6 CHEMILUMINESCENCE AS AN INDICATOR OF PHAGOCYTE ACTIVATION

6.6.1 The natural history of white blood cells

There are about $4–11 \times 10^9$ white cells in each litre of human blood, some one-thousandth that of the red cell population. White cells are also found in tissues, and in extracellular fluids such as the synovial, alveolar, and peritoneal fluids. Some of these cells form the reticuloendothelial system. Microscopically, the white cells of the blood can be separated into three groups: ones with a polymorphic nucleus, the monocytes, and the lymphoctyes. The first of these are sub-classified into neutrophils, eosinophils, and basophils. Monocytes are transformed into 'activated' macrophages when attracted into tissues by chemotactic substances, and lymphocytes are divided in B-cells and plasma cells, responsible for antibody production, together with various types of T-cell (helper, suppressor, cytotoxic). The neutrophil is the predominant white cell in the blood, accounting for 40–75%. Lymphocytes account for some 20–45% of the white cell population. Eosinophils and monocytes account for about 5% each, whereas basophils are less than 1% of the total. They are all formed from stem cells in the bone marrow, branching out into the three major cell lines. All white cells are concerned with the body's defence against invading organisms, lymphocytes being responsible for antibody production, and for killer T cells; whereas neutrophils, eosinophils, and macrophages kill invading pathogenic organisms after phagocytosing them.

Neutrophils are the first cells to invade a site of inflammation following, for example, an infection with bacteria. Their half time in the blood is only 6–7 hours, but in tissue may be some 1–4 days. In the fridge, they also keep for several days. In an inflammatory response the neutrophils are followed some hours later by the 'activated' macrophages. These two cell types ingest invading organisms and viruses, particularly if they are coated with antibody or complement since these white cells contain Fc and C3b receptors.

The question arises, how do these cells actually kill a bacterium, or other microorganism, after ingesting it? The phagolysosomes, formed after phagocytosis, contain degradative enzymes which are ultimately responsible for degrading the macromolecules in the bacterium. However, the respiratory burst (see section 6.2.2) also plays a crucial role. It leads to the production of reactive oxygen metabolites. These metabolites produced within the phagolysosomes, oxidise proteins, carbohydrates, and lipids on the surface of the bacterium, enabling the proteases and other degradative enzymes to work quickly and efficiently to destroy the invading organism. The degradative enzymes of the neutrophil are mainly stored within two types of granule. The large electron dense azurophil, or primary, granules contain acid phosphatases, β glucuronidase, cationic proteases, some lysozyme, and the chloro-peroxidase myeloperoxidase. The smaller, less dense 'specific' granules contain alkaline phosphatase, lysozyme, lactoferrin, vitamin B binding protein, and various hydrolases, but no peroxidase. There is also a third, intermediary, granule, whose precise function remains to be established. Genetic defects in the respiratory burst, e.g. chronic granulomatous disease, or myeloperoxidase deficiency, leave the individual susceptible to certain infections. This is consistent with the multi-level defence system in mammals involving both humoral and cellular components. Impairment or loss of one does not necessarily lead to death of the host, since another mechanism is ready to take over the defence role. Perhaps this range of different defence systems is one of the keys to the evolutionary success of mammals.

In contrast to their beneficial role, recently attention has been focused on the possible hazards of neutrophil infiltration. They occur in large numbers in the joints of patients with rheumatoid arthritis, but not in osteo-arthritis. They infiltrate the myocardium within minutes of an infarction. In animal models, prevention of this infiltration increases survival rate and reduces complications such as arrythmias. They also infiltrate the kidney in certain types of glomerulonephritis, and the lung in several pathological conditions. Are these neutrophils producing, inappropriately, degradative enzymes or reactive oxygen metabolites which may lie close to the origins of the disease? Do these cells interact with platelets, or the atheromatous plaque, and thereby trigger the initial infarction in the onset of a heart attack?

Reactive oxygen metabolites, including O_2^-, H_2O_2, $\cdot OH$, OCl^-, and $^1O_2$, are certainly damaging to isolated cells and macromolecules, causing oxidation of –SH groups to S–S and SO_3, halogenation of tyrosine and histidine residues, chloramine formation from lysines, and oxidation of purine and pyrimidine bases, sugars, and fatty acids. This can lead to destruction of soluble proteins, degradation of hyaluronic acid and collagen, cleavage of nucleic acids, and the breakdown of membrane permeability. These potentially lethal properties of oxygen metabolites are why we have several scavenging mechanisms designed to protect cells, and the extracellular matrix, against them. Within the cell, superoxide dismutases in the cytoplasm or mitochondria convert O_2^- to H_2O_2·H_2O_2 can be removed by catalase or by glutathione + glutathione peroxidase. Vitamin E acts as an oxygen metabolite scavenger within membranes, whilst ascorbate and albumin act as scavengers outside the cell. To establish whether any of these protection mechanisms break down in disease, and whether reactive oxygen metabolites play a central role in diseases of the lung, heart, kidney, joint, brain, and other tissues, it is essential that sensitive methods are available to measure them, and that the mechanism of their production and release by cells is fully characterised. Chemiluminescence provides such a method.

6.6.2 Some questions
Chemiluminescence has been applied to a number of key questions concerning the natural function, the pathogenic role, and the evolutionary significance of reactive oxygen metabolite production by neutrophils and other phagocytes:

(1) What is the pathway responsible for producing reactive oxygen metabolites?
(2) What are the primary stimuli and secondary regulators responsible for controlling this pathway?

(3) What is the intracellular mechanism of cell activation, and its modification by secondary regulators?

(4) Where, and in what quantity, are the metabolites produced, and how does this relate to the nature of the particular cell stimulus and mechanism of cell activation?

(5) Is the activation of this pathway a threshold phenomenon, a stimulus acting by 'switching on' individual cells?

(6) What is the natural function, or pathological significance, of the metabolites produced within, or outside of, the cell?

(7) Are there any drugs which can modify the cell response, or scavenge any of the oxygen metabolites?

(8) What are the differences and similarities between types of phagocyte, namely neutrophils, eosinophils, monocytes, and macrophages?

(9) Does measurement of reactive oxygen metabolite production by neutrophils or any other cell, particularly using chemiluminescence, have any application as a tool in the clinical laboratory for diagnosis, prognosis, or assessing the management of a disease?

6.6.3   The pathway generating reactive oxygen metabolites

The reduction of oxygen to water requires four electrons. Only one is needed to produce superoxide anion, the initial product of the 'respiratory burst'. Thus, there must be both a source of electrons, together with a mechanism for transferring them to oxygen. The electron source in phagocytes appears to be NADPH, supplied through the pentose phosphate pathway, which is activated concomitantly with reactive oxygen metabolite production. During the 1960s, three groups of Japanese workers discovered an unusual cytochrome in neutrophils. In the latter 1970s, Segal and Jones showed, using spectral criteria, that this cytochrome was a 'b' type and had the very negative redox potential of -245 mV, sufficient to act as an electron donor to oxygen. Further evidence that this cytochrome b_{245} is essential for the generation of O_2^- has come from studies of chronic granulomatous disease. In this disease, bacteria are ingested by neutrophils but the respiratory burst fails to occur, resulting in inefficient microbial killing in the cells. Two biochemical forms of the disease have been characterised. In one, the larger X linked group, the cytochrome b_{245} appears to be absent altogether, whereas in the other autosomal recessive, it is present but fails to become reduced following cell activation. O_2^- fails to form in either case, and indicator-dependent chemiluminescence is greatly reduced or absent. This b cytochrome, unlike cytochrome oxidase (a/a_3) in mitochondria, is not inhibited by azide or cyanide. The link between the NADPH oxidase and the b_{245} cytochrome may be direct. However, an FAD flavoprotein and/or a quinone have been proposed as intermediates. The NADPH oxidase is very unstable once out of a membrane, and has been proved very difficult to isolate and purify from broken cells. A solubilised preparation of the oxidase, but with little FAD flavin, can be isolated which is capable of generating O_2^-. However, Cross and Jones have isolated a complex with a one-to-one stoichiometry of FAD to b cytochrome, with the necessary kinetic properties for transfering electrons from NADPH to O_2. Oxidation of NADPH is a two electron process, whereas reduction of O_2 to O_2^- requires only one. Flavoproteins are common in Nature as intermediates between such electronic transitions.

The pathway responsible for O_2^- formation in the cell is membrane-bound. It can be found in the cell membrane, in the membrane of the primary granules containing myeloperoxidase, and in granules formed by fusion of cell membrane vesicles or phagosomes with the primary granules. Once O_2^- is formed, however, it leads to formation of other radicals or active metabolites in the aqueous phase:

(1) Formation of hydrogen peroxide by spontaneous dismutation

$$2O_2^- + 2H^+ \rightarrow H_2O_2 + O_2 \tag{6.25}$$

(2) Formation of OCl^-, catalysed by myeloperoxidase

$$H_2O_2 + Cl^- \rightarrow H_2O + OCl^- \tag{6.26}$$

(3) Formation of hydroxyl radical via the Haber-Weiss reaction

$$H^+ + O_2^- + H_2O_2 \xrightarrow[\text{iron salt}]{} H_2O + \cdot OH + O_2 \tag{6.27}$$

(4) Formation of singlet oxygen

$$OCl^- + H_2O_2 \rightarrow H_2O + Cl^- + {}^1O_2. \tag{6.28}$$

The uptake of oxygen per stimulated neutrophil at the peak of the respiratory burst is about 0.1 fmol per second, i.e. about 60 million oxygen molecules per second per cell. This compares with an ultraweak chemiluminescence of $< 10^{-4}$ photon per second per cell, and 1–100 photons per second per cell in the presence of luminol. If all of the O_2 uptake were converted into H_2O_2 within the cell (0.45 pl in volume) then this would rise by some 0.2 mM per second. In fact, with a suspension of neutrophils at 10^5–10^6 per ml μM concentrations of H_2O_2 can be detected in the medium under some conditions. The conclusions from these calculations are that the production of reactive oxygen metabolites is substantial, yet the chemiluminescence, even indicator-dependent, is very small. Luminol therefore appears to be acting as a true indicator, consuming only a very small quantity of the metabolites produced. This is confirmed by the lack of significant loss of luminol at an initial concentration of 10 μM during a 15–30 minute experiment. Exposure of luminol to myeloperoxidase is required for maximum cellular chemiluminescence, but luminol also reacts with O_2^- and probably some other species. The oxidase system producing O_2^- is much less well characterised in eosinophils and monocytes/macrophages. The former have a peroxidase, similar, but not identical, to the neutrophil myeloperoxidase, which generates OBr^-. The latter have little or no peroxidase unless they have phagocytosed a neutrophil.

6.6.4 How are phagocytes activated?
To fully rationalise the mechanisms underlying activation of the pathway generating reactive oxygen metabolites, the answers to six specific questions are required:

(1) What are the natural primary cell stimuli, together with the secondary regulators which modify their action? Can we find experimental stimuli which will help elucidate how these natural stimuli work?
(2) Does cell activation involve a change in affinity of the 'oxidase' for its substrates, namely NADPH or O_2, or an increase in maximum activity? These two

parameters are the familiar K_m and V_{max} of the Michaelis–Menton (or Briggs–Haldane) equation.

(3) Since V_{max} consists of two components; k, the turnover rate of the enzyme, and E, the enzyme concentration, the question arises, is any increase in the V_{max} of the oxidase a reflection of increased turnover rate (i.e. enzyme activation) or enzyme concentration (e.g. by recruitment of components via membrane–membrane fusion or aggregation within the same membrane)?

(4) Are intracellular messengers involved in mediating the effects of the primary stimuli or secondary regulators? Obvious candidates are Ca^{2+}, H^+, cyclic AMP, cyclic GMP, diacyl glycerol, inositol trisphosphate, and members of the prostaglandin–thromoboxane-leukotriene pathway?

(5) What is the molecular basis of the action of any intra-cellular regulators? Do the secondary regulators act by modifying their concentration or do they act independently?

(6) What is the relationship between 'oxidase' activation and the other phenomena associated with phagocyte activation — amoeboid cell movement, secretion, cell aggregation, internal vesicular movement and membrane fusion, and activation of intermediary metabolism (glycogenolysis, pentose phosphate pathway, arachidonate metabolism)?

The stimuli

Two groups of naturally-occurring primary stimuli can provoke the respiratory burst in neutrophils and other phagocytes — phagocytic and chemotactic (Table 6.4). All of these provoke both indigenous ultraweak and indicator-dependent chemiluminescence. On the outside surface of the cells are receptors specific for (a) complement fragments C3b (the 'b$_i$' fragment of complement component C3) and C5a (this and desarg C5a are proteolytic fragments of complement C5); (b) the Fc site exposed in an antibody–antigen complex; and (c) chemotactic peptides released by infecting bacteria. The N terminus, N formyl met leu phe, of these latter peptides has been widely used as an experimental stimulus over a wide range of concentrations (nM–μM). There is some evidence for two classes of receptor for N formyl met leu phe: one — high affinity (K_d $ca.$1–10 nM) for chemotaxis and oxidase activation, the other — low affinity (K_d $ca.$ μM) for secretion. This is quite sensible since the cells will thus only release reactive metabolite enzymes once they have reached the source of the infection, from where the chemotactic factors originate. Release of the oxygen metabolites may then inactivate the chemotactic factors and soluble toxins from the bacteria, which would otherwise have deleterious effects on other cells. Down regulation and movement of receptors to one end or side of the cell ('capping') can also occur. This may result in different degrees of sensitivity to various stimuli in cells isolated from different sites, e.g. blood, synovial or peritoneal fluids. Certain sugars, e.g. mannose and sialic residues, in glycoproteins on the cell surface also act as 'receptors' for certain unopsonised particulate stimuli. Concanavalin A is a model soluble experimental stimulus mimicking these.

One of the most potent experimental stimuli of the 'oxidase' and luminol chemiluminescence is the tumour promoter phorbol myristate acetate, also causing secretion of proteins from the secondary granules. Its structural resemblance to diacyl glycerol gives a clue to one of the molecular mechanisms mediating one group of cell stimuli (phagocytic), through the activation of protein kinase C by diacyl glycerol. In contrast, activation of the respiratory burst by the Ca^{2+} ionophores A23187 or ionomycin supports the case of intracellular free Ca^{2+} mediating the other group — chemotactic. The major site of production of luminol chemilumines-cence is different for the two groups of stimuli, within intracellular vesicles and phagolysosomes in the case of phagocytic stimuli, and extracellular via a secretory pathway in the case of chemotactic stimuli. However, this is not necessarily a true reflection of the site of O_2^- production from the 'oxidase', but rather the means by which luminol can be exposed to myeloperoxidase and its products. Some phagocytic stimuli can release this enzyme from the primary granules, thereby generating extracellular luminol-dependent chemiluminescence. Cytochalasin B stimulates primary granule secretion in the presence of both chemotactic factors and phorbol esters, thereby enhancing the chemiluminescence in the presence of luminol. Inhibition of phagocytosis by cytochalasin B inhibits activation of oxidase and prevents access of luminol to myeloperoxidase; luminol-dependent chemilumines-cence is thus virtually abolished (Fig. 6.6). Cytochalasin B treatment enables 'cytoplasts' to be isolated, plasma membrane vesicles encapsulating cytoplasm, but containing no granules and thus no myeloperoxidase. The respiratory burst in these cytoplasts measured directly by O_2 uptake induced by phagocytic or chemotactic stimula is high, virtually the same as in the intact cells, confirming the plasma membrane as a major side of the cytochrome b_{245} oxidase. However luminol dependent chemiluminescence is greatly reduced in cytoplasts compared with intact cells, because of the lack of myeloperoxidase. Addition of horseradish peroxidase to intact cells or cytoplasts enhances the luminol chemiluminescence response. How-ever, in the absence of cytochalasin B, phorbol esters, and large particles unable to be ingested, can stimulate secretion from secondary granules, indicated by release of vitamin B12 binding protein. This would be expected also to result in O_2^- release, since a large fraction of the cytochrome b_{245} is in the plasma membrane, and thus produces some extracellular luminol chemiluminescence. This chemiluminescence is not detected in the bulk medium. Rather it occurs very close to the cell surface, because of the short lifetime of O_2^-. In contrast longer lived oxidants released from both primary and secondary granules do appear to react with pholasin in the external medium, well away from the cells. A particular problem in the interpretation of chemiluminescence data can be the presence of extracellular myeloperoxidase released by cells prior to experimental stimulation.

The membrane attack complex of complement ($C5b6789_n$, $n = 12$–18) conventio-nally kills cells. However, neutrophils are able to protect themselves within minutes of attack by complement, by removal of the complex containing the key component C9, via budding and endocytosis. Prior to this Ca^{2+} has leaked into the cell via the complex, thereby activating the oxygen metabolite pathway in the absence of cell lysis.

During the 1970s a number of prostaglandins, catecholamines, and histamine, acting through cyclic AMP and cyclic GMP, were shown, under experimental conditions, either to potentiate or to inhibit enzyme secretion from neutrophils. Less is known of the secondary regulators of the reactive oxygen metabolite pathway, Leukotriene B_4 (5G, 12R) 12,20-trihydroxy-6, 14-*cis*-8, 10-*trans*-icosatetraenoic acid is a potent chemo–reactant of neutrophils down to 1 nM, via a receptor with a K_d *ca*.

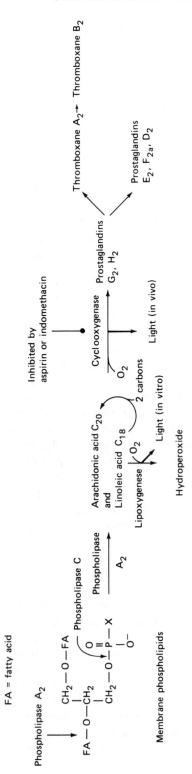

Fig. 6.5 — Chemiluminescence through the prostaglandin generating pathway.

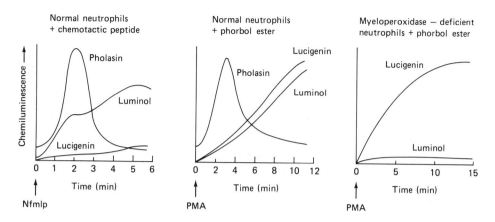

Fig. 6.6 — Comparison between luminol, lucigenin and pholasin as indicators of neutrophil reactive oxygen metabolites. These figures are approximate and intended only to illustrate the differences in time of onset, time to peak, and magnitudes of response when using different stimuli and cells.

10 nM, and it activates the respiratory burst. It can be considered as a *secondary stimulus* since it is produced by neutrophils themselves. Primary chemotactic stimuli activate a phospholipase A_2, probably through an increase in intracellular Ca^{2+}, leading to production of free arachidonate. This is metabolised along the 5-lipoxygenase pathway to leukotriene B_4. Inhibitors of this pathway, such as indomethacin, inhibit the respiratory burst induced by the chemotactic peptide N formyl met leu phe suggesting that leukotriene B_4, or a metabolite on this pathway, may be an intermediary between the primary stimulus and part of the 'oxidase' activation. Similarly, prior exposure of macrophages to interferon increases their luminol chemiluminescence response to phorbol esters, possibly via the lipoxygenase pathway, though ·OH production is increased rather than that of O_2^- and H_2O_2. Interferon also acts on monocytes and macrophages to induce protein synthesis and allow cells, previously with no 'oxidase', to produce O_2^-. Similar effects on gene expression have been observed with certain viruses. Adenosine, on the other hand, inhibits the chemiluminescence response provoked by chemotactic stimuli, but not that provoked by phagocytic ones which are dependent on protein kinase C activation. Nor does adenosine inhibit neutrophil activation induced by phorbol esters. In constrast experimentally, cytochalasin B, a product of the fungus *Helminthosporium dematioideum*, virtually abolishes (>95% inhibition) phagocytic activation and potentiates chemotactic stimuli. These observations are consistent with our hypothesis that there are two routes for 'oxidase' activation (Fig. 6.9). Both involve membrane fusion, and probably recruitment of components for the oxidase.

Characteristics of the respiratory burst and its associated chemiluminescence
Resting neutrophils respire at about 0.5–5 amol O_2 per second per cell. Addition of a stimulus, after a lag of 20–30 seconds, causes a 5 to 10-fold increase in O_2 uptake (Fig. 6.7). This can be detected either by measuring a decrease in oxygen concentration or, in a steady-state gassed system, by measuring the increase in oxygenation required to

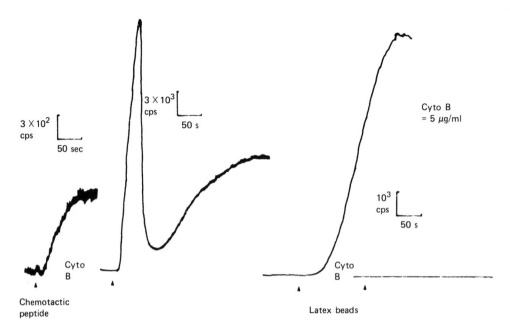

Fig. 6.7 — Effect of cytochalasin on neutrophil luminol dependent chemiluminescence. Reprinted by permission from Hallett & Campbell (1983). Copyright ®1983 The Biochemical Society, London.

maintain a steady concentration of O_2. It means that 10^7 cells in 2 ml would consume cell the oxgyen within a few minutes. Phagocytic-activated respiration may continue for 10–30 minutes fully stimulated, whereas with chemotactic stimuli, the respiratory burst tends to be biphasic, consisting of a highly elevated first phase lasting only 1–3 minutes, followed by a slower, less active phase which again may last more than 30 minutes. Concomitantly, and with a similar time course, there is an increase in ultraweak and indicator-dependent chemiluminescence (Figs 6.1 and 6.6). The maximum increase in light emission compared with resting cells is some 10^3–10^4 fold with luminol or 10^2–10^3 with lucigenin, considerably greater than the fold increase in respiration. There are three possible reasons for this. Firstly, there could be a substrate cycle, resulting in an amplification of the initial O_2^- signal to produce the oxygen metabolite responsible for either the ultraweak or indicator-dependent chemiluminescence signal. Secondly, recruitment of components may lead to a greater amplification of the production of the key oxidant from O_2^- required for maximum chemiluminescence compared with O_2 uptake. Thirdly, access of the chemiluminescent substrate to a catalytic component, for example myeloperoxidase, may be necessary for the maximum chemiluminescence signal. The latter is the main explanation for the large fold increase in luminol-dependent chemiluminescence. As with respiration, there is a 20–30 second lag before light emission increases, the chemiluminescence reaching a maximum within 3–5 minutes and then decaying slowly. The initial lag is due to intracellular or membrane events necessary for 'oxidase' activation since removal of oxygen prior to addition of the cell stimulus, followed by re-oxygenation, abolishes the lag.

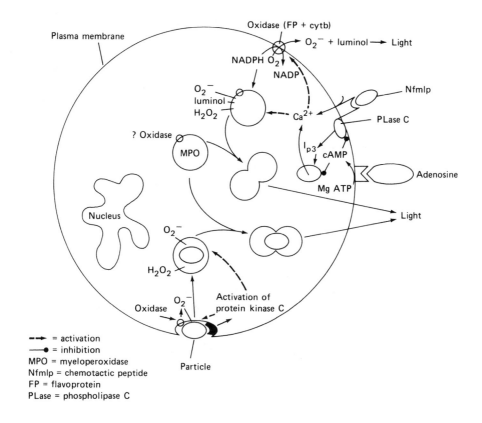

Fig. 6.8 — Mechanism of neutrophil activiation.

By placing an oxygen electrode into a chemiluminometer (Fig. 2.6) respiration and chemiluminescence can be monitored simultaneously, and related to the concentration of respiratory substrate, oxygen (Fig. 6.12). Cell activation results in an increase in maximal activity (V_{max}) and an increase in the apparent affinity of the 'oxidase' for its primary substrate, O_2. There is also an increase in affinity for NADPH in broken cell preparations. Although K_m is not strictly the affinity of the enzyme it provides a measurable indicator of affinity within the cell. For example, using N formyl met leu phe + cytochalasin B as the stimulus, the apparent K_m for O_2 measured from respiration decreases from 10 μM to 4 μM, and measured from chemiluminescence decreases from 40–50 μM to 20–30 μM. Although these are not strictly K_m measurements, but rather $K_{0.5}$, three important conclusions can be drawn from such experiments. Firstly, more than 95% of the chemiluminescence, both ultraweak and luminol-dependent, requires oxygen, being lost rapidly by gassing with N_2. The possibility exists that a small residual component of the light emission could be caused by a metastable metabolite, but the majority of the chemiluminescence must be caused by a rapidly turning over oxygen metabolite. Secondly, the relatively high apparent K_m for O_2 of the oxidase, compared for example with the terminal cytochrome oxidase, in mitochondria and bacteria (which have K_m app O_2 around 1 μM), means that oxygen concentration may be an important factor limiting metabolite production *in vivo*. Although the oxidase would be nearly saturated at arterial or venous oxygen concentration, 50–100 μM (Fig. 6.12), at an inflammatory site O_2 may be as low as 1–10 μM. Thirdly, the 'oxidase' adapts to low oxygen concentration, becoming more sensitive to its substrate O_2, after exposure to hypoxic or anaerobic conditions.

The role of intracellular regulators
The primary stimuli of the 'oxidase', and thus the agents responsible for initiating chemiluminescence, act at the cell membrane, yet they activate processes within the cell. There must, therefore, be an intracellular signal connecting events at the cell membrane with those activated internally.

Indirect experiments such as inhibition of cell activation by removal of extracellular Ca^{2+} or the calmodulin 'inhibitor' trifluoperazine, and stimulation of cells by the Ca^{2+}-ionophore A23187, have lead to the suggestion that intracellular Ca^{2+} plays a key role in phagocyte activation. Rapid changes in cyclic AMP have also been detected, induced by the chemotactic peptide N formyl met leu phe. However, the story remained confused until it became possible to measure directly the concentration of free Ca^{2+} in the living cell (see Chapter 7). The first published report by Hallett and Campbell in 1982 used the Ca^{2+}-activated photoprotein obelin in fused neutrophil–erythrocyte ghost hybrids. Other workers have used the fluorescent indicators quin2 and fura2 incorporated into the cell as the tetra-acetoxy methyl ester, being hydrolysed back to the acid indicator by an intracellular esterase. Within less than a second of stimulation of cells with N formyl met leu phe or concanavalin an increase in the cytoplasmic free Ca^{2+} from 0.1–0.3 μM in the resting cell to 0.6–1 μM in the stimulated cell occurs. However, somewhat surprisingly, neither phagocytic stimuli nor phorbol esters cause a detectable increase in intracellular Ca^{2+}, despite the fact that removal of external Ca^{2+} or addition of trifluoperazine inhibit the cell response induced by these two stimuli. Furthermore, prevention or reduction of the increase in cytosolic Ca^{2+} by a chelating agent, e.g. EGTA incorporated inside the cell, inhibits the response of cells to the chemotactic stimulus but has no effect on the phagocytic stimulus. This has been shown in both neutrophils and macrophages.

The two classes of stimuli (Table 6.4) therefore fall also into two mechanistic groups (a) phagocytic stimuli and phorbol esters are independent of a rise in cytosolic Ca^{2+}, and (b) chemotactic (N formyl met leu phe, C5a and LTB_4), and secretory stimuli releasing primary granules, are dependent on a rise in cytosolic Ca^{2+} (Fig. 6.9). Adenosine inhibits group (b) but not group (a), and abolishes the intracellular Ca^{2+} transient (Fig. 6.10), but may also affect directly the intracellular mechanism activated by Ca^{2+}. Adenosine acts via an A_2 receptor which activates adenylate cyclase. The role of cyclic GMP, however, is less well established.

Where do the intracellular signals come from?
There are two potential sources of a rise in cytosolic free Ca^{2+}: increase through the cell membrane via receptor operated channels, or release from an internal vesicular store. Both appear to be involved in neutrophil activation. Acute removal of extracellular Ca^{2+}, or depletion of the internal store, inhibit the stimuli dependent

on a rise in cytoplasmic free Ca^{2+}. The significance of the two sources of free Ca^{2+} in other cells reflects the site of the intracellular events which are to be activated, as well as the time course and the need to maintain an elevated free Ca^{2+} for a particular time. Chemotactic stimuli activate a membrane bound phosphatidyl inositol phospholipase, analogous to phospholipase C. This stimulates the breakdown of a minor membrane component phosphatidyl inositol 4,5 bisphosphate (PIP_2) to diacyl glycerol and inositol 1,4,5 trisphosphate (IP_3). The former remains within the membrane, and has the potential to activate the enzyme protein kinase C; the latter is soluble. Once in the cytoplasm IP_3 releases Ca^{2+} from its internal store, which then leads to activation of the oxidase and the enormous-fold increase in chemiluminescence. There are four main pieces of evidence supporting this mechanism:

(1) Cytoplasmic Ca^{2+} rises prior to chemiluminescence, and prevention of the rise stops cell activation by chemotactic stimuli.
(2) Chemotactic stimuli raise cytoplasmic Ca^{2+} and IP_3.
(3) The effects of chemotactic stimuli are mimicked by Ca^{2+} ionophores, but these do not release internal IP_3.
(4) Protein phosphorylation is consistent with the Ca^{2+} mechanism.

There are five pieces of evidence supporting the hypothesis that most phagocytic stimuli (in particular C3b, unopsonised particles and possibly Fc) and phorbol esters act via protein kinase C, independently of a rise in cytoplasmic free Ca^{2+}:

(1) These stimuli provoke no detectable rise in cytoplasmic Ca^{2+} prior to oxidase activation and chemiluminescence, and depletion of internal Ca^{2+} stores has no effect on them.
(2) These stimuli can act synergistically with the Ca^{2+} dependent ones.
(3) Activators of protein kinase C, such as oleoyl acetyl glycerol, mimic these stimuli.
(4) Phosphorylation of internal proteins is consistent with protein kinase C activation.
(5) Inhibitors of protein kinase C, such as retinal and tamoxifen, inhibit the increase in O_2 uptake and chemiluminescence induced by phorbol esters. Tamoxifen is a drug which causes 30% remission in women in breast cancer. The precise signal in the cell for protein kinase C activation is not known. The only known means of producing diacyl glycerol, to activate this kinase, is from PIP_2 breakdown which would also release IP_3 and thus Ca^{2+}.

How do the intracellular signals activate the 'oxidase'?
There is, as yet, little direct evidence for this in neutrophils. Nevertheless, from what is known in other cells, we can make some guesses. The calmodulin concentration in neutrophils is about 7 μM, ca. 0.04% total protein. The K_d Ca^{2+} for this protein is about 2 μM. So a rise in cytosolic free Ca^{2+} in the μM range will activate calmodulin-dependent kinases leading to phosphorylation of the myosin light chain (mol. wt 20 000) in the cytoskeleton. The cytoskeleton also contains other calcium binding proteins such as gelsolin and actinogelin. As we have seen, the other group of stimuli, independent of an increase in intracellular Ca^{2+}, are likely to be mediated via the protein kinase C discovered by Nishizuka in 1977. This protein is a monomer

of molecular weight 77 000. It requires Ca^{2+} and phosphatidyl serine for maximal activity. It can be fully activated by diacylglycerol, which increases the sensitivity of the enzyme for Ca^{2+}, enabling it to be fully active at 0.1 μM Ca^{2+}, the cytosolic Ca^{2+} concentration in the resting cell. Diacylglycerol is formed in the membrane from phospholipid hydrolysis. It can be by-passed by phorbol esters, which activate protein kinase C directly, perhaps acting as structural analogues of diglyceride. Protein kinase C acts independently of calmodulin, but is also inhibited by trifluoperazine which may be a source of misinterpretation of cell experiments with this drug. Protein kinase C phosphorylates different proteins from the calmodulin-activated kinase, in particular a 40 000 or 47 000 molecular weight protein. The mechanism by which the phosphorylation of the 20K, 40K, or 47K proteins, together with interactions with protein kinase A activated by cyclic AMP, cause activation of the oxidase pathway, recruitment of its components and membrane fusion remains to be elucidated. The precise role of phosphorylation of other proteins induced by Ca^{2+} or protein kinase C is not known.

Evidence for the mechanism by which secondary regulators such as adenosine act comes from a combination of pharmacological experiments, measurements of intracellular free Ca^{2+} and the use of bacterial toxins which ADP ribosylate, using NAD^+ as substrate, GTP regulatory binding proteins (G proteins) associated with adenylate cyclase in the cell membrane. Pertussis toxin causes ADP riboxylation of the so-called N_i, or inhibitory subunit, of adenylate cyclase. This inactivates the N_i component, preventing the cyclase from being switched off. Pertussis toxin inhibits secretion from neutrophils induced by chemotactic factors and leukotriene B_4, but not that by Ca^{2+} ionophores. Cholera toxin, on the other hand, increases cyclic AMP by causing ADP ribosylation of the N_a, or activating subunit, of adenylate cyclase. It also inhibits granule secretion. As expected forskolin, which bypasses the adenosine receptor and activates adenylate cyclase, inhibits the Ca^{2+}-dependent chemiluminescent stimuli, but not the Ca^{2+}-independent ones. Furthermore, adenosine and pertussis toxin both appear to reduce or abolish the increase in intracellular free Ca^{2+} induced by adenosine. The possibility exists that adenosine, acting via cyclic AMP, inhibits the phospholipase responsible for IP_3 production, thereby preventing release of internal Ca^{2+}. However, complete inhibition of the chemiluminescence repsonse by adenosine cannot be achieved.

Is neutrophil activation a 'threshold' phenomenon?

All measurements of the 'respiratory burst' and chemiluminescence in neutrophils and other phagocytes have, so far, been carried out on populations of thousands or millions of cells. A major problem in elucidating the molecular basis of cell activation, and injury by agents such as the immune system, bacterial toxins or viruses, is the inability to take into account the heterogeneity in the responses of individual cells within the population. Using 2,7 dichlorofluorescein as an intragranular indicator of reaction oxygen metabolites, and pholasin to measure their release, we have shown there are three principle sources of this heterogeneity which must now be taken into account (Fig. 6.11). Firstly individual cells switch on their oxidase at different times after addition of the stimulus. Secondly, there is a critical concentration of stimulus for each neutrophil at which a 'threshold' for oxidase activation and secretion occur. Below this threshold concentration no activation

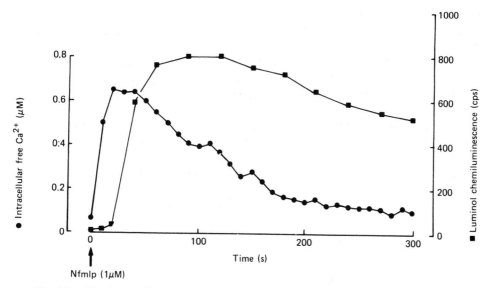

Fig. 6.9 — Correlation of intracellular free Ca^{2+}, measured by obelin, and luminol chemilumi-nescence in human neutrophils. Nfmlp = chemotactic peptide. Campbell and Patel unpublished.

occurs. Thirdly, the magnitude of the maximal oxidase activity, and thus rate of oxygen metabolic production, varies from cell to cell. I have proposed that a chemisymbiosis between the intracellular signal is required to provoke the threshold in each cell (see chapter 10, section 10.3.2). Three possible molecular mechanisms underlying each chemical threshold in neutrophils and other cells are; firstly, the concerted action of a combination of molecular reactions leads to a critical event, such as granule-plasma membrane fusion, or a gel-sol transition necessary for activation to occur. Secondly, a gradient O_2, which exists in all cells, may, through a limiting O_2 concentration at a particular site in the cell, determine both the initiation time, and magnitude, of the threshold. This may be particularly relevent *in vivo* where O_2 concentrations are much lower than *in vitro*. (Fig. 6.12). Thirdly, the concentration of intracellular signals have to reach a critical concentration at the necessary location inside a cell to switch on the 'oxidase'. The size, and distribution of a cytoplasmic cloud of IP_3 or Ca^{2+}, or diacyl glycerol, will be detemined by receptor occupancy, internal buffering e.g. from mitochondria or pumps, and degradation. Whether adenosine acts to alter the number of cells crossing the 'threshold', or the magnitude once this has been achieved, or both, remains to be established. Substances such as interferon, or internal programming, which induce the proteins required for 'oxidase' activation in precursor cells such as promyelocytes and monocytes, will also control the number of cells in a given population capable of producing oxygen metabolites.

Other pathways
Several other pathways are activated concomitantly with the 'respiratory burst': glycogen breakdown for energy supply and reducing equivalents, the pentose phosphate pathway for NADPH, the substrate of the 'oxidase' and required by glutathione reductase, glutathione oxidation, ion pumps (Na$^+$ and Ca^{2+}).

6.6.5   Clinical applications

Neutrophils, and other phagocytes, can do a lot of things. They can move, aggregate, secrete, change their shape, phagocytose, produce reactive metabolites and chemiluminesce. Most of these are required for the cells to fulfil their main function, to protect the host from invading micro-organisms. Neutrophils are the first line of defence, eosinophils deal with some of the more unusual invasions, for example from protozoa, and macrophages are part of the reinforcement and cleaning-up operation. Phagocytes also remove dead or living cells orginating from the host. Their numbers vary greatly, particularly during or following an infection. They do not function properly if certain key proteins are genetically abnormal, and they can be affected by drugs or toxins, to the detriment of the host. When they invade the joints of patients with rheumatoid arthritis or the myocardium following a heart attack, are they a friend or a foe? Assessment of neutrophil function involves testing the efficiency of three main properties:

(1) Chemotaxis.
(2) Phagocytosis.
(3) Biocidicity i.e. microbial killing.

The production of reactive oxygen metabolites is involved in all of these processes. However, it does not seem to be *essential* for bacterial killing. Rather, the oxygen metabolites help the phagocytes to kill the microbes more quickly. Two interesting applications of chemiluminescence for testing the efficiency of the killing process are to use either lipid vesicles containing luminol or luminous bacteria. The former emit light after they have been taken up and the membrane of the vesicle damaged within the phagosome. The latter stop emitting light when they are killed. Biochemical assessment of neutrophils involves examining four parameters:

(1) The sensitivity of the cells to primary stimuli and secondary regulators in relation to the numbers of chemotactic factor receptors (e.g. C5a and N form met leu phe, leukotriene B4), phagocytic receptors (Fc, C3b/b_i, glycoprotein), and adenosine receptors (A_2).
(2) The components involved in reactive oxygen metabolite production, in particular the 'oxidase + b cytochrome' and peroxidase, together with secretable products including O_2^-, H_2O_2, OX^- ($X = Cl$ or Br), chloramines, and lipid peroxides.
(3) Secretable proteins from the primary or secondary granules, e.g. peroxidase, β glucuronidase, esterases, elastase, vitamin B12 binding protein.
(4) Eicosanoid and leukotriene production.

Chemiluminescence, in conjunction with other tests, provides a means of assessing phagocyte function, particularly the production of reactive oxygen metabolites, oxidants, and (per) oxidases. It also provides some insights into the past history of the cells.

A number of potential applications of neutrophil chemiluminescence in the clinical laboratory have been explored (Table 6.9). They can be divided into two groups. The first (group A) involves the detection of reduced or increased reactive oxygen metabolite production by neutrophils isolated from the patients, stimulating the cells with standard stimuli. The responsiveness of the cells is related to the clinical

Table 6.9 — Clinical investigation of granulocyte chemiluminescence

Clinical application	Specific disease or test
A. Evaluation of abnormalities intrinsic to the cells	
1. Congenital and inherited	Susceptibility to infection in the new born
	Myeloperoxidase deficiency
	Chronic granulomatus disease
2. Acquired	Rheumatoid arthritis
	Pregnancy
	Diabetes mellitus
	Carcinoma
	Burns
B. Evaluation of the potency of humoral abnormalities	
1. Defective opsonic capacity in extracellular fluids (serum, synovial fluid)	Inflammatory disease (rheumatoid arthritis)
2. Pathogenicity of infectious agents	Bacteria
	Viruses
	Higher protists
	Activators of complement
3. Pathogenicity of immune system	Immune complexes (serum, synovial fluid)
4. Pathogenicity of ingested particles	Pollen (allergic response)
	Dusts (lung disease)
5. Inhibitory action of drugs (*in vivo* or on cells)	Penicillamine
	Non-steroidal anti-inflammatories
	Methimazole

problem; the second (group B) uses activation of normal neutrophils by a sample from the patient to assess the stimulatory or inhibitory capability of factors in extracellular fluids such as plasma or synovial fluid.

For example, using the group A approach the stimulation of indicator-dependent chemiluminescence induced by opsonised particles may be reduced in cells isolated from burns patients and neonates. This may be related to their susceptibility to sepsis and infections. That the cells have already been stimulated can be assessed by comparing the response to opsonised particles with that to phorbol esters, which by-pass 'down regulated' receptors. The problem with these studies is standardisation and the lack of, or too variable, a normal range.

Results are more clear-cut with cells from patients with genetic abnormalities, and who are prone to certain infections. Cells from patients with chronic ganulamatous disease produce little or no O_2^-; the chemiluminescence responses are therefore lost. In contrast, cells from patients with myeloperoxidase deficiency produce O_2^-, sometimes at a higher concentration than normal cells. These cells therefore show a 'respiratory burst' but take some four to five times as long to kill bacteria such as *Staphylococcus aureus* compared with normal cells. They are also unable to kill the cells of the yeast *Candida albicans* at all, hence the clinical expression of myeloperoxidase deficiency, disseminated Candidiasis. The lack of myeloperoxidase can be detected, in the presence of a normal or high respiratory burst, by a reduced, but detectable, ultraweak chemiluminescence coupled with the virtual absence of luminol-dependent chemiluminescence.

In contrast, normal neutrophil indicator-dependent chemiluminescence can be used to identify a plethora of stimuli and inhibitors in extracellular fluids which may be pathogenic (group B). For example, the tumorigenicity of certain dusts may be related to their potency at stimulating chemiluminescence. When searching for potential stimuli in body fluids it is essential to remove all the cells from the fluid being investigated, and to prevent any cell breakage. These are not always easy to achieve. Many workers have used synovial fluid centrifuged only in a bench centrifuge. This does not remove all of the neutrophils, centrifugation at $> 10\ 000$ g being required to do this in this viscous fluid. Alternatively the synovial fluid can be diluted in buffer to reduce viscosity and enable any cells to be centrifuged easily.

Cellular chemiluminescence can be provoked and detected using unseparated, whole blood samples. A dilution of more than one in a hundred is required to prevent significant absorption of light by haemoglobin from the erythrocytes. However, none of these investigations has yet to find routine use in a laboratory where clincial decisions are taken. There are four main reasons for this:

(1) Lack of establishment of a normal range for neutrophil chemiluminescence.
(2) Wide variability in response from individuals, and effects of infections such as influenza, caused particularly by cellular hetereogeneity.
(3) The relationship between potency of stimulation, or inhibition of a pathogen or a drug, and its pathogenicity or therapeutic efficiency has yet to be established for most substances.
(4) Susceptibility of neutrophil chemiluminescence to a wide range of experimental conditions, and interfering substances in clinical samples (Table 6.10). A particular problem is the presence of proteins such as albumin which can reduce the chemiluminescence response by more than 90%.

A further problem, in comparing results from various laboratories using purified cells, is the variation in method of isolation and storage (Table 6.10, section 2). There are four particular hazards:

(1) Density separation and the number of osmotic shocks to remove red cells affects the response, in magnitude and shape, of the final preparation of neutrophils.
(2) Storage temprature and time. Cell responses decrease after an hour or two even when stored at 0°C, whereas most workers store neutrophils for 1 h, prior to measurement of chemiluminescence, to allow the cell to recover from the stresses of the isolation procedure.
(3) Protein is required to prevent aggregation and secretion, but it will reduce the chemiluminescence unless removed or diluted out.
(4) Release of lactate (white cells are essentially glycolytic) may dramatically reduce the pH of the medium containing the stored cells.

Table 6.10 — Experimental conditions affecting luminol-dependent neutrophil chemiluminescence

Experimental variable	Parameter to standardise
1. Detector (commercial or home-built luminometer, scintillation counter)	Calibration in photon. s^{-1} Photomultiplier spectral response
2. Preparation of cells	Source and species Time of day (circadian rhythm) Separation method (dextran, Ficoll, Percoll) Number of osmotic shocks to remove red cells Temperature during prepartion Media during preparation a (e.g. albumin)
3. Characterisation of final cell suspension	Identification of cell type (polymorph, neutrophil) Purity (red cells, platelets, other white cells) Storage conditions (cell concentration, temperature Medium-pH, ions, glucose and albumin

4. Conditions for assay of chemiluminescence	Cell concentration–stimulus ratio
	Type of stimulus and concentration
	Luminescent indicator (type, concentration, pre-incubation)
	Oxygenation and mixing
	Medium —buffer (HCO_3^-, CO_2, Hepes)
	  pH
	  ions (Ca^{2+}, Mg^{2+}
	  substrates (glucose, aminoacids)
	  protein (albumin, gelatin and globulin)
	  pH indicator (phenol red)
	  DMSO (from luminol stock) or NFMLP
	  Presence of free –SH groups
	  Temperature (20–40°C)

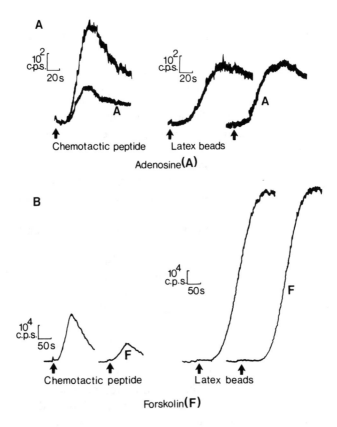

Fig. 6.10 — Effects of adenosine on luminol chemiluminescence in neutrophils (11 μM luminol). A. Adenosine inhibits chemotactic peptide but not latex bead (unopsonised particle) stimulation. B Forskolin increases cyclic AMP and mimics adenosine.

6.7 CONCLUSIONS

Ultraweak and indicator-dependent chemiluminescence provide a continuous monitor, in intact organs, isolated cells and organelles, and in enzymatic and non-enzymatic reactions, of oxidative processes involving oxygen radicals and other metabolites of oxygen. A few reports describe visualising indicator-dependent signals using image intensification. But is chemiluminescence any better than other techniques? Why should we bother to study such processes anyway?

It is acceptable to investigate a genuine natural phenomenon like biolumines-cence, since the major objective of science is to provide a depth of understanding of all such natural phenomena. However, it could be argued that cellular ultraweak chemiluminescence is just an interesting curiosity, a minor by-product of more interesting and significant processes. I believe this not to be the case. The increasing awareness of the importance of oxygen derivatives in tissue physiology and patho-logy (Table 6.11 A and B) means that any non-invasive method of detecting them and assessing their potential potency cannot be ignored. Singlet oxygen is a 'natural phenomenon' in itself. Its real significance in biology is only just being appreciated.

Table 6.11 — Possible roles for reactive oxygen metabolites

Role and source of metabolites	Example
A. Natural Function	
1. Intra or extracellular second messenger	Insulin action
2. Spermicide prevention of double fertilisation	Egg fertilisation
3. Microbial killing by phagocytes	Neutrophils, macrophages
4. Predator killing in certain insects	Exothermic excudate
B. Pathology	
1. Inappropriate release by phagocytes	Rheumatoid arthritis
	Dust diseases of the lung
	Myocardial infarction
2. Induction by transition metals	Fe^{2+}/Fe^{3+} iron overload
	Haemachromatosis
	Cu^{2+}, Wilson's disease
3. Induction by toxins	CCl_4 in liver
	Paraquat in lung
	Some chemical carcinogens
	(O_3, NO, NO_2, smoking)
4. Irradiation	UV
	Porphyria
	X-rays
5. Scavenger deficiency	Vitamin E
	Selenium
	Ascorbate (scurvy)
6. Genetic effect	Chronic granulomatous disease
	Myeloperoxidase deficiency
C. Therapy	
1. Generators	Porphyrin killing of tumours
2. Scavengers	Penicillamine

Furthermore, the importance of a small number of excited-state molecules, produced at a critical time in a cellular process such as division, is only just being realised. Chemiluminescence, together with energy transfer (see Chapter 9), provides a unique probe for such events.

6.7.1 Potential importance of reactive oxygen metabolites in biology

Reactive oxygen metabolites have been implicated in many diseases and in the action of several drugs (Table 6.11B). In some cases, O_2^- has been identified as the probable pathogen, in others ·OH or $^1O_2$. Several chemical changes, such as

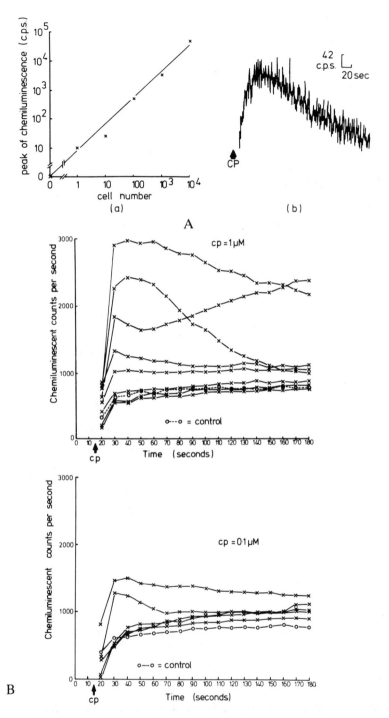

Fig. 6.11 — Pholasin as an indicator of single neutrophil activation. A. (a) Cell dose response and (b) A single cell trace. B. Pholasin signals from several individual cells. Each trace represents one cell x = CP = Chemotactic peptide, o = no CP from Roberts *et al.* (1987) with permission of Academic Press.

increased vascular and membrane permeability, degradation of hyaluronic acid, and loss of scavengers such as ascrobate, $-$ SH groups (albumin is the major source of free $-$ SH in plasma), or vitamin E, are certainly consistent with this hypothesis. However, definitive evidence for their primary role in pathogenesis has been lacking. They do play a key role in microbial killing by phagocytes. However, lack of oxygen metabolite production is not apparently lethal in humans. People lacking the ability to produce them from neutrophils are simply more prone to certain infections. There is strong epidemiological evidence for the beneficial effects of unsaturated fatty acids in the diet in preventing ischaemic heart disease. The possible link with oxygen metabolites requires investigation.

An exciting possibility is an alternative role for reactive oxygen metabolites as short-lived intra- or extra-regulators (Table 6.11A). They would thus cause irreversible covalent modifications of proteins or other molecules associated with cell regulation.

In order to study these two potential biological roles of oxygen metabolites, it is essential that a sensitive method for measuring them within, and near to, living cells is available, enabling production to be correlated with their effects on cells and tissues. Removal of the oxygen metabolites, demonstrated directly by such an indicator, should then alleviate or prevent the phenomenon thought to be caused by them.

A complete understanding of the biological significance of reactive oxygen metabolites requires the answer to eight major questions:

(1) Which cells or conditions produce them *in vivo*?
(2) What are the naturally-occurring primary stimuli, both in healthy and diseased individuals, that provoke their production?
(3) What is the pathway or reaction responsible for their production *in vivo*, and how is it activated?
(4) What are the natural intra- and extra-cellular mechanisms for scavenging reactive oxygen metabolites, normally responsible for protecting cells and tissues against attack?
(5) What are the naturally-occurring secondary regulators and drugs which can modify the production or occurrence of oxygen metabolites?
(6) What is the relationship between the concentration of oxygen metabolite and the concentration of its primary substrate, oxygen, *in vivo*?
(7) What proportion of oxygen metabolites is released by the cells, necessary for them to cause damage to, or act on, surrounding cells and the extracellular matrix?
(8) What is the chemical basis of their physiological or toxic action, and can evidence be obtained that they are a primary cause of a pathological condition *in vivo*?

6.7.2   Is chemiluminescence a good indicator of reactive oxygen metabolites?

Chemiluminescence is but one of several methods available. Others are based absorbance, fluorescence, and electron spin resonance or electron paramagnetic (spin) resonance (Table 6.12). None is 'ideal' nor do they necessarily monitor the same oxygen metabolite in the same part of the cell. But chemiluminescence offers some particular advantages over the other methods. Ultraweak chemiluminescence

Table 6.12 — Some indicators of reactive oxygen metabolites

Indicator	Metabolite detected
A. Absorbance	
Endogenous absorbance spectrum at 77°K in DMF	O_2^-
Reduction of ferricytochrome c (acetylated)	O_2^-
Reduction of nitro blue tetrazolium to formazine	O_2^-
Reduction of tetranitromethane in dimethyl-formamide	O_2^-
Oxidation of chromophores, e.g. o-dianisidine	H_2O_2 + peroxidase
GLC detection of ethene from methional, B methyl mecaptopropanal, methionine or 4-methylthio-2-oxobutanoic acid	·OH
HPLC detection of hydroxylated phenols	·OH
Furan oxidation to diketone	O_2
B. Fluorescence	
Loss of scopoletin fluorescence	H_2O_2 + peroxidase
Formation of fluorescent product of 2,7-dichlorofluorescin	H_2O_2 + peroxidase
C. Chemiluminescence	
Ultraweak (indigenous)	Dioxetan cleavage, $^1O_2$
Luminol	O_2^-, H_2O_2 + peroxidase, OCl^-
Lucigenin	O_2^-, $^1O_2$
Pholasin	O_2^-,
Rubrene	$^1O_2$
Anthracene	$^1O_2$
Methoxyvinylpyrene	$^1O_2$
D. Electron spin resonance	
Endogenous frozen at 77K, in DMF	$·O_2^-$, ·OH
Nitrone spin traps, e.g. 5,5′ dimethylpyroline-N-oxide	·OH
Various aromatic compounds	O_2^-, $^1O_2$

is the only method available at present for detecting, non-invasively, excited carbonyls and $^1O_2$ in intact organs, whereas indicator-dependent chemilumines-cence has four particular advantages:

(1) *Sensitivity*
This arises because of the inherent sensitivity of chemiluminescence over other analytical techniques, and because of the large amplification of the signal relative to the fold increase in oxygen uptake, or that produced by absorbing or fluorescent

indicators with linear relationships. This sensitivity has three important implications. Firstly, comparison of photon yield and luminol consumption with oxygen consumption and total 'peroxide' formation shows that $< 0.1–1\%$ luminol is consumed in most experiments. In fact, O_2 uptake and H_2O_2 formation are often greater than the luminol concentration used. Thus, luminol appears to be acting as an 'ideal' indicator, though reaction with a minor oxidant resulting in a 'scavenging' action cannot be entirely ruled out. Secondly, a signal can be detected from a small number of cells e.g. from a thumb prick. A time course in a single cell is possible using a more sensitive chemiluminescent compound than luminol, e.g. pholasin (Fig. 6.11). Thirdly, oxygen metabolite production can be monitored concomitantly with an effect of the oxygen metabolites on cells or substances in the surrounding fluid.

(2) *Oxygen dependence*

More than 95% of luminol-dependent chemiluminescence requires O_2 (Fig. 6.12b). This has rarely, if ever, been shown for other indicators. Also, stoichiometric O_2^- indicators should 'inhibit' the respiratory burst, since they work by re-oxidising it back to the O_2. The fact that oxidised cytochrome c, and the more specific acetylated form, do not appear to do this in many instances suggests they are only reacting with a small fraction of the metabolites, or are acting as an electron acceptor directly from cytochrome b245, and not all of them as has been claimed.

(3) *Site of oxidant production*

It is possible to identify the site of oxygen metabolite production (i) by allowing uptake, followed by subcellular fractionation to isolate luminescent organelles, (ii) by using solid phase reagents to measure extracellular metabolites, (iii) to synthesise lipophilic indicators for measurements within membranes, and (iv) to incorporate indicators into the cell cytoplasm by microinjection lipsome fusion or micropinocytosis (see Chapters 4 and 7).

(4) *Homogeneous*

The method is continuous and homogeneous, not requiring separation of cells from medium. Unlike absorbance and fluorescence measurements in cells measurement of chemiluminescence is little affected by the problem of light scattering, arising from the particulate nature of the material being analysed.

Nevertheless, there are problems.

(1) None of the chemiluminescent indicators is specific for one oxygen metabolite. This can be circumvented to some extent by the use of inhibitors and scavengers (Table 6.13), superoxide dismutase for O_2^-, catalase for H_2O_2, azide for peroxidase, β carotene for $^1O_2$, and benzoate or mannitol for $\cdot OH$. However, virtually any biological molecule including amino acids, carbohydrates, and nucleic acids can act as a hydroxyl radical scavenger, with rate constants similar to that for benzoate ($k = 2 \times 10^9 \, M^{-1} \cdot s^{-1}$). The specificity of the other scavengers is also questionable. For example, superoxide dismutase is usually used at around $10 \, \mu M$, enormous if it is simply acting catalytically! An alternative approach is to use a different indicator. Lucigenin may be relatively selective for O_2^-, but its kinetics may be complicated by the tendency of acridinium salts to form pseudo base (see Chapter 8). Pholasin seems very good for O_2^-.

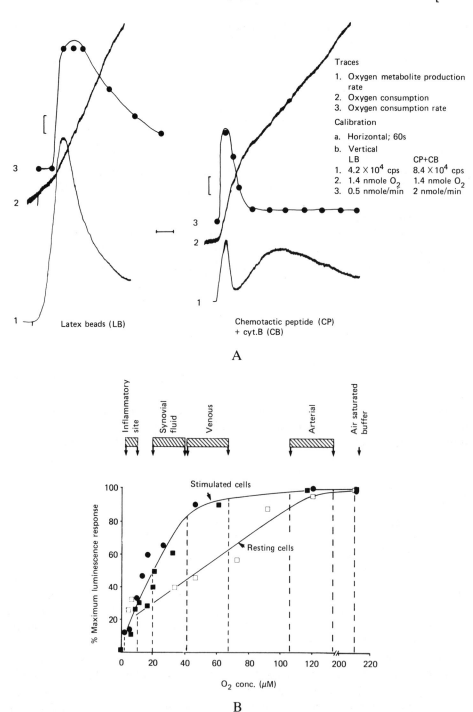

Fig. 6.12 — Correlation between respiratory burst (traces 2 and 3) and luminol chemiluminescence (trace 3) in rat neutrophils. A. Time course (Hallett, Edwards & Campbell
unpublished). B. Effect of O_2 concentration.

Table 6.13 — Inhibitors of neutrophil chemiluminescence

Class of compound	Example
1. Remove substrate (O_2)	N_2 or argon
2. Receptor blockers	Boc leu phe
3. Enzyme inhibitors	N_3^- (myeloperoxidase)
	cytochalasin B (? membrane phospholipase)
4. Oxygen metabolite scavengers	Superoxide dismutase, adrenaline (O_2^-)
	catalase, glutathione + peroxidase (H_2O_2)
	mannitol, benzoate ($\cdot OH$)
	RNH_2 (OCl^-)
	N_3^- (? O_2^-, ?OCl^-)
	albumin (?O_2^-, ?OCl^-)
	free ^-SH groups

(2) The exquisite sensitivity of chemiluminescence requires highly pure reagents and a pure cell preparation. For example, most people work with preparations of polymorphs which may be at best 95% neutrophils but will also contain some eosinophils and basophils.

(3) The source, preparation, and storage of cells, together with the conditions for studying their chemiluminescence, have marked effects on the time course and magnitude of the response (Table 6.10). The response is affected by animal species, or whether cells are taken from the blood or another fluid such as synovial fluid or the peritoneal cavity. The latter is very convenient when using rats. An overnight intraperitoneal injection of caseinate produces an infiltrate of large numbers of polymorphs within 16 hours. This is followed some hours later by macrophage invasion, if these are required. Purification using dextran and gradient separation on Ficoll or Percoll affects their responsiveness, particularly since Percoll itself consists of particles. Neutrophils can be stored on ice, in a protein medium at low density to prevent aggregation or attachment to the vessel. The response of the cells stored on ice usually decays over a period of a few hours. The cells are very unstable if stored at 37°C. Freshly prepared cells are usually stored for an hour to allow them to stabilise before use.

(4) Substances exist in biological systems which may quench the excited state product of the chemiluminescent reaction, thereby suggesting that a reactive oxygen metabolite scavenger is present when it is not.

6.7.3 Consequences of using chemiluminescence
In spite of the problems associated with the interpretation of indigenous ultraweak and indicator-dependent chemiluminescence, several important observations have been made by workers using them:

(1) Red or infrared ultraweak chemiluminescence has provided some of the most convincing data that singlet oxygen does occur in biological systems.

(2) Ultraweak and indicator-dependent chemiluminescence have led to the identification of a large number of cell types producing reactive oxygen metabolites (Tables 6.1 and 6.8). Many others, particularly in the haemolymph and tissues of invertebrates, remain to be discovered.

(3) Using indicator-dependent chemiluminescence, the relationship between O_2 concentration and the production of reactive oxygen metabolites by neutrophils has, for the first time, been defined. Three important conclusions have been made. Cell activation increases the apparent affinity of the oxidase for O_2, the oxidase adapts to low oxygen concentration and O_2 may limit *in vivo* the amount of oxygen metabolites produced.

Consider a sphere consisting of packed, activated neutrophils. Let r (cm) be the radius of the sphere, where O_2 concentration is zero at the centre:

$$r = \sqrt{(6CD/A)} \tag{6.29}$$

where A = rate of oxygen consumption (ml O_2/tissue/min),
  D = diffusion constant for O_2 = 1.98×10^{-5}
  C = external oxygen concentration = 120 μM in an artery.
  A = 20 nmol O_2/min/10^7 neutrophils

Volume of one neutrophil = 0.45 pl, therefore

  1 ml packed cells = 2.22×10^9 neutrophils

From this simple equation it is possible to show that neutrophil oxygen uptake itself could reduce local O_2 concentration to sub-μM levels.

(4) A wide range of naturally-occurring and experimental primary stimuli and secondary regulators have been identified (Table 6.4).

(5) Two distinct molecular mechanisms for neutrophil activation have been identified, one a chemotactic pathway depends on an increase in intracellular free Ca^{2+}; the other, a phagocytic pathway, does not.

(6) A number of drugs have been identified which affect reactive oxygen metabolite production (Table 6.4).

(7) New reactive oxygen metabolite scavengers have been identified, including albumin, the major protein and source of free –SH groups in the blood.

(8) The respiratory burst has been separated from the production of reactive oxygen metabolites formed after O_2^-, for example, via myeloperoxidase and loss or reduction in genetic deficiency diseases.

(9) A lack of correlation between ultraweak or indicator-dependent chemiluminescence and covalent changes in biological molecules, for example SH group oxidation, has highlighted the existence of novel oxidants released by activated phagocytes.

Ideally, an indicator of oxygen metabolite production should be specific for a particular metabolite, sensitive enough to detect production from one cell without significant scavenging, be capable of monitoring and visualising both intra- and extra-cellular production, be non-toxic and readily available. No indicator currently satisfies all these criteria.

No doubt the controversy over which is the best indicator to use for monitoring reactive oxygen metabolites will continue. As is so often the case in science, discoveries are not restricted entirely to one technique, but rather will depend on the creativity of the experimenter. The increasing number of ingenious experiments using chemiluminscence speaks for itself, even if, as is the case, important discoveries have been made by alternative methods and approaches.

6.8   THE FUTURE

New cells and biological reactions producing ultraweak chemiluminescence are being discovered regularly. The cloning of bioluminescent proteins will not only extend the range and sensitivity for such indicators of oxygen metabolites but also, through site directed mutagenesis and chemical modification of the luciferins, enable more specific indicators to be developed. There is a particular need to detect oxygen metabolite production inside living cells, in the cytoplasm, nucleus and mitochondria, and in whole tissues *in situ*. This is necessary if the role of these toxic species in the pathogenesis of inflammatory diseases, reperfusion, injury and cancer is to be established. The development of fibre optics probing the lung or the gastrointestinal tract, and perhaps even the heart during a catheterisation, might provide, for the first time, a way of relating oxidative processes in tissues to the clinical manifestations of particular diseases.

6.9   RECOMMENDED READING

Books

Barenboim, G. M., Domanskii, A. N. & Turoverov, K. K. (1959) *Luminescence of biopolymers and cells*. transl. Chen, R. F. Plenum Press, New York.

DeLuca, M. A. (1978 and 1987) *Methods in Enzymology* volumes 57 and 133. Bioluminescence and chemiluminescence (several articles).

Halliwell, B. & Gutteridge, J. R. C. (1985) *Free radicals in biology and medicine*. Clarendon Press, Oxford.

Packer, E. (1984) (ed) *Methods Enzymol* **105** Oxygen radicals.

Schölmerich, J., Andreesen, R., Kapp, A., Ernst, M. & Woods, W. G. (1987) (eds) *Bioluminescence and chemiluminescence: new perspectives*. John Wiley, Chichester and New York (several papers).

Van Dyke, K. & Castranova, V. (1987) *Cellular chemiluminescence* vol. I. CRC Press, Florida.

Zhuravlev, A. I. (1972) (ed) *Ultraweak luminescence in biology*. Nauka Publishing House, Moscow (in Russian).

Discovery of ultraweak chemiluminescence

Allen, R. C., Strijernholm, R. C., & Steel, R. H., (1972) *Biochem. Biophys. Res. Commun.* **47** 679–684.

Colli, L., Facchini, U., Guidotti, G., Dugnani Lonati, R., Orsenigo, M., & Sommariva, O. (1955) *Experimentia* **12** 479–481 Further measurements on the bioluminescence of the seedlings.

Gurvich, A. G. & Gurvich, L. D. (1934) *Mitogenic radiation* (in Russian) Leningrad, also Medgiz, 1945.

Tarusov, B. N., Polivida, A. I., & Zhuravlev, A. I. (1961) *Radiobiologiya* I 150–151. Ultraweak luminescence of animal tissues (in Russian).

Vassil'ev, R. F. & Vichutinskii, A. A. (1962), *Nature* (Lond.) **194** 1276–1277. Chemiluminescence and oxidation.

Reviews

Allen, R. C. (ed) (1982) In: *Chemical and biological generation of excited states*, pp 309–344. Biochemical excitation: chemiluminescence and the study of biological oxygenation reactions.

Barenboim, G. M., Domaskii, A. N., & Turoverov, K. K. (1969), *Luminescence of biolymers and cells*. Chapter 4. Chemiluminescence of Cell and Organisms. Transl. Chen, R. F., Plenum Press, New York and London.

Boveris, A., Cadenas, E., & Chance, B. (1981) *Fed. Proc.* **40** 195–198. Ultraweak chemiluminescence: a sensitive assay for oxidative radical reactions.

Campbell, A. K., Hallett, M. B., & Weeks, I. (1985), *Meth. Biochem. Anal.* **31** 317–416. Chemiluminescence as an analytical tool in cell biology and medicine.

Chance, B., Sies, H., & Boveris, A. (1979) *Physiol Rev.* **59** 527–610. Hydroperoxide metabolism in mammalian organs.

Respiratory burst in cells

Ado, A. D. (1933) *Z. Ges. Exp. Med.* **87,** 473–480. Ubër den Verlauf der oxydativen and glykolytischen Prozesse in den Leukocyten des ertzündeten Gewebes während der Phagocytose.

Baldridge, C. W. & Gerald, R. W. (1933) *Am. J. Physiol.* **103** 235–236. Extra respiration of phagocytosis.

Klebanoff, S. J. & Clark, R. A. (1978) *The neutrophil: function and clinical disorders*. North-Holland, Amsterdam.

Segal, A. W. & Jones, O. T. G. (1978) *Nature* (*Lond*) **276** 515–517. Novel cytochrome b system in phagocytic vacuoles of human granulocytes.

Warburg, D. (1980) *Physiol. Chem.* **57** 1–16. Beobachtugen über die Oxydations prozesse in Seeigelei.

Cellular and organ ultraweak chemiluminescence

Aniansson, H., Stendahl, O., & Dahlgren, C. (1984) *Acta path. microbiol. immunol. Scand sect. C* **92** 357–361. Comparison between luminol- and lucigenin-dependent chemiluminescence of polymorphonuclear leukocytes.

Barak, M., Ulitzur, S., & Merzbach, D. (1983) *J. Immunol.* **64** 353–363. The use of luminous bacteria for determinations of phagocytosis.

Barsacchi, R., Pelosi, G., Comici, P., Bonalso, L., Maiorino, M., & Ursini, F. (1984) *Biochim. Biophys. Acta* **804** 356–360. Glutathione depeletion increases chemiluminescence emission and lipid peroxidation in the heart.

Boveris, A., Cadenas, E., Reiter, R., Filipkowski, M., Nakase, Y., & Chance, B. (1980) *Proc. Natl. Acad. Sci. US* **77** 347–351. Organ chemiluminescence: non-invasive assay for oxidative radical reactions.

Cadenas, E., Brigeluis, R., & Sies, H. (1983) *Biochem. Pharmacol.* **32** 147–150. Paraquat-induced chemiluminescence of microsomal fractions.

Cooke, E. & Hallett, M. B. (1985) *Biochem. J.* **232** 323–327. The role of C-kinase in the physiological activation of the neutrophil oxidase. Evidence from using pharmacological manipulation of C-kinase activity in intact cells.

Dahlgren, C. (1987) *Agents and Actions* **21** 1–4. Polymorphonuclear leukocyte chemiluminescence induced by formyl-methionyl-leucyl-phenylanine and phorbol myristate acetate: Effects of catalose and superoxide dismutase.

de Chatelet, L. R., Long, G. D., Shirley, P. S., Bass, D. A., Thomas M. J., Henderson, F. W., & Cohen, M. S. (1982) *J. Immunol.* **129** 1589–1593. Mechanism of luminol-dependent chemiluminescence of human neutrophils.

de Nascimento, A.-L, T. O., & Cilento, G. (1985) *Biochim. Biophys. Acta* **843** 254–260. Induction of chemiluminescent processes in the fungus *Blastocladiella emersonii* by exposure to enzyme-generated triplet benzaldehyde.

Eschenbach, C. & Adrian, U. (1985) *Klin. Wochenschr* **63** 1218–1225. DMNH Ein neuerkompetenter Indikator für die Messung der Chemilumineszenz von neutrophilen Granulocyten und Monocyten.

Esser, A. & Stauff, J. (1968) *Z. Naturforschng.* **23** 1554–1555. Lumineszenz von Hefe.

Foerder, C. A., Klebanoff, S. J., & Shapiro, B. M. (1978) *Proc. Natl. Acad. Sci. US* **75** 3183–3187. Hydrogen peroxide production, chemiluminescence, and the respiratory burst of fertilisation: inter-related events in early sea urchin development.

Haber, M. J. & Topley, N. (1986) *J. Biolum. Chemilum.* **1** 15–27. Factors affecting the measurement of chemiluminescence in stimulated human polymorphonuclear leucocytes.

Hallett, M. B. & Campbell, A. K. (1983). *Biochem. J.* **216** 459–465. Two distinct mechanisms for activation of polymorphonuclear leucocytes.

Hallett, M. B. & Campbell, A. K. (1984) *Cell Calcium* **5** 1–19. Is intracellular calcium the trigger for neutrophil activation?

Houk, M. S. (1974) *Exp. Cell Res.* **83** 200–206. Respiration of starfish oocytes during meiosis, fertilization, and artifical activation.

Ito, M., Karmali, R., & Krim, M. (1985) *Immunology* **56** 533–541. Effect of interferon on chemiluminescence and hydroxyl radical production in murine macrophages stimulated by PMA.

Jenner, D. E., Holt, M. E., & Campbell, A. K. (1987) *J. Biolum. Chemilum.* **1** 165–171. Luminol dependent chemiluminescence and thiol group oxidation provoked by neutrophils is attributable to different oxidising speices.

Lew, D. C., Monod, A., Waldvogel, F. A., & Pozzan, T. (1987) *Eur. J. Biochem.* **162** 167–168. Role of cytosolic free calcium and phospholipase C in leukotriene B4- stimuated secretion in human neutrophils. Comparison with chemotactic peptide formyl-methionyl-leucyl-phenylalanine.

Marnett, L. J., Wlodawer, P., & Samuelsson, B. (1974) *Biochem. Biophys. Res. Commun.* **60** 1286–1294. Light emission during the action of prostaglandin synthetase.

Mills, E. L., Gerrard, J. M., Filipovich, D., White, J. D., & Quil, P. G. (1978). *J. Clin. Invest.* **61** 807–814. The chemiluminescence response of human platelets.

Minkenberg, I. & Ferber, E. (1984) *J. Immunol. Metabol.* **71** 61–67. Lucigenin-dependent chemiluminescence as a new assay for NAD(P)H-oxidase activity in particular fractions of human polymorphonuclear leukocytes.

Patel, A., Hallett, M. B., & Campbell, A. K. (1987) *Biochem. J.* **248** 173–180. Threshold responses in reactive oxygen metabolite production in individual neutrophils, detected by flow cytometry and microfluorimetry.

Pirrene, M. H. & Denton, E. J. (1952). *Nature (Lond.)* 1039–1042. Accuracy and sensivity of the human eye.

Posner, G. A., Lever, J. R., Miura, K., Lisek, C., Seliger, H. H., & Thompson, A. (1984) Biochem. Biophys. Res. Commun. 123 869–873. A chemiluminescence probe specific for singlet oxygen.

Pozzan, T., Lew, D. L., Wollheim, C. B., & Tsien, R. Y. (1983) Science **221** 1413–1415. Is cytosolic ionised calcium regulating neutrophil activation?

Quickenden, T. I. & Que Hee, S. S. (1974). *Biochem. Biophys. Res. Commun.* **60** 764–770. Weak luminescence from the yeast *Saccharomyces cerevisiae* and the existence of mitogenetic radiation.

Quickenden, T. I. & Que Hee, S. S. (1976) *Photochem. Photobiol.* **23** 201–204. Spectral distribution of luminescence emitted during yeast growth and relationship to mitogenetic radiation.

Quickenden, T. I., Comarmond, M. J., & Tilbury, R. N. (1985) *Photochem. Photobiol.* **41** 611–615. Ultraweak bioluminescence spectra of stationary phase *Saccharomyces cerevisiae* and *Schizosaccharonyes pombe*.

Roberts, P. A., Knight, J., & Campbell, A. K. (1987) *Anal. Biochem.* **160,** 139–148. Pholasin — a bioluminescent indicator for detecting activation of single neutrophils.

Roberts, P. A., Newby, A. C., Hallett, M. B., & Campbell, A. K. (1985) *Biochem. J.* **127** 669–674. Inhibition by adenosine of reactive oxygen metabolite production by human polymorphonuclear leucocytes.

Roberts, J. & Quastel, J. H. (1964) *Nature (Lond).* **202,** 85–86. Oxidation of reduced triphosphopyridine nucleotide by guinea pig polymorphonuclear leucocytes.

Roder, J. C., Helfand, S. L., Werkmeister, J., McGarry, R., Beaumont, T. J., & Duwe, A. (1982). *Nature (Lond)* **298** 569–572. Oxygen intermediates are triggered early in the cytolytic pathway of human NK cells.

Rosen, H. and Kebanoff, S. J. (1976) *J. Clin. Invest.* **58** 50–60. Chemiluminescence and superoxide production by myeloperoxidase-deficient leucocytes.

Sbarra, A. J. and Karnovsky, M. L. (1959) *J. Biol. Chem.* **234** 1355–1362. The Biochemical basis of phagycotosis. I. Metabolic changes during the ingestion of particles to polymorphonuclear leucocytes.

Seim, S. (1983), *Acta. path. microbiol. immunol. scand. sect. c.* **91** 123–128. Role of myeloperoxidase in the luminol-dependent chemiluminescence response of phagocytosing human monocytes.

Suematsu, M., Oshio, C., Miura, S. and Tsuchiya, M., (1987). *Biochem. & Biophys. Research Communications* **149** No. 3, 1106–1110. Real-time visualization of oxyradical burst from single neutrophil by using ultrasensitive video intensifier microscopy.

Uchida, T., Kanno, T., & Hosaka, S. (1985) *J. Immunol. Meth.* **77** 55–62. Direct measurement of phagosomal reactive oxygen by luminol-binding microspheres.

Worner, P. & Patschke, H. (1980), *Thrombosis Res.* **19.** 277–282. Chemiluminescence in wasted human platelets during prostaglandin-thromboxane synthesis
induced by N-ethylmaleimide and thiomersal.

Singlet oxygen
Gorman, A. (1981). *Chem. Soc. Rev.* **10,** 205–231. Singlet molecular oxygen.
Khan, A. N., (1976). *J. Phys. Chem.* **80** 2219–2228. Singlet molecular oxygen: a new
kind of oxygen.
Nakano, M., Noguchi, T., Sugioka, K., Fukuyama, H., Sato, M., Shimizu, Y.,
Tsuji, Y., & Inaba, H. (1975). *J. Biol. Chem.* **250** 2404–2406. Spectroscopic
evidence for the generation of singlet oxygen in the reduced nicitinamide
adenine dinucleotide phosphate-dependent microsomal lipid peroxidation
system.
Williams, D. & Chance, B. (1983). *J. Biol. Chem.* **258** 3628–3631. Spontaneous
chemiluminescence of human breath. Spectrum, lifetime, temporal distribution,
and correlation with peroxide.

Acellular Ultraweak Chemiluminescence
Gurwitsch, A. A., Eremeyer, V. F., & Karabchievsky, Y. A. (1965). *Nature* (*Lond*)
206 20–22. Ultraweak emission in the visible and ultra-violet regions in oxidation
of solutions of glycine by hydrogen peroxide.
Slawinska, D. & Slawinski, J. (1982) In: *Luminescent assays*: perspectives in
endocrinology and clinical chemistry. pp 221–227. Eds Serio, M. & Pazzagli, M.
Raven Press, New York. Chemiluminescence in the peroxidation of noradrena-
line and adrenaline.
Trusch, M. A., Mimnaugh, E. G., Siddik, Z. H., & Gram, T. E. (1983) *Biochem.
Biophys. Res. Commun.* **112** 378–383. Bleomycin-metal interaction: ferrous
iron-initiated chemiluminescence.

7

Chemiluminescent indicators for inorganic ions

It may seem something of a paradox to be considering inorganic substances in a book aimed principally at the biology and biomedical applications of the *organic* reactions involved in chemiluminescence. Certainly the uniqueness of life depends on the synthesis and degradation of carbon compounds. However, animals, plants, and unicellular organisms, both pro- and eu-karyote, cannot survive without inorganic substances. The characterisation of their biological role requires their measurement, not only in the fluids bathing cells but within the living cell itself.

Many methods for measuring free ions, and their complexes with other elements or ligands, have been developed during the past hundred years, based mainly on precipitation and colorimetric indicators, as well as on flame emission and atomic absorption spectrophotometry. Chemiluminescent assays have also been developed for a number of cations and cation complexes, and for a limited number of anions. To see whether these have any real value in the research or clinical laboratory, it is necessary first to remind ourselves of the major functions of inorganic ions in living systems.

7.1 THE PRINCIPAL BIOLOGICAL ROLES OF INORGANIC IONS

7.1.1 Cations

Apart from the non-metal cations H^+ and NH_4^+, biologically significant cations occur in three major metallic groups of the periodic table, the alkali metals, the alkaline earths, and the transition metals. They can be found within cells or in the fluids bathing cells, and are essential nutrients for all living cells. The metallic cations may be found free or complexed to inorganic anions, either in solution or in precipitates. They also form complexes with many small and large organic ligands, including porphyrins, nucleotides, phospholipids, carbohydrates, nucleic acids, and proteins. Their biological roles (Table 7.1) can be conveniently divided into six main categories:

(1) Structural role: e.g. calcium phosphate in bone and teeth, calcium carbonate in the shells of invertebrates, and calcium in the maintenance of membrane integrity.

Table 7.1 — Major biological roles of inorganic cations

Metal group	Cation	Functions
Group I (alkali metals)	Na^+, K^+	Osmotic balance across membranes Electrical activity across membranes Activation of some enzymes
Group II (alkaline earths)	Mg^{2+}	ATP Mg^{2-}, the substrate for all kinases and many pumps Enzyme activator Mg porphyrin in chlorophyll
	Ca^{2+}	Biomineralisation (Ca phosphate, $CaCO_3$, $CaSO_4$) Structure of soft tissues (cell-cell junctions, cell adhesion, cell permeability, electric stability) Current carrier in some cells Intracellular regulator Intracellular pathogen
Transition metals	Fe^{2+}/Fe^{3+} Cu^+/Cu^{2+}	Redox catalysts, particularly containing haems (respiratory chain, reactive oxygen metabolite generators and scavengers, peroxidases)
	Fe^{2+}, Cu^{2+}	Oxygen-carrying pigments (haemoglobin, haemocyanin)
	Fe^{2+}, Cu^{2+}	Cofactor or metalloprotein in some luciferases
	Co^{2+}	Vitamin B_{12}
	Fe^{2+}, Cu^{2+}, Co^{2+}, Mn^{2+}, V^{2+}, Zn^{2+}	Catalytic function at the active centre of some enzymes
Non-metal	NH_4^+, H^+	pH and ionisation state of acid groups, source of N for amino acid synthesis

(2) Osmotic role: e.g. Na^+ and K^+ in the maintenance of the osmotic balance across biological membranes.

(3) Electrical role: e.g. K^+ in the maintenance of the resting potential of many cells, and Na^+ or Ca^{2+} in the generation of action potentials in excitable cells.

(4) Role in oxido-reduction reactions (electron transfer), e.g. Fe^{2+} in the haems of cytochromes in the respiratory chain, and in phagocytes the unique cytochrome responsible for the univalent reduction of 0_2 to 0_2^-.

(5) Catalytic role: e.g. the activation of substrates, such as ATP by Mg^{2+} (ATP Mg^{2-}); or the activation of enzymes, such as by transition metals at the active centre; or the activation and inhibition through binding at allosteric sites or binding to special regulatory proteins such as calmodulin.

(6) Carrier or transporter role: e.g. O_2 on haems, anions such as phosphate by Ca^{2+} across membranes, or cation exchange across membranes such as Na^+ for Ca^{2+}.

These categories embrace and control the physical (1,2), electrical (3,4) and chemical (5,6) properties of all living cells. Disturbances in cations both within and outside cells also play an important role in the pathogenesis of many diseases, for example renal failure (Na^+, K^+), diabetes (K^+), parathyroid disease (Ca^{2+}), haemachromatosis (Fe^{2+}), and Wilson's disease (Cu^{2+}). If we wish to characterise the static and dynamic aspects of cations in the physiology and pathology of tissues they have to be measured. But how sensitive a technique is required?

The range of cation concentrations varies from as much as 500 mM for Na^+ in sea water to less than 100 nM for cytosolic free Ca^{2+}. For *free* transition metals concentrations may be as low as pM–fM. Where the concentration of free cation, or cation in a tissue extract, is in the μM–mM range, there is usually a simple chemical method available, or one utilising an ion specific electrode. Flame emission and atomic absorption spectroscopy can also measure cations in this range. Many cations can be estimated in tissue sections by X-ray microprobe analysis or electron energy loss. However, for free cations $<0.1\,\mu$M a more sensitive technique is required, particularly when the measurement must be carried out in the presence of a large amount of bound cation. We also need to measure cation concentrations in live cells. Chemiluminescence enables free Ca^{2+} inside cells, and transmission metals in tissue extracts, to be measured in the required concentration range.

7.1.2 Anions

Chloride is the major extracellular anion in metazoans. However, inside the cell organic anions such as glutamate and isethionate often assume the role of major anion. Other important intracellular organic anions include aspartate, nucleotides, phospholipids, nucleic acids, and proteins. Most proteins, with a few notable exceptions (e.g. histones), have greater glu + asp than lys + arg contents and are thus negatively charged at physiological pH. Apart from Cl^-, the anions within and outside the cell are much less permeable across biological membranes than are their balancing cations, and may be virtually impermanent in the case of macromolecular anions. Hardly surprisingly, therefore, several inorganic ions in addition to Cl^- have important biological roles (Table 7.2). As with the metallic cations, these inorganic anions also play a key role in determining the physical, electrical and chemical properties of cells. Their roles (Table 7.2) can be conveniently divided into six main categories:

(1) Structural role, e.g. phosphate in bone and teeth and carbonate in the shells of invertebrates.
(2) Osmotic role, e.g. Cl^- as the counter ion for Na^+ extracellularly.
(3) Electrical role, e.g. Cl^- in the resting and spike potentials of some cells.
(4) Role in oxido-reduction reactions (electron transfer) e.g. NO_3^-/NO_2^-; $SO_4^{2-}/SO_3^{2-}/H_2S$; O_2/O_2^-.
(5) Catalytic role, e.g. regulation of enzymes by Cl^-, or covalent phosphate on ser, thr and tyr residues in proteins, Se in glutathione peroxidase.
(6) Source of inorganic elements for the synthesis of organic molecules, e.g. HCO_3^-, SO_4^{2-}, NO_3^-, HPO_4^{2-}, I^-.

Table 7.2 — Major biological roles of inorganic anions

Anion	Function
Cl^-	Major extracellular anion in Metazoans
	Osmotic and electrical balance across cell membranes
I^-	Thyroid hormones
HCO_3^-	Acid-base balance
	CO_2 source in fixation reactions
NO_3^-	Nitrogen source in plants and micro-organisms
O_2^-	Bactericidal activity in phagocytes
OCl^-	Bactericidal activity in phagocytes
	Spermicidal activity in some eggs
$OSCN^-$	Bactericidal activity in milk
$HPO_4^{2-}/H_2PO_4^- (PO_4^{3-})$	Biomineralisation in bones and teeth
	Substrate for ATP synthesis
	Product of phosphatases and pumps
CO_3^{2-}	$CaCO_3$ in shells of invertebrates
SO_4^{2-}	Sulphur source in plants and micro-organisms
	Buoyancy in some invertebrates
Se	Active centre of glutathione peroxidase

Major anions such as Cl^-, phosphate, and HCO_3^- in the mM range can be measured by titration to form coloured compounds or precipitates, and Cl^- can also be measured by using an electrode. However, many of the other anions exist in the nM–μM range. A few of these can be measured by using chemiluminescence.

But let us begin by examining one of the most significant applications of chemiluminescence in biology during the last twenty years, the use of coelenterate photoproteins to measure intracellular free Ca^{2+}. Chemiluminescence has played a key part in establishing the unique role of this cation in cell activation and cell injury.

7.2 PHOTOPROTEINS AND INTRACELLULAR FREE CALCIUM

7.2.1 The first direct measurement of intracellular free Ca^{2+}

'Rasa ligno, pareum adeo in tenebris splendet', being roughly translated 'struck by wood to such an extent it shines brightly in the darkness'. Thus, for the first time, Forskål in 1775 described in his *Fauna Arabica* the luminescence of the jelly fish *Aequorea forskalea* (Fig. 7.1), or *Medusa Aequorea* as he called it. Forskål would no doubt have been surprised to learn that this jelly fish contains its own calcium indicator, a protein now known as aequorin which emits light when it binds calcium. Calcium does not, however, provide the energy for light emission; this arises from the oxidative chemiluminescent reaction of the prosthetic group covalently linked to the protein (Fig 7.2). Calcium acts as the trigger allowing the reaction to proceed. It was this unique property which enabled Ridgway and Ashley in 1967 to develop the first method, generally applicable to a wide range of cell types, for measuring the

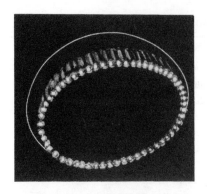

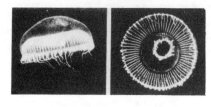

A

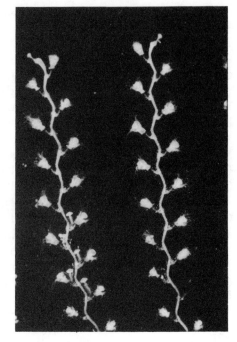

B

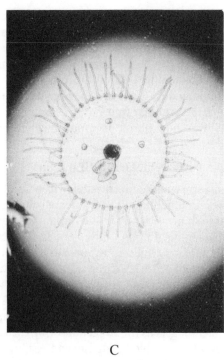

C

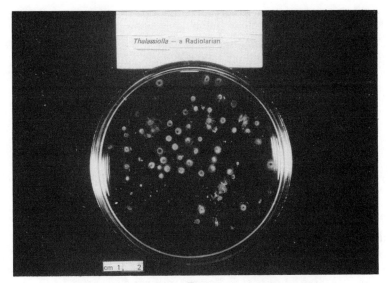

D

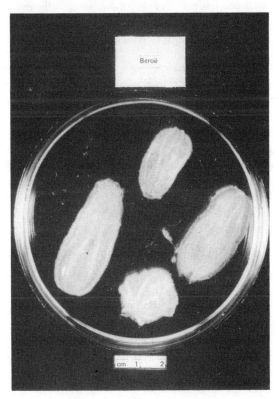

E

Fig. 7.1 — Sources of Ca^{2+}-activated photoproteins. A. *Aequorea forskalea*. kind gift from
Professor John Blinks. B. *Obelia geniculata*. C. *Obelia lucifera*. D. *Thalassicolla spp*. E. *Beroë*
(NB ctenophore photoproteins are light sensitive).

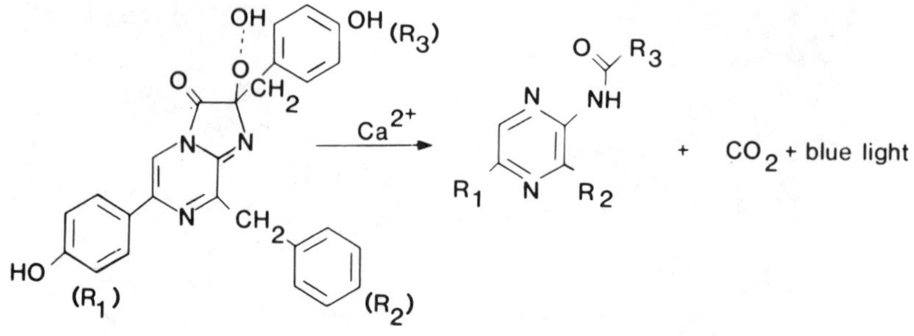

Fig. 7.2 — Reaction of photoproteins with Ca^{2+}.

concentration of free calcium in the cytoplasm of a living cell. They injected aequorin into the giant muscle fibre of the barnacle *Balanus nubilus* and were able to detect an increase in light emission from inside the cell just before the onset of muscle tension, after the fibre had been excited electrically. The light emission returned virtually to the level in the resting cell within a few seconds, being followed by relaxation of the myofibrils within the muscle fibre. This experiment provided two crucial pieces of evidence in the study of muscle contraction. Firstly, if intracellular Ca^{2+} is to be the trigger for muscle contraction, then it is *necessary* that changes in cytoplasmic free Ca^{2+} should precede events in the myofibrils. Secondly, it enabled, for the first time, free Ca^{2+} to be measured directly in a resting and a stimulated cell. It rose from about $0.3\,\mu M$ at rest to $5\,\mu M$ after electrical stimulation at the peak of activation. The light emitted from aequorin within the cell was thus able to provide evidence sufficient to *prove* that Ca^{2+} was the internal trigger for muscle contraction since prevention of the increase in light signal by injecting a Ca^{2+} chelator, EGTA, into the cell also prevented contraction.

Sixty years ago, A. B. Macullum in a correspondence with a Dr Myra Sampson in *Science* drew attention to two papers published by Grandis and Mainani at the turn of the century suggesting that purpurin, now known as murexide, could be used to localise calcium in animal and plant cells. Macullum concluded 'a sensitive microchemical reagent for calcium in tissues and cells is a great desideratum . . .'. Some 23 years later, Pollack injected the red dye alizarin sulphate as a 'calcium indicator' into an amoeba and observed a 'shower of red crystals near to the site pseudopod formation'. Farsighted as these early studies were, it was not until the experiments of Ridgway and Ashley, and those of John Blinks and co-workers, at the end of the 1960s, using the photoprotein aequorin, that a method with the necessary specificity and sensitivity for detecting and quantifying cytosolic free Ca^{2+}, generally applicable to all eukaryotic cells, became available. This long gap between awareness and realisation occurred not only because of the relatively slow acceptance of intracellular calcium as a regulator, in spite of L. V. Heilbrunn's pioneering arguments (see section 7.6), but also because the crucial importance of measuring cytosolic free Ca^{2+} was not generally appreciated. Nor was it appreciated how low the cytosolic free Ca^{2+} was, compared with both the total cell calcium and the concentration of Ca^{2+} outside the cell.

Why then do we need to measure free Ca^{2+} in living cells? Have we really learned anything new about how cells work as a result of striving to put aequorin not only into giant cells but also into small mammalian cells? How does aequorin measure up to the other indicators of free Ca^{2+}, in particular the new fluorescent indicators quin2, fura-2, and indo-1 (Fig. 7.6), developed thanks to the ingenuity of Roger Tsien?

Let us therefore first examine why those of us who have tried so hard to measure the concentration of free Ca^{2+} in living cells believe that it is crucial to an understanding of the biological function of this very special cation.

7.2.2 Why measure intracellular free Ca^{2+}?

The calcium ion has an unusual importance in biological phenomena, and the literature concerning its effects is extremely voluminous . . . no other ion exerts such interesting effects of protoplasmic viscosity as the calcium ion'

L. V. Heilbrunn, 1937.

All eukaryotes, both animal and plant, and many prokaryotes, require Ca^{2+} in the medium bathing their cells in order to survive, to grow, and to carry out their specialised functions. In 1883 Sydney Ringer demonstrated that the normal beating of a frog heart requires external Ca^{2+}. Such a requirement for extracellular Ca^{2+} has since been demonstrated for several phenomena involving cell activation, as well as several examples of cell injury. The problem facing the early physiologists was to define precisely what the calcium was doing.

Imagine you are in the laboratory and that you are still using a glass pipette. It snaps and you cut yourself. You rush over to the cold water tap, but fortunately within a few minutes the bleeding stops. Ca^{2+} has played a major part in this scenario, both within and outside your cells. Its diverse roles are illustrated particularly well in the formation of the blood clot. Ca^{2+} in the blood is essential for the activity of at least five enzymes in the blood clotting pathway. However, Ca^{2+} is not a regulator of this pathway in a 'physiological' sense. Certainly, removal of Ca^{2+} prevents clotting, but the stimulus for activating the proteolytic cascade leading to formation of the fibrin clot is not a change in Ca^{2+} concentration, but rather the release of substances from damaged tissues and platelets. In contrast, secretion of substances from platelets, together with platelet aggregation, require an increase in cytosolic Ca^{2+} within the platelet, initiated by the primary stimulus, e.g. thrombin. Here, intracellular Ca^{2+} is acting as a 'physiological' regulator since it is the change in Ca^{2+} itself, within the cell, which actually initiates the chemical and morphological changes associated with platelet aggregation and secretion. The only way to demonstrate definitively this unique role for Ca^{2+} inside the cell is to measure the free Ca^{2+} in the cytosol of the living cells, and to correlate the time course and magnitude of any increase with the physiological event.

The phenomena we are dealing with, therefore, involve a change of state in a cell, or a population of cells, initiated by a primary stimulus, and modified by a secondary regulator (Table 7.3). The primary stimulus may be physical, for example an action potential, touch or light, or it may be chemical, for example a neurotransmitter or hormone. The primary stimulus usually acts at the cell membrane. There has, therefore, to be a mechanism for transmitting a signal, that the cell membrane has been exposed to a stimulus, to the chemical processes within the cell responsible for the phenomenon concerned. Many of these phenomena involved thresholds within

Table 7.3 — Phenomena initiated by a rise in intracellular free Ca^{2+}

Phenomenon	Example	Primary stimulus	Secondary regulator
Cell movement	Skeletal muscle	Acetylcholine	?
	Heart muscle	Action potential	β adrenaline (a) Acetyl choline (i)
	Smooth muscle	Noradrenaline	Eisosanoids
	Neutrophil chemotaxis	Complement fragment ($C5_a$)	Eicosanoids
Vesicular secretion	Neurotransmitter from nerve terminal	Action potential	Adenosine (i)
	Insulin from β cell	Glucose	Glucagon (a) Adrenaline (i)
Cell aggregation	Platelet	Thrombin	Adenosine (i)
Cell division/ transformation	Lymphocyte	Antigen	?
Cell fertilisation	Egg	Sperm	?
Vision†	Retinal rod	Light	Dark adaptation (a)
Bioluminescence	*Aequorea*	Touch	Light (i)
Ultraweak chemilumines cence	Neutrophil	Chemotactic stimuli	Adenosine (i)
Intermediary metabolism	Glycogenolysis in hepatocytes	Adrenaline	Insulin (i)
Cell protection	Vesiculation of 'lethal' pore-formers on nucleated cells	Membrane attack complex of complement	Adenosine (i)

† The role of Ca^{2+} in vision remains controversial; some believe a reduction in cyclic GMP is the primary intracellular signal for vision.

individual cells where the cell rapidly transforms from a resting to stimulated state. Thus, a heart muscle cell beats or it does not, two platelets aggregate or they do not, a luminous cell flashes or it does not. Similarly, threshold responses occur in several types of cell injury. The membrane attack complex of complement ($C5b6889_n$) either kills a cell or the cell recovers from attack. The time taken to reach any threshold may vary from cell to cell, this time together with the magnitude of the eventual response being capable of modification by a secondary regulator. Ca^{2+} can also damage macromolecules and organelles within the cell, yet it also activates mechanisms enabling cells to protect themselves from attack by pore-forming agents such as the membrane attack complex of complement, T cell perforins, bacterial toxins, and viral proteins. See Chapter 10, section 10.3.2 for a fuller explanation through the chemi-symbiotic hypothesis.

How does one show whether a rise in cytoplasmic free Ca^{2+} is the mediator of such events? How does one distinguish types of cell activation dependent on a rise in intracellular free Ca^{2+} from those which are not, for example ones mediated solely through a cyclic AMP or cyclic GMP dependent protein kinase, or through protein kinase C? How can the importance of intracellular Ca^{2+} in a cellular event be rationalised with the importance of other intracellular regulators such as cyclic AMP, cyclic GMP, diacyl glycerol, and inositol trisphosphate? Is intracellular Ca^{2+} a friend or foe? Direct measurement of cytoplasmic free Ca^{2+} plays a central role in answering these questions, and in particular to solving five particular problems (Fig. 7.5):

(1) Is a rise in intracellular Ca^{2+} directly responsible for cell activation?
(2) If so, is the source of the Ca^{2+} for cell activation extra or intra-cellular, and how is it released into cytoplasm?
(3) How and where does the Ca^{2+} act?
(4) Do secondary regulators act by altering the free Ca^{2+} change induced by the primary stimulus, or on the mechanism by which Ca^{2+} acts, or independently of Ca^{2+}?
(5) Is a rise in intracellular free Ca^{2+} a cause or a consequence of cell injury, and is it responsible for activating protection mechanisms within the cell?

Indirect experiments based on manipulation of extra- and intra-cellular Ca^{2+} provide clues that Ca^{2+} may be involved in a cellular phenomenon. However, such experiments are unable to distinguish an 'active' role for intracellular Ca^{2+}, from a 'passive' role in the integrity of the cell or maximal activity of enzymes. Nor can the electrical role of Ca^{2+} through the opening of voltage-sensitive Ca^{2+} channels in some cells be distinguished from the activation of intracellular processes by Ca^{2+}.

There are five major reasons why direct measurement of cytoplasmic free Ca^{2+} is crucial to understanding its role in cell regulation. Firstly, correlation of an increase in cytoplasmic free Ca^{2+} with the onset, duration, and location of intracellular events responsible for the phenomenon (Table 7.3), together with prevention of the phenomenon by stopping the Ca^{2+} increase, provides definitive evidence for intracellular Ca^{2+} as the cell 'trigger'. Secondly, measurement of the magnitude of the Ca^{2+} rise identifies the sensitivity of Ca^{2+} necessary in the Ca^{2+} binding protein(s) responsible for mediating the effect of Ca^{2+} within the cell. Thirdly, it enables the source of Ca^{2+} for cell activation to be identified, be it intra- or extra-cellular. Fourthly, it enables $^{45}Ca^{2+}$ flux data to be interpreted correctly, avoiding artefacts caused by changes in specific activity during cell activation. Fifthly, measurement of the cytosolic free Ca^{2+} is essential if this cation is to be held responsible for cell injury, as opposed to any increase being a consequence of such injury.

What sort of sensitivity is required, then, for such intracellular measurements?

In 1956 Keynes and Lewis reported that the calcium content inside an invertebrate nerve axoplasm was about 0.5 mmole per kg wet weight, some 20-fold lower than that outside the cell. The following year, Hodgkin and Keynes showed that the mobility of $^{45}Ca^{2+}$ within the axon of a giant nerve of the squid was more than 45 times slower than that in free solution. They concluded that most of the calcium must be bound within the axoplasm, and that the free ionized calcium must be less than

10 μM. Studies by Portzehl, Caldwell, and Ruëgg in 1964, using the then recently introduced Ca^{2+} buffer EGTA in crab muscle, reduced the estimate for cytosolic free Ca^{2+} in resting cells to less than 0.3 μM. The validity of this surprisingly low value was supported by experiments on isolated myofibrils which contracted in solutions containing sub-μM free Ca^{2+}, together with the isolation of vesicles from sarcoplasmic reticulum capable of lowering the surrounding free Ca^{2+} to 0.01–0.1 μM Ca^{2+} in the presence of MgATP. Thus, by the mid 1960s it had become clear that eukaryotic cells were capable of maintaining an electrochemical gradient of Ca^{2+} of some 10 000-fold across the cell membrane (Fig. 7.4 pp 393), in contrast to the 10 to 20-fold gradients of Na^+ and K^+. More than 99% of the Ca^{2+} within the cell is therefore bound within organelles such as endoplasmic reticulum, mitochondria, specialised vesicles, and possibly the nucleus. Less than 0.005% of the cell's Ca^{2+} is free in the cytoplasm. An indicator is therefore required which can detect free Ca^{2+} approximately in the range of 0.1–10 μM. Atomic absorption spectrometry and X-ray microprobe analysis are neither sufficiently sensitive nor do they distinguish between free and bound Ca^{2+} within the cell.

7.2.3 The ideal indicator

The ideal indicator of cytosolic free Ca^{2+} should satisfy the following eight criteria:

(1) *Specificity*
   Ca^{2+} should be the only ion generating a signal from the indicator, particularly in the presence of Mg^{2+}, K^+, and Na^+ (1–100 mM).
(2) *Sensitivity*
   To cover the entire range of free Ca^{2+} in resting, stimulated, and injured cells, the indicator should be capable of quantifying free Ca^{2+} in the range of 10 nM–100 μM. Furthermore, the signal from the indictor should be readily detected over any background signal arising from the cell itself, or from the apparatus.
(3) *Speed of response*
   Ca^{2+} transients occurring within 10 ms, and lasting many minutes, must be detectable.
(4) *Incorporation into cells*
   It must be possible to incorporate the Ca^{2+} indicator into the cytoplasm of the living cell without significant disturbance to cell structure and function. Once there, it should be stable for hours, ideally days in a cell culture. It should be non-toxic.
(5) *Effect on cell Ca^{2+} balance*
   The indicator should not significantly disturb the Ca^{2+} balance of the cell.
(6) *Diffusion*
   The diffusion of the indicator should not be a factor affecting the signal from the Ca^{2+} transient.
(7) *Distribution of free Ca^{2+} within the cell*
   All cells exhibit some polarity. To establish Ca^{2+} as the intracellular mediator of a primary stimulus, the free Ca^{2+} increase must ultimately be localised within the cell, and this localisation correlated with chemical and structural changes associated with cell activation or injury.

(8) *Availability*

The Ca^{2+} indicator should be readily available and inexpensive.

Three groups of indicators satisfy several, if not all, of these criteria and have been used inside cells (Fig. 7.6 pp 394):

(1) Ca^{2+}-activated photoproteins.
(2) Indicator dyes
    (a) spectrophotometric (murexide, antipyryl azo III, arsenazo III)
    (b) fluorescent (quin2, fura-2, indo-1)
    (c) nuclear magnetic resonance (FBAPTA).
(3) Ca^{2+} sensitive microelectrodes.

Two have found wide application in both large and small cells in the plant and animal kingdoms, namely the Ca^{2+}-activated photoproteins discovered by Shimomura and colleagues in 1962 and the fluorescent dyes invented by Tsien. But before identifying the valuable properties of the photoproteins as intracellular Ca^{2+} indicators we first need a source, and a method of isolating the photoprotein in a form suitable for incorporation into cells.

7.2.4 Source of Ca^{2+}-activated photoproteins

Luminous coelenterates have been described by naturalists since Pliny observed in the first century A.D. his luminous marine lung, 'Pulmo marinus', probably the pink jelly fish *Pelagia noctiluca*. They include a wide variety of hydroids, hydrozoan and scyphozoan jelly fish, sea pens, sea pansies, and sea combs and sea jellies. They emit blue to blue–green light (max 460–508 nm) when touched. The chemical puzzle presented by the experiments of McCartney in 1810 and Harvey in 1926 was that it seemed impossible to extract a conventional O_2 requiring luciferin–luciferase reaction from any but the luminous anthozoans (see Chapter 3). The problem was solved on 1962 by Shimomura and co-workers who showed that the jelly fish *Aequorea forskalea*, or *Aequorea aequorea* as they call it, contains a protein of molecular weight about 20 000 with the chemiluminescent chromophore covalently linked to it and already precharged with oxygen. All that is necessary to trigger the chemiluminescent reaction is to add calcium. The imidazolopyrazine ring (Fig. 3.11; Fig. 7.2) ultimately responsible for coelenterate chemiluminescence has since been found in some luminous squid, shrimp, copepods and fish (Table 3.19). However, in only two groups (Fig. 7.1), luminous coelenterates and luminous radiolarians, the latter being a group of protozoa, have Ca^{2+}-activated photoproteins been found.

There are about 10 000 individual species making up the coelenterates, now regarded as two phyla, Cnidaria (*ca.* 9500 species) and *Ctenophora* (*ca.* 200 species). They are found in the deep oceans, in surface waters, and in rock pools. Many are luminous (Figs. 3.3B, 7.1; Table 7.4). Fresh or brackish water coelenterates are rare, and none has been reported as being luminous. For any organism to provide a practical source of a Ca^{2+} indicator three criteria must be satisfied:

(1) It must be luminous and contain an extractable Ca^{2+}-activated photoprotein.
(2) It must be regularly available in large quantities. Usually many thousands are required to provide mg quantities of photoproteins.

Table 7.4 — Some luminous coelenterates

Class	Group common name	Species
PHYLUM: *CNIDARIA*		
Hydrozoa	Hydroids	*Obelia geniculata, Obelia dichotoma, Obelia longissima, Obelia bicuspidata, Obelia australiensis, Obelia commisuralis, Gonothyraea loveni, Clytia edwardsi, Clytia johnstoni*
	Medusae	*Obelia lucifera, Aequorea forskalea (Aequorea aequorea)? = Aequorea victoria, Aequorea vitrina, Tima bairdi, Mitrocamella polydiademata, Eutonnia indicans, Mitrocoma (Halistaura* or *Thaumantias) cellularia*
	Siphonophores	*Praya cymbriformis, Hippopodius gleba*
Scyphozoa	Medusae	*Phialidium hemisphaericum, Pelagia noctiluca, Atolla* spp., *Periphylla* spp.
Anthozoa	Sea pansies and pens	*Renilla reniformis, Renilla köllikeri, Ptilosarcus guerneyi*
PHYLUM: *CTENOPHORA*		
	Sea combs and jellies	*Mnemiopsis leideyi, Beroë albens, Beroë ovata, Pleurobrachia* spp.

(3) The protein should be easily extractable in high yield. Whilst the chemiluminescence reaction itself appears to be identical in all luminous coelenterates (Fig. 7.2), thanks to the elegant work of Cormier and his colleagues we know that the luminous anthozoans contain a conventional O_2 requiring luciferin–lucferase reaction. No photoprotein can be isolated containing the precharged prosthetic group, nor does Ca^{2+} directly affect the luciferase, but rather acts on the luciferin-binding protein (see Chapter 3). Most, but not all, ctenophores are luminous, but the yield of Ca^{2+}-activated photoprotein is low, a problem exacerbated by its inactivation by light. Luminous scyphozons such as *Pelagia* can sometimes be found in very large numbers many hundreds of metres below the surface of the sea. However, these are only available on specially organised research cruises, and are often difficult to extract cleanly because of the problem of removing large amounts of mesoglea (the jelly) in these species.

The conventional life cycle of a hydrozoan involves a fixed stage, the hydroid, and a free-floating stage, the jelly fish or medusa. The hydroid *Obelia geniculata* (Fig. 7.1), some 1–2 cm long, occurs in very large numbers in both the Northern and Southern hemispheres growing on a variety of brown seaweeds, particularly *Laminaria*. It is one of the commonest luminous organisms of the British coasts during the summer, particularly near Plymouth in Devon. It is found in rock pools, where it is often confused with its non-luminous relative *Laomedea* (formerly *Campanularia*) *flexuosa*, but grows best on seaweed never exposed at low water. It has therefore to be collected by divers working in 5–10 feet of water. The jelly fish which it releases is *Obelia lucifera* (Fig. 7.1C), only a millimetre or so in diameter, is found in large numbers in the plankton during the summer months, but it is difficult to obtain in large enough quantities for practical extraction. Both the hydroid and jelly fish stages are difficult to find outside the summer months of June to September. Two other smallish jelly fish which also occur in large numbers in coastal waters are *Phialidium* (*hydroid Clytia*) and *Mitrocoma* (formerly *Halistaura*).

Only two species have provided sufficient photoprotein for physiological experiments, the jelly fish *Aequorea forskala* (Fig. 7.1A), also known as *Aequorea aequorea*, whose hydroid has not been definitively identified but is probably *Campanulina paracuminata*, and the hydroid *Obelia geniculata* (Fig. 7.1B). In Puget Sound, USA, *Aequorea* is known as *Aequorea victoria*. It has been collected in large quantities in Friday Harbor, Washington, where one person can collect several thousand jelly fish in a few days. *Aequorea* has also been collected in slightly smaller quantities from the inland sea of Japan. It occurs spasmodically elsewhere throughout the world, including the European coasts. *Obelia* has been collected in sufficient quantities at Plymouth, England, and Port Philip Bay, Victoria, Australia, where it is known as *Obelia australiensis*. It is not clear whether these two species of *Obelia* are identical. Although the quantities potentially available are very large, it is much more difficult to obtain the sort of yield and purity of photoprotein obtained by Blinks and colleagues or Shimomura and Johnson with aequorin. For me, the real value of working with obelin was two-fold. Firstly, it provided a source of Ca^{2+}-activated photoprotein in the mid 1970s when it was difficult to obtain aequorin. Secondly, it provided the stimulus for my exploration of chemiluminescence as an analytical tool, together with energy transfer for detecting chemical events inside cells.

Before leaving the sources of Ca^{2+}-activated photoproteins, the recent application of cloning techniques by Cormier's group and a group in Japan (see Section 7.6 for references) to produce large amounts of apoaequorin in *E. coli*, now provides for the possibility of large, commercially available supplies of aequorin. It has been estimated that one bacterial culture, transformed with the expression plasmid containing apoaequorin cDNA, may yield aequorin equivalent to several years collection.

The apoprotein can, however, be converted to aequorin by incubation with the prosthetic group, synthesised artificially, in the presence of oxygen. This takes several hours to complete and requires an SH group protectant for optimal yields. The gene product has a modified amino acid structure from aequorin as normally extracted and isolated from *Aequorea* itself. There is an extra peptide on the N terminal end of the protein (Fig. 3.18b). The effects of this on the thermodynamic properties of the Ca^{2+}-indicator have yet to be fully determined. This strengthens the use of aequorin as the photoprotein of choice for physiological experiments. An acrylated derivative of aequorin has been produced by Shimomura which is more sensitive to Ca^{2+} than the natural aequorin preparations normally used and is consumed faster. Thus the cloned aequorin is likely also to have a different sensitivity to Ca^{2+}. Furthermore extracted aequorin is really a family of isoaequorins, perhaps six or more, varying in just a few amino acid residues from each other. The use of cloned aequorin is thus likely to lead to a better standardised reagent. I have established a method for cloning obelin which should now make this photoprotein also available to a larger number of workers.

7.2.5 Preparation of reagents

Two methods are now available for preparation of Ca^{2+}-activated photoproteins, one based on extraction and purification of native photoproteins from the organism, the other on reactivation of apophotoprotein prepared either by cloning the DNA or by heat inactivation of crude extracts of the organism. The advantage of the latter is that the apophotoprotein itself, like calmodulin, seems to be highly heat stable. Thus, contaminating proteins can be easily removed under conditions where the reacted prosthetic group (coelenteramide) drops off the protein in the absence of Ca^{2+}. Since the apoprotein is an oxygenase it can be transformed back into photoprotein by incubation with the 'luciferin' (coelenterazine) in the presence of O_2 (Fig. 7.3). The yield can be high since with pure luciferin, either native or a synthetic benzyl derivative, > 80% reactivation is possible. If the photoprotein is to be inactivated, by heat or Ca^{2+}, prior to purification of the apoprotein, the conditions require very careful examination. Irreversible loss can occur if the overall protein concentration is low, if the pH becomes very acid, if the ionic strength is low, or if $(NH_4)_2SO_4$ is present. An –SH group protectant is also required. The pH optimum for reactivation is broad, maximum approximately pH 7.5. For apoobelin preparation 0.5 M Na, 0.1% gelatin and 5 mM mercaptoethanol are required for high yields. Coelenterazine is very unstable, being susceptible to autoxidation and photooxidation, and must be stored dry and in the dark over periods of more than an hour or so. The affinity of apoobelin for coelenterazine is extemely high, as little as 0.1–1 μM being adequate for maximum reactivation. This enables a very sensitive assay for coelenterazine, down to 10–100 amol in 1–10 μl, to be established, to detect it in luminous and non-luminous organisms. The coelenterazine from one copepod leg can easily be detected. Coelenterazine can be purified, to remove contaminating coelenteramide and coelenteramine formed spontaneously, by thin layer and reverse phase chromatography on silica in methanol: ethyl acetate (5:1 v/v); R_F coelenterazine (yellow fluorescent at 366 nm) 0.84, coelenteramide (blue fluorescent) 0.61. Coelenterazine concentrations in stock solutions are estimated from its absorbance peak at 423 nm ($\varepsilon = 8900$ $M^{-1}cm^{-1}$), which disappears on autoxidation.

Although the purification procedures for native photoproteins contain variations between different groups of workers, they all consist of four main stages:

(1) Isolation of the luminous cells in as pure a form as possible.
(2) Crude extraction, and storage as a precipitate in ammonium sulphate.
(3) Purification by ammonium sulphate fractionation, gel filtration, and DEAE or QAE ion exchange chromatography.
(4) Desalting and freeze-drying for storage of the final preparation.

A vital component of all stages is to minimise Ca^{2+} contamination. Once triggered, the active photoprotein is gone. Sea water contains about 10–11 mM Ca^{2+}, but even double glass distilled water may contain up to 10 μM Ca^{2+}. Even Analar reagents are contaminated with sufficient Ca^{2+} to trigger photoproteins in the absence of a protecting Ca^{2+} chelator.

Free Ca^{2+} contamination can be minimised by the following procedures:

(1) Use plastic vessels and pipettes. Fresh, acid-washed, glassware can be used, but Ca^{2+} can leach off even this.
(2) Spectroscopically pure reagents containing very low, defined Ca^{2+} contamination.
(3) Ca^{2+} chelators. Ethylene diamine tetra acetic acid, EDTA (50 mM), is alright for purification, being cheaper than its more Ca^{2+}-specific relative EGTA. Remember that at pH 7, $>90\%$ of the chelator is in the $EDTA^{2-}$ form. Fully ionised EDTA is $EDTA^{4-}$, since it is a tetra basic acid. Addition of Ca^{2+} will therefore lead to displacement of $2H^+$ per mol of chelator. Will your buffer cope with 200 mM H^+?! BAPTA has a lower pKa, so is better at physiological pH.
(4) Removal of Ca^{2+} from H_2O or buffers by resins such as chelex 100 or Ca^{2+} ligand columns (e.g. parvalbumin).

Ag contamination from pH electrodes should also be avoided since it increases the Ca^{2+}-independent light signal from these photoproteins.

The most useful method of preparation of aequorin has been developed by Professor John Blinks and his co-workers. Rings at the edge of the *Aequorea* medusa bell, containing the luminous cells, are carefully cut off. The luminous cells and particles are dislodged by shaking. This removes the bulk of the jelly, the mesoglea. Unfortunately, this procedure is more difficult, though not impossible, to do in high yield with *Obelia*. Usually the whole hydroid is scraped off the seaweed on which it grows, and used for extraction.

Crude aequorin is extracted in hypotonic EDTA and precipitated with $(NH_4)_2SO_4$ (75% saturation is just sufficient). It can be stored at this point, ideally at $-70°C$. However, I have found that $(NH_4)_2SO_4$ precipitation of crude obelin causes large losses of activity. It is therefore better to load the initial homogenate on to DEAE cellulose at the appropriate ionic strength and elute with NaCl >300 mM. The crude protein is purified by $(NH_4)_2SO_4$ fractionation (25–75% cut), gel filtration on Sephadex G 50, ion exchange on QAE Sephadex A 50 (pH step + salt elution), Sephadex G 50, and DEAE Sephadex A 50. The final preparation is de-salted on Sephadex G 25 or Bio-gel P 50 using Ca^{2+} free buffer (and EDTA free). The final pure preparation is lyophilised in the presence of a small quantity of chelex 100. The freeze-drying step is notoriously irreproducible, and not just because of Ca^{2+} contamination. Obelin is stabilised by using gelatin during freeze-drying. Photoproteins continuously chemiluminescence even when freeze-dried, but active material can be stored for years at $-70°C$.

The overall yield of pure aequorin by the Blinks method is 20%, 4.5×10^{15} photon/mg protein (*ca*. 10 mg per 1000 jelly fish). Although pure from other contaminants, a puzzling feature of aequorin is that it is not one protein. A family of at least eight photoproteins can be isolated, several even from a single jelly fish. Some, but not all, are caused by proteolysis during isolation. These isoaequorins can be separated and purified by using phenyl Sepharose CL-4B and h.p.l.c. They have molecular weights ranging from 20 100 to 22 000 and isoelectric points from 4.2–4.9, saturating rate constants of $0.45–1.33 s^{-1}$ and chemiluminescence emission maxima from 460–472 nm. The evolutionary significance of this microheterogeneity is not known. However, the individual amino acid changes can be identifed using the cDNA sequence. This approach will provide unique sequence aequorin since each cDNA probe contains only one of the genes from the family.

To clone the aequorin genes, mRNA is extracted from the 'rings' containing the luminous cells. mRNA with poly A attached to it is purified on oligo dT cellulose. Double stranded cDNA is then synthesised from the *Aequorea* poly (A⁺) RNA. The cDNA library is then produced by incorporating the cDNA into a plasmid, e.g. pBR322. This can then be grown in *E. coli*, e.g. strain SK1592. Expression of the cDNA to form active apoprotein is low, unless the cDNA is removed from the cloned plasmid using restriction enzymes, and then inserted into an expression vector with a good RNA polymerase promoter. The use of expression vectors will then yield large amounts of apoaequorin produced by the *E. Coli* (e.g. 10^4–10^6 mols aequorin per cell). The cDNA library can then be screened for clones producing apoaequorin in one of three ways:

(1) Oligonucleotide probe, e.g. $3'$ACC AT^A_GTG^G_TTACC T^A_GGG $— 5'$ coding for a known amino acid sequence in aequorin (in this case residues 173–178, trp. tyr.thr.met.asp.pro).

(2) Antibody to apoaequorin if the cDNA is expressed in the *E. coli*.

(3) Reactivation using coelenterazine if the cDNA is expressed in the *E. coli*.

mRNA from *Obelia* can be detected by using a rabbit reticulocyte lysate translation assay, followed by reactivation of the photoprotein, using coelenterazine (Fig. 7.3). Total RNA is extracted and purified, using guanidinium thiocyanate, with centrifugation through CsCl, and phenol extraction. The mRNA is then purified on an oligo dT column, the cDNA prepared using reverse transcriptase, and incorporated into an expression vector. Clones producing apoobelin are detected by reactivation with coelenterazine, followed by measurement of Ca^{2+}-activated chemiluminescence. Our chemiluminometer can detect at least down to 10 tipomol (10^{-20} mol) of apoobelin.

7.2.6   Properties suitable as a Ca^{2+} indicator (Table 7.5)

Ca^{2+} is the only cation known which triggers photoprotein chemiluminescence in cells. Sr^{2+}, Ba^{2+}, and rare earths such as La^{3+} and Yb^{3+} do stimulate photoproteins but do not occur in sufficient concentrations under physiological conditions. Mg^{2+}, Na^+, K^+, and H^+ do not stimulate coelenterate photoproteins but do reduce their apparent affinity for Ca^{2+}. Mg^{2+} also reduces the Ca^{2+}-independent light signal. The protein has four domains (Fig. 3.18c); three are Ca^{2+} sites having the so-called E–F hand conformation, the other is where the luciferin (coelenterazine) and O_2 bind. Three Ca^{2+}'s must therefore bind to initiate maximum chemiluminescence, with an apparent affinity for each site of about $10\,\mu M$ at physiological concentrations of Mg^{2+}, H^+, K^+, Na^+. This places an upper detection limit of around $100\,\mu M$ free Ca^{2+}, above which changes in free Ca^{2+} cannot be accurately detected since the photoprotein is $>95\%$ saturated with Ca^{2+}. These Ca^{2+} indicators are therefore unsuitable for quantification of extracellular free Ca^{2+}, in the mM range. The lower limit for free Ca^{2+} is determined ultimately by the light from photoprotein with no

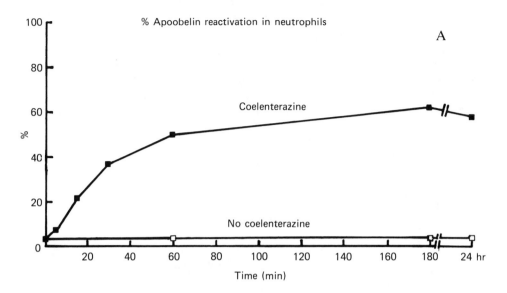

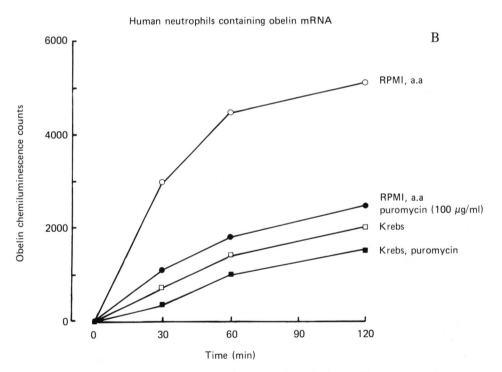

Fig. 7.3 — Reactivation of obelin in living cells. Apoobelin (A) or mRNA (B) from Obelia incorporated into human neutrophils by liposome-cell fusion. Puromycin or lack of amino acids inhibit formation of apoobelin from mRNA inside cells. $\frac{RPMI}{a.a.}$ = tissue culture medium + amino acids. Reactivation to form obelin by adding coelenterazine (1 μM). Reprinted from Campbell *et al.* (1988). *Biochem J.* **252**, 143–149. Copyright © 1988, The Biochemical Society, London.

Table 7.5 — Properties of aequorin and obelin

Property	Aequorin	Obelin
Mol.wt	Approx. 20 000	Approx. 20 000
Prosthetic group = coelenterazine	2-(p-hydroxybenzyl)-6-(p-hydroxy phenyl)3,7-dihydro-8benzyl-imidazo[1,2-a)pyrazin-3-one]	
λ_{max} (nm) of chemiluminescence	460	475
ϕ_{CL}	0.18 (15°C)	—
Number Ca^{2+} sites	3	3
Isoelectric point	4.2–4.9	ca. 4.5
Solubility	ca. 40 mg ml$^-$ (0.2 mM	—
Absorbance $\varepsilon_{1cm}^{1\%}$ 280 nm	27.0	—
460 nm	0.81	—
Activating cations	Ca^{2+}, Sr^{2+}, rare earths	
Inhibiting agents	Mg^{2+}, K$^+$, Na$^+$, H$^+$, alcohols, local anaesthetics	
Rate constant + Ca^{2+} (s^{-1}) rise	100	300
Rate constant + Ca^{2+} (s^{-1}) decay	1.4	4
$K_d^{Ca^{2+}}$ of each Ca^{2+} site (μM)	16	ca. 20

Ca^{2+} bound to it, assuming you have enough photoprotein present to detect a light signal. In practice, the Ca^{2+}-independent light signal from aequorin and obelin places a lower detection limit of about 10 nM for free Ca^{2+}. The Ca^{2+}-independent chemiluminescence of obelin preparations appears to be slightly greater than that of aequorin.

When saturated with Ca^{2+} (Ca^{2+} > 100 μM) aequorin chemiluminescence decays exponentially with a rate constant of 1–1.4 s^{-1} ($t_{\frac{1}{2}} \doteq 0.6$ s). The range of rate constants for the various isoaequorins is 0.95–1.33 s^{-1}. Thus different preparations may have different isoaequorin composition, and thus slightly different kinetics. The saturating rate constant for obelin is 4 s^{-1} ($t_{\frac{1}{2}} \doteq 0.18$ s).

In the absence of Ca^{2+} the decay constants for photoproteins are only 10^{-6} to 10^{-7} s^{-1}, depending on ionic conditions and temperature. This is equivalent to a half time of some 12 days at room temperature. Thus, within the range of 10 nM–100 μM free Ca^{2+} concentration can be quantified. Acetylation of the protein may increase the sensitivity of aequorin for Ca^{2+}. The speed of response of aequorin to a change in free Ca^{2+} produces little or no distortion of the Ca^{2+} signal, provided that this occurs no faster than 100 s^{-1} (i.e. rise time ca 6 ms). Obelin is two–three times faster than this. This means that although millisecond Ca^{2+} transients may be detectable by the Ca^{2+} indicators, they will introduce a time constant resulting in the recorded signal being a slight distortion of the true Ca^{2+} signal. The diffusion rate of aequorin in cells is about 10^{-7} cm^{2}s^{-1}, sufficiently fast not to distort free Ca^{2+} responses in most

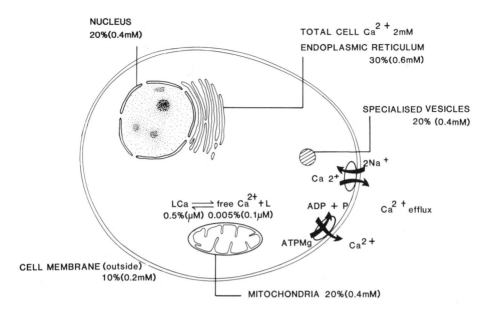

Fig. 7.4 — The calcium balance of a typical cell.

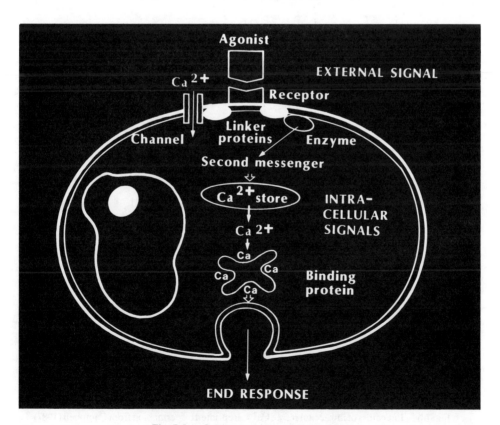

Fig. 7.5 — Calcium as an intracellular signal.

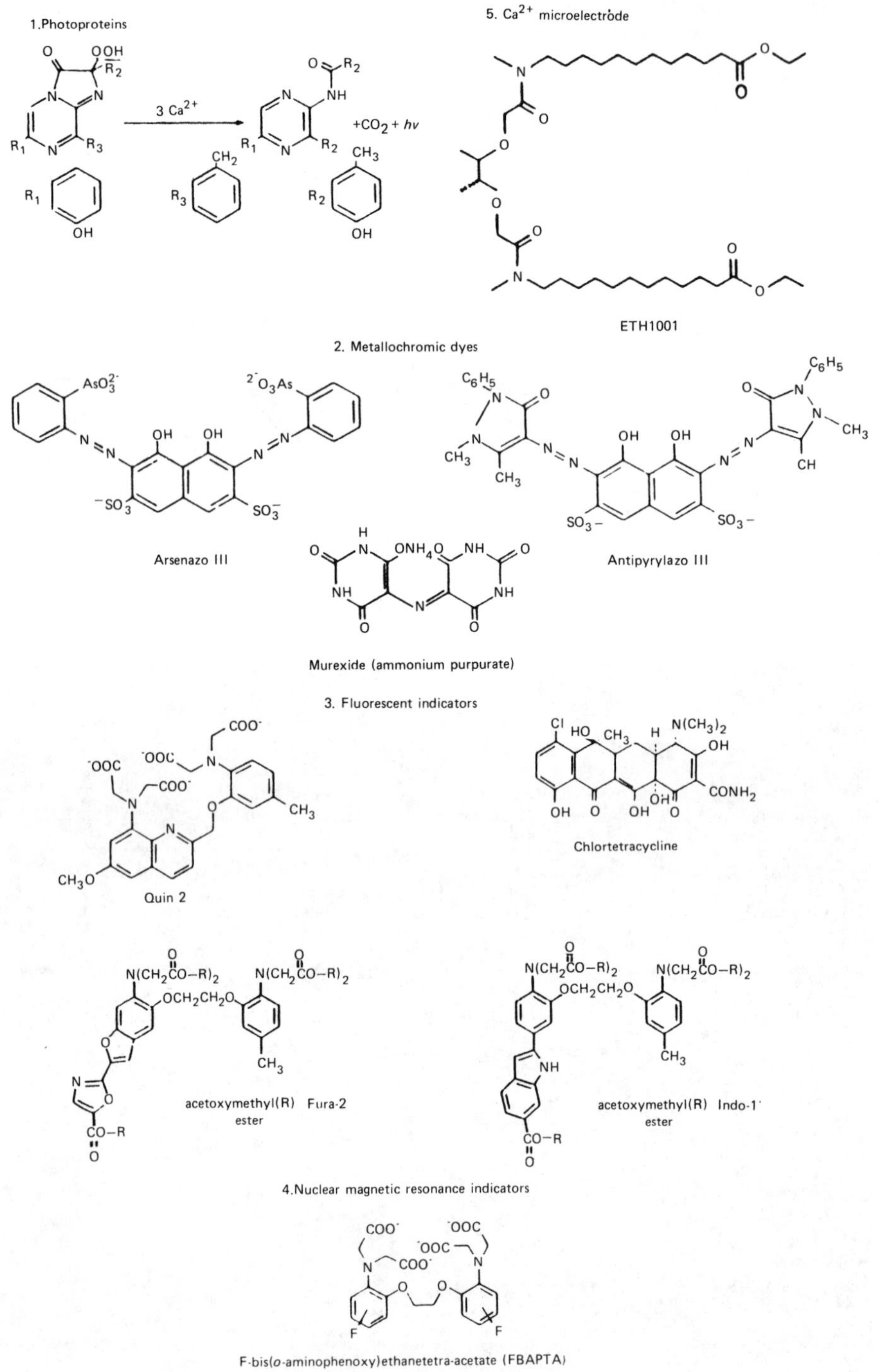

Fig. 7.6 — Indicators of intracellular free Ca^{2+}. Reprinted by permission from Campbell (1987) *Clin. Sci.* **72** 1–10. Copyright © The Biochemical Society, London.

cells. These photoproteins do not appear to bind significantly to internal structures within the cell, though this has never been investigated extensively.

Aequorin and obelin have been incorporated into a wide range of small and large cells from the plant, animal, and higher protist kingdoms (Tables 7.6 and 7.7) without serious impairment of cell function or cell Ca^{2+} balance. Axons still conduct action potentials, muscles still contract, cells still secrete or divide. Most experiments use intracellular concentrations of photoprotein in the nM–μM range. With a Ca^{2+} dissociation constant of approximately $20\,\mu$M and free Ca^{2+} in the range of 0.1–10$\,\mu$M, the amount of Ca^{2+} bound by the indicator is thus insignificant. Visualisation of the photoprotein signal using an image intensifier enables any free Ca^{2+} changes within the cell to be localised.

The photoproteins aequorin and obelin therefore satisfy most of the criteria for an ideal indicator of intracellular free Ca^{2+} (section 7.2.3). The two key questions now are: how does one get a protein of molecular weight 20 000 into the cytoplasm of a living cell, and how can the light signal from within be converted to establish what is the absolute free Ca^{2+} concentration once it is there?

Table 7.6 — Incorporation of Ca^{2+}-activated photoproteins into cells

Method	Example (approx. cell volume)	Estimate intracellular free Ca^{2+}	
		Resting	Stimulated (agent)
1. Micro-injection	Barnacle muscle (10 μl)	0.3 μM	5 μM (electrical)
	Single rat myocyte (5 pl)	0.25 μM	1–3 μM (spontaneous)
2. Cell permeation	Pigeon erythrocyte ghost (0.5 pl)	0.3 μM	10 μM (complement)
3. Cell fusion	(a) erythrocyte ghost cells rat neutrophil (0.5 pl)	0.3 μM	0.8 μM (chemotactic peptide)
	human neutrophil (0.5 pl)	0.1–0.3 μM	0.7 μM (chemotactic peptide)
	(b) mRNA from liposome human neutrophil (0.5 pl)	0.1–0.3 μM	0.7 μM (chemotactic peptide)
4. Micropinocytosis	rat macrophage (1pl)	0.1–0.3 μM	ca 1 μM (chemotactic peptide)

Table 7.7 — Some cells loaded with Ca^{2+}-activated photoproteins

Cell type†	Example
A. Monera	
Bacteria	*E. coli*
B. Plantae	
Alga	*Cara cariolensis, Fucus*
C. Animalia	
Protozoa	*Spirostomum, Amoeba*
Slime mould	*Physarum*
Muscle (invertebrate)	Barnacle (*Balanus*)
Muscle (vertebrate) smooth	Ferret
skeletal	Frog, human
heart	Rat, frog
Nerve (invertebrate)	Squid, snail, sea hare
Nerve (vertebrate)	Frog
Photoreceptor (invertebrate)	Horseshoe crab
Photoreceptor (vertebrate)	Rabbit
Secretory cells (veterbrate)	Rat pancreatic acini
Secretory cells (invertebrate)	Midge salivary gland
Eggs (invertebrate)	Starfish, sea urchin
Eggs (vertebrate)	Mouse, toad
Blood cells	Platelets, erythrocytes, neutrophils, macrophage
Tissue culture cells	LT3, monkey kidney, fibroblasts, oligodendrocyte

† These are only intended to illustrate the diversity of cell types and organisms which have been studied.
  For further examples see references in section 7.6.

7.2.7   Incorporation of photoproteins into cells
There are five methods now available to incorporate Ca^{2+}-activated photoproteins into cells (Table 7.6; Fig. 4.4):

(1) Micro-injection.
(2) Cell permeation, including reversible osmotic swelling and specialised media.
(3) Cell fusion, with an erythrocyte ghost or liposome containing the photoprotein, apoprotein or mRNA.
(4) Release from internalised, micropinocytotic vesicles.
(5) Intracellular synthesis from mRNA or cDNA.

Following the pioneering experiments of Ashley and Ridgway, and Blinks, in the late 1960s, aequorin has been injected into a variety of giant cells with internal volume of anything from $10\,\mu l$ to $10\,nl$. These include invertebrate giant nerves such as squid axons, single invertebrate and vertebrate striated muscle, eggs, slime mould, some protozoa, and amoebae. Pressure injection must be used; there is too much salt in the preparation to use iontophoresis successfully. The problem facing cell physiologists in the mid 1970s was how to incorporate photoproteins into mammalian cells, some with volumes less than a picolitre. Skilful filling and handling of micropipettes has enabled Cobbold and his co-workers to inject single oocytes, myocytes, hepatocytes and fibroblasts, using concentrations approaching saturating aequorin (0.1–0.3 mM). Other workers, including ourselves, have developed three alternative methods (Table 7.6; Fig. 4.4) for 'micro-injecting' populations of small cells, thereby enabling changes in free Ca^{2+} within the cells to be correlated with other physiological and chemical parameters in whole population. Exposure of the photoprotein to Ca^{2+} must be avoided at all times otherwise the active photoprotein will be lost.

The first of these methods depends on transient permeabilisation of the cell membrane. This can be achieved by using hypotonic solutions or specially designed permeabilisating solutions containing substances such as EGTA, ATP, and high K^+, or mechanically by disruption of the cell membrane using a glass rod with a rubber end, known as a 'rubber policeman', the so-called scrape-loading technique. Following permeabilisation and photoprotein entry, the cell membrane reseals. Erythrocytes are surprisingly easy to load with photoprotein, using reversible cell swelling to form 'ghosts'. The resulting ghosts contain 1–50% of the original haemoglobin, contain ATP, and are capable of maintaining an internal Ca^{2+} concentration of 0.1–0.3 μM in the presence of 1 mM outside. The Ca^{2+} pump within the cell membrane is therefore still functional. Other cells loaded this way include pancreatic acini, tissue culture cells, smooth muscle, and platelets. The scrape-loading method works with attached cells, for example fibroblasts grown in culture dishes. The electropermeation method developed by Zimmerman based on a very high potential lasting a few μs to punch a hole in the cell membrane has yet to be used. Two problems need to be solved if this method is to be used successfully. The holes are not big enough and are difficult to reseal.

The second method depends on the fusion of a membrane vesicle, for example an erythrocyte ghost or liposome containing photoprotein, with the cells being studied. Sendai virus is able to induce fusion betweeen two or more cells to produce a population of hybrids where more than 50% are 1:1 erythrocyte ghosts cell hybrids. These can be purified on a Percoll gradient. Other fusogens, such as polyethylene glycol, are either insufficiently efficient at introducing fusion, or result in too much loss of internal photoprotein through cell lysis or release of intracellular Ca^{2+}. Conventionally prepared liposomes result in negligible uptake of cytoplasmic photoprotein. Cells do take up plenty of photoprotein entrapped within these liposomes, but more than 99% of this is either within secondary lysosomes or still in liposomes attached to the cell's surface.

These liposomes normally consist of phosphatidyl choline (lecithin), and other lipids such as phosphatidyl serine, with or without cholesterol. However, Huang and Connor have developed an ingenious way of making liposomes release entrapped material into the cytosol, which we have used successfully to incorporate obelin into the cytosol of intact human neutrophils. The liposomes are composed of dioleoyl-phosphatidyl ethanolamine + palmitoyl-homocysteine (8:2). They fuse very efficiently with other membranes at acid pH (pH <6.5). By coupling a monoclonal antibody to palmitate the liposomes can be targetted on to the surface of a cell, though this is not essential. They are then endocytosed. Once the liposomes are in the acid environment of the lysosomes they fuse, internally, with the lysosomal membrane, releasing their contents into the cytosol. A simpler (dipalmitoyl phosphatidyl choline) liposome composition results in their being temperature-, rather than pH-sensitive. Addition of these liposomes, containing obelin, to non-phagocytic cells on ice causes them to bind to the plasma membrane, and then fuse on increasing the temperature to 41°C, thereby releasing obelin into the cytosol.

The pH sensitive liposomes have been applied successfully to the incorporation of apoobelin, and its mRNA, into human neutrophils (Fig. 7.3). Apoobelin is converted into obelin on incubation of the cells with coelenterazine. Translation of mRNA to apoobelin within the cells, could therefore be detected, and was inhibited both by removal of amino acids from the incubation medium or by the presence of puromycin, an inhibitor of protein synthesis.

The third method relies on the fact that all eukaryotic cells continuously take up fluid into micropinocytotic vesicles. When this is done in hypertonic medium containing photoprotein, and the cells then returned to isotonicity, the material within the micropinocytotic vesicles is released into cytoplasm without the cell membrane rupturing. The reason why the cell is able to withstand such extreme conditions is that invaginations within the cell membrane enable the cell to swell transiently and then recover from the osmotic jump. A tiny vesicle is unable to do this, so it 'pops', releasing its contents inside the cell. This method works best with cells having a relatively rapid micropinocytotic uptake. For example, macrophages can take up to 15 fl per cell in 15 minutes. Thus 10^7 cells in 100 μl will take up to 0.15% of the medium within this time.

These three methods have been used successfully to incorporate photoproteins into neutrophils, macrophages, smooth muscle, erythrocytes, kidney cells, platelets, and tissue culture cells. Before using these experimentally 'injected' cells three questions must be answered:

(1) How much photoprotein has been taken up into the cell?
(2) Is it all in the cytosol?
(3) Has the loading procedure damaged the viability or function of the cell so much that it is no longer possible to correlate intracellular changes in free Ca^{2+} with physiological events within the cell?

An individual cell will stand up to about 10% of its volume being increased rapidly by micro-injection. Consider a population of 10^6 cells, of individual volume 10 pl (volume = $4/3\pi r^3$, $r \doteqdot 13 \mu$m) suspended in a volume of 100 μl. 10% of the total cell volume is therefore 0.1 μl, or 0.1% of the whole suspension. When making erythrocyte ghosts uptake from the medium may be as much as 10%, but in most other cells it varies between 0.01 and 1% depending on the precise conditions. Low percentage entrapments necessitate high photoprotein concentrations, either in the microsyringe or in the medium surrounding the cell. Thus, with an aequorin concentration of 100 μg ml^{-1} (5 μM) in a volume of 100 μl our hypothetical cells taking up 0.1% of the fluid would take up a total of 10 ng of aequorin per cell. This is equivalent to an intracellular aequorin concentration of about 5×10^{-8} M, insufficient to cause any Ca^{2+} buffering problems. How easy it is to detect a signal from such low photoprotein concentrations we shall see in the next section.

The total incorporation of photoprotein inside the cells can be measured by lysing them either in a hypotonic medium or with a detergent such as triton (the commercial preparation Nonidet NP40 is best), both containing a saturating Ca^{2+} concentration (mM). All of the photoprotein will then luminesce within a few seconds (Fig. 7.7). But is it all in the cytoplasm? Has any of it been entrapped within other compartments in the cell? It is vital to find this out, otherwise both the interpretation of changes in light emission from the cell, together with the conversion to free Ca^{2+}, will be invalid.

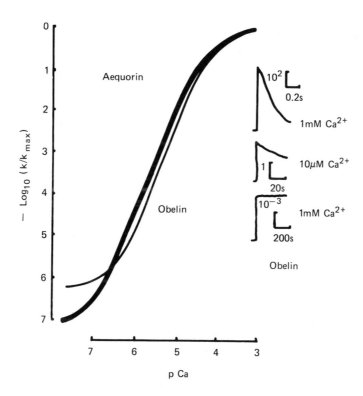

Fig. 7.7 — Relationship between free Ca^{2+} and the rate constant of photoprotein chemiluminescence.

It is claimed that digitonin only lyses the plasma membrane, and not organelle membranes. This has yet to be fully tested using photoproteins.

Four methods have been used in attempts to solve the problem:

(1) Careful disruption of the cell, followed by subcellular fractionation. This is not satisfactory as secondary lysosomes are easily broken during homogenisation.

(2) Correlation with a fluorescent indicator, e.g. fluorescein-labelled albumin (Fig. 4.4). Visualising the fluorescence shows that it originates from the cytoplasmic compartment. The main difficulty with this criterion is that tiny vesicles ($<0.1\,\mu$m diameter) may not be detectable. Entrapment of a nuclear fluorescent stain, such as ethidium bromide or propidium iodide, only leads to fluorecent nuclei in those cells where the dye has been released from liposomes into the cytosol.

(3) Specific stimulation of cytosolic photoproteins. This cannot be definitely achieved by using Ca^{2+} ionophores, but can be by using complement. Activation of complement by an antibody bound to a cell surface antigen results in the formation, within a few seconds, of the membrane attack complex. This contains five proteins C5b, C6, C7, C8, and up to 12–18 C9's. Binding of C9 lets Ca^{2+} into cell, an increase in cytosolic free Ca^{2+} in the μM range being detectable within 5s (Fig. 7.8). Thus, a rapid luminescence from cytoplasmic photoprotein occurs, but not from photoprotein within intracellular vesicles, nor from within vesicles such as liposomes, attached to the cell membrane and which are unable to activate the complement pathway.

(4) Translation of mRNA to apoprotein (Fig. 7.3) will occur only in the cytosol, to any significant degree.

Criteria of cell viability include a cell count, trypan blue exclusion, lack of ethidium bromide or propidium iodide nuclear fluorescence (when added to the extracellular fluid as opposed to within liposomes), electron microscopy, measurement of ATP content, and release of the cytosolic enzyme lactate dehydrogenase into the external medium. But perhaps one of the best indications of the intactness of a cell is its ability to maintain a large electrochemical gradient of Ca^{2+} across its cell membrane. Thus, cells with sub-μM cytoplasmic free Ca^{2+} will be the only ones with any active photoprotein left, after a few minutes. The response of the cells to stimuli is aiso vital. Even the severe treatment of fusion of neutrophils with erythrocyte ghosts leaves hybrids with an apparently intact cytoskeleton and a response 50–80% that of the original cells when stimulated.

7.2.8 How is a light signal converted to free Ca^{2+}?

No one would contend the importance of being quantitative in science, but conversion of the photoprotein light signal from inside cells to the absolute free Ca^{2+} concentration is important not only because we want to define the range of free Ca^{2+} inside the cell, but also because a qualitative description of the photoprotein chemiluminescence may lead to two misinterpretations. One, that a particular fold-increase in light signal is equivalent to the same fold-increase in free Ca^{2+}, and the other, that a decrease in light intensity with time is always indicative that the free Ca^{2+} is also decreasing.

Because photoproteins bind three Ca^{2+}'s, the mathematical relationship between light intensity and free Ca^{2+} contains a term $(Ca^{2+})^3$. Thus a doubling in the free Ca^{2+} can in theory lead to an eight-fold change in light intensity. In other words, the photoprotein light signal is not linearly related to the free Ca^{2+}. This is actually an advantage under some circumstances since it amplifies the Ca^{2+} signal. The second possible misinterpretation can arise because, like any chemiluminescent substrate, the prosthetic group on the photoprotein only reacts once and is consumed during the reaction. As little as 0.001% of the intracellular photoprotein may be consumed during a muscle twitch lasting less than a second. However, prolonged elevation, e.g. several minutes, of the free Ca^{2+} in the μM range results in a significant loss of photoprotein during the experiment. The result is a decrease in light intensity even when the free Ca^{2+} is not dropping (Fig. 7.7). The free Ca^{2+} may even be increasing!

Both of these potential artefacts can be avoided by converting the light trace to absolute free Ca^{2+} concentration.

The reaction of photoprotein (P) with Ca^{2+} can be considered in four strages, only the first of which is reversible:

Stage 1 Ca^{2+} binding

$$P + 3Ca^{2+} \underset{}{\overset{K}{\rightleftharpoons}} PCa_3 \tag{7.1}$$

(N.B. The three Ca^{2+}'s are unlikely to bind simultaneously; see equation (7.11)).

Stage 2   The rate limiting step

$$PCa_3 \xrightarrow{k_{RL}} \text{intermediate X} \tag{7.2}$$

Stage 3   The generation of the excited state

$$X \xrightarrow{k_{ex}} Y^* \tag{7.3}$$

Stage 4   Photon emission

$$Y^* \xrightarrow{k_{em}} Y + h\nu \tag{7.4}$$

What then are the relative rates of these four stages?

Ca^{2+} binding is thought to be limited principally by diffusion, and is therefore fast and not thought to be rate limiting. Though this is not entirely proven at low free Ca^{2+} concentrations the 'on' rate for Ca^{2+} binding (k binding) will therefore be of the order of $500 s^{-1}$ ($t_{\frac{1}{2}} \doteqdot 1.4 ms$). The rate constant of the rate limiting step (eqn (7.2)) is easily measured by addition of a saturating Ca^{2+} concentration, e.g. $1-10 mM$, when all the photoprotein will form into PCa_3 within a few milliseconds, and then decay exponentially as for any first order reaction. Therefore, at saturating free Ca^{2+}:

$$\frac{dPCa_3}{dt} = -k_{RL} (PCa_3) = -k_{sat} (PCa_3) \tag{7.5}$$

since 'all' the photoprotein is (PCa_3).

This is the classic exponential equation which on integration becomes:

$$(PCa_3)_t = (PCa_3)_0 \exp(-k_{sat}t) \tag{7.6}$$

$(P Ca_3)_0 = Po = $ *amount* of photoprotein at time 0. Since we measure the reaction by photon emission, the quantum yield (ϕ_{CL}) of the reaction must be incorporated into the equation.

$$\phi_{CL} = \frac{\text{number of photons emitted}}{\text{number of photoprotein molecules reacted}} \qquad (7.7)$$

ϕ_{CL} for aequorin $\doteq 0.18$ at 15°C.

The reaction is followed by measuring light intensity, which as we saw in Chapter 4 (section 4.7), unlike absorbance in spectrophotometry, has no 'concentration' term in it. $\phi_{CL}.Po$ is therefore the total photon yield of photoprotein, and is independent of its molar concentration. Equation (7.6) therefore becomes:

$$I = dh\nu/dt = k_{sat}\phi_{CL}.Po.\exp(-k_{sat}t) \qquad (7.8)$$

since $dh\nu/dt = -\phi_{CL}(dPCa_3/dt)$.

No correction for the efficiency of the apparatus is necessary in this equation since it is incorporated both into I and Po and therefore cancels out.

$$k_{sat} = \ln 2/t_{\frac{1}{2}} = 0.693/t_{\frac{1}{2}} \qquad (7.9)$$

$$k_{sat} \text{ quoted for aequorin} = 1 \text{ to } 1.4\,s^{-1} \, (t_{\frac{1}{2}} = 0.50\text{--}0.69\,s)$$

$$\text{for obelin} = 4\,s^{-1} \, (t_{\frac{1}{2}} = 0.18\,s)$$

k_{sat} is relatively insensitive to temperature over the range of 0–40°C. Thus, for obelin at mM Ca^{2+} more than 98% of the photoprotein will be consumed within 1 sec (*ca.* 6 half times) or 3–5 sec for aequorin.

The rate constant (k_{ex}) for stage 3 (eqn (7.3)) is fast ($100\,s^{-1}$, $t_{\frac{1}{2}} = 6.9\,ms$ for aequorin, and $250\text{--}280\,s^{-1}$, $t_{\frac{1}{2}}$ 2.5–2.8 ms for obelin). k_{ex} can be measured in a stopped-flow apparatus by measuring the up phase in light emission on addition of Ca^{2+}. Alternatively k_{ex} can be measured from the decay constant on rapid removal of Ca^{2+}. This will leave X, but $P Ca_3$ will revert rapidly to $P + Ca^{2+}$. The final stage (eqn 7.4) is extremely fast, 1–10 ns for a normal decay of a single excited state to ground state.

What then happens at non-saturating free Ca^{2+}?

$$dh\nu/dt \text{ will always} = k_{sat}.\phi_{CL}. \, (P Ca_3)_t \qquad (7.10)$$

$(P Ca_3)_t$ can be calculated from the binding constants for Ca^{2+} in stage 1, but this requires an assumption about the mechanism of Ca^{2+} binding to form $P Ca_3$.

Two kinetic models have been proposed. One (model 1, eqn (7.11)) by Moisescu, Ashley, and Campbell assumes an ordered binding of the three Ca^{2+}'s. The other (model 2, eqn (7.14)) proposed by Blinks and co-workers assumes that the photoprotein exists in two principal forms, tight (T) with low chemiluminescence (Ca^{2+}-independent) and relaxed (R). Only R can bind Ca^{2+}, binding being random. Our model (model 1) assumes the Ca^{2+}-independent signal to be negligible relative to the Ca^{2+}-activated signal. This appears to be true inside cells where micro-injection of Ca^{2+} chelators reduces light emission by more than 90%. Model 1 takes account of inhibition by other cations (Mg^{2+}, K^+, Na^+, H^+) of Ca^{2+} activated photoprotein chemiluminescence by assuming a 'competition' with the Ca^{2+} binding sites, direct or indirect, which reduces the apparent affinity for Ca^{2+}. Model 2 takes cation inhibition into account by assuming that any individual cation pulls the equilibrium over to the tight (T) form. Both models assume three Ca^{2+} sites, confirmed by 3 EF hand domains in the amino acid sequence (see Chapter 3; Fig. 3.18c), and that PCa_3 is the only significantly chemiluminescent form, though this latter point has never been proven.

Model 1

$$P + Ca^{2+} \underset{}{\overset{K_1}{\rightleftharpoons}} PCa + Ca^{2+} \underset{}{\overset{K_2}{\rightleftharpoons}} PCa_s + Ca^{2+} \underset{}{\overset{K_3}{\rightleftharpoons}} PCa_3 \overset{k_{sat}}{\longrightarrow} hv \quad (7.11)$$

where K's = equilibrium dissociation constants, at any time t.
When the free Ca^{2+} is constant:

$$(PCa_3)_t = P_t K_1 K_2 K_3 (Ca)^3 / (1 + K_1 \, Ca + K_1 K_2 (Ca)^2 + K_1 K_2 K_3 (Ca)^3)$$

and

$$- dP/dt = (dhv/dt)/\phi_{CL} = k_{app} P_t \quad (7.12)$$

Thus from eqns 7.10 and 7.12:

$$d(hv/dt) = k_{app}\phi_{CL}Po \exp(-k_{app}t) = I \quad (7.13)$$

where $k_{app} = k_{sat}K_1 K_2 K_3 \, (Ca)^3 / (1 + K_1 Ca + K_1 K_2 (Ca)^2 + K_1 K_2 K_3 (Ca)^3)$.

Model 2

$$T \overset{K_{TR}}{\rightleftharpoons} R + 3Ca \overset{K_{RCa}}{\rightleftharpoons} RCa_3 \overset{k_{sat}}{\longrightarrow} hv \quad (7.14)$$

$$k_{app} = k_{sat} \, [(K_{RCa}Ca)/(K_{RCa} + K_{RCa}.K_{TR} + Ca)]^3 \ .$$

The only real difference between the two models, kinetically, is whether Ca^{2+} binding is assumed to be ordered or random.

In principle, if the binding constants for Ca^{2+} in the presence of physiological concentrations of K^+, Na^+, Mg^{2+} and H^+, and k_{sat}, are known, then the free Ca^{2+} inside the cell could be estimated simply by measuring k_{app} and fitting it in either to (7.13) or (7.14). In practice, however, it is simpler to construct a standard curve of k_{app} versus free Ca^{2+} *in vitro* using standard solutions prepared for Ca^{2+} buffers, e.g. EGTA or BAPTA in media mimicking conditions inside the cell (Fig. 7.7).

In vitro, where free Ca^{2+} is constant the decay of photoprotein light emission should be a pure exponential. Thus from (7.13)

$$\log_e I = -k_{app}t + \log_e (k_{app}\phi_{CL}Po) \quad (7.15)$$

the slope of a plot of $\log_e I$ versus t gives k_{app}. An alternative, but less precise, method is to measure peak height on addition of Ca^{2+} to the photoprotein:

$$\text{peak height} = k_{app}\phi_{Cl}Po = L \quad (7.16)$$

$$\text{at saturating } Ca^{2+}, \ L = L_{max} = k_{sat} \, \phi_{CL}Po \quad (7.17)$$

therefore

$$L/L_{max} = k_{app}/k_{sat} \quad (7.18)$$

Three precautions are therefore necessary for accurate measurement of k_{app}, for calibration outside of the cell:

(1) The decay should be a pure exponential down to $< 1\%$ peak height. $\log_e I$ versus t is more precise than using L/L_{max} to estimate k_{app}.
(2) Ca^{2+} should be added to the photoprotein and not vice versa. This mimics the *in vivo* situation and is particularly important at high temperatures (e.g. 37°C), where low protein concentrations lead to spurious k_{app} if photoprotein is added to the Ca^{2+} buffer. This artefact is the result of inactivation of photoprotein via a low quantum yield reaction.
(3) Protein carrier is necessary at 37°C.

In the cell Ca^{2+} may be changing. As long as the amount of photoprotein remains constant (i.e. consumption $< 1\%$) throughout the Ca^{2+} transient, then k_{app} can be estimated at any time from the intensity (I_t) at this time, and a knowledge of the total photoprotein light present ($P_T.\phi_{CL}$). This latter parameter can be estimated at the need of the experiment by exposing all the active photoprotein using Nonidet P40 to lyse rapidly the cell membrane. This exposes the photoprotein to saturating Ca^{2+} within a few hundred milliseconds. Thus,

$$k_{app} = \text{fractional photoprotein consumption rate at time } t$$

$$= I_t/P_T.\phi_{CL} \tag{7.19}$$

However, in many cell experiments, particularly where the free Ca^{2+} rises to μM for more than a few seconds (e.g. Fig. 7.8), photoprotein consumption is significant, and thus both Ca^{2+} and P will be changing. Yet at any time t equation (7.13) can be rewritten

$$I_t = k_{app}\phi_{CL}P_t \tag{7.20}$$

$P_t = $ *active photoprotein remaining at time t*. P_t can be estimated at the end of the experiment. If the end of the experiment occurs at time t_2 and P_t is required at time t_1, then

$$P_t = \sum_{t_1}^{t_2} I + Ca^{2+} \text{ exposable photoprotein left at } t_2 \text{ (i.e. with detergent)} \tag{7.21}$$

$\sum_{t_2}^{t_2} I$ can be estimated most easily if the chemiluminometer is connected to a computer (see Chapter 2), which can convert chemiluminescence traces to free Ca^{2+} automatically.

The Ca^{2+}-independent light signal of 10^{-6}–$10^{-7} s^{-1}$ places a detection limit on free Ca^{2+} of about 10 nM (Fig. 7.7). Consumption rates may rise to $10^{-2} s^{-1}$ at 1–10 μM free Ca^{2+}. The practical range for free Ca^{2+} using photoproteins therefore depends on four main parameters:

(1) Ca^{2+}-independent light signal.
(2) Amount of photoprotein incorporated into the cell.
(3) The consumption rate.
(4) Timing of the Ca^{2+} transient.

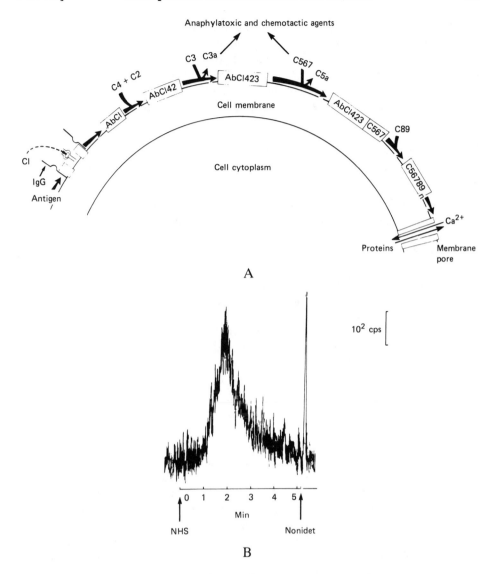

Fig. 7.8 — Complement activation as a criterion of cytoplasmic photoprotein. A. Complement activation. B. Free Ca^{2+} NHS = normal human serum as source of complement. Cells = human neutrophils, containing obelin and coated with antibody to activate complement. Nonidet = detergent to expose remaining obelin to Ca^{2+}.

Free Ca^{2+} cannot be measured once the photoprotein signal becomes insignificantly different from either the Ca^{2+}-independent signal or the machine background.

Consider a cell 10 pl volume containing 10 ng aequorin = 0.5 pmol: aequorin produces 4.4×10^{15} photons mg^{-1}, therefore if the cell contains 4.4×10^{10} photon potential = Po. If the apparatus efficiency = 1% and machine background = 10 cps total photoprotein counts = 4.4×10^{8}. If the resting cell has a free $Ca^{2+} = 10^{-7}$ M with $k_{app} = 10^{-6} s^{-1}$ then $I = 440$ cps, easily detectable over machine background.

The practical Ca^{2+} range for photoproteins is from $10\,nM$–$100\,\mu M$. Calibrations inside giant cells can be checked by injection of EGTA and Ca/EGTA buffers.

Photoproteins have been used to quantify free Ca^{2+} in cells from more than 40 species representing at least seven different phyla, including plant cells, unicellular protozoa, invertebrate and vertebrate cells (Table 7.7). What then has been learnt about the biology of these cells as a result of measuring free Ca^{2+} in them with photoproteins?

7.2.9 Discoveries arising from the use of photoproteins

There have been at least seven major consequences of using photoproteins to measure free Ca^{2+} in cells:

(1) *Identification of primary stimuli dependent on an increase in free Ca^{2+}* (Fig. 7.9)
    In muscle excited electrically or by neurotransmitter, in eggs fertilised by sperm, in oocyte meiosis provoked by hormones, in neutrophils activated to produce reactive oxygen metabolites by chemotactic peptide, the primary stimulus provokes an increase in free Ca^{2+} prior to the physiological event. Prevention of the photoprotein light signal by Ca^{2+} chelation prevents the cell response. Events mediated, for example, by protein kinase C, not apparently requiring an increase in cytoplasmic free Ca^{2+}, have been identified. These include phagocytosis of unopsonised and opsonised particles by phagocytes. Important information has also come from the correlation between the time course of intracellular events and the Ca^{2+} transient. In several cell types, including some muscles, cell activation remains even when the free Ca^{2+} has virtually returned to resting levels, implying either another mechanism not dependent on Ca^{2+} is operating, or that there has been a change in the senstivity of the intracellular mechanisms to Ca^{2+}.

(2) *Four ranges of free Ca^{2+} in cells*
    Measurement of absolute free Ca^{2+} has established that resting cells have free Ca^{2+} of the order of 30–$300\,nM$, stimulated cells 1–$5\,\mu M$, reversibly injured cells 5–$50\,\mu M$, and dying cells $> 100\,\mu M$ free Ca^{2+}. These latter two high free Ca^{2+}'s can only be detected by photoproteins for a few seconds because of the rapid consumption of the Ca^{2+} indicator. In some cells, e.g. fertilised eggs, the free Ca^{2+} after activation may be as high as 10–$30\,\mu M$, at least transiently.

    Correlation of the absolute free Ca^{2+} provoked by different naturally occurring stimuli, or artificial substances such as ionophores, has highlighted the fact that an increase in Ca^{2+} may be insufficient in itself to provoke the maximal cell response. This is well illustrated in neutrophil and platelet activation.

(3) *Identification of secondary regulators working via Ca^{2+}* (Fig. 7.10).
    Secondary regulators could, in principle, act in one of three ways, by altering (a) the time of onset or magnitude of the Ca^{2+} transient, (b) the mechanisms by which Ca^{2+} acts, or (c) independently of Ca^{2+}, thereby potentiating or inhibiting the effect of the primary stimulus. Photoproteins have established that a major action of catecholamines in heart muscle is to increase the magnitude of the Ca^{2+} transient, and to speed up removal of Ca^{2+} on recovery after each action potential. CO_2 in barnacle muscle can also modify cytosolic free Ca^{2+} by affecting the sarcoplasmic reticulum. Adenosine abolishes the free Ca^{2+} transient induced by N formyl met leu phe in neutrophils.

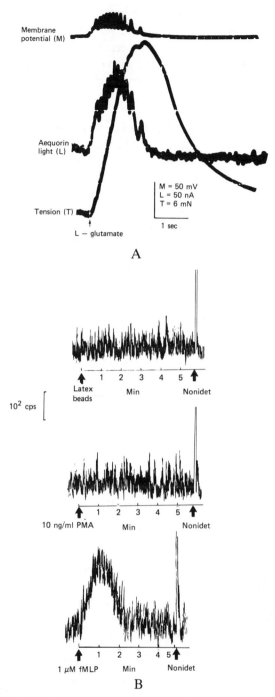

Fig. 7.9 — Primary stimuli and intracellular free Ca^{2+}. A. Barnacle muscle containing aequorin, stimulated by L glutamate from Ashley & Campbell (1978) *Biochim. Biophys. Acta* **512,** 429–435 with permission of Elsevier Science Publications Bv. B. Human neutrophils containing obelin and stimulated by chemotactic peptide (Nfmlp), note: particles or phorbol ester (PMA) cause no rise in cytosolic free Ca^{2+}.

(4) *Localisation of Ca^{2+} transients* (Fig. 7.11)

Visualisation of aequorin signals from salivary glands have established that only when Ca^{2+} hits the gap junction is it electrically sealed. Similarly in fish and sea urchin eggs fertilised by sperm, a Ca^{2+} wave has been visualised beginning at the sperm attachment site which precedes several chemical events within the cell, including exocytotic granule release.

(5) *Identification of free Ca^{2+} source*

There are two questions, firstly is the Ca^{2+} coming from outside or inside the cell, and secondly how is it released into the cytosol.

In neutrophils the intracellular free Ca^{2+} increase induced by chemotactic stimuli is reduced but not abolished by removal of external Ca^{2+}. Thus as internal release and a receptor-activated Ca^{2+} channel are required.

Whether the endoplasmic reticulum or mitochondria are the intracellular source of Ca^{2+} for cell activation has long been a source of controversy. The use of mitochondrial uncouplers, together with visualisation of aequorin signals in sea urchin eggs, has established that mitochondria are not the source of Ca^{2+} during cell activation but rather act as 'sinks' buffering the cytoplasm and restricting the Ca^{2+} transient to localised areas within the cell. Photoproteins have played an important role in identifying the endoplasmic reticulum as the major source of Ca^{2+} for cell activation, and mitochondria as an internal buffer localising the free Ca^{2+} change.

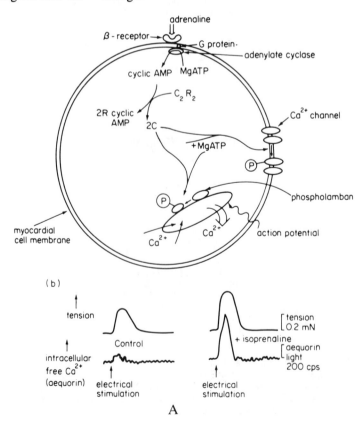

A

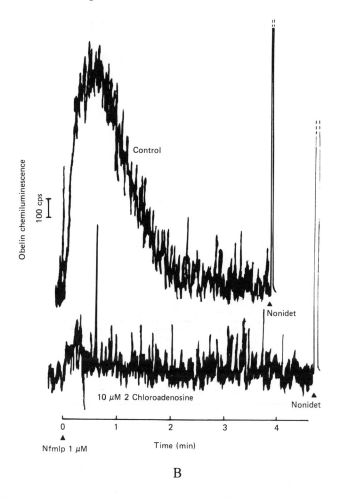

B

Fig. 7.10 — Secondary regulators and intracellular Ca^{2+}. A. β adrenergic agonists (e.g. isoprenaline) increase the Ca^{2+} transient (from Campbell (1983) with permission). B. 2Cl adenosine inhibits or abolishes the Ca^{2+} transient in neutrophils induced by chemotactic peptide (Nfmlp). Campbell *et al. Biochem. J.* **252**, 143–139 (1988) with permission.

(6) *The role of Ca^{2+} in cell injury*

Is an elevation of cytoplasmic free Ca^{2+} a cause or a consequence of cell injury? By using aequorin in isolated myocytes, injury to the cell has been detected, caused by anoxia or metabolic poisoning, prior to detectable increases in intracellular free Ca^{2+}. In contrast, the earliest detectable intracellular event following C9 binding to the membrane attack complex of complement is a large increase in cytoplasmic free Ca^{2+} (Fig. 7.8). This rise in intracellular Ca^{2+} can activate cell responses before, and even in the absence, of cell lysis. This has led to the discovery of a mechanism by which nucleated cells, long known to be more resistant to killing by complement compared to 'aged' erythrocytes, can protect themselves against attack. They do this by removal of the lethal membrane attack complexes through vesiculation. This may have important implications in the mechansims underlying several immune-based diseases including rheumatoid arthritis and multiple sclerosis. Furthermore, there may be a similar protection mechanism when other membrane pore-formers, e.g. T cell perforins, bacterial toxins and viruses, attack cells.

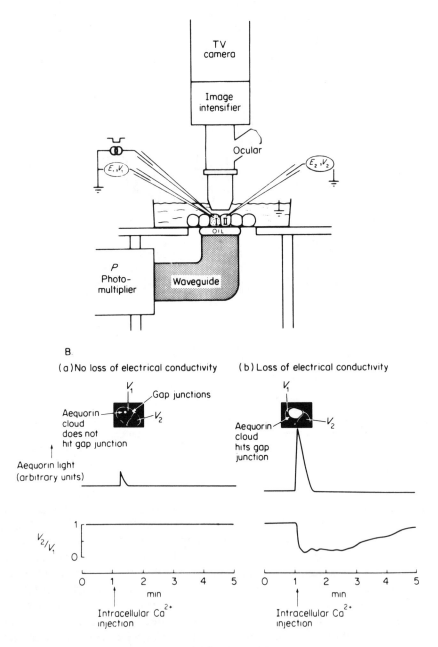

Fig. 7.11 — Localisation of cytosolic free Ca^{2+} with aequorin from, Rose & Loewestein (1976) *J. Membr. Biol.* **28**, 97–119, with permission of Springer Verlag.

(7) *Threshold phenomena* (see Chapter 10)

Examples of cell activation or cell injury occurring through quantal events in individual cells could be explained in several ways. A critical level of stimulus could be required to provoke an increase in free Ca^{2+} within the cell, or to distribute the Ca^{2+} far enough from its site of release to reach its site of action. Alternatively, the concerted action of groups of molecules or structures within the cell could be responsible for 'all or none' cellular events. In starfish oocytes a Ca^{2+} threshold induced by the hormone 1 methyl adenine is necessary to provoke meiosis. Similar Ca^{2+} thresholds appear to occur in fertilised eggs, certain invertebrate nerves, the sealing of gap junctions, and in individual hepatocytes activated by hormones such as vasopressin. Cobbold and colleagues have observed oscillations in cytosolic free Ca^{2+} in individual hepatocytes which may allow a cell to remain activated without depleting itself of Ca^{2+}. Similar oscillations in free Ca^{2+} also occur in other cells particularly at sub-optimal concentrations of stimulus and their frequency may play a key role in determining whether the end-response of an individual cell crosses the activated threshold.

It might be argued that several of these important observations might have been made while using one of the other types of indicators, dyes or micro-electrodes (Fig. 7.6). Some have been repeated, using metallochronic indicators such as arsenazo III. Ca^2 microelectrodes have been used to detect Ca^{2+} release from leaky cells incubated with the key intracellular metabolite inositol trisphosphate, which releases Ca^{2+} from internal stores. Yet until the invention by Tsien in the late 1970s of a family of fluorescent indicators, derivatives of tetracarboxylic acids, nearly all the new information about intracellular free Ca^{2+} up to 1980 came from using photoproteins. Are these new fluors, then, real competitors of the Ca^{2+}-activated proteins? Are photoproteins now redundant?

7.2.10 Photoproteins versus fluors

Photoproteins and Tsien's fluors (Fig. 7.6) compare favourably on five counts:

(1) Specificity — they all show high specificity for Ca^{2+}.
(2) Quantification — both provide quantification of intracellular free Ca^{2+}.
(3) Oxygen independence — both are independent of oxygen concentration; important in stimulated cells which may markedly reduce the cytoplasmic O_2.
(4) Free Ca^{2+} distribution — both can be used to study free Ca^{2+} distribution in cells when using image intensification.
(5) Problems of endogenous signals — both can have problems in the endogenous signals. On the one hand reactive oxygen metabolites in phagocytes, platelets, and some eggs give rise to ultraweak chemiluminescence; on the other hand endogenous fluors such as NADH or NADPH have fluorescence emissions close to that of quin 2. Concentrations of quin 2 > *ca.* $100\,\mu M$ are necessary to swamp this endogenous fluorescence. The more fluorescent fura-2 and indo-1, with shifts in excitation or emission spectra respectively, were designed to circumvent this problem since the free Ca^{2+} is measured from a ratio of fluorescent emissions.

However, seven important differences have been defined:

(1) *Loss of indicator*

Photoproteins are comsumed by Ca^{2+}, fluors are not. However, photobleaching or chemical bleaching (e.g. by oxygen metabolites) causes problems when studying cells for long periods using fluors.

(2) *Linearity of calibration*

Fluorescence increases linearly with free Ca^{2+}, whereas photoprotein chemiluminescence shows a power law relationship (7.12). Whilst the latter complicates slightly quantification of photoprotein signals, it has the advantage that small, localised Ca^{2+} transients, are more easily detectable by photoproteins owing to this amplification factor.

(3) *Signal detection*

Chemiluminescence is usually a more sensitive method of analysis than fluorescence. This is true so long as the chemical background for chemiluminescence is low. Fluorescence, in the absence of time resolution, is always limited by endogenous fluorescence, including that of the solvent itself. However, individual cells can be examined in the fluorescence activated cell sorter, since the cell becomes a microcuvette. A relatively simple chemiluminometer can detect tipomol (10^{-21}) quantities of photoprotein. In cells the photoprotein concentration is usually $\ll 1 \mu M$, whereas quin 2 must be at a concentration of 0.1–1 mM to be easily detectable in cells, a concentration which buffers much of the intracellular Ca^{2+}. Even fura-2 and indo-1 have to be used at 10–50 μM. Photoproteins can be used at concentrations which do not significantly alter the Ca^{2+} balance of the cell, and are detectable by much simpler apparatus. Fura-2 needs, for example, a dual wavelength excitation fluorimeter.

(4) Ca^{2+} *range*

Photoproteins have a wider working range for free Ca^{2+} than the fluors. This is because of the sensitivity of chemiluminescence, and its amplification by Ca^{2+}, enabling very small fractional saturations of photoprotein to be detectable. In contrast, the lower apparent affinity of photoproteins for Ca^{2+}, compared with the fluors, extends the upper limit of the range for free Ca^{2+}. Fura-2, for example, is not much use above 2–3 μ on Ca^{2+}, the interesting range pathologically.

(5) *Interference by other ions*

Whilst in theory photoprotein signals could be affected by changes in intracellular K^+, Na^+, Mg^{2+}, or H^+, this does not seem to have led to any major misinterpretations. The fluors are less susceptible to interference by H^+ or Mg^{2+}, but may be affected by transition metals such as Zn^{2+}, common in all eukaryotic cells.

(6) *Incorporation into cells*

The elegant method of incorporating fluors into small cells is based on the use of membrane permeant acetoxy methyl esters which hydrolyse back to the impermeant tetracarboxylate indicator, a reaction catalysed by a cytoplasmic esterase. Incorporation of fluors into small cells is thus much simpler than with photoproteins. However, there are two snags. Firstly, the ester hydrolysis generates toxic formaldehyde, secondly some cells do not contain the esterase, whilst others have it within granules resulting in incorporation of indicator into a non-cytosolic compartment. This can lead to misinterpretation of results, e.g. in neutrophils where fusion of granules containing extracellular Ca^{2+} may occur, or secretion.

(7) *Release of indicator*

Fluors, being of much lower molecular weight than photoproteins, are more readily released by activated or injured cells. With injurious agents like the membrane attack complex of complement small molecule release may occur within seconds, making interpretation of fluorescent signals very difficult. Furthermore we have shown that the thresholds for such permeability changes occur at different times in individual cells. The fluors also leak out of healthy cells during an experiment, and this must be corrected for. It has always been assumed that photoproteins do not. This can only be proven by visualisation or by measuring spent photoprotein at the end of the experiment. A further problem with the fluors, such as fura-2, is leakage into intracellular organelles, because of the lipophilic nature of the fluor. This may be a particular problem in long term studies.

We have compared fura-2 with obelin by loading them into the same population of human neutrophils. Fortunately there was a good measure of agreement between the two Ca^{2+} indicators. Both identified stimuli dependent (N formyl met leu phe) and independent (latex particles and phorbol ester) of a rise in cytosolic free Ca^{2+}. The time of onset of any rises in cytosolic free Ca^{2+} were rapid (within < 1 second) from a resting free Ca^{2+} of 10–100 nM to a maximum of 0.5–1 μM. Extracellular EGTA (i.e. no external Ca^{2+}) decreased the Ca^{2+} transient measured by either indicator by 75%, but did not abolish it. Two important discrepancies were, firstly the time courses were different. Fura-2 showed a faster rate of rise, time to peak and rate of decay than obelin. Secondly 2 Cl adenosine could abolish the obelin signal, but only reduce that from fura-2. Furthermore, obelin loading in cells via liposomes appeared to increase the cytosolic free Ca^{2+}, compared with fura-2 loaded cells without liposome treatment. At least some of these discrepancies could be accounted for by heterogeneity of loading of the two Ca^{2+} indicators into particular cells. Some discrepancies between quin2, inferior to fura-2 anyway, and aequorin in platelets loaded with both indicators have also been observed, by Salzman's group.

In summary the Ca^{2+}-activated photoproteins aequorin and obelin still have much to offer in the measurement of intracellular free Ca^{2+}. They cover a wide range of free Ca^{2+}, can now be incorporated into any mammalian or other eukaryotic cell, do not significantly disturb the Ca^{2+} balance, and are not prone to leakage from healthy or injured cells, unless these are dead. The major problems of consumption and single cell analysis can now be circumvented by reactivating apoobelin within the cell, and by increasing the incorporation of photoprotein into individual small cells using amplification from mRNA or cDNA (Fig. 7.3). The beautiful Ca^{2+} localisation studies initiated by Tsien using fura-2 have been made possible not only by the ease of incorporation of fura-2 into large numbers of small cells but also by the power of the software, enabling pseudo-colour images of absolute intracellular free Ca^{2+} to be displayed and photographed temporally. Such software can equally be applied, with a little modification, to photoprotein signals, provided enough protein is incorporated into each cell.

From this discussion it can be seen that whilst the fluors have made free Ca^{2+} measurements in cells apparently available to large numbers of scientists, accurate interpretation of their signals from cells is not always straightforward. The two main groups of indicators, fluor and photoproteins, are thus likely to continue as complementary techniques. Cloning of the photoprotein genes should lead to equal availability relative to synthetic fluors, and will enable some of the superior features of obelin over aequorin to be realised. Furthermore, chemiluminometers are inherently simpler in construction than fluorometers, particularly if dual excitation is required as is the case with fura-2. The development of continuously turning over, synthetic chemiluminescent Ca^{2+}-indicators offers an exciting challenge to the synthetic organic chemist.

7.3 MEASUREMENT OF TRANSITION METAL IONS AND THEIR COMPLEXES

7.3.1 Which chemiluminescent systems are sensitive to transition metals?

In 1916 Harvey discovered that peroxidases, or 'oxidase' as he called them, from plants such as the potato, were capable of catalysing the chemiluminescence of pyrogallol in the presence of H_2O_2. He noted 'that an extract dilution of less than 1:10,000 could be detected'. Twelve years later, Albrecht discovered the chemiluminescence of luminol, when it was oxidised by hydrogen peroxide or hypochlorite. He also found that this chemiluminescence too could be provoked or enhanced by a number of transition metal-containing compounds, including potassium ferricyanide, ammoniacal solutions of copper salts, manganese dioxide, colloidal platinum, as well as blood and potato peroxidase.

As early as 1937 Specht suggested that this phenomenon could be used as a sensitive assay for blood spots in forensic medicine. The list of 'oxidants' and 'catalysts' of the luminol reaction is now quite extensive (Table 7.8). In principle, any could be detected sensitively through their ability to stimulate light output. Assays can also be established for enhancing or inhibiting components which modify the quantum yield or rate of the luminol reaction. A number of bioluminescent systems have also been found which can be stimulated by transition metals such as Fe^{2+} or Cu^{2+}, or when complexed to protein in haems, e.g. *Chaetopterus* (Fig. 7.12).

Many hundreds of papers have now been published showing both standard curves, and values for tissue extracts, using several of these chemiluminescent detectors for transition metals. But are they of any practical use to the biologist trying to tackle real problems in the research or clinical laboratory? The answer to this lies in the response to three supplementary questions. Firstly, how specific are these assays? Secondly, how sensitive are they? Thirdly, why do we want to measure the substance affecting the chemiluminescence anyway?

7.3.2 Some assay characteristics

Bright chemiluminescence from acridinium salts such as lucigenin and acridinium esters can be detected simply by addition of H_2O_2, particularly under alkaline conditions which favours the formation of HO_2^-, their most active oxidant. The chemiluminescence can be enhanced by addition of some iron and other transition metal compounds, including peroxidases. However, the enhancement is considera-

Table 7.8 — The triggering of luminol chemiluminescence

Condition	Specific agent
1. Oxidant — no other substances required	O_2 (DMSO), O_3, H_2O_2, O_2^-, OCl^-, KO_2, $K_2S_2O_8$, NIO_4, NOCl OO_4, BO_4^-
2. Oxidant + O_2	Fe^{2+}, Co^{2+}, t-butoxide (DMSO), Br_2, I_2, Fe $(CN)_6^{3-}$, MnO_4^-
3. 'Catalyst' + H_2O_2/OH^- (pH 10–13)	
(a) free transition metal cations	Fe^{2+}, Fe^{3+}, Co^{2+}, Ni^{2+}, Cu^{2+}, Hg^{2+}, Cr^{3+}, Mn^{4+}, V^{2+}, V^{4+}
(b) transition metal anion complexes	$Fe(CN)_6^{3-}$, MnO_4^-, $AuCl_4^-$, $SbCl_6^-$
(c) transition metal cation complexes	UO_2^+, $Cu(NH_3)_6^{2+}$, VO_2^+
(d) free porphyrins	coproporphyrin
(e) porphyrin Fe^{3+}	protohaem, haematin, deutero-haematin and other modified haematins
(f) haemoproteins	haemoglobin, myoglobin, catalase, peroxidases (myelo-, lacto-, ovo-, micro-,), cytochrome c
(g) other Fe^{3+} proteins	ferritin
(h) corrins	vitamin B_{12}
4. Enhancers	benzothiazoles (e.g. firefly luciferin on peroxidase), iodophenol at pH 8
5. Inhibitors	Ce^{4+}, Vn, Th, Zr^{2+}, cysteine, tris, CN^-, bicine, albumin, serum, EDTA

Buffers may be 0.1 M, OH^- alone, barbitone, borate, or bi-carbonate.

bly less than with phthalazinediones such as luminol. The much greater signal:noise achieved with these latter chemiluminescent compounds results in lower detection limits for transition metals, and their complexes. Luminol, with its greater quantum yield compared with isoluminol, has been the most popular choice of chemiluminescent compound for the assay of such compounds. However, other chemiluminescent reactions should not be ignored since some involve metal ions which do not catalyse the luminol reaction. For example, Pb^{2+} can catalyse acridinium salt chemiluminescence.

The principle of these assays is, therefore, that addition of an analyte, containing a transition metal, to a chemiluminescent substrate and an oxidant, enhances the light emission. This may be detected in one of three ways: (a) an increase in peak height, (b) an increase in the rate of reaction and thus the decay rate constant, and (c) an increase in integrated light output. The transition metal may be free, complexed to inorganic or organic ligands, or be within a protein (Table 7.8). To establish a working assay the following decisions must be taken:

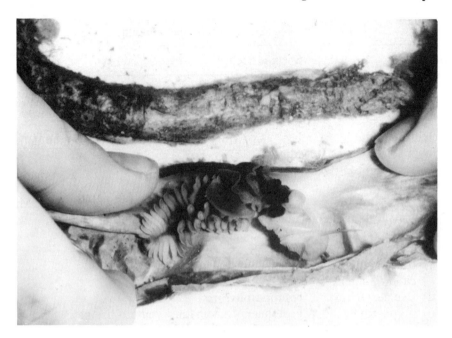

Fig. 7.12 — The luminous polychalate Chaetopterus. It secretes a blue luminous slime.

(1) Choice of chemiluminescent substrate.
(2) Choice of oxidant.
(3) Assay conditions, particularly buffer, pH, temperature, substrate and oxidant concentrations, assay volume, and when to add the analyte.
(4) Parameter to relate to analyte concentration, i.e. peak height, $k(t_{\frac{1}{2}})$, or total light output in a defined time or when the reaction is complete. The last is only valid if there is an increase in overall quantum yield.
(5) Light detector, particularly photographic film versus a photo-multiplier tube or photodiode.

Several transition metal cations or complexes, e.g. $Fe(CN)_6^{3-}$, can enhance luminol chemiluminescence with O_2 as the oxidant (Table 7.8). However, the most sensitive assays, down to p–f mol $(10^{-12}–10^{-15})$ utilise H_2O_2. The buffer chosen depends not simply on its pKa being compatible with the pH of the assay, but on whether it generates a high chemical blank in the absence of the analyte. Barbitone is thus much preferred to borate on these grounds. The higher the chemical blank without addition of the transition metal the poorer the detection limit of the assay. Luminol reacts fastest and with greatest apparent quantum yield at alkaline pH (>9, optimum 11–13). This pH optimum may not be compatible with maximum activity of the analyte. Biological 'catalysts' may prefer a lower pH, and luminol will work at least down to pH 5.

The reaction is usually initiated either by addition of the oxidant, H_2O_2, or sometimes the analyte. The reaction may be $>95\%$ complete within a few seconds or may continue for many minutes. Peak height or light output in a defined interval are the most popular parameters to relate to analyte concentration. Since some analytes also affect the total light output when the reaction is complete, i.e. they affect the apparent quantum yield, this can be monitored instead, but is convenient only if the reaction is over within a few seconds. Most assays use $1-100 \mu M$ luminol, but since a few assays use mM concentrations of the chemiluminescent substrate the light emitted on addition of the analyte may be bright enough to see with the naked eye, and to record on photographic film. However convenient this may be there is no convincing evidence that such assays have better precision, signal:noise, or detection limits than those producing much lower light intensities detected by a photomultiplier tube–quite the reverse. In fact, once a chemiluminescent assay becomes visible one has to question whether alternative methods using precipitation, absorbance, fluorescence, or chromatography are perfectly adequate.

Luminol has been used to establish assays for individual transition metals such as Fe, Cd, Co, Cu, Vn, Mn, Cr, and Ni down to one part per 10^9, sufficiently sensitive for analysis of these cations as pollutants or food contaminants. Complexes such as the Co^{2+} corrin in vitamin B_{12}, Fe^{2+} or Fe^{3+} in ferritin, haem, and haem-proteins such as haemoglobin, cytochromes, catalase, and peroxidase can be readily detected in range $1 nM-1 \mu M$ ($0.1 pmol-0.1 nmol$ in $0.1 ml$ sample), and in several cases down to $1 pM$ ($0.1 fmol$ in $0.1 ml$). One assay for haematin has been produced with a detection limit of $0.1 pM$. These assays approach the sensitivity necessary for quantification of haem proteins in single entharyotic cells (Table 4.6), and for use as an indicator of bacterial biomass (down to $1-100$ bacteria ml^{-1}), though this appears no better than ATP when using the firefly system (see Chapter 5).

Impressive as some of these sensitivities may sound, these assays have yet to find wide application, mainly because of five problems associated with them:

(1) *Lack of specificity*

   Luminol chemiluminescence is not provoked by any of the alkali metal or alkaline earth cations. However, any metal capable of existing in more than one valency state appears capable of 'catalysing' the reaction. Thus, an assay using luminol will not be able to select out one particular substance in a sample containing a mixture of transition metals, or their complexes, unless the analyte required is first isolated, for example by ion exchange or high pressure liquid chromatography. Several tricks such as the use of O_2 instead of H_2O_2 as oxidant, differences in kinetics, or the rate of binding to inhibitory ligands such as EDTA have been used in an attempt to circumvent this problem of lack of specificity. None has proved ideal, though Cr(III) has been determined in water, leaves, liver, and blood by using the fact that the Cr–EDTA complex is slow to form relative to many other transition metals. CrIII can thus be left in free solution, able to catalyse the luminol reaction, whereas the others bound to EDTA are relatively impotent.

(2) *Signal:noise*

   H_2O_2 causes a chemical blank with many organic contaminants in biological samples. Furthermore, the luminol:H_2O_2 blank can be very high, many 10000's photons, restricting the detection limit of the assay.

(3) *Interference from the sample*

There are three principal sources of interference from biological samples:

(a) other sources of chemiluminescence apart from luminol, e.g. substances which react with H_2O_2;

(b) inhibition or enhancement of the chemical blank;

(c) enhancement or quenching of the luminol reaction by oxygen intermediate scavenging, or excited state quenching or enhancement.

The last of these is a major problem with crude biological extracts, particularly containing proteins. Albumin in serum samples causes significant quenching of the luminol reaction at $>0.2\%$ (w/v) i.e. 5% serum, when using haem proteins $+ H_2O_2$ as the oxidant, but may enhance if other oxidants are used.

(4) *Non-linearity*

Linear assays for cations, complexes, or even organisms can, in principle, be established. However, many of the chemiluminescence traces are non-exponential because the reaction is not first order with respect to the chemiluminescent substrate. This problem may arise, for example, if either the oxidant or the 'catalyst' are consumed significantly during the reaction. The result is that many standard curves are non-linear with respect to analyte concentration. This makes precision difficult to assess. 'Hooking' at high doses can also occur, producing a bell-shaped standard curve. Shimomura has suggested that the photoprotein from the marine worm *Chaetopterus* (Fig. 7.12) might provide a useful, linear, assay for Fe^{2+}. Since this may be important in the generation of highly toxic hydroxyl radical (.OH) this is an important cation to measure. But this assay has yet to be used with a biological sample.

(5) *The chemistry of the reaction*

In spite of the fact that the luminol reaction has been known for more than 50 years to be chemiluminescent, the precise chemical mechanism generating the excited state is still not fully understood (see Chapter 1). Any assay utilising such an inadequately characterised reaction is unsatisfactory and likely to be prone to artefacts. There are three main sources of controversy:

(a) *light versus dark reactions*

There appear to be at least two routes from luminol to aminophthalate, one with high quantum yield, the other with a low one. Different analytes, as well as different assay conditions (e.g. oxidant of pH), may alter the balance between them, thereby changing the apparent quantum yield.

(b) *the precise oxidant is often unclear*

The luminol reaction can be provoked by many oxygen-containing oxidising agents, as well as several which do not contain oxygen themselves (Table 7.8). These latter substances use the O_2, present at about $240\,\mu M$, in air-saturated water at room temperature. An additional complication is the generation of reactive oxygen metabolites by metal cations, for example:

$$M^{n+} + O_2 \rightarrow M^{(n+1)} + O_2^- \tag{7.22}$$

or

$$M^{n+} + H_2O_2 = M^{(n+1)} + .OH + OH^- \tag{7.23}$$

Thus, in some assays, more than one oxidant may be involved in the reaction, leading to confusing kinetics reflected in the chemiluminescence trace.

(c) *the mechanism of 'catalysis', if it is true catalysis, is unclear*

A particularly puzzling aspect is the ability of more than one valency state of an individual cation to trigger the luminol reaction. For example, both Fe^{2+} and Fe^{3+} will provoke luminol chemiluminescence, though Fe^{3+} is much less potent. Fe^{2+} complexed to EDTA is very poor, yet the haem in haemoglobin or myoglobin, where iron is also in the Fe^{2+} state, is a reasonable 'catalyst' enabling fmol quantities to be determined. Yet the best known catalysts are peroxidases, particularly microperoxidase, a proteolytic fragment of 11 amino acids from cytochrome c, where the iron is in the Fe^{3+} state! Catalase with Fe^{3+} is also a good catalyst. Furthermore, when many haem proteins are exposed to OH^-, as they are in many luminol assays, the haem dissociates to form protoporphyrin IX Fe II Cl, the iron rapidly autoxidising to the Fe III state (protohaemin). OH^- is then exchanged for Cl^- to form haematin, a good catalyst of the luminol reaction.

A proposed intermediate is $Fe^{3+} - O$, ($Fe^3 + H_2O_2 \rightarrow Fe^{3+}\text{-}O + H_2O$) or the perhydroxy radical $Fe^{2+}\text{-}OOH$ formed by the followed reaction:

$$\text{porphyrin } Fe^{3+} + HO_2^- \rightarrow \text{porphyrin } Fe^{3+}\text{-}OOH^- \rightarrow \text{porphyrin } Fe^{2+}\text{-}OOH$$

↑	|	|
(Haematin) OH	OH	OH
I	II	III

(7.24)

(the primary oxidant for
luminol)

The naturally-occurring porphyrin in haem contains vinyl groups at positions 2 and 4. Changing these groups has established that there is a correlation between the electronegativity and hence the ability to generate III in (7.24)) and the potency in provoking a high luminol flash:background ratio. The order of potency is:

deutrohaemin (H) > mesohaemin(ethyl) > protohaemin(vinyl) >
> diacetyl > haemin(Ac) > nitrated haemin (OH,NO_2)

(7.25)

Groups at positions two and four on the porphyrin ring are in brackets.

(EDTA Fe)$^{2-}$ is a poor 'catalyst' as the complex is non-planar, compared with the planar haem complex.

Do these observation explain the different potencies of the various haems found in haemoglobin and cytochromes, in catalase, or in peroxidase? Not entirely, since there are at least four factors which can influence the potency of porphyrin–metal complexes in triggering luminol chemiluminescence:

(1) The readiness with which the haem dissociates from the respective protein exerts a considerable influence on its catalytic potential.

(2) Its ability to autoxidise rapidly to the Fe^{3+} state.

(3) Its tendency to aggregate and thus reduce its potency.

(4) The electronegativity of the porphyrin ring and thus its readiness to form the perhydroxy radical intermediate (7.24) and (7.25)).

The possibility exists that the reason why microperoxidase is such a good catalyst of luminol chemiluminescence is that the haem can remain attached to a peptide, in a soluble, Fe^{3+}, unaggregated form which readily forms Fe^{2+}–OOH with H_2O_2 and alkali.
$$\underset{\overset{|}{OH}}{}$$

Further problems requiring explanation are the autooxidation of vinyl groups on natural haem and the ability of porphyrins, without metals complexed to them, to provoke luminol chemiluminescence by generating reactive oxygen metabolites, for example following irradiation.

(d) *The mechanism by which inhibitors and enhancers (Table 7.8) act is unclear*
For example, 6 hydroxy benzothiazoles, including firefly luciferin and iodophenol enhance up to 1000-fold luminol chemiluminescence provoked by horseradish peroxidase at a pH around 8 (Fig. 7.13 see section 7.3.3), whilst reducing the chemical blank in the absence of the peroxidase. The aminophthalate anion is still the emitter, as shown by the chemiluminescence spectrum. So, such an effect cannot, it appears, be explained by energy transfer. Perhaps the thiazoles, like micro-peroxidase, optimise conditions for formation of the OH–Fe^{2+}–OOH as the primary oxidant.

The analytical reason for wanting to know the cause of differences in potencies between luminol 'catalysts' is that the better the 'catalyst' the more sensitive the assay for the analyte.

7.3.3 Peroxidases

Peroxidases are enzymes capable of oxidising substances (AH_2), with H_2O_2 as the oxidant, to form $A + 2H_2O$. They appear to be relatively non-specific compared to many other enzymes. The familiar concept of K_m may not be relevant, as the substrate may not actually bind to the enzyme in the conventional way.

Peroxidases are found in all eukaryotes, both plant and animal, often occurring in discrete organelles called peroxisomes. Some cells have evolved a special use for them. In neutrophils a myeloperoxidase, and in eosinophils a similar peroxidase, play a key role in the killing of phagocytosed microorganisms. In milk, lactoperoxidase has a similar biocidal function. Certain invertebrate eggs, for example in the sea urchin and starfish, release an ovoperoxidase on fertilisation, which appears to prevent attachment of a second sperm by being spermicidal, and which may oxidise proteins on the outer membrane of the egg. In the thyroid gland a peroxidase is responsible for iodination of tyrosines on thyroglobulin as a precursor in T_4 and T_3 synthesis. Peroxidases may also be released extracellularly in disease, e.g. myocardial infarction and rheumatoid arthritis.

An important property of all peroxidases is their ability to form oxyanions, which may lead to singlet oxygen formation via the reactions:

$$H_2O_2 + X^- \rightarrow OX^- + H_2O \tag{7.26}$$

$$OX^- + H_2O_2 \rightarrow {}^1O_2(\Delta g) + H_2O + X^- . \tag{7.27}$$

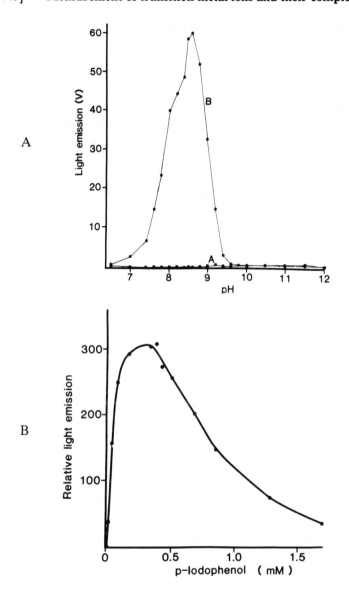

Fig. 7.13 — Enhanced peroxidase mediated luminol chemiluminescence. A. pH optimum A unenhanced. B. enhanced by 336 μM p-iodophenol, 59 μM luminol, 1.87 mM H_2O_2, 0.1 M tris or glycine buffer. B. Effect of p-iodophenol pH 8.5, 56 μM luminol, 1.87 mM H_2O_2. From Thorpe *et al.* (1985) *Anal Biochem.* **145** 96 with permission of Academic Press.

They exert some selectivity for X^-. Myeloperoxidase forms OCl^-, thyroid peroxidase OI^-, and lactoperoxidase $OSCN^-$. They also catalyse halogenation of tyrosines and histidines in proteins.

Peroxidases are very high turnover enzymes, and having the ability to produce products which are coloured or precipitate, have been widely used as labels in immunoassay and in immunocytochemistry, and more recently in recombinant DNA technology. Thus, their biological, pathological, and analytical potential can be realised only if there are good assays for them. Assays with greater sensitivity than spectrophotometric or precipitation assays which can be visualised would greatly enhance their potential.

Peroxidase can generate chemiluminescence in two ways: one by producing $^1O_2$, the other by catalysing the oxidation of a chemiluminescent substrate. With the exception of glutathione peroxidase, which contains selenium at its active centre, peroxidases are haemoproteins with the iron in the Fe^{3+} state. They are good catalysts of the luminol reaction, though their relative potency in terms of kinetics and quantum yield varies considerably, as do their pH optima. The pH optimum for horseradish peroxidase with a colorimetric substance is 5–6, whereas with luminol it is 8–9. The active intermediate, the primary oxidant, is presumably the same as that for haem (7.24), i.e.

$Fe^{3+}-O$

or

$Fe^{2+}-OOH$
|
OH

When using luminol, horseradish peroxidase can be detected down to 1–10 fmol. The rate of reaction is much slower than with the artificial peroxidase, microperoxidase used widely as the catalyst for isoluminol derivatives in immunoassays (see Chapter 8).

The ratio of luminol, or another phthalazine dione, to peroxidase varies greatly in different assays. In an immunoassay using the isoluminol derivative ABEI (see Chapter 8), microperoxidase concentrations will be much greater than the chemiluminescent compound. The concentration of microperoxidase may be as high as $1–10\,\mu M$, whereas the antibody or antigen labelled with ABEI is likely to be at a concentration well below $1\,\mu M$, perhaps even nM. The kinetics are fast, 95% of the chemiluminescence being complete within 10 s. The question arises whether microperoxidase is behaving as a true catalyst, or acting as the oxidant. In contrast, when peroxidase itself is being assayed the chemiluminescent substrate is in excess. In an immunoassay using horseradish peroxidase as label this may be at a concentration of nM–μM, or less, whereas luminol will be present at a concentration of some 0.1–1 mM. Under these conditions chemiluminescence is detected for many minutes, since the rate of reaction, and thus loss, of luminol is relatively slow. The oxidant concentrations, usually H_2O_2, is always much greater than either luminol or peroxidase, mM at least and sometimes as high as 100 mM. The signal:noise is different for each peroxidase–phthalazine dione pair, and varies with the oxidant used, e.g. H_2O_2 versus perborate.

H_2O_2 + luminol or other phthalazinediones produce a significant chemical blank in the absence of peroxidase, reducing the sensitivity of peroxidase detection. The purity of the water used to prepare solutions is a critical factor in reducing this chemical blank. In 1983 Whitehead and collaborators made the unexpected discovery that addition of firefly luciferin and other 6-hydroxy benzothiazole derivatives, not only enhanced peroxidase catalysed luminol chemiluminescence several hundredfold, but also reduced the chemical blank from luminol + H_2O_2 alone. The 6-hydroxy group is essential. 6-methyl, 6-ethyl, 2-amino or 2-hydroxy derivatives or benzothiozole, itself are inactive. A screen of other phenols has shown that *p*-iodophenol and *p*-phenylphenol are even better 'enhancers' than firefly luciferin.

p-phenylphenol > 1,6 dibromonapth-2-ol > p-iodophenol >
> p-chlorophenol > 6 hydroxybenzothiazole

p-iodophenol increases the luminol signal, and that from several other cyclic diacyl hydrazides, up to 2500-fold. pH and indicator concentration are critical. Significant enhancement is only observed between pH 7 and 9.5 (optimum *ca.* pH 8.6; Fig. 7.13A), and within a particular concentration range of enhancer. The result is a prolonged chemiluminescence with a very slow decay. Enhancement does not appear to alter the chemiluminescence spectrum (δ_{max} 425 nm). The use of enhancers, such as firefly luciferin and *p*-iodophenol, which increase the light yield and reduce the chemical blank, improves the detection limit for peroxidase, enabling simple, nonisotopic immunoassays and DNA assay to be established (see Chapter 8).

A variety of reaction pathways have been proposed to explain the peroxidase-provoked chemiluminescence of luminol (Fig. 7.15) both in aprotic and protic solvents. There is general agreement that the excited state emitter is the dianion of 3-aminophthalate. However proposed intermediates leading to its formation include luminol radicals, a diazoquinone, an endoperoxide and a dioxethanone. At alkaline pH one route is for the luminol anion to react with a peroxidase-oxidant product (e.g. $Fe^{3+} - O$ or $OH - Fe^{2+} - OOH$ Eqn. 7.24) to form the luminol radical anion ($L^{-\cdot}$). Some of this then reacts with O_2 to form O_2^- and a diazoquinone. O_2^- reacts with other, unreacted luminol radicals ($L^{-\cdot}$) to form the endoperoxide $L\overset{O^{2-}}{\underset{O}{\diagdown\diagup}}$), which then breaks down to form the 3-aminophthalate dianion in an excited state. An alternative route to 3-aminophthalate, but with low chemiexcitation quantum yield, is via a hydroxy diazo derivative of the monocarboxylate anion (Fig. 7.15). Thus the enhancers of peroxidase-provoked luminol chemiluminescence could act to increase the efficiency of oxidant supply to form the luminol radical anion, they could activate the high quantum yield route via an endoperoxide or diazoquione intermediate, or they could reduce reactions through 'dark' or low-quantum yield routes. One possibility to be excluded is hydroxyl radical ($\cdot OH$) or singlet oxygen scavenging. Both of these appear to oxidise luminol by low quantum yield routes.

Several luminous organisms, including the bivalve mollusc *Pholas*, the acorn worm *Balanoglossus* and the earthworm *Diplocardia* contain luciferases which act as peroxidases (see Chapter 3). Hatchet fish light organs also contain an active luminol 'peroxidase'. Their luciferins are thus ideal substrates for other peroxidases. Pholasin, the protein-bound luciferin from *Pholas*, enables horseradish peroxidase to be detected down in the attomol range, and potentially lower. The assay is linear with respect to peroxidase over six orders of magnitude. This is sufficiently sensitive to detect the release of myeloperoxidase from a single human neutrophil.

7.4   MEASUREMENT OF ANIONS

Relatively few useful anions can be measured by using chemiluminescence. Oxyanions such as OCl- provoke luminol and pholasin chemiluminescence, providing potentially sensitive assays. Several simple anions inhibit some chemiluminescent reactions. Cl^-, for example, inhibits the firefly luciferin–luciferase reaction. A concentration of 150 mM Cl^- can reduce the light signal from this chemiluminescence reaction over 90%. This reduces the sensitivity of the assay for ATP. It can lead to artefacts if samples for ATP assay contain high anion concentrations, as Cl^- is not the only anion to inhibit. Some are even more potent ($SCN^- > I^- > Br^- > Cl^- >$ acetate), hence the use of acetate as the buffer in ATP assays (see Chapter 5). None of these provide really useful assays for anions in biological samples because they are not specific enough. We saw in chapter 5 (section 5.4) that nitrate in sea water can be measured using O_3 provoked chemiluminescence.

Phosphate is an important metabolite, both clinically in bone disorders and in other disturbances of body calcium. It is also a key metabolite in the energy metabolism of the cell. Phosphate is formed after hydrolysis of nucleotides involved in ion pumps, and forms after hydrolysis of phosphate covalently linked to proteins, as a regulator of their biological activity. Colorimetric assays are very insensitive. ^{31}PNMR, whilst being an exciting non-invasive tool, is also limited by sensitivity. Formation of ATP from inorganic phosphate (Pi) produces a highly sensitive assay:

$$\text{glycerine 3 phosphate} + \text{Pi} + \text{NAD} \rightarrow \text{1:3 diphosphoglycerate} + \text{NADH} \tag{7.28}$$

$$\text{1:3 diphosphoglycerate} + \text{ADP} \rightarrow \text{ATP} + \text{3 phosphoglycerate} \tag{7.29}$$

or

$$\text{Mg ATP} + \text{firefly luciferin-luciferase} \rightarrow h\nu \tag{7.30}$$

7.5   CONCLUSIONS AND THE FUTURE

Chemiluminescence has two main practical applications in the study of biological chemistry of inorganic ions.

Coelenterate photoproteins for detecting intracellular Ca^{2+} have had a major impact on the experimental approach to cell biology. In spite of recent advances in fluorescent Ca^{2+} indicators, the chemiluminescent photoprotein aequorin, and its British relative obelin, have, and continue to have, important uses in the study of physiology and pathology of cells. The possibility of improved photoprotein reagents in high yield using cloned apoprotein plus synthesised prosthetic groups, coupled with visualisation of the light signal using image intensification offer exciting possibilities for examining individual cells where Ca^{2+} events occur. Now that methods for getting photoproteins into small cells have been developed, studies need no longer be restricted to large cells, an easy target for the microsyringe of the physiologist. Furthermore, the methods for incorporating photoproteins into small cells are also applicable to enzymes, protein activators, and antibodies, providing a powerful approach to the dissection of the molecular biology of cells whilst they remain intact. Incorporation of mRNA or cDNA coding for the apoprotein into cells will enable them to make their own Ca^{2+} indicator. This will also enable free Ca^{2+} to be measured, for the first time within live bacteria. Transgenic cells or organisms produced by retrovirus insertion of the cDNA into the host genome offer exciting prospects for the future.

The importance of trace metals in tissue physiology and pathology, their role in vitamins, their use as a measure of biomass, their toxic action as pollutants, and their ability to generate toxic metabolite of oxygen in disease, certainly provide a convincing argument for the need to have sensitive and specific assays for transition metals. In principle, chemiluminescence should be a valuable analytical approach. However, lack of specificity, together with interference from substances in biological samples, have led to disappointingly few genuine biomedical applications of all but a few of such chemiluminescence assays established with pure standard solutions. The improvement in specificity and quantum yield with bioluminescent reagents provides a stimulus to search among the many groups of luminous organisms whose chemistry is unknown for new indicators. Some squid, fish, and decapod shrimp appear to have peroxide-sensitive systems. The value of a peroxidase assay sensitive in the tipomol (10^{-21}) region would be substantial to those using antibodies and recombinant DNA techniques in both research and clinical laboratories.

If only there were chemiluminescent assays for Na^+ and important anions!

7.6 RECOMMENDED READING
Intracellular calcium

Ashley, C. C. & Campbell, A. K. (eds) (1979). Detection and measurement of free Ca^{2+} in cells. Elsevier/North–Holland, Amsterdam.

Ashley, C. C. & Ridgway, E. B. (1968) *Nature (Lond.)* **219** 1168–1169. Simultaneous recording of membrane potential, calcium transient and tension in single barnacle muscle fibres.

Blinks, J. R., Wier, W. G., Hess, P., & Prendergast, F. G. (1982) *Prog. Biophys. Mol. Biol.* **40** 1–114. Measurement of Ca^{2+} concentrations in living cells.

Blinks, J. R., Mattingly, P. H., Jewell, B. R., van Leeuwan, M., Harrer, G. C. & Allen, D. G. (1978) *Methods Enzymol.* **57** 292–328. Practical aspects of the use of aequorin as a calcium indicator: assay, preparation, microinjection, and interpretation of signals.

Borle, A. B. & Snowdowne, K. W. (1982) *Science* **217** 252–254. Measurement of intracellular free calcium in monkey kidney cells with aequorin.

Campbell, A. K. (1983) *Intracellular calcium: its universal role as regulator.* John Wiley, Chichester.

Campbell, A. K. (1986) *Cell Calcium* **7** 287–296. Lewis Victor Heilbrunn: Pioneer of calcium as an intracellular regulator.

Campbell, A. K. (1987) *Clin. Sci.* **72** 1–10 Intracellular calcium: friend or foe?

Campbell, A. K., Dormer, R. L. & Hallett, M. B. (1985) *Cell Calcium* **6** 69–83. Coelenterate photoproteins as indicators of cytoplasmic free ca^{2+} in small cells.

Campbell, A. K. & Dormer, R. L. (1978) *Biochem. J.* **176** 53–66. Inhibition by calcium of cyclic AMP formation in sealed pigeon erythrocyte 'ghosts': a study using the photoprotein obelin.

Campbell, A. K., Patel, A. K., Razavi, Z. S., & McCapra, F. (1988) *Biochem. J.* **252**,143–149 Formation of the Ca^{2+} activated photoprotein obelin from apo-obelin and mRNA in human neutrophils.

Cobbold, P. H. & Bourne, P. K. (1984) *Nature (Lond.)*, **312** 446–448. Aequorin measurements of free calcium in single heart cells.

Cobbold, P. H., Cheek, T. R., Cuthbertson, K. S. R., & Burgoyne, R. D. (1987) *FEBS Lett.* **211** 44–48. Calcium transients in single adrenal chromaffin cells detected with aequorin.

Cobbold, P. H. & Rink, T. J. (1987) *Biochem. J.* **248,** 313–328. Fluorescence and bioluminescence measurement of cytoplasmic free calcium.

Connor, H. & Huang, L. (1985) *J. Cell Biol.* **101** 582–589. Efficient cytoplasmic delivery of fluorescent dye by pH-sensitive immunoliposomes.

Eisen, A. & Reynolds, G. T. (1985) *J. Cell Biol.* **100** 1522–1527. Source and sinks for the calcium released during fertilisation of single sea urchin eggs.

Gilkey, J. C., Jaffe, L. F., Ridgeway,, E. B., & Reynolds, G. T. (1978) *J. Cell Biol.* **76** 448–466. A free calcium wave traverses the activating egg of the medaka.

Hallett, M. B. & Campbell, A. K. (1980) *Biochem. J.* **192** 587–596. Uptake of liposomes by rat isolated fat cells: adhesion, endocytosis or fusion?

Hallett, M. B. & Campbell, A. K. (1982) *Nature (Lond.)* **295** 155–158. Measurement of changes in cytoplasmic free Ca^{2+} in fused cell hybrids.

Hallett, M. B. & Campbell, A. K. (1983) *Immunology* **50** 487–497. Direct measurement of intracellular free Ca^{2+} in rat peritoneal macrophages: correlation with oxygen-radical production.

Inouye, S., Noguchi, M., Sakaki, Y., Takagi, Y., Miyata, T., Iwanaga, S. & Tsuji, F. I. (1985) *Proc. natl. Acad. Sci. US* **82** 3154–3158. Cloning and sequence analysis of cDNA for the luminescent protein aequorin.

Inouye, S., Sakaki, Y., Goto, T. & Tsugi, F. I. (1986) *Biochem.* **25** 8425–8429. Expression of apoaequorin complementary DNA in *Escherichia coli*.

Jaffe, L. F. & Sardet, C. (1986) *Biol. Bull.* **171** Calcium pulses and waves through ascidian eggs.

Johnson, P. C., Ware, J. A., Clivedon, P. B., Smith, M., Dvorak, A. M., & Salzman, E. W. (1983) *J. Biol. Chem.* **260** 2069–2076. Measurement of ionised calcium in blood platelets with the photoprotein aquorin.

Morgan, B. P., Luzio, J. P. & Campbell, A. K. (1986) *Cell Calcium* **7** 399–411. Intracellular Ca^{2+} and cell injury: role of Ca^{2+} in complement membrane attack.

Morgan, J. P. & Morgan, K. G. (1982) *Pflügers Archiv.* **295** 75–77. Vascular smooth muscle: the first recorded calcium transient.

Nakano, T., Terawaki, A., & Arita, H. (1986) *J. Biochem.* **99** 1285–1288. Measurements of thromboxane A_2-induced elevation of ionised calcium in collage-stimulated platelets with the photoprotein, aequorin.

Neering, I. R. & Fryer, M. W. (1988) *Biochem. Biophys. Acta* **882** 34–43. The effect of alcohols on aequorin luminescence.

Prasher, D. C., McCann, R. O., & Cormier, M. J. (1985) *Biochem. Biophys. Res. Commun.* **126** 1259–1268. Cloning and expression of the cDNA coding for aequorin, a bioluminescent calcium-binding protein.

Prasher, D. C., McCann, R. O., Longiaru, M. and Cormier, M. J. (1987) *Biochem.* **26,** 1826–1332. Sequence comparisons of complementary DNAs encoding Aequorin isotypes.

Ridgway, E. B. & Ashley, C. C. (1967) *Biochem. Biophys. Res. Commun.* **29** 229–234. Calcium transients in single muscle fibres.

Rose, B., & Loewenstein, W. R. (1976) *J. Memb. Biol.* **28**, 87–119. Permeability of cell junction depends on local calcium activity.

Shimomura, O. (1986) *Biochem. J.* **234** 271–277. Isolation and properties of various molecular forms of aequorin.

Shimomura, O. & Shimomura, A. (1985) *Biochem. J.* **228** 745–749. Halistaurin, phialidin and modified forms of aequorin as Ca^{2+} indicator in biological systems.

Shimomura, O., Johnson, F. H., & Saiga, Y. (1962) *J. Cell Comp. Physiol.* **59** 223–239. Extraction, purification and properties of aequorin, a bioluminescent protein from the luminous hydromedusan Aequorea.

Stephenson, D. G. & Sutherland, P. J. (1981) *Biochim. Biophys. Acta* **678**, 65–75. Characterisation of obelin from the hydroid *Obelia australiensis*.

Woods, N. M., Cuthbertson, K. S. R., & Cobbold, P. H. (1987) *Cell Calcium* **8** 79–100. Agonist-induced oscillations in cytoplasmic free calcium concentration in single rat hepatocytes.

Other calcium indicators

Grynkiewicz, G., Poenie, M., & Tsien, R. Y. (1985) *J. Biol. Chem.* **260**, 3440–3450. A new generation of Ca^{2+} indicators with greatly improved fluorescence properties.

Tsien, R. Y., Pozzan, T., & Rink, T. J. (1982) *J. Cell Biol.* **94** 325–334. Calcium homeostasis in intact lymphocytes: cytoplasmic Ca^{2+} monitored with a new, intracellularly trapped fluorescent indicator.

Poinie, M., and Tsien, R. Y. (1986) *TIBS* **11** 450–455. Fluorescence ratio imaging: a new window into intracellular ionic signalling.

Smith, G. A., Hesketh, T. R., Metcalfe, J. C., Feeney, J., and Morris, P. G. (1983) *Proc. Nat. Acad. Sci. US* **80** 7178–7182. Intracellular calcium measurements by F NMR of fluorine-labelled chelators.

Transition metals

Albrecht, H. O. (1928) *Zeitschrift für physikalischen Chemie* **136** 321–330. Über die Chemiluminescenz des Aminophthalsäurehydrazids. (translation available).

Campbell, A. K. & Simpson, J. S. A. (1979) *Tech. Metabol. Res.* **B213**, 1–56. Chemi- and bioluminescence as an analytical tool in biology.

Carter, T. J. N. & Kricka, L. J. (1982) *Clinical and biochemical luminescence*, 135–151. Eds. Kricka, L. J. & Carter, T. J. N. Marcel Dekker, New York and Basel.

Ewetz, L. and Thore, A. (1976) *Anal. Biochem.* **74** 564–570. Factors affecting the specificity of the luminol reaction with hematin compounds.

Harvey, E. N. (1916) *Science* **44** 208–209. The mechanism of light production by animals.

Harvey, E. N. (1916) *Am. J. Physiol.* **41** 449–454. Studies on bioluminescence II. On the presence of luciferin in luminous bacteria.

Harvey, E. N. (1926) *Biol. Bull.* **51** 89–97. Oxygen and luminescence with a description of methods for removing oxygen from cells and fluids.

Isacsson, U. and Wettermark, G. (1974) *Anal. Chim. Acta.* **68** 339–362. Chemiluminescence in analytical chemistry.

Klopf, L. L. & Nieman, T. A. (1983) *Anal. Chem.* **55** 1080–1083. Effect of iron (II), cobalt (II), copper (II), and manganese (II) on the chemiluminescence of luminol in the absence of hydrogen peroxide.

Longeon, A., Henry, J. P., & Henry, R. (1986) *J. Immunol. Meth.* **86** 103–109. A solid-phase luminescent immunoassay for human chorionic gonadotropin using *Pholas dactylus* bioluminescence.

Michelson, A. M. (1978) *Methods Enzymol.* **57** 385–406. Purification and properties of *Pholas dactylus* luciferin and luciferase.

Morrison, M. & Schonbaum, G. R. (1976) *Ann. Rev. Biochem.* **45** 861–888. Peroxidase-catalysed halogenation.

Rosewell, D. F. & White, E. H. (1978) *Methods Enzymol.* **57** 409–423. The chemiluminescence of luminol and related hydrazides.

Shimomura, O. & Johnson, F. H. (1966) *Bioluminescence in action.* pp. 495–521. Ed. Johnson, F. H., & Haneda, Y. Princeton University Press, Princeton.

Shimomura, O. & Johnson, F. H. (1968) *Science* **159** 1239–1240. Chaetopterus photoprotein: crystallisation and cofactor requirements for bioluminescence.

Thorpe, G. H. G. & Kricka, L. J. (1987) *Methods Enzymol.* **133** 331–353. Enhanced chemiluminescent reaction catalysed by horseradish peroxidase.

Thorpe, G. H. G., Kricka, L. J., Gillespie, E., *et al.* (1985) *Anal. Biochem.* **145** 96–100. Enhancement of the horseradish peroxidase catalysed chemiluminescent oxidation of cyclic diacyl hydrazides by 6-hydroxy-benzothiazoles.

Thorpe, G. H. G., Kricka, L. J., Moseley, S. B., & Whitehead, T. P. (1985) *Clin. Chem.* **31**, 1335–1341. Phenols as enhancers of the chemiluminescent horseradish peroxidase luminol-hydrogen peroxide reaction: application in luminescence-monitored enzyme immunoassays.

Vasileff, T. P., Svarnas, G., Neufeld, H. A., & Spero, L. (1974) *Experientia* **30** 20–22. Hemin as a catalyst for chemiluminescence.

Whitehead, T. P., Thorpe, G. H. G., Carter, T. J. N., Groucutt, C., & Kricka, L. J. (1983) *Nature (Lond).* **305** 158–159. Enhanced luminescence procedure for sensitive determination of peroxidase-labelled conjugates in immunoassay.

8

An alternative to radioactive labels

8.1 DO WE NEED RADIOACTIVE PROBES?

8.1.1 Some of their successes

Radioisotopes are atoms of the same element having different nuclear masses but identical atomic number and chemical properties, whose disintegration is accompanied by the release of α, β, or γ particles enabling the isotope to be detected and quantified.

Considering how much we rely on radioactivity in both the research and clinical laboratories nowadays, it is remarkable by how much biochemistry managed to advance in the first half of this century, before radioactive probes became widely available. Remember glycolysis, the urea and Krebs cycles were elucidated without the availability of ^{14}C or ^{3}H labelled substrates. Yet radioisotopes have had an immense impact on the advancement of medical science. Their availability meant that no longer were studies on metabolism limited by insensitive assays for metabolites. Fluxes of whole pathways could be studied easily, as well as fluxes of substances crossing biological membranes. Would it really have been possible to elucidate the pathways of proteins and nucleic acid synthesis without the use of labelled amino acids and nucleotides? The quantification of ligand–ligand interactions such as antibody–antigen binding, and agonist– or antagonist–receptor binding, has relied heavily on ^{3}H and ^{125}I labelled reagents. The development of the new biotechnologies using monoclonal antibody and recombinant DNA techniques equally have depended on the use of radioisotopes, ^{125}I and ^{32}P respectively.

The strength of radioactive isotopes is that they provide highly sensitive probes (Table 4.7) for following the fission and fusion of covalent and non-covalent chemical bonds in biological systems, with minimum disruption to the physicochemical properties of the molecules or atoms to which they are attached (Table 8.1). Furthermore, isotope disintegration is not affected by physical or chemical factors in the sample, and small interferences with detection systems, such as quenching in scintillation fluids, can be easily compensated for. One only has to see how the

Table 8.1 — Some important analytical applications of radiosiotopes in biology

	Application	Example	Isotopes commonly used
1	Metabolism	intermediary, protein, nucleic acid, phospholipid	^3H, ^{14}C, ^{32}P, ^{35}S
2.	Fluxes across membranes	cations	^{23}Na, ^{45}Ca, ^{42}K, ^{86}Rb
		anions	$^{32}PO_4$
		organic substrates	^3H, ^{14}C
3.	Ligand-ligand interactions	immunoassay	^3H, ^{125}I
		receptors	^3H, ^{125}I
4.	Recombinant DNA technology	gene sequencing, Southern and Northern blots	^{32}P
5.	Localisation of reactions	autoradiography in light and electron microscope	^3H, ^{14}C, ^{125}I

investigation of the biochemistry of oxygen and its metabolites is severely hampered at present by the lack of a suitable radioactive isotope to realise how valuable such probes are in the research laboratory

In spite of these successes for radioisotopes in biology and medicine, the last ten years or so have seen major efforts to find an alternative. Why?

8.1.2 Problems and limitations of radioisotopes

There are five main problems with radioisotopes, placing some practical limitations on their application in the biological laboratory:

(1) *Preparation of the isotope*

This requires a special, costly apparatus such as a nuclear reactor or particle accelerator together with the safety facilities for handling highly radioactive compounds. Preparation of isotopes can therefore only be carried out by large industrialised units or special research facilities, which are not routinely available to the biologist.

(2) *Hazard*

Radioactive isotopes are dangerous, not only at the high levels involved in their preparation but also even at the levels in reagents used in research and clinical laboratories. ^{125}I readily concentrates in the thyroid and ^{32}P in bone. Safety precautions, including special 'hot' rooms and cabinets, together with personnel screening are therefore essential. Special facilities are also really required for disposal of radioactive waste, though all too often it is discarded down the sink. Public pressure through the political lobby has already lead to legislation in many countries restricting the use of radioisotopes such as ^{125}I and ^{32}P to specialised laboratories. If you are not in one, you need an alternative.

(3) *Sensitivity and speed of detection*
The detection limit for a radioisotope can be estimated from the half time ($t_{\frac{1}{2}}$) of its decay. The shorter the half life the lower the detection limit (Table 4.7):

$$A = A_{\exp}(-kt) \qquad (8.1)$$

$$k = 0.693/t_{\frac{1}{2}} \qquad (8.2)$$

Whilst ^{125}I and ^{32}P can be detected in a few minutes in the 1–10 attomol (10^{-17}–10^{-18}) region, ^{3}H and ^{14}C are considerably less easy to detect, and often require counting times of many minutes. Autoradiography requires hours or days.

(4) *The consequences of isotope decay*
The most sensitively detectable isotopes by their very nature decay rapidly. Some decay is so fast as to make the isotope practically useless. Others like ^{32}P have lost sufficient radioactivity within a few weeks to be undetectable. A further problem is the effect of the decay on the molecule to which the isotope is attached. Biological molecules, such as proteins, are particularly susceptible to radiolytic damage. The immunological activity of some ^{125}I-labelled antibodies may thus last only a few weeks. Monoclonal antibodies, often stored at much lower protein concentrations than polyclonals which have high concentrations of protective protein carrier, are particularly susceptible. They may not, for example, survive freeze-thawing. Also it is very difficult, if not impossible, to prepare immunologically-active and stable antibodies with more than one radioactive iodine per molecule, because of this damage.

(5) *The need for a separation step*
Virtually all analytical applications of radioisotopes require a separation step in order to isolate the appropriate material for radioactive counting. Individual intermediates on a metabolic pathway must be isolated and separated from the initial substrate, cells must be separated from extracellular fluid in flux studies, antibody–antigen complexes must be separated from free antibody and antigen in an immunoassay, drugs on receptor much be separated from unbound drugs, DNA recombinants must be separated before and after addition of a ^{32}P labelled probe and in autoradiography unreacted or unbound label must be washed away from the microscope section or nitrocellulose sheet. These separation steps introduce imprecision, are sometimes laborious, and complicate automation for clinical application. An enormous variety of separation procedures have been developed for immunoassay, involving precipitation, gel filtration, chromatography, filtration, electrophoresis, solid phase centrifugation, large beads, or magnetic particles. Different DNA or RNA molecules of varying size are separated by electrophoresis. Related to this problem is the fact that several ligand–ligand interactions of biological interest cannot be studied using separation techniques because the dissociation rate is too fast. Suppose the binding of a substance (S) to its ligand (L) is limited principally by diffusion. Then the 'on rate' (k_1) for $S + L \xrightarrow{k_1} SL$, will be about $10^8 \text{M}^{-1}\text{min}^{-1}$. The 'off rate' ($k_{-1}$) for $SL \rightarrow S + L$ is then related to the equilibrium dissociation constant (K_d):

$$k_{-1} = K_d \cdot k_1 \text{ min}^{-1} \tag{8.3}$$

$$SL \rightleftharpoons S + L \tag{8.4}$$

$$K_d = S.L/SL \quad (K_a = 1/K_d). \tag{8.5}$$

For practical purposes we cannot really handle a ligand–ligand interaction by separation if the half time for dissociation is <5 min. This places restrictions on k_{-1}, and thus on K_d:

If $t_{\frac{1}{2}} = 5$ min then

$$k_{-1} = \ln 2/t_{\frac{1}{2}} = 0.693/t_{\frac{1}{2}} = 0.14 \text{ min}^{-1},$$

then $K_d = k_{-1}/k_1 = 1.4 \times 10^{-9}$ M i.e. 1.4 nM.

Thus, separation assays using binding proteins with K_d's >1–10 nM (i.e. $K_a <$ $10^9 - 10^8 \text{ M}^{-1}$) will not work. This is well below the K_m for most enzymes, and the K_d for several receptors. An analytical method not requiring a separation step, that is 'homogeneous' in contrast to the 'heterogeneous' assay requiring separation, would open up new possibilities for investigating the molecular biology of living systems.

These human, economic, and very real practical problems are reason enough to search for an alternative to radioisotopes. But there is another, perhaps more compelling, scientific one. There are several analytical problems in cell and molecular biology which cannot be solved by using radioisotopes. Yet these still require particular molecules to be tagged so that their interactions with other species can be quantified. Ligand–ligand interactions where a separation step is impossible, for example within a cell, or because the 'off' rate (k_{-1}) is too fast, require a homogeneous assay. These ligand–ligand interactions include antibody–antigen and receptor–agonist/antagonist interactions, enzyme–substrate binding and DNA–DNA or RNA–DNA hybridisation in intact cells. Radioactive tags cannot be used to measure distances within and between molecular complexes either *in vitro* or inside cells. For example, when adrenaline binds to its β receptor what are the distances between the receptor, the G protein complex with its α, β and γ sub-units and adenylate cyclase before and after binding. When the membrane attack complex of complement forms C5b678 and bind the last components C9 activating the polymerisation of up to 12–18 C9 molecules, does this polymer still remain attached to the C5b-8 complex? Visualisation of molecular movement in living cells, for example capping in lymphocytes and neutrophils, or the movement of receptors to one part of a cell, cannot be followed by using radioactive tags. Some of these problems have been solved by using photoluminescent labels, i.e. fluorescence. Yet the solution to many of them ultimately requires labels with the equivalent sensitivity of the best radioisotopes, such as ^{125}I and ^{32}P. Chemiluminescence offers the possibility of achieving this, and potentially of doing better.

8.1.3 The alternatives
On what criteria should we base our selection if we are to search for an alternative to a radioisotopic label in established assays, and to develop new approaches to advance our knowledge of how cells work? We can list nine, beginning with the important scientific criteria:

(1) *Sensitive, with fast detection and quantification (high signal:noise)*
The non-isotopic label should be easily detectable in the femtomol–attomol $(10^{-15}–10^{-18})$ range, and potentially in the attomol–tipomol range $(10^{-18}–10^{-21}$ mol), using ideally a relatively simple and inexpensive apparatus. The signal should be complete within a few seconds for application in techniques such as immunoassay, yet capable of following reactions for many minutes or even hours when used in conjunction with living cells. It is worth remembering that the sensitivity of many analyte assays is not limited by the detection limit of the label, but rather the properties of the other reagents, e.g. the affinity of an antibody. Nevertheless, a sensitive label is still required if the final, analytical step is to be fast, and able to cope with large numbers of samples quickly.

(2) *Visible*
For certain applications it should be possible to see the native signal, after amplification if necessary, and signals from within cells using both the light, and the electron microscope.

(3) *Couplable to biological molecules*
The label must be capable of attachment to a wide range of small molecular weight substances, and macromolecules of biological and clinical interest. Coupling must not seriously affect either the detection of the label or the chemical and biological properties of the molecule to which it is attached.

(4) *Homogeneous*
A change in signal from the label, induced when the molecule to which it is attached binds to another, should enable the number of molecules, together with the distance between them, to be quantified. This must be done without separation of bound and free ligand.

(5) *Stable*
The label, both initially and when attached to the molecule of biological interest, should be stable, having a shelf life of at least several years.

(6) *Unsusceptible to environmental factors*
The signal from the label should not be interfered with by physical or chemical factors present in biological samples.

(7) *Clinically applicable*
For use in assays of patient samples it should be possible to automate both the initial assay steps and the detection of the label, whilst retaining the necessary assay precision for the clinical range of the analyte in question. Common tests, such as thyroid function, may involve many samples per day in a clinical laboratory; others only arrive in ones and twos. Large genetic screening programmes can require hundreds of samples per day for a national laboratory.

(8) *Safe*
The label itself should be safe to humans, and safe to prepare. It should also be non-toxic to animals, or the experimental systems in which it is to be used.

(9) *Easily available*
The label should be easily available, easy to make in gram quantities, and cheap.

A large number of compounds have been investigated based on absorbance, spin response, and various types of luminescence (Table 8.2), as well as particulate labels including bacteriophage, erythrocytes and latex particles, in an attempt to satisfy these criteria. Many fail on the first criterion alone. They have nowhere near the sensitivity of radioactive iodine, though they may be usable for analytes present in the μM–mM range (Table 4.3). Enzyme labels including peroxidase, oxidase, dehydrogenases, kinases, amylase and β galactosidase have been used to establish many workable and commercial immunoassays. They can be used in any of the immunoassay strategies, competitive, immunometric, two site and homogeneous (see section 8.2). When used in conjunction with a solid phase it is often known as ELISA (enzyme linked immunosorbent assay). An homogeneous assay strategy has been established known as EMIT (enzyme multiplied immunoassay technique). Here hapten–enzyme conjugates bind to their respective antibodies in such a way that steric hindrance or conformational change in the enzyme inhibits the substrate interacting with the enzyme. Addition of sample displaces the hapten-enzyme conjugate, thereby enhancing enzymatic activity. ELISA or EMIT assays use colorimetric indicators to detect the enzyme. This is much less sensitive than chemiluminescence. However several of the enzymes, e.g. peroxidase, oxidase, kinases and dehydrogenases can be detected with much greater sensitivity using a chemiluminescent substance. A number of luminescent labels can be readily detected in the fmol–amol range (10^{-15}–10^{-18}) and appear to satisfy many of the other criteria. Let us therefore examine the application of chemiluminescent labels

Table 8.2 — Some non-radioactive labels

Class of label	Example	Method of detection
Biological	erythrocyte	rosetting
	bacteriophage	plaques, ^{125}I label, lux gene expression
Enzyme	Peroxidase	absorbance
		chemiluminescence
	β galactosidase	absorbance
	dehydrogenase	absorbance, chemiluminescence
Organic	fluorescein	fluorescence
		fluorescence polarisation
	Acridinium ester	chemiluminescence
	ABEI	chemiluminescence
	AENAD	enzyme cycling
		chemiluminescence
	tetramethyl-piperiodinoxyl	electronspin resonance
	di-π-cyclopentadienyl-iron (ferrocene)	atomic absorption
Inorganic	Eu^{2+} chelate	time resolved fluorescence

in an essential technique for any research or clinical laboratory — immunoassay, and see how they match up to their rivals, the photoluminescent (fluorescent) labels.

8.2 CHEMILUMINESCENCE IMMUNOASSAY
8.2.1 Principles
Radioimmunoassay was introduced by Yalow and Berson in 1959, and has since become one of the most powerful techniques to have developed in biology and medicine in the last 30 years (Table 8.3). It is based on using an antibody as a reagent, which specifically can select out from a biological sample, for example blood, one substance, the antigen, which is the one we want to measure. The antigen can be a hormone, e.g. thyroid stimulating hormone; a therapeutic drug, e.g. a cardiac glycoside or an antibiotic; a drug of abuse such as heroin, cocaine, or an anabolic steroid; or it can be a pathogen such as *Salmonella* or HIV, the AIDS virus. The amount of antibody–antigen complex formed is related directly to the concentration of substance in the original blood sample by including in the assay a known amount of antigen, tagged with a label, ^{125}I in the case of radioimmunoassay. This will 'compete' with the antigen in the sample for the antibody. Alternatively the antibody can itself be labelled with ^{125}I. The antibody–antigen complex, now labelled with ^{125}I, is then separated from the mixture and assayed for radioactivity. The technique has to cover a wide range of analyte concentrations (Table 4.3), from μM in the case of certain marker proteins and drugs to pM in the case of free hormones such as those from the thyroid gland. These provide the clinician with essential information of

Table 8.3 — The importance of immunoassay

Analyte	Application
Clincial medicine	
Hormones	endocrine disorders
Embryonic proteins	cancer detection
	pregnancy testing
	congenital defects
Abnormal proteins	genetic defects
Microorganisms (bacteria, Viruses, protozoa)	identification of infectious agent
Drugs therapeutic	patient treatment
abuse	identification of abuser
Veterinary medicine	
Proteins and pathogens	diagnosis and treatment
Forensic medicine	
Proteins, toxins and microorganisms	cause and causer of crime
Food industry	
Pathogens	detection of dangerous food contaminants (microorganisms and toxins)

direct value at the patient's bedside, enabling tumour types to be identified and located, endocrine disorders to be diagnosed, infections to be identified, inherited and congenital defects to be picked up in the newborn and in 'carriers', and drug therapy to be followed. Yet the value of immunoassay is not restricted to human medicine. It also has a vital role to play in veterinary medicine, agriculture, forensic medicine, and the food industry (Table 8.3).

Since the introduction of radioimmunoassay by Yalow and Berson, a plethora of assays have appeared using a wide range of different procedures. The first most obvious variables in these are:

(1) The nature of the analyte, be it small molecular weight (hapten) or a macromolecule, and whether either free or total analyte is to be measured.
(2) The label, and whether it is attached to the antigen or antibody.
(3) The method of detection of the label. Even with radioisotopes one may have a choice between scintillation or γ counting.
(4) Whether a polyclonal or monoclonal antibody is used.
(5) The method of separation of the antibody–antigen complex from the free components, or whether it is 'homogeneous'.

However, one feature transcends these variables, and that is whether the assay belongs to the class known as *radioimmunoassay* (RIA), when a radioisotope is used, the original type, or whether it is known as *immunoradiometric* (IRMA), invented by Miles and Hales in the late 1960s (Fig. 8.1). In non-isotopic immunoassays simply replace R by F for fluorescence or fluoro- and C for chemiluminescence or chemilumino-, i.e. FIA and IFMA or CIA and ICMA.

Let us examine these in a little more detail. *Radioimmunoassay* conventionally uses a labelled antigen (Ag*) mixed with unlabelled antibody (Ab), usually at a concentration necessary to produce about 50% binding of the labelled Ag.

$$Ab + Ag^* \overset{K}{\rightleftharpoons} AbAg^* \tag{8.6}$$

Unlabelled antigen (Ag), will also bind to antibody. In other words, it 'completes' for the limited amount of antibody available:

$$Ab + Ag^* + Ag \overset{K}{\rightleftharpoons} AbAg^* + AbAg \tag{8.7}$$

Since the amount of antibody is fixed and the equilibrium constant, K, is invariant, as more analyte (unlabelled) is added less of the antibody is able to bind labelled antigen, so AbAg* decreases and free Ag* increases. By separating antibody bound Ag* from free Ag* a standard curve can be constructed (Fig. 8.2), enabling the concentration of analyte from a biological sample to be measured.

With the introduction of solid phase reagents it became possible to design a 'competition' strategy using labelled antibody instead of labelled antigen. For example, consider a reaction mix containing solid phase antigen, free antigen (the analyte), analyte antibody, and a labelled anti-antibody produced in a different animal species from the analyte antibody, e.g. sheep anti-rabbit IgG where the analyte antibody was produced in a rabbit. The amount of this labelled anti-antibody bound to the solid phase through an Ag-Ab-antiAb* sandwich will depend on how much free Ab-analyte is left in free solution. This will increase as the concentration of free analyte is increased.

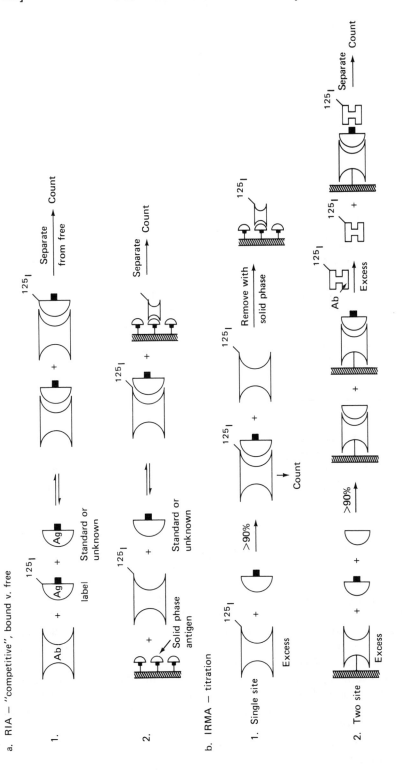

a. RIA — "competitive", bound v. free

b. IRMA — titration

Fig. 8.1 — Principles of immunoassay.

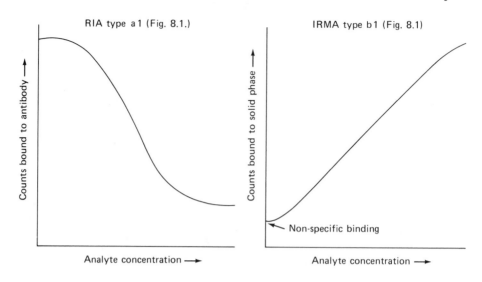

Fig. 8.2 — Typical standard curves in RIA and IRMA.

Immunoradiometric assays, on the other hand, are non-competitive, and always use labelled antibodies (*Ab):

$$^*Ab + Ag \overset{K}{\rightleftharpoons} {}^*AbAg \qquad\qquad (8.8)$$

The assay is thus essentially a titration. Labelled antibody is added in large excess over the analyte, thereby pulling the equilibrium over to the right. The excess antibody not bound to antigen is then removed, for example by adding solid phase antigen, and the *AbAg remaining measured. As the concentration of analyte increases so does the amount of measured antibody–antigen complex (Fig. 8.2).

The relative merits of these two assay strategies, together with what limits their ultimate sensitivity, have long been the source of heated debate amongst immunoassayists. Is it the antibody? Is it the label? Is it the strategy? The consensus from a combination of mathematical analysis and empirical observation, is that under optimal conditions greatest sensitivity can be achieved when using the non-competitive IRMA approach. Furthermore, a label of greater sensitivity than ^{125}I would enable analytes such as certain viral antigens, beyond the present detection limits of radioassays to be measured. Nevertheless, the 'competitive' type assays are still in wide use and perfectly adequate for many analytes, such as drugs, where the ultimate sensitivity of immunoassays is not necessary.

This then is the background to the use of chemiluminesent labels in immunoassay. Before examining how they fit into the argument about immunoassay sensitivity let us first examine whether they can be used at all to establish immunoassays capable of detecting useful analytes in patient samples, and other biological material. First, what chemiluminescent labels could we use, and second, how can they be applied to an immunoassay?

8.2.2   The labels

Chemiluminescent reactions require up to five components: the chemiluminescent substrate which reacts to form the actual light emitting moiety, oxidants, cofactor(s), inorganic ions, and a catalyst. In principle, any one of these components, even inorganic ions within organic complexes, can be coupled to an antibody or antigen. The labelled reagent can then be used in a 'competitive' or 'non-competitive' binding assay, and the chemiluminescence initiated by addition of the remaining components of the reaction. If a substance is to be used as a chemiluminescent label, it must satisfy four requirements:

(1) It must be capable of participation in a chemiluminescence reaction.
(2) It must be possible to attach it to the antigen or antibody, not normally chemiluminescent, to form a reagent stable until the reaction is triggered.
(3) The label should retain high quantum yield and the necessary reaction kinetics after being coupled.
(4) It should not significantly alter the physico-chemical properties of the molecule to which it is attached, in particular its immunological activity.

Three groups of compounds (Table 8.4; Fig. 8.3) have been investigated extensively in an attempt to satisfy these requirements, together with the criteria for an ideal non-specific isotopic label (section 8.1.3):

(1) Synthetic organic compounds: phthalazine diones and acridinium esters.
(2) Cofactors in bioluminescent reactions: NAD and ATP.
(3) Enzymes: peroxidase, oxidases, kinases, and luciferases.

These have to be covalently linked either to the antigen or antibody. Such a linkage does not occur spontaneously, but any one of three procedures can be carried out to facilitate this (Table 8.5, Figs 8.3 and 8.4):

(1) Chemical modification of the label, e.g. diazotisation, isothiocyanate, N-hydroxysuccinimide, hemisuccinate, imidoesters.
(2) Chemical modification of the antigen or antibody, e.g. hemisuccinate, glutaraldehyde.
(3) Conjugation using bifunctional reagents, e.g. mixed anhydride, carbodiimide, bis N-hydroxy succinimides, azido-succinimides.

Isothiocyanate, N-hydroxy succinimide, and imidoester groups readily label primary amines over the pH range 7–10, whilst carbodiimide, and mixed anhydride coupling using isobutyl chloroformate, enable amide bonds to be formed between $-NH_2$ and $-CO_2H$ groups. The reaction can be carried out in organic media if the substances are soluble, or alternatively aqueous media. On the other hand, pyridyl disulphides, maleimides, thiophthalimides, and active halogens label free SH groups over the pH range 7–8. Nitrenes and carbenes formed by photo-activation are much less specific, even going for C-H bonds.

What then is the best chemiluminescent label to use and the best way of coupling, for example, a synthetic organic chemiluminescent label to a hapten, or to a protein antigen or to an antibody?

Early experiments using diazoluminol, succinyl luminol, or luminol + glutaraldehyde, resulted in a dramatic loss, some 10 to 100-fold, in total light yield

Table 8.4 — Some chemiluminescent labels

Compound	Assay
A. Synthetic organic compounds	
luminol	NaOCl or peroxidase + H_2O_2
isoluminol	lactoperoxidase + H_2O_2
amino butyl ethyl or isoluminol (ABEI) or AHEI (H = hexyl)	microperoxidase + H_2O_2
ABENH	microperoxidase + H_2O_2
acridinium esters	OH^-/H_2O_2
adamantylidene adamantane 1,2 dioxetan	heat
B. Bioluminescence cofactors	
aminoethyl NAD (AENAD)	bacterial luciferase after AENADH formation
aminoethyl ATP (AEATP)	firefly luciferin-luciferase
FAD	bacterial luciferase or fluorescence
C. Enzymes and photoproteins	
peroxidase	pyrogallol
	luminol
	luminol + enhancer
	pholasin
bacterial luciferase	NADH/FMN/long chain aldehyde + oxidoreductase
firefly luciferase	ATP Mg + luciferin
aequorin	Ca^{2+}
glucose dehydrogenase	bacterial luciferase/oxido-reductase

ABENH =
7[N-4-aminobutyl)-N-ethylamino]naphthalene-1,2-amino-2,3-dihydrophthalazine-1,4-dione

from the luminol chemiluminescent reaction. Substitution on the heterocyclic ring of luminol results in complete loss of chemiluminescence, whereas substitution of any electron withdrawing group on the aromatic $-NH_2$ results only in a reduction in quantum yield. It might be thought, therefore, that substitution of electron donating groups on the $-NH_2$, such as ethyl, should enhance the quantum yield of luminol. However, because of steric hindrance with the heterocyclic ring, this still results in a reduction in quantum yield. This led Schroeder to investigate a similar substitution on isoluminol (Fig. 8.3) when the shift of the $-NH_2$ by one position on the benzene ring allows this without causing steric hindrance.

The quantum yield of luminol in DMSO is about 5%, but only 1.5% in aqueous media. Isoluminol, however, only has a quantum yield of about 5–10% that of luminol. Substitution of two ethyl groups on the aromatic $-NH_2$ of isoluminol raises its quantum yield virtually to that of luminol. But this derivative still has no linking

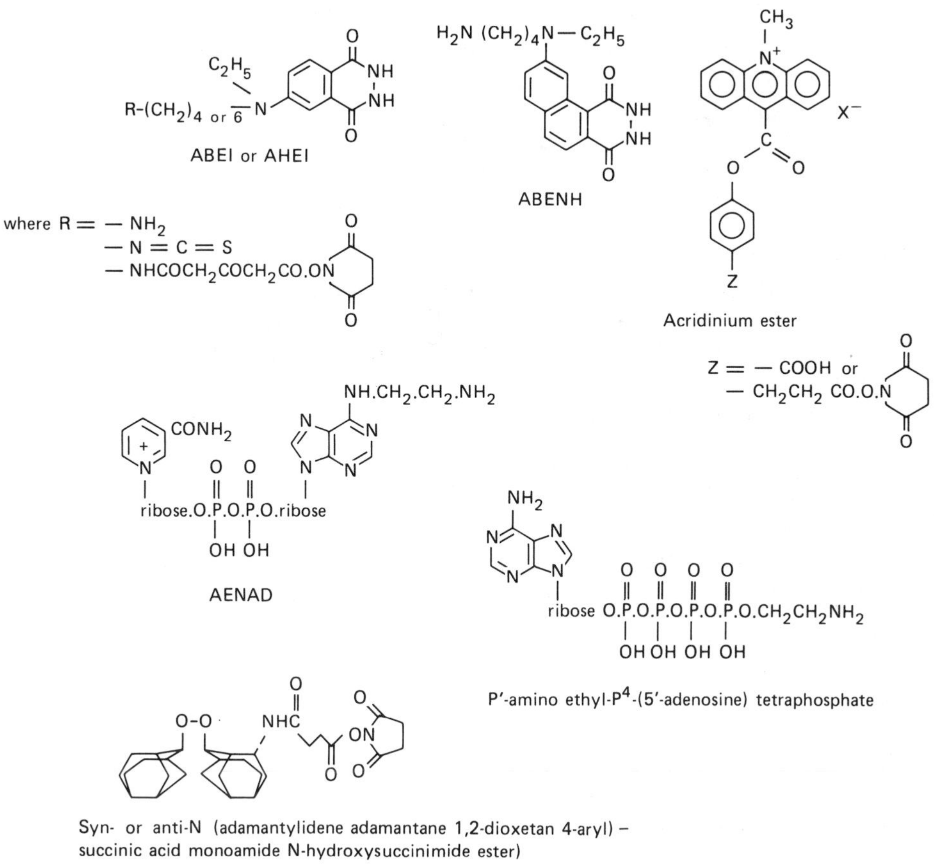

Fig. 8.3 — Chemiluminescent labels and their linking groups.

group on it. However, substitution with an ethylamine solves this. Two popular compounds have been ABEI (aminobutylethylisoluminol) and AHEI (H = hexyl instead of butyl) (Fig. 8.3). Their quantum yields are 80–94% that of luminol. The free primary amine can be covalently linked to carboxyl groups by mixed anhydride or carbodiimide coupling, or converted to an N-hydroxy succinimide or isothiocyanate which avidly react with free –NH$_2$ groups, such as lysine side chains on the surface of proteins. If you are lucky, the coupling of ABEI to a hapten or protein has little or no effect on its quantum or reaction kinetics. It can then generate as many as 10^{19} photon counts in 10 seconds in a sensitive chemiluminometer, with a detection limit of about 5×10^{-17} mol. However, decreases in quantum yield, with an inevitable reduction in the detection limit of the label, have been reported. ABEI coupled to succinyl cyclic AMP suffers about a two-fold drop, whereas the heavy iodines acting as efficient excited state quenchers on thyroxine (T$_4$) result in ABEI T$_4$ with only 1–2% of the light yield of ABEI itself.

Table 8.5 — Examples of coupling strategies

---

1 Modification of parent chemiluminescent compound
   luminol → succinyl luminol
   isoluminol → aminobutylethyl isoluminol (ABEI)
   NAD → aminoethyl NAD (AENAD)
   ATP → aminoethyl ATP (AEATP)

2. Modification of the label
   luminol + HNO_2 → diazoluminol + IgG → IgG conjugate
   ABEI + thiophosgene → ABEI NCS + RNH_2 → Ig G conjugate
   acridinium ester + Nhydroxysucinimide → NOH succacridinium ester +
   RNH_2 → IgG conjugate
   ABENH + dimethyladipimate → imidate + RNH_2 → protein conjugate

3. Modification of the substance to be labelled
   steriod → steroid CO_2H + ABEI → ABEI steroid
   digoxin + $NaIO_4$ → aldehyde + luminol → conjugate
   peroxidase + gluteraldehyde or bifunctional + label → conjugate
   cyclic AMP → succinyl-cyclicAMP → ABEI succ cAMP

4. Bifunctional reagents
   luminol + glutaraldehyde → insulin
   ABEI + carbodiimide → sterod CO_2H
   $AECO_2$ + carbodiimide → IgG
   succinyl luminol + carbodiimide → IgG
   ABEI + isobutylchloroformate → succinyl cAMP
   AENAD + carbodiimide → biotin
   AENAD + carbodiimide → steroid
   AEATP + carbodiimide → fluorodinitrobenzene

These problems, and those of the high chemical blank associated with the stimuli for a phthalazinedione chemiluminescence (see section 8.2.3) led us with Professor McCapra's group to investigate the potential of his acridinium salts (Fig. 8.3). These are half a lucigenin molecule with an ester linkage, providing a good leaving group and thus a fast, high quantum yield reaction from the efficient generation of excited state N-methyl acridone. Not only does this product detach itself as a result of the chemiluminescent reaction, and is thus unlikely to be interfered with by haptens or proteins to which the acridinium ester is orginally attached, but also the triggering conditions for these esters are simpler than those for phthalazine diones, requiring only alkaline hydrogen peroxide. No additional catalyst is required, as is necessary for phthalazine diones. The result is a similar total photon yield per mol with both classes of compound, but a redution of 10 to 100-fold in the reagent blank when using acridinium ester. These labels can therefore be detected down to 10^{-17}–10^{-18} mol and can be attached to NH_2 groups in proteins by using an *N*-hydroxy succinimide derivative (Fig. 8.3) with little or no effect on their quantum yield. By adjusting the

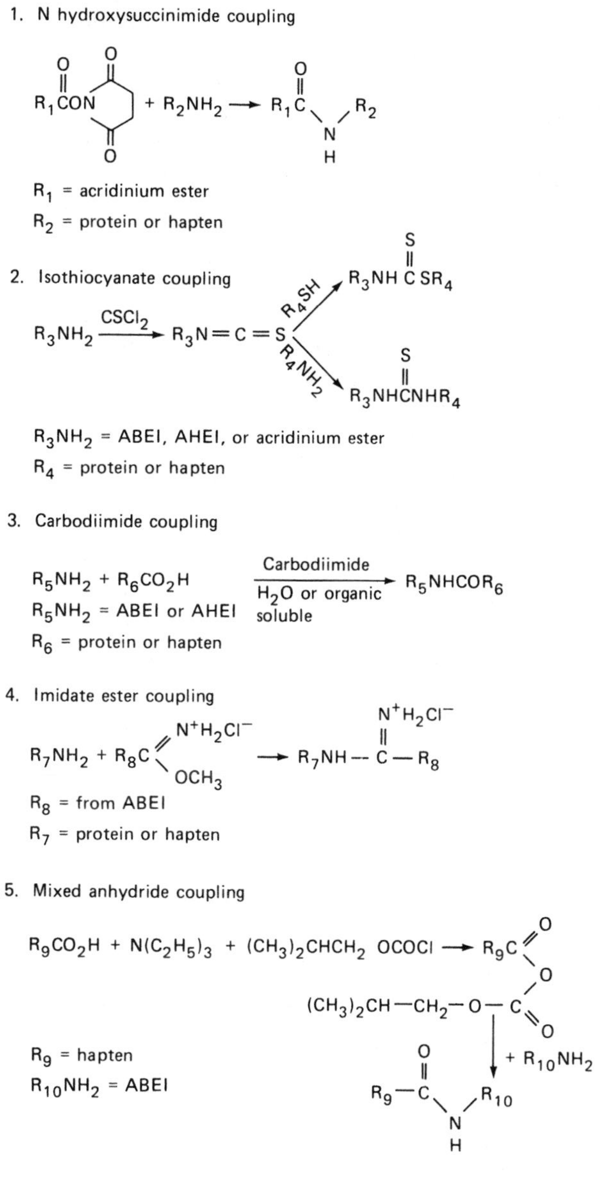

1. N hydroxysuccinimide coupling

R_1 = acridinium ester

R_2 = protein or hapten

2. Isothiocyanate coupling

R_3NH_2 = ABEI, AHEI, or acridinium ester

R_4 = protein or hapten

3. Carbodiimide coupling

$$R_5NH_2 + R_6CO_2H \xrightarrow[\substack{H_2O \text{ or organic} \\ \text{soluble}}]{\text{Carbodiimide}} R_5NHCOR_6$$

R_5NH_2 = ABEI or AHEI

R_6 = protein or hapten

4. Imidate ester coupling

R_8 = from ABEI

R_7 = protein or hapten

5. Mixed anhydride coupling

$$R_9CO_2H + N(C_2H_5)_3 + (CH_3)_2CHCH_2\ OCOCl \longrightarrow$$

R_9 = hapten

$R_{10}NH_2$ = ABEI

Fig. 8.4 — Some reactions for coupling labels to antigens and antibodies.

conditions to optimise the signal:noise, after only one second 10^{-18}–10^{-19} mol of acridinium ester can be detected.

There has been less success in the coupling of acridinium esters to haptens. Though this has been achieved with some steroids, our attempts to couple them directly to succinyl cyclic AMP highlighted a danger with them: the ester link is susceptible to hydrolysis resulting in formation of *N*-methyl acridylate, which is poorly chemiluminescent. This problem may also reduce the shelf life of some proteins labelled with acridinium esters. Chemical modification of the leaving group, or the acridinium moiety, will increase the shelf life.

Thus, of the synthetic and cofactor labels, ABEI and acridinium esters are the ones with the necessary chemical and luminescent properties to provide a wide range of labelled haptens, protein antigens, and antibodies. The labelled reagents themselves can be readily detectable in the fmol–amol (10^{-15}–10^{-18} mol) range. Antibodies containing at least three molecules of label per molecule of IgG can be prepared, therefore, having a higher 'specific activity' than is achievable with radioactive iodine. However, increasing the specific activity of protein antigens still further often leads to precipitation, because of loss of surface charge and increased hydrophobicity. It may be possible to circumvent this problem by using imidoester coupling. This retains the positive charge on the coupled nitrogen or the protein surface. As many as 20–30 label molecules can be coupled to one antibody molecule without precipitation, and with the retention of biological activity. The labelled derivatives can be detected within a few seconds, yet without the necessary triggering conditions can have a shelf life of many years.

Peroxidase has been a popular antibody label in enzyme-linked immunoassay and immunochemistry for many years. Since it 'catalyses' phthalazinedione chemiluminescence (see Chapter 7), peroxidase, usually horseradish, can be used as a 'chemiluminescent' label. However, the detection limit using luminol alone is only about 10 fmol, owing to the high chemical blank with luminol + H_2O_2 alone, μM–mM concentrations being required. Using enhancers such as p-iodophenol (see Chapter 7 and section 7.3.3), or bioluminescent substrates such as the luciferin from *Pholas*, this detection limit may be improved considerably, down to 0.1 fmol.

The chemiluminescent label eventually chosen should be characterised by the following six parameters:

(1) *Purity*
The label should give a single spot on tlc or hplc, and each antigen or antibody molecule should have at least one molecule of label attached to it.

(2) *Chemical identity*
This includes: the R_F value of the original label on chromatography, chemically modifying the labelled molecule (e.g. ABEI succinyl cyclic AMP + KOH → succinyl ABEI + cyclic AMP), UV, and mass spectrometry for hapten labels, and identification of amino acid residues in proteins which have been labelled. The final structure of the label should also be checked, e.g. acridinium salts sometimes convert to acridylate.

(3) *Number of mols of label per mol of antigen or antibody*
This should be measured by absorbance, fluorescence of the label and the reaction product, and by the chemiluminescence yield. Extinction coefficients, quantum yields, and spectra may however be altered by conjugation of label to antigen or antibody. The molar coupling ratio should be 1:1 for a hapten, and 1–10 mols label per mol for proteins.

(4) *Chemiluminescence*
The assay conditions required for maximum photon yield, with minimum reagent blank, for the labelled derivative must be defined. Also the detection limit of both

the label and the labelled conjugate free or bound to antibody must be determined. This should be 10^{-16}–10^{-18} mol to better ^{3}H, and be equivalent to ^{125}I. The effect of antibody binding on quantum yield, kinetics, spectra and polarisation of the light emission should also be characterised.

(5) *Immunoactivity*
The ability of the labelled conjugate to displace radioactive label from antibody, and to bind to antibody to give a chemiluminescence Scatchard plot, should be established. The affinity constant should be similar to that of the unlabelled analyte, e.g. K_a for ABEI succinyl cAMP is similar to that for cyclic AMP binding to antibody, even though succinyl cAMP has a 10 times greater affinity. Too many molecules of label on an antibody may destroy its ability to bind to antigen, either through labelling of crucial amino acid side chains or precipitation. As was discussed earlier, the latter problem may be circumvented by using imidate esters, which retain the charge on lys residues.

(6) *Stability*
The chemical stability, e.g. label remains attached to antigen or antibody, should be checked by tlc or hplc. This, together with its chemiluminescence and immunological reactivity, should show that the labelled reagent will last for years. As always with any luminescent reagent it is important to check that it is not sensitive to light, as photo-inactivation would lead to loss of activity under normal laboratory conditions.

The value of chemiluminescent labelled antigens or antibodies is critically dependent on whether assays can be established to quantify analytes of any real interest. How then have these been optimised, and simplified, to try to make them as easy to use and detect as a radioactive label?

8.2.3 Assay of the label
To get the most out of a chemiluminescent label it is necessary to find conditions which provide the most sensitive and convenient way of quantifying it. In particular, there are four things to optimise:

(1) Maximum signal:noise, i.e. highest photon emission, or maximum intensity, with lowest reagent blank when no label is present.
(2) Rapid detection, at most a few seconds.
(3) Non-interference from assay reagents or substances in the biological sample.
(4) Simplicity and convenience.

To satisfy these conditions it is necessary first to find the best way of presenting the sample. Interference from serum (Fig. 8.5) for example may necessitate solid phase assays, where all the interfering substances can be washed away before triggering the chemiluminescence. Quenching of the chemiluminescence when labels are bound to solid phase may require their release by hot, strong alkali before assay. The key parameters to optimise in the assay itself are usually:

(1) Selection of the oxidant.
(2) Selection of the 'trigger' or 'calatyst', if required.
(3) Concentration of reagents and order of addition.
(4) Buffer and pH.
(5) Method of light detection.

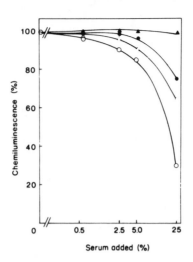

Effect of serum on ABEI chemiluminescence. To 50 μl of 10 nM ABEI in phosphate buffer, pH 7.4, containing 0.01% BSA was added 50 μl of varying dilutions of either rabbit serum (△, ▲) or human serum (○, ●). The chemiluminescence of the ABEI was then measured using the microperoxidase–H_2O_2 assay at pH 9.0 (▲, ●) or at pH 13.0 (△, ○). The percentage inhibition of the chemiluminescence without any serum added is shown

Fig. 8.5 — Interference of ABEI chemiluminescence by serum from Chemiluminescence energy transfer. Patel & Campbell (1985) in Alternative Immunoassays pp 153–183. Ed. Collins, W. P. © 1985. Reprinted by permission of John Wiley & Sons Ltd.

Phthalazine diones
Luminol works best at pH 9–13. Several early immunoassays with phthalazinedione labels used alkali + NaOCl ± H_2O_2 as trigger. Though the reaction is very fast, being complete within a few seconds, the apparent quantum yield is poor; the detection limit being only about 10^{-15} mol luminol. This, together with the damaging effect of NaOCl on auto-injectors in several commercial chemiluminometers, led Schroeder to systematically investigate alternative oxidants and triggers. These included NaOCl + $CoCl_2$, $NaIO_4$, $NaBO_3$, potassium *t*-butoxide, KO_2, $K_2S_2O_8$, hypoxanthine + xanthine oxidase to generate O_2^-, and H_2O_2 + $K_3Fe(CN)_6$, deuterohaemin, haematin, catalase, lactoperoxidase and microperoxidase (11 amino acid N terminus + haem of cytochrome c). All bar one (O_2^-) were capable, at the appropriate concentration, of fast kinetics, all the photon counts being generated within 10–100 s. The most efficient stimulus was microperoxidase, with a quoted detection limit for luminol in the range of 1 pM (10^{-16} mol in 100 μl) at pH 9–12, whereas lactoperoxidase and potassium *t*-butoxide were poor, with detection limits several hundredfold worse. Reproducibility with microperoxidase was good, being ±5%. The vinyl groups of deuterohaemin and haematin are prone to autoxidation and hence aggregation. However, under rapid conditions, this does not appear to affect seriously their potency as 'catalysts'.

Microperoxidase + H_2O_2 is thus the 'trigger' of choice for optimal detection of luminol and ABEI. However, the precise concentrations, together with the pH for optimal signal:noise, vary not only between luminol and ABEI (Fig. 8.6), but also

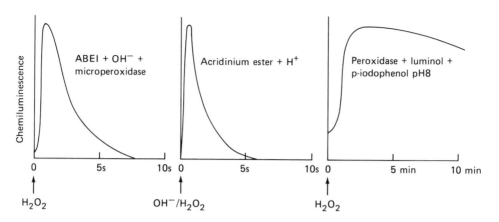

Fig. 8.6 — The detection of chemiluminescent labels. These traces are only approximate and intended to illustrate the flash of ABEI or acridinium ester versus the glow of luminol + peroxidase.

often between the unconjugated and conjugated label. Unlike luminol, ABEI produces most counts at pH 13, but has a long $t_{\frac{1}{2}}$ of 15 s at this pH requiring > 90 s for completion of the reduction (i.e. $6 \times t_{\frac{1}{2}}$ see chapter 2 section 2.1.3). The chemical blank of H_2O_2 + microperoxidase also increases with pH (Fig. 8.6). A typical assay for ABEI therefore uses 5 μM microperoxidase at pH 9 or 13, and is initiated by addition of 175 mM H_2O_2. This generates $> 10^{19}$ recorded photon counts per mol ABEI, compared with 7×10^{17} using NaOCl, and a detection limit of 5×10^{-17} mol. The latter is determined by the reagent blank, which at best is never less than 2000–3000 recorded photon counts in 10 s. This chemical blank is highly susceptible to 'organic' contaminants in the water used to prepare the buffers, which can increase the blank 10-fold. It can also be affected by the type of buffer used. Borate, for example, a common buffer in immunoassay, gives a very high blank with microperoxidase + H_2O_2 alone, whereas barbitone and phosphate are good buffers, generating much lower blanks.

Acridinium esters
These react with HO_2^-. Since the pKa for H_2O_2 is *ca.* 13, acridinium esters work best in strong alkali, e.g. 0.1 mM NaOH, and H_2O_2 *ca.* 50–200 mM. No 'catalyst' is required. Thus the reagent blank is much less, and more reproducible, than with phthalazine diones, and the detection limit for acridinium esters is about 10^{-17}–10^{-18} mol. One problem is the formation of 'pseudobase' (Fig. 8.7). OH^- will attach to the same carbon as that attacked by HO_2^-. At alkaline pH the reverse reaction, necessary to produce the moiety for H_2O_2 to initiate rapid chemiluminescence, is slow. Thus acridinium esters are best assayed by pre-incubation at pH 5–7 and then triggering them rapidly by jumping the pH with H_2O_2. Counting times are 1–10 s. The time to peak intensity, as with phthalazine diones, is usually just a few milliseconds, and may require adjustment if the injector has any light leak. Some assays, not requiring the maximum sensitivity of acridinium ester detection, slow down the reaction by reducing the concentration of H_2O_2, thereby improving

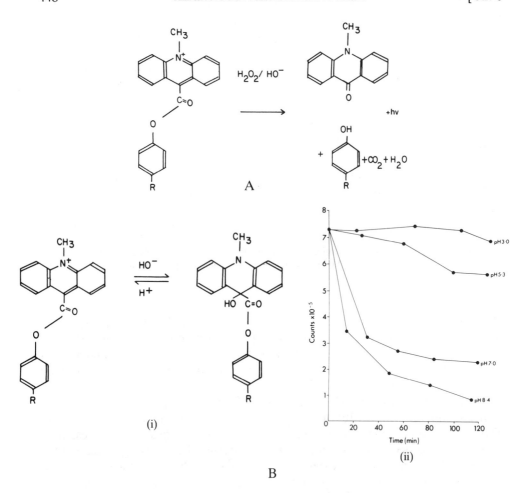

Fig. 8.7 — The chemical reaction of acridinium esters. A. Chemiluminescence; B. Pseudobase formation, graph = pH dependency of reaction. cts = recorded in 10 secs on adding H_2O_2.

precision. It is important to monitor the exponential decay rate constant when using phthalazine dione or acridinium ester labels to ensure that there are no substances present which interfere with the kinetics of the chemiluminescence.

Acridinium esters are the labels used in the MagicLite immunoassay systems produced by Ciba–Corning, using single tube assays and detecting chemiluminescence using a highly sensitive chemiluminometer.

Peroxidase
Peroxidase labels have been detected by using luminol over the wide pH range of 5–13, though best signal to noise seems to be achieved at alkaline pH. Since these assays are carried out with 'excess' luminol, usually 10–1000 μM, the chemiluminescence trace is quite different from that obtained with the organic labels (Fig. 8.6), being slow to decay. Counts are integrated over a 10–60 s interval. However, detection below 10^{-14}–10^{-15} mol peroxidase is difficult because of the 'blank' light

emission from luminol + H_2O_2 alone. As we saw in the last chapter (section 7.3.3) a puzzling, but significant, discovery by Whitehead, Kricka, and colleagues was that addition of μM concentrations of firefly luciferin not only increased the peroxidase induced luminol light intensity by 2–3 orders of magnitude but also reduced the reagent blank (i.e. luminol + H_2O_2 alone). A number of 6 hydroxybenzothiazoles and other phenols also have this property. Dehydro firefly luciferin ($\geqslant 10 \mu$M) and 6-hydroxybenzothiazole are most potent, increasing the signal some 3000-fold, whereas 2-cyano-6-hydroxybenzothiazole is some 10-fold less potent. Substituted phenols and naphthols also are potent enhancers of peroxidase, p-iodo and p-phenyl-phenol being particularly potent. Being more easily available, and thus cheaper, these tend to be used in commerical assays. The condition should be optimised for each particular enhancer. p-iodophenol and firefly luciferin enhancement has a sharp pH optimum of 8.6, with little or no effect below pH7 or above pH 9.5. At the optimum pH and using about 50 μM luminol with 0.2 mM H_2O_2 as oxidant, the full enhancement by p-iodophenol was 2500 with luminol, 120 with isoluminol and 90 with ABEI. In a typical immunoassay using peroxidase as a label the concentration of luminol will be 50–200 μM, with 1–2 mM H_2O_2 and pH 8–8.5, the optimal p-iodophenol concentration being about 340 μM. At higher concentrations of enhancer the potentiation decreases. Under similar conditions the optimal concentration for p-phenyl phenol is 11 μM and for firefly luciferin 55 μM.

A particular advantage of peroxidase as a label is that the chemiluminescent substrate concentration can be high, so that the light intensity remains elevated for several minutes, unlike the acridinium esters where the reaction is all over in a few seconds. In the enhanced chemiluminescence immunoassays available in the Amer-lite system, produced by Amersham International, assays are carried out in 96 well microtitre plates, darkened and read with a 'mask' to prevent cross-talk (i.e. light), between wells. Light intensity stabilises within 2 minutes or so and remains elevated for as long as 20–30 minutes, plenty of time to scan a whole microtitre plate . The chemiluminescence from such plates can also be recorded on high sensitive photographic film e.g. Polaroid type 612 (ASA 20 000) or X-ray film (OG1). The unknown sample can be matched by eye to a standard or more accurately using circles of step neutral density filters (Fig. 8.10). A photographic record can be obtained in seconds to minutes. However it is not clear precisely what is the detection limit for peroxidase using luminol-enhanced chemiluminescence nor whether it makes horseradish peroxidase better than micro-peroxidase. It appears to be 0.1–1 fmol, far below the detection limit of the acridinium esters. It is also unsatisfatory not knowing the mechanisms of the enhancement, as this might result in unsuspecting interference from substances in biological samples. Four possible mechanisms have been considered: energy transfer, oxidant processing, influence on low ('dark') versus high ('light') quantum yield pathways, effect on hydrophobicity which is known in miscelles or through antibody binding to increase chemiluminescence quantum yield. Energy transfer to the thiazole or phenol has been ruled out since the spectrum is unchanged (λ_{max} luminol 425 nm). Peroxidase labelled antibodies detected by luminol-enhanced chemiluminescence also have important application in Western blotting, the detection of specific protein antigens following electrophoresis and transfer to nitrocellulose. Recording the light signal on high speed Polaroid or X-ray film improves detection of peroxidase compared with absorbing dyes.

One possibility is that the benzothiazoles or phenols stabilise the Fe^{2+}–OOH or Fe^{3+}-O intermediate oxidant in a soluble form, thereby bringing horseradish peroxidase to the same efficiency as microperoxidase. Nevertheless luminol-enhanced peroxidase chemiluminescence has been used to record an immunoassay for ferritin (2.5–20 ng ml^{-1}) on Polaroid film with a 30–60 s recording time, and with 35 s development. Whilst the potential of chemiluminescent labels is usually only achieved using a sensitive photomultiplier as detector (see Chapter 2), recording on photographic film may have valuable applications at the patient's bedside, in the surgery, in the Third World and for 'in the field' assays. Furthermore, a label producing a long-lived chemiluminescence trace can be very useful since reagents can be mixed away from the light detector, and have potential for use inside cells.

An interesting approach to the problem of triggering the chemiluminescent label has been developed by Hummelen and colleagues. They have synthesised stable 1,2-dioxetans linked to adamantylidene adamantane (Fig. 8.3). These can be coupled to antigen and antibodies. At 25°C these dioxetans have a half life of 10 000 years, but on heating to greater than 200°C the dioxetans breakdown within 0.1–1 s producing a chemiluminescence with a λ_{max} at 425 nm and a ϕ_{CL} of about 10^{-4} (i.e. 6×10^{19} photons/mol). The excited state product is also capable of energy transfer to rubrene or 9,10-dibromoanthracene. Thermo-chemiluminescent immunoassays for substances such as IgG have been established using these labels but their clinical usefulness remains to be established, particularly in view of the relatively poor quantum yield. Of course, the chemical blank is virtually zero, in the absence of the label.

Finally, a word about signal to noise. Measurement of total ($>98\%$) chemiluminescence counts requires a counting time of $> 6 \times t_{\frac{1}{2}}$. The optimal counting time will, however, depend on the time course of the trace relative to the reagent blank (Fig. 8.6). If the label produces a peak within a few seconds and then decays, but the blank does not decay or increase with time, then the optimal signal to noise will be achieved at the peak in light intensity.

We now have the reagents and the detection system. Let us see some working examples of their use in real assays.

8.2.4 Some actual immunoassays

Immunoassays using chemiluminescence have now been established for more than 50 haptens and polypeptides (Table 8.6) These include hormones, drugs, embryonic proteins and bacterial or viral antigens. Peroxidase has been the most popular label, but ABEI and acridinium esters have also been applied to a wide range of analytes. Several assays are being applied in laboratories where clinical decisions are taken, as well as being used as a routine research tool. Chemiluminescence immunoassay kits from several major companies are now available (see Appendix V). The most successful range of assays use either acridinium esters or peroxidase detected by luminol-enhanced chemiluminescence. Some use polyclonal, others monoclonal, antibodies. Some are the chemiluminescence equivalent of radioimmunoassay, using labelled antigen in a competition assay, CIA (Chemiluminescence ImmunoAssay) for short. Others use the non-competitive, labelled antibody approach, ICMA (ImmunoChemiluminoMetric Assay). Whilst these two alternative strategies are very similar in principle to their radioisotope counterparts, the special requirement

Table 8.6 — Some heterogeneous chemiluminescence immunoassays

Label	Detection	Analyte
A. Synthetic		
luminol	H_2O_2 + peroxidase	testosterone
diazoluminol	H_2O_2 + peroxidase	IgG
isoluminol	H_2O_2 + peroxidase	biotin
ABEI	H_2O_2 + microproxidase	progesterone, cortisol, oestradiol, testosterone, steroid glucuronides, cyclic nucleotides, C9, IgG, digoxin
AHEI	H_2O_2 + peroxidase	progesterone, hepatitis B
ABEN	H_2O_2 + peroxidase	gentamicin, thyroglobulin, ferritin, transferrin, $\alpha_2$macroglobulin, caeruloplasmin
acridinium ester	H_2O_2/OH^-	IgG, αfetoprotein, TSH, ferritin, parathyroid hormone, T_4, 17β oestradiol
AAD	heat (thermochemiluminescence)	IgG
B. Cofactor		
AENADH	bacterial oxido-reductase/luciferase	biotin,2,4-dinitrofluorobenzene
ATP	firefly luciferin – luciferase	2,4-dinitrofluorobezene
FAD	theophylline	bacterial luciferase
C. Enzyme		
peroxidase	luminol	cortisol, insulin
peroxidase	luminol + enhancer	progesterone, digoxin, T_4, ferritin, antirubella IgG, anticytomegalo virus IgG, IgE, αfetoprotein, choriogonadotropin, carcinoembyronic antigen, factor VIII-related antigen,
peroxidase	pyrogallol	anti-sinbis virus, anti SEB IgG
peroxidase	pholasin	anti IgG
bacterial luciferase	NAD, FMN, oxidoreductase, long chain aldehyde	IgG (cell surface)
firefly luciferase	ATP, luciferin	methotrexate, IgG
aequorin	Ca^{2+}	progesterone
GbP dehydrogenase	NAD, G6P, bacterial luciferase	progesterone

Note: in principle both the competitive and immunometric strategies are applicable with any label. However these examples incorporate a mixture of the two strategies.
ABEI = aminobutylethyl isoluminol	AHEI = aminohexylethyl isoluminol
ABEN (or ABENH) = 7-[(N-4-aminobutyl)-N-ethylamino]naphthalene-1,2-dicarboxylic acid hydrazide
AAD = adamantylidene adamantane 1,2-dioxetan

for detection and quantification of chemiluminescent labels imposes two important practical restrictions in the assay protocols:

(1) Solid phase separation is the most practical method of separating bound complexes from free antibody and free antigen (Fig. 8.8). These may be coupled covalently by diazotisation, CNBr, carbodiimide or other chemical agents. However many plastics bind proteins, particularly antibodies, readily. Thus plastic tubes, beads or microtitre plates can be incubated with a solution containing antibody, at the appropriate pH, and sufficient protein will remain stuck on the surface of the plastic to be used as the solid phase reagent in an immunoassay. Isolation of immune complexes by filtration, e.g. in automated machines, not only makes assay of chemiluminescence difficult, but the filters often introduce contaminants resulting in

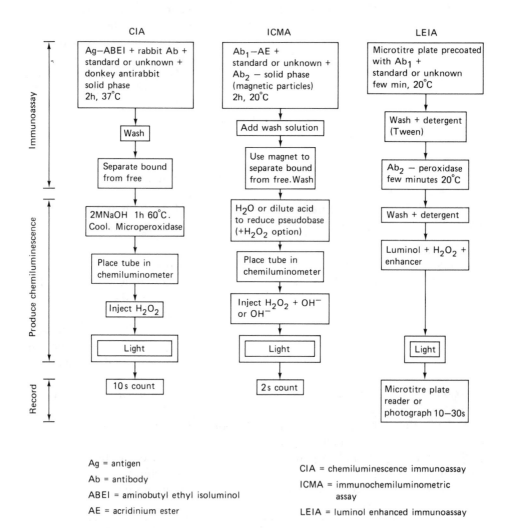

Fig. 8.8 — Protocols of CIA and ICMA.

increased chemiluminescence blanks. Furthermore, a solid phase reagent such as antibody or antigen on aminocellulose, or antibody or antigen coated on to particles, or antibody coated tubes or microtitre plates, makes it easy to wash away not only free labelled reagents but also substances from serum, urine, or other biological material which may interfere with the chemiluminescence. Antibodies bound to magnetic particles makes separation easy and rapid using a magnet. This is used in the MagicLite system. However, the particles may need to be removed prior to the chemiluminescence assay as they may quench the luminescence.

(2) A pH jump may be required to trigger the chemiluminescence. Both ABEI and acridinium esters are most sensitively detectable at pH 11–13. Antibody–antigen complexes dissociate rapidly, within seconds, at pH > 11. Thus, the assay is usually set up at pH 7–9, all manipulations prior to triggering the label being conducted at this pH. Some ABEI and peroxidase assays are carried out at pH 7–9, in which case the pH jump is unnecessary. In contrast peroxidase-labelled antigen or antibody detected by luminol enhanced chemiluminescence at pH 8.6 will not require a pH jump.

Chemiluminescence assays for a range of steroids have been established, using ABEI-antigens by the groups of Kohen and Collins. Assays for several polypeptides have also been established (Table 8.6). Five steps are involved in these assays. Competition for free rabbit antibody between a standard or an unknown quanitity of analyte and ABEI-antigen is the first step. Labelled antigen bound to a rabbit antibody is then bound by solid phase donkey anti-rabbit IgG. This incubation takes only 2 h at 37°C to equilibrate. After washing, the ABEI-Ag must be dissociated from the solid phase by strong, hot alkali if reproducible, accurate chemilumines-cence signals are to be obtained. The reason for this is not clear since in several homogeneous assays (see section 8.2.6 and Chapter 9) antibody binding does not quench ABEI chemiluminescence. Free ABEI-Ag is then assayed with microperoxi-dase and H_2O_2. An alternative protocol, still based on 'competition', involves the use of a 'universal' reagent, ABEI sheep anti IgG. In this assay, a standard amount of antigen is on a solid phase; this is incubated with a limiting amount of rabbit antibody, together with a standard or unknown sample of analyte (the same antigen as the solid phase reagent) and ABEI sheep anti-rabbit IgG. As the free analyte concentration increases less rabbit antibody binds to the solid phase, and thus the amount of ABEI-sheep anti-rabbit IgG bound to the solid phase decreases.

ICMA (immunochemiluminometric assays) using both ABEI and acridinium ester antibodies have also been established. Here, excess labelled ABEI antibody is added to standard or unknown sample of analyte. The antibody–antigen reaction is allowed to equilibrate, and excess solid phase antigen is added to mop up any free ABEI-antibody not bound to antigen. After removal of the solid phase, the amount of chemiluminescent label in the remaining fluid is determined. The problem here is that interfering substances from the biological sample are still present. Urine and serum ($> 5\%$) markedly reduce ABEI chemiluminescence (Fig. 8.5). Thus, most ICMAs in use, with either ABEI or acridinium esters, use the so-called two-site approach. This is based on the principle that a polyclonal antiserum contains antibodies to more than one epitope on the antigen. The resulting assay is then more likely to select the complete amino acid sequence of any polypeptide, and less likely to detect peptide fragments. A more specific way of establishing a two site assay is to use either two monoclonal antibodies selected for the distinct epitopes, or one monoclonal and one polyclonal antibody. The principle of these two-site ICMAs therefore is to incubate standard or unknown analyte with solid phase antibody to which it will bind. The solid phase is washed and the second antibody, chemiluminescently labelled, is then added. The amount of chemiluminescence bound to the solid phase, after removal of excess free, labelled second antibody, will be directly proportional to the concentration of analyte orginally present. The two-site approach will of course, only work in general with protein and macromolecular antigens. Small molecules such as drugs or steroids can only bind one antibody molecule at at time.

In some assays all three reagents, solid phase, analyte, and labelled antibody, can be added together in the first step. For example, in a two-site ICMA for α-fetoprotein we have used a solid phase (cellulose) polyclonal antibody incubated for only one hour with α-fetoprotein plus acridinium ester labelled monoclonal to α-fetoprotein (0.3–3 mols label per mol antibody) (8.9). The cellulose is centrifuged and unbound antibody-label in the supernatant is removed. The solid phase is then washed once and acridinium ester on the solid phase, directly proportional to the α-fetoprotein originally added assayed by injection of 0.1 M NaOH containing 0.1% H_2O_2 (Fig. 8.9B):

$$\alpha FP + \text{solid phase } Ab_1 + Ab_2 - AE \rightarrow \text{solid phase } Ab_1 - \alpha FP - Ab_2 - AE$$
$$+ Ab_2 AE \quad (8.9)$$

Ab_1 = polyclonal, Ab_2 = monoclonal, αFP = α-fetoprotein, AE = acridinium ester.

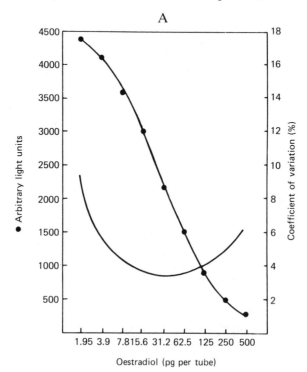

Oestradiol (pg per tube)

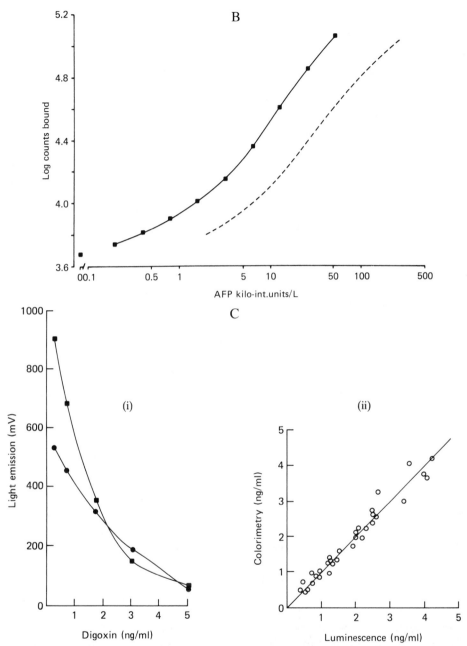

Fig. 8.9 — Some examples of CIA and ICMA. A. CIA or oestradiol using ABE1, with precision profile, from Barnard *et al.* (1985) Chemiluminescence immunoassays and immunochemilu-mino metric assays In: *Alternative immunoassays* pp 132. ed. Collins, W. P., Wiley, Chichester, © 1985. Reprinted by permission of John Wiley & Sons Ltd. B. ICMA for α-fetoprotein using acridinium ester from Weeks *et al.* (1983) see section 8.5 of The American Association for Clinical Chemistry — 3 mols acridinium ester per mol antibody; --- lmol acridinium ester per mol antibody. C. Peroxidase enhanced immunoassays for digoxin from Coyle *et al.* (1986) *Anal. Clin. Biochem.* **23** 42–46 with permission of Blackwell Scientific Publications Ltd.; (i) Standard curve A 45m B 15 min incubation and   (ii) correlation with colorimetric competitive enzyme immunoassay.

The counting time is 10 s. Non-specific binding of acridinium ester–antibody to the solid phase, the factor limiting sensitivity, is usually $<0.5\%$. This low non-specific binding, together with the high sensitivity of detection of the label ($ca.$ 10^{-17}–10^{-18} mol), are sufficient to provide an assay for α-fetoprotein at least as sensitive as the ^{125}I equivalent IRMA, but with all the advantages of a non-isotopic label. The assay is capable of measuring α-fetoprotein in patient samples both accurately and precisely in the detection of neural tube defects in unborn children. More recently, magnetic particle solid phase assays have been developed, and the counting time reduced to 1–2 s.

A similar two-site approach has been adopted to establish assays for α-fetoprotein and ferritin using peroxidase-labelled polyclonal antibodies detected by 6 hydroxybenzo-thiazole or iodophenol enhanced luminol chemiluminescence. Polystyrene beads or the wells of microtitre plates are coated first with rabbit anti-α-fetoprotein or antiferritin. The analyte is incubated with this solid phase for about two hours, and rabbit anti-analyte antibody labelled with peroxidase added. Excess labelled antibody not bound to the solid phase is then removed and the peroxidase, attached to the beads or the plates, is assayed by addition of 200 μl of assay cocktail containing luminol (1–2 μmol), firefly luciferin, 6 hydroxybenzothiazole or iodophenol (10–50 nmol), and H_2O_2 (2–5 μmol) in tris buffer (0.1 M pH 8). The continuous glow from each well can then be detected with a chemiluminometer, or in some assays, is bright enough to be seen and recorded on fast Polaroid film (Fig. 8.10), film exposure 30–60 s, development $ca.$ 35s. Because of the prolonged chemiluminescence all the manipulations can be carried out outside the chemiluminometer. The whole microtitre plate (96 wells) can be read in just a few minutes.

Thus, the isoluminol derivatives, acridinium esters, and peroxidase can be used as labels in virtually any immunoassay strategy currently available. They work in 'competitive' or 'non-competitive' assays, in one- and two-site assays, and with polyclonal or monoclonal antbodies. They can be used to prepare 'universal' reagents, e.g. labelled anti-antibody. They provide assays for haptens and polypeptides, though hapten assays are difficult, if not impossible, to establish using the two-site approach, but this is true whether you are using chemiluminescent labels or radioactive ones. They can be used to measure 'free' analytes in plasma, for example, free T_3 or T_4, or free drugs and steroid hormones where a large fraction of the analyte is bound to thyroglobulin in the case of T_3 or T_4, or albumin in the case of drugs.

Are chemiluminescent labels then a real alternative to radioisotopes in immunoassay? To answer this we must examine three parameters, firstly their performance in assays on which clinical decisions are to be taken, using conventional immunoassay criteria; secondly how well they satisfy the criteria of an ideal non-isotopic label laid down in section 8.1.3; and thirdly, what factors really limit detection in immunoassays to see whether the detection of the label itself really matters anyway.

8.2.5　Are chemiluminescence immunoassays a real alternative?

Chemiluminescent labels, based on isoluminol derivatives, acridinium esters, and peroxidase, do satisfy the nine criteria for an ideal non-isotopic label (section 8.1.3).

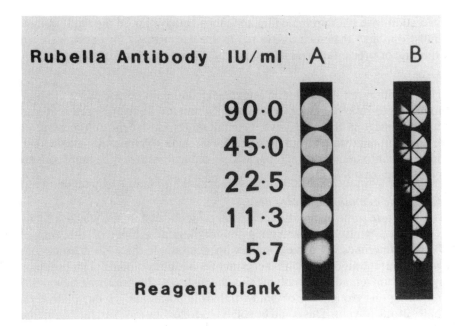

Fig. 8.10 — Detection of Rubella using peroxidase enhanced immunoassay with Polaroid film. From White *et al.* (1987) In *Bioluminescence and chemiluminescence* pp 335 Ed. Scholmerich *et al.* John Wiley, Chichester, © 1987. Reprinted by permission of John Wiley & Sons Ltd.

(1) *Sensitivity*

As with any immunoassay this is determined by three factors: antibody affinity, specific activity of the label, and the precision of the separation step. All three chemiluminescent labels can be readily detected within a few seconds in the fmol–amol range. They are all therefore superior to ^3H (Table 4.7), used in many of the early steroid assays. However, the chemical reagent blank for isoluminol derivatives and peroxidase limit their detection to about 10^{-15}–10^{-16} mol. Only the acridinium esters, at present, appear to approach the detection limit for ^{125}I of 1–10 amol (10^{-17}–10^{-18} mol). However, the improvement of peroxidase 'enhancers', the use of luciferins such as pholasin with a higher quantum yield and lower chemical 'noise' than luminol, and the use of a better peroxidase than horseradish, for example a luciferase peroxidase, may enable tipomol (10^{-21}) sensitivity to be achieved.

Chemiluminometers are simpler electronically than scintillation counters, though automation will be necessary for large number assays. Detection of isoluminol derivatives and acridinium esters, however, requires an injection into the sample chamber, whilst the sample is directly in front of the photomultipier tube. This introduces another factor of imprecision, and also must not allow any light leak into the chamber. It is a complication which can be avoided by using enzyme labels such as peroxidase or luciferase which produce a steady glow, initiated before placing the sample in front of the light detector. Highest signal:noise, and thus the best sensitivity, for a simple organic label, e.g. ABEI or acridinium esters, requires the fastest chemiluminescence reaction possible.

Detection with photographic film appears attractive for 'in the field' assays, but one must establish that such assays really are better than, for example, a simple colorimetric or precipitation assay, to justify the use of chemiluminescence.

(2) *Visible*
Image intensifiers are available for detecting chemiluminescence as sensitiviely as a photomultiplier tube (Chapter 2), beyond the intensity per unit area necessary to produce an image in the naked eye. Chemiluminescent labels are therefore applicable to light microscopy, but not yet to electron microscopy. Again, a label producing a continuous glow, e.g. peroxidase or luciferase, may be more convenient than one producing a flash.

(3) *Couplable to biological molecules*
This may require some ingenuity from the organic chemist, but as we have seen, has been achieved whilst still preserving immunological activity of the antigen or antibody. Sometimes, as for example with oestradiol, the easiest group on the antigen to couple to is one which is necessary for antibody binding. This problem can easily be circumvented by modifying the chemistry, or by preparing new antibody not requiring this group. A problem with thyroid hormones is their relatively poor solubility in any solvent. This could be solved by following the example of the thyroid gland, use thyroglobulin and then cleave off the labelled thyroxine.

(4) *Homogeneous*
As we shall see (section 8.2.6) several types of homogeneous assay, not requiring a separation step, have been established using chemiluminescent labels.

(5) *Stable*
So far, all the labelled reagents have been found to have a shelf life of many years, though the ester bond in the acridinium labels is relatively labile and care must be taken to minimise spontaneous cleavage from the antigen or antibody. Modified compounds are now available which are much more stable than the original acridinium esters.

(6) *Environmental factors*
Chemiluminescent labels are more susceptible to interference than radioactive labels, for example by proteins or small molecules in serum and urine. The reaction kinetics may change because of 'scavengers' or 'enhancers' of active oxygen intermediates. Haemoglobin absorbs light, and catalyses phthalazine dione chemiluminescence. There may be 'excited state' quenchers present, e.g. drugs. These problems can be easily circumvented by using an assay protocol which removes all interfering substances before assaying for chemiluminescence. Solid phase, such as magnetic particles, may preclude measurement at high concentrations on them. But microtitre plates, cellulose, or sepharose and low particle concentrations seem all right.

(7) *Clinically applicable*
Many clinically useful assays have been established (Table 8.6) which satisfy the normal criteria for a good immunoassay. The working range can be established by evaluation of the precision profile (Fig. 8.11A) and a good correlation ($r > 0.95$) with its radioactive counterpart checked (Fig. 8.11B). At least one has been reported by

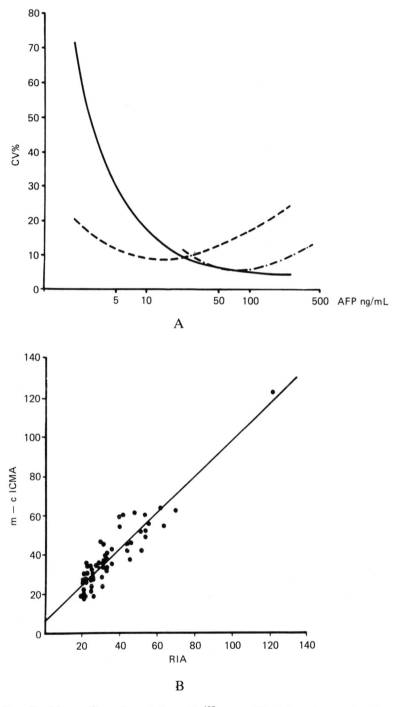

Fig. 8.11 — Precision profile and correlation with ^{125}I assay. ICMA for α-fetoprotein using acridinium ester labelled antibody, from Weeks *et al.* (1983) see Section 8.5 with permission of The American Association for Clinical Chemistry. A. Precision Profile. B. Correlation with [125]I 8.5 RIA.

Weeks and colleagues, for thyroid stimulating hormone (TSH), which has superior performance to its ^{125}I equivalent, due to the combination of a good, well characterised monoclonal to TSH and the merits of acridinium ester labels. This is a two-site ICMA using solid phase polyclonal and a labelled monoclonal to TSH. The nonspecific binding to the solid phase is $< 0.1\%$ of the label added. The whole assay takes only three hours, and produces a linear dose response up to 60 mU TSH per litre with an absolute sensitivity of 0.004 mU per litre. This assay is therefore not only able to measure TSH in normal and hypothyroid patients, 0.4–4 mU TSH per litre, but also in hyperthyroid patients with circulating TSH levels < 0.03 mU per litre. Thus a chemiluminescence assay has enabled an important clinical parameter to be detected in patients where previous assays using radioisotopes have failed.

It has been reported that both ^{125}I based and peroxidase enhanced luminol chemiluminescence (Amerlite) assays for TSH may be subject to interference, particularly when patients sera contain anti IgG or anti TSH antibodies. Using an increased sensitivity of antibody detection in the assay e.g. by using acridinium esters, it may be possible to circumvent this problem by increasing the dilution of the patient sample.

The concentration of *free* hormones and drugs (hapten) can be several orders of magnitude less than the total present in the plasma. By adding a fixed amount of hapten linked to a protein antigen and a labelled antibody to the hapten, the amount of hapten-protein conjugate bound to the labelled antibody increases as the free hapten concentration decreases. This can be measured by adding a solid-phase antibody to the protein antigen in the hapten conjugate, enabling it to be separated from unbound label. The amount of labelled hapten antibody is insufficient to disturb the equilibrium between bound and free hapten in the plasma. Weeks and colleagues have established such an assay for free thyroxine (T_4) in human plasma. A T_4 rabbit IgG conjugate and a mouse monoclonal antibody to T_4 labelled with acridinium ester (1 mol per mol) are mixed with a serum sample or standard for about 30 minutes. Excess sheep anti rabbit IgG attached to magnetic particles, is then added which binds all the T_4-IgG conjugate, some of which will have acridinium ester labelled monoclonal antibody bound to it. The greater the free T_4 in the serum sample the less label will be bound to the magnetic particles. These are separated by a magnet, washed and the chemiluminescence attached to them assayed. The standard curve for free T_4 ranged from 2 pM to 100 pM sufficient for normal subjects (10–30 pM) and patients with chronic renal failure (2–15 pM).

A second advantage of chemiluminescent labels in immunoassay over ^{125}I has again been highlighted by the acridinium esters. As predicted by Miles and Hales in the late 1960s, by increasing the number of moles of label on an antibody to more than one (3 mols per mol in Fig. 8.9B) the sensitivity of the immunassay is increased. This is not possible with ^{125}I, owing to radiolytic damage.

When examining the claims for immunoassay sensitivity always check firstly whether the detection limit was established with a biological sample or pure standard, and secondly whether the value quoted refers to the amount or concentration in the sample or in the assay tube.

(8) *Safe*

So far as we know, the chemiluminescent labels are safe to handle, though I would not like to take a spoonful in my coffee! Thus, unlike radioisotopes, no special precautions appear necessary. However, proper toxicity tests have not been fully carried out. Many organic compounds, including acridines, the precursors of the acridinium esters, are carcinogenic. This needs to be checked. Care may be particularly necessary during synthesis, where several other reagents are also dangerous.

(9) Easily available

Isoluminol derivatives and acridinium esters are relatively easy to synthesise in gram quantities. Remember, 1 μg, just a speck, of ABEI or acridinium ester is capable of generating as many as 10^{11} photon counts in your apparatus! Keep the synthesisers away from the assayists if you want to avoid contamination of your reagents. Many assays using ABEI, acridinium ester or peroxidase are now commercially available (Appendix V).

Empirically, therefore, our three labels hold up against the criteria. But do they also hold up against the theoretical arguments?

What determines the ultimate sensitivity of an immunoassay has long been a source of argument. Is it the antibody; is it the label; or is it the overall strategy itself? Ignoring the label for a minute, it has been recognised since the 1960s that two factors limit the sensitivity of a 'competition' immunoassay, be it an RIA or CIA; (i) the overall experimental error (ε) including pipetting and radioisotope counting, and (ii) the affinity constant (K_a l.mol^{-1}) of the antibody. Ekins has shown that if the specific activity of the label is not limiting, i.e. infinite in mathematical terms, the best sensitivity attainable is given by the relationship:

$$\text{sensitivity} = \varepsilon/K_a \tag{8.10}$$

Thus, if the experimental errors including non-specific binding in the absence of added analyte are 1%, perhaps the best one can hope for, and the antibody affinity is 10^{12} l.mol^{-1}, the greatest normally found, then the detection limit will be of the order of 10 fM, or in an assay of volume 100 μl, 1 amol (10^{-18} mol). It seems obvious that with an isotope like ^{3}H only detectable down to about 10^{-16} mol this theoretical sensitivity can never be achieved. Thus, chemiluminescent labels win out over ^{3}H. ^{125}I and acridinium esters, detectable down to about 10 amol, enable the theoretical detection limit of an immunoassay to be approached. Many antibodies available, however, have much lower affinity than 10^{12} l.mol^{-1}. The affinity of a typical monoclonal may be only 10^9–10^{10} l.mol^{-1}, as cloning procedure tends to select out the more prevalent low affinity–producing hybridoma clones. This reduces the detection limit to 0.1–1 fmol in 100 μl (1–10 pM). In practice, radioactive counting, and other experimental errors, are greater than 1%. The triggering of the chemiluminescent reaction introduces a further step, with a consequent increase in error.

The empiricists argue that because 'non-competitive' immunoassays, IRMAs and ICMAs, use *excess* labelled antibody, the affinity constant should no longer be a limiting factor. Non-specific binding can be reduced by using a labelled monoclonal antibody and the 'specific activity' of this labelled reagent can be increased by using chemiluminescence some 3 to 5-fold relative to ^{125}I. Mathematical analysis has shown that, over simplified as these views may be, under optimal conditions the sensitivity of an immunochemiluminometric assay could be anything from 5–10 times more sensitive than is ^{125}I equivalent, and 10–50 times more sensitive than a 'competitive' radioimmunoassay using an antibody with the same affinity. This means that, given the right antibody, not only can ICMAs measure analytes such as free hormones in the pM region (Table 4.3), the bottom end of the present radioisotope assays, but also with a chemiluminescent label detectable in the tipomol region (10^{-12}) analytes not measurable at present in the fM–aM (attomol–tipomol in 100 μl) range should become detectable.

Of the other labels used to establish chemiluminescence immunoassays (Fig. 8.3 and Table 8.4) few, if any, have yet to reach the clinical laboratory. The main reason for this is inferior performance compared with ABEI, acridinium ester or peroxidase based assays. However the development of cloned bioluminescent reagents based on luciferase or photoproteins may allow the superior sensitivity of detection of these chemiluminescent systems to be exploited clinically and commercially.

8.2.6 Homogeneous immunoassays

Homogeneous immunoassays are ones not requiring a separation step. Reagents and samples are added sequentially, or together, into one tube, and a signal is detected after an appropriate time interval. It is virtually impossible to establish such immunoassays using radioisotopes, though a homogeneous assay has been reported using a change in radioluminescence when a labelled drug bound to its receptor, using β quenching. Homogeneous assays, therefore, represent an area of great potential for chemiluminescent labels to exploit. But why bother? What would be the value of a universal strategy for homogeneous immunoassays? There are five important advantages of a homogeneous method, compared to a heterogeneous one requiring a separation step:

(1) It would provide a method for detecting chemical events inside living cells.
(2) Free substances in biological fluids could be detected without significant disturbance of equilibria.
(3) Ligand–ligand interactions could be quantified with $K_d < 10^{-8}$M, too fast to detect a bound ligand in the absence of free ligand, since on separation dissociation will occur within seconds, e.g. enzyme–substrate, receptor–agonist interactions.
(4) It would remove imprecisions inherent in the separation step.
(5) Being simpler, it would make assays cheaper, less laborious, and easier to automate.

The first study of an antibody–antigen interaction without a separation step was carried out by Arquilla and Stavitsky in 1956. Using erythrocytes labelled with insulin, they detected insulin antibodies in diabetics by their ability to activate complement, and thereby lyse the erythrocytes. This was a remarkable achievement, being published three years prior to Yalow and Berson's first accredited immunoassay paper. It is interesting to note tht Arquilla and Stavitsky's paper was actually submitted in December 1954! Perhaps they should be given more credit for establishing really the first immonoassay.

A homogeneous immunoassay was established for penicillin by Dandliker and Feigen in 1961, by measuring a change in fluorescence polarisation when fluorescein-labelled penicillin bound to its antibody. The advent of electron spin resonance (esr) labels enabled a change in rotational correlation to be used by Leute and co-workers in 1972 for a homogeneous immunoassay for morphine. Since then a variety of enzyme, fluorescent, and chemiluminescent labelled haptens and protein antigens have been used to establish homogeneous immunoassays on the basis of enhancement or quenching of the signal when bound to antibody (Table 8.7). These can be classified on the basis of the method of detection of the label, together with the effect on the signal caused by antibody binding to antigen (Table 8.8).

Homogeneous immunoassays have, however, only occasionally found real application in the clinical laboratory, for example, for certain drugs. Their wide-ranging potential has yet to be realised because of three major problems:

(1) Many homogeneous assays depend on the unique property of a particular antibody–antigen interaction, found by chance with one antiserum and therefore not generally applicable for all antibody-antigen complexes.
(2) The detection limits of the labels used in homogeneous assays to date are inferior to ^{125}I and the best chemiluminescent labels, namely acridinium esters.
(3) Many homogeneous immunoassays are interfered with by components of serum and other biological fluids.

In heterogeneous assays this last problem is circumvented by removal of interfering substances during the separation step. Heterogeneous assays are thus still preferred

Table 8.7 — Homogeneous immunoassays

Basis of assay	Type (see Table 9.10)	Example of analyte
enzyme inhibition	A1 + Bla	several drugs and proteins
erythrocyte lysis by complement	A2 + Bla	insulin
change in esr		morphine
liposome lysis inhibition	A1 + Bla	dioxin
nephelometry (formation of immune complexes detected by light scattering)	A6 + Bld	IgG, IgA, α, antitrypsin complement component C9
fluorescence quenching	A2 + Bla	human albumin, IgG, placental lactogen, dinitrophenol
fluorescence enhancement	A2 + Blb	thyroxine (T_4)
fluorescent substrate	A2 + Bla	drugs, IgG
change in fluorescence polarisation	A3	penicillin
fluorescence excitation energy transfer	A2 + Bla or b	T_4, drugs, IgG, various proteins
chemiluminescence enhancement	A4 + Bla, b or c	steroids
chemiluminescence energy transfer	A4 + B2a	cyclic nucleotides, progesterone, IgG

Table 8.8 — Classification of homogeneous interactions

A. Method of detection of the label
1. Spectrophotometry — enzyme label
2. Fluorescence emission
3. Fluorescence polarisation
4. Chemiluminescence
5. Electron spin resonance
6. Light scattering (nephelometry)
7. Radioactivity
8. Biological e.g. cell lysis

B. Change in signal induced by the antigen-antibody interaction
1. Enhancement or reduction of signal without change in spectrum
   (a) alteration in ability of label to interact with another molecule (e.g. a substrate, an enzyme or antibody directed towards the label as opposed to the analyte)
   (b) excited state quenching (direct or by energy transfer)
   (c) enhancement of quantum yield or kinetics
   (d) change in light scattering or polarisation
2. Change in spectrum
   (a) excitation energy transfer (see chapter 10)
   (b) rotational correlation in esr

---

by most immunoassayists. However, the potential sensitivity of chemiluminescent labels offers great promise that they may yet provide a universal, homogeneous strategy.

A homogeneous immunoassay based on chemiluminescence would require the binding of the labelled antigen to the antibody, or conversely a labelled antibody to the antigen, to change some parameter associated with the light signal when the chemiluminescence is triggered (8.11):

$$Ab + Ag^* \rightleftharpoons AbAg^* \xrightarrow{\text{trigger}} \text{different light signal from } Ag^* \text{ alone} \qquad (8.11)$$

This could be a change in one of three parameters:

(1) Light intensity, through a change in quantum yield or reaction kinetics.
(2) Colour, through an alteration in the chemiluminescence spectrum.
(3) Polarisation.

Very little investigation has been carried out on chemiluminescence polarisation. Fluorescence polarisation is usable because the incident exciting light can be produced as plane or circularly polarised. Whether the apparent random orientation of chemiluminescent labels could allow polarisation of the emitted light to be detected is a question of considerable fundamental interest, but yet to be investigated. It might also be possible on solid phase.

The other two parameters, namely colour and intensity, have been exploited to establish the principle of homogeneous chemiluminescence immunoassays (Table 8.9). The binding of several isoluminol-steroid conjugates to their respective antibodies can result in up to a 10-fold enhancement in maximum light intensity and has been used to establish working assays with reasonable sensitivity, though usually lower than the equivalent heterogeneous assay. Changes in the kinetics of acridinium esters on antibody–antigen binding may also provide the basis for homogeneous immunoassays. However, with both of these, it is not clear how general the phenomenon is. We have never seen a change in ABEI chemiluminescence when a range of antigens from 315–150 000 molecular weight bind to their respective unlabelled antibodies. Furthermore, homogeneous experiments highlight two particular problems associated with using chemiluminescence as a homogeneous detector of antibody-antigen binding. Firstly, the pH optimum for isoluminol derivatives and acridinium esters occurs where antibody–antigen complexes dissociate within seconds. Conditions, therefore, have to be found for triggering chemiluminescence and yet retain antibody–antigen complexes intact, i.e. pH < 9. This can be done, but it may reduce the sensitivity of detection of the labels currently available. Secondly, serum, urine, and other substances in biological fluids reduce chemiluminescence (Fig. 8.5). The homogeneous chemiluminescence immunoassays can therefore only be carried out with serum concentrations < 5–10%.

Table 8.9 — Some homogeneous chemiluminescence immunoassays and other binding protein assays

Chemiluminescent label	Homogeneous principle	Analyte
isoluminol or ABEI	quenching or enhancement of chemiluminescence (i.e. ϕ_{CL}) by antibody or binding protein	steroids, biotin
ABEI	resonance energy transfer (colour shift)	nucleosides, cyclic nucleotides, steroids, C9, IgG
ABEI	quenching by energy transfer	biotin
G6P dehydrogenase	local production of NADH, close to bacterial luciferase	progesterone

A further chemical problem with acridinium esters is that the excited state N-methyl acridone dissociates from the molecule to which it is attached, as a result of the chemiluminescent reaction (Fig. 8.7). Good as they are as labels in heterogeneous assays, we have yet to find the condition for acridinium esters in a working homogeneous immunoassay.

In contrast, by using ABEI linked to haptens or proteins and then bound to their respective antibodies labelled with fluorescein, a change in chemiluminescence spectrum towards the green has been detected relative to unbound ABEI-antigen.

By measuring the ratio of blue to green light simultaneously, homogeneous immunoassays have been established for a range of useful analytes, including cyclic nucleotides, steroids, polypeptides, and immunoglobulin. This apparently universal method is based on chemiluminescence energy transfer between the excited state product of the ABEI reaction and the fluor on the antibody. It is a radiationless transfer, not involving the direct transfer of a photon. It occurs by dipole–dipole resonance through the so-called Förster mechanism. This phenomenon forms a central part of the next chapter, which focuses on the principles and applications of different types of chemiluminescence energy transfer.

Another, apparently universal, method of homogeneous immunoassay has been developed in France, using the bacterial oxidoreductase–luciferase system as the source of the chemiluminescence. The principal reagents are the antigen labelled (Ag) with a dehydrogenase (DH) (e.g. NAD dependent Gb dehydrogenase from a bacterium), free antibody (Ab), and a solid phase (e.g. Sepharose) containing an anti-antibody (AAb) as well as luciferase (L) and oxidoreductase (OR) coupled to it. In the presence of unlabelled antigen a certain amount of the labelled antigen will bind to the solid phase (●):

$$L \overset{OR}{\underset{\bullet}{\rule{1.2cm}{0.4pt}}} AAb + Ab + Ag + AgDH \rightleftharpoons L \overset{OR}{\underset{\bullet}{\rule{1.2cm}{0.4pt}}} AAb–AbAg + L \overset{OR}{\underset{\bullet}{\rule{1.2cm}{0.4pt}}} AAb–AbAg–DH. \qquad (8.12)$$

The less free antigen, the more dehydrogenase (DH) is bound. When glucose is added, NADH is generated. If FMN and long chain aldehyde are also present then the oxidoreductase–luciferase will generate light. However, because these are on the solid phase, only the NADH generated close to the solid phase will generate light. To ensure minimum light emission from NADH generated by dehydrogenase on antigen not bound to the solid phase, an NADH 'scavenging' system, e.g. pyruvate + lactate dehydrogenase, is also added. Thus increasing antigen concentration decreases light intensity. No separation step is required. This system has already been used to establish working assays for hapten and protein analytes. How rugged it is in a clinical setting remains to be established.

8.3 OTHER PROTEIN LIGAND—LIGAND ASSAYS

Several other binding proteins apart from antibodies are used in biology to assay substances using radio-labelled ligands. Protein kinase, the regulatory subunit, has been widely used to assay cyclic AMP using ^{3}H cAMP as label. The binding of ^{3}H biotin to its high affinity binding protein avidin can be used to assay this B vitamin, of particular interest in paediatric biochemistry and in inherited or congenital carboxylase deficiency where biotin is a cofactor.

Biotin-isoluminol was one of the first successful chemiluminescent ligands to be synthesised, by Schroeder in 1976. He found a 10-fold enhancement in light intensity on triggering the chemiluminescence when bound to avidin. As with the antibody-enhanced isoluminol chemiluminescence the cause is unknown, but may involve a stabilisation of the excited state product and thus an increase in quantum yield, or it may involve a greater efficiency in isoluminol reacting with the active oxygen species responsible for oxidation. Standard curves for biotin in the 50 nM range were established. This is not sufficiently sensitive for plasma levels. We have established a chemiluminescence energy transfer assay using ABEI-biotin bound to fluorescein–avidin resulting in both a quenching and spectral shift (see Chapter 9; section 9.10). This assay can detect biotin at least down to 1 nM, is homogeneous, and is capable of detecting plasma biotin. As with the ABEI energy transfer immunoassays, the concentration of serum in the assay tube must be < 5% to prevent serious interference with ABEI chemiluminescence.

8.4 CONCLUSIONS AND PERSPECTIVES

Inevitably laboratories committed to radioisotopes, and with many years' experience in handling them, will be reluctant to change to a new technology, particularly if it appears simply to replace, rather than improve, the old one. Yet in several European countries, many laboratories wishing to carry out immunoassay are not licensed to use radioactivity. With the availability of several commercial chemiluminometers coupled to reagent kits, many are taking up chemiluminescence in earnest. In the Third World or, nearer home, a general practitioner's surgery or a special clinic, scintillation or γ counters are expensive and thin on the ground. A simple assay protocol, with a small chemiluminometer, has much potential here. Many specialised laboratories throughout the world are now using chemiluminescent assays for cancer screening, detection of endocrine disorders, identification of foetal and neonatal abnormalities, genetic screening programmes, pregnancy testing, and drug assays.

Chemiluminescent labels, therefore, are real alternatives to radio-labels in 'competitive' and 'non-competitive' ligand–ligand assays. They are at least as sensitive, and potentially more sensitive, than their radioactive counterpart. They can generate a light signal lasting just a few seconds or one lasting a hour or more. They can be detected very quickly, within a few seconds and perhaps even down to 0.1 s. It is like having an isotope with a controllable half-life. Without the triggering conditions it is completely stable; with them the whole lot decays within a second or so. They are therefore stable, safe, and they can be used to establish homogeneous assays.

It might be thought at first sight that enzyme labels would be more sensitive than simple organic ones. As was pointed out in Chapter 4, using absorbing or fluorescent substrates, the turnover number of most enzymes, together with the blank rate or signal without the enzyme, does not often enable their catalytic potential to be realised. Use of a chemiluminescent substrate may enhance the use of enzyme labels, particularly peroxidase. Cloning and solid phase reagents may open much scope for using luciferases or photoproteins as lables.

The greatest competitor to chemiluminescence in photoluminescence. Let us see, then, what advantages and disadvantages enable us to select which one to use in a particular analytical application.

8.4.1 Photoluminescence versus chemiluminescence

Photoluminescence is the emission of light from atoms or molecules in electronically exicted states produced by the absorption of electromagnetic radiation, usually ultraviolet or visible light (see Chapter 1). Fluorescence, first defined by Stokes in 1853, involves emission from singlet spin states, phosphorescence from triplet spin states. The latter usually requires the so-called 'forbidden' intersystem crossing from excited singlet to excited triplet. The formation of the initial excited singlet takes only a few femto seconds. Intersystems crossing is also fast, perhaps ps. However, the decay from triplet back to ground singlet is very slow and may be milliseconds to seconds compared with nanoseconds for fluorescence decay.

The ideal fluor or phosphor has a high extinction coefficient, making it easy to excite a large proportion of the molecules, a high fluorescence quantum yield, and a big Stokes shift. The latter is the shift towards the red end of the emission maximum relative to excitation maximum. The bigger this is the less interference from Rayleigh and Raman scatter of the incident light. Fluorescein is a popular fluor because of its high extinction coefficient of some $70\,000\,l.mol^{-1}.\,cm^{-1}$ (NADH is only 6600) and high quantum yield of about 70%. It has a relatively small Stokes shift, however, from *ca.* 495 nM excitation to *ca.* 525 nM emission. Phycobiliproteins from blue-green, and red algae (see Chapter 9 for structures of chromophores) have even larger extinction coefficients of $10^5-10^6\,lM^{-1}\,cm^{-1}$, high quantum yield, and large Stokes shift (phycocyanin $395 \rightarrow 630$ nmh; phycoerythrin $488 \rightarrow 575$ nm). They have evolved as very efficient energy transfer proteins. Fluorescein can be coupled to proteins by using its isothiocyanate, and phycobiliproteins coupled by using bifunctional coupling reagents. Both have important applications in cell biology, in immunofluorescence, and in the fluorescence activated cell sorter for studying single cells. The other group of useful fluors used in conjunction with antibody–antigen reactions are the umbelliferones. When coupled to a sugar such as β galactose they are virtually non-fluorescent. The enzyme β galactosidase, coupled to an antibody, can be sensitively detected by its ability to release fluorescent umbelliferone from the sugar.

Conventional fluorescence using these fluors is limited by the noise signal, which itself cannot be reduced below that of contaminants in the reagents or the solvent, and even water itself. The fluorescence life time of fluorescein is 4 ns, that of dansyl chloride 14 ns, very close to both protein fluorescence (*ca.* 4 ns), and that of the solvent (*ca.* 10 ns). This limits fluorescent label detection to the nM–pM range (0.1 pmol–0.1 fmol in 0.1 ml), still far short of the best chemiluminescent labels. However, there are two ways round these limitations of fluorescent labels.

(1) A reduction in volume to nl–pl, whilst maintaining fluor concentration. This is, to all intents and purposes, what happens in the fluorescence activated cell sorter, the narrow high energy beam being produced by a laser. Just a few hundred molecules of phycobiliproteins on an antibody attached to the surface of a cell can be recognised. It may become possible to detect one antibody molecule on a cell passing through the cell sorter, if this antibody is labelled with some 50–100 phycoerythrin molecules. The photon yield from a chemiluminescent label is independent of volume, but its reagent blank is not. This too, like the fluorescence blank, is reduced by reducing the assay volume. Similarly, by coupling a highly efficient chemiluminescent molecule, e.g. a photoprotein, in large numbers onto an antibody, one antibody molecule will be detectable.

(2) Slow decaying photoluminescence, detected by a time resolved luminometer.

Tetraisofluorescein (erythrosin) is phosphorescent, with a lifetime of several hundred milliseconds. It has a large Stokes shift from 540 nm (excitation) to 690 nm (emission), and can be coupled to proteins, using its isothiocyanate. This, therefore, circumvents the problem of the background fluorescence. However, because of the long life time of phosphors, their energy is easily quenched, particularly by oxygen in the solvent. The quantum yield of erythrosin is only 1%. This, together with its long decay, reduces considerably the sensitivity of detection.

However, a much more attractive group of substances for time resolution are the lanthanide chelators. Europium (Eu^{3+}), terbium (Tb^{3+}) and samarenium (Sm) when bound to β diketones, with a Stokes shift from 340 nm excitation to 613 nm emission, where a red sensitive photomultiplier can have an efficiency approaching 1% (see Chapter 2). The extinction coefficient is some 150 000 $l.mol^{-1}.cm^{-1}$. Some elegant chemistry has been developed to use these Eu^{3+} chelates to develop working time-resolved fluorescence immunoassays. Commercial kits are now available for this method.

The europium fluorescence appears to involve four steps; first, excitation of the organic ligand to an excited singlet; second, energy transfer involving intersystem crossing within the organic ligand thereby generating a triplet excited state; third, and the particularly slow step, energy transfer from the triplet excited state ligand to the europium ion; and finally, fluorescence decay of the excited europium to ground state. The result is a fluorescence decay over a period of several hundred microseconds, instead of the usual nanosecond decay. The europium fluorescence can therefore be separated from the much faster fluorescence of the background signal. Unfortunately, water is a strong quencher of the fluorescence, and the β diketones are difficult to couple to the antibodies whilst maintaining the complex with europium stable. So europium is coupled to antibodies by binding to EDTA, linked covalently using an isothiocyanate or a phenyl group. After the antibody–antigen reaction on solid phase followed by washing, the europium is released by acid pH. Its fluorescence is then assayed in an enhancing solution consisting of 2 napthoyl trifluoroacetone plus tri(n-octyl) phosphine oxide as enhancer. Triton X 100 is also added to form micelles, in acetone-potassium hydrogen phthalate pH 3.2 as solvent.

The fluorescence is measured by summing the emission of 1000 single ms pulses, where the exciting laser pulse lasts only 0.4 μs. Background fluorescence is therefore very low. The result is working assays for analytes, including polypeptides such as TSH.

Unfortunately, solvent quenching still reduces the quantum yield to only 6%. The approximate detection limit of the europium chelates is about 10^{-16} mol in an assay, but perhaps down to 10^{-17} mol (10^{-13} M) in pure solution. The immunoassays are, however, very cumbersome, involving many additions and washing steps, and the apparatus required is very expensive. However, improvements in the chemistry and simplification of electronics, together with a homogeneous system, may improve their clinical viability. These time-resolved fluorescence assays are still not as good in performance as the best chemiluminescence immunoassays.

8.4.2 Recombinant DNA technology

Recombinant DNA technology is poised to have a major impact on clinical and forensic medicine, and is already a very powerful tool in the research laboratory. Clinically, five obvious applications have already been identified:

(1) Identification of the gene and its mutation in inherited disorders, such as familial hypercholesterolaemia, muscular dystrophy, and cystic fibrosis, enabling early foetal diagnosis and the definitive detection of carriers.

(2) Identification of polymorphisms, including those highlighted by restriction enzymes (restriction fragment length polymorphisms-RFLP's), and the relationship of these to both inherited disease and the susceptibility of individuals to diseases such as ischaemic heart disease, the arthritides, multiple sclerosis, diabetes, and other auto-immune based diseases.

(3) Early detection and identification of cancer cells, including cells expressing oncogenes, or endocrine tumours identifiable through the mRNA for a particular polypeptide hormone.

(4) Definitive identification of microorganisms including free bacteria and viruses, as well as ones latent in the host's DNA.

(5) DNA fingerprinting. In forensic medicine this enables the individual source of a blood spot or semen sample to be definitively identified for criminal identification and paternity disputes.

In the research laboratory cloning enables the intron-exon sequence of protein families, and the amino acid structure-function relationships to be established. Furthermore, cloning, together with alteration of individual amino acids by site-directed mutagenesis, enable the molecular biology of binding sites, e.g. for substrates, cations, and cofactors, to be identified. This also has potential for extending the source of bioluminescent indicators themselves. In this way the affinity of a luciferase for ATP or O_2, its colour and kinetics could be tailor-made for an analytical need either in a tissue extract or within a cell.

Where does chemiluminescence come into all this?

To identify specific pieces of DNA or RNA, a probe complementary to the piece being searched for is used. It may be short, just a few nucleotides long, e.g. an oligonucleotide probe, or it may be equivalent to the complete sequence, e.g. cDNA, or cloned mRNA. To search for an abnormal gene or a polymorphism, DNA is isolated from a few cells, e.g. white blood cells. It is broken into fragments with enzymes, particularly restriction endonucleases which cleave usually at unique palindromic sequences in the DNA. These fragments are then separated by gel electrophoresis. But to pick out the particular one we are looking for the DNA must be melted, i.e. separated into single strands, in order to react with the probe complementary to it. To do this the DNA is transferred to nitrocellulose, the process of Southern blotting, where it sticks firmly, after 'melting' by using heat, or a denaturing agent. The probe is then added and binds only to DNA complimentary to it (i.e. where the A–T/G–C bases match). To identify the bands containing the hybridised probe-DNA fragment, the probe has first to be labelled, prior to hybridisation. The most sensitive label is ^{32}P, which is detected by autoradiography. A similar procedure, using a family of species-specific cDNA or RNA probes, can be used to identify particular microorganisms, e.g. *Salmonella*. The technique can be applied to RNA (Northern blots), and can be used in tissue sections, to pick out cells containing particular pieces of DNA or expression of specific genes to make mRNA.

There are two major problems inherent in these procedures, which may restrict the widespread application of this exciting technology to clinical medicine.

(1) ^{32}P is hazardous, more short lived than ^{125}I (Table 4.7). Its detection by autoradiography is cumbersome, often requires exposure times of hours or even days to produce a visible image on the X-ray film, and is difficult to quantify.

(2) The blotting procedure is cumbersome and introduces imprecision.
To circumvent these problems, a non-isotopic replacement for ^{32}P is required, together with a solid phase or a homogeneous method for detecting and quantifying DNA–DNA and DNA–RNA hybrids in solution, without the need for gel separation. Three chemiluminescent systems have so far been used to label DNA probes: ABEI, acridinium esters and peroxidase. Peroxidase can be attached to DNA directly by using glutaraldehyde, or indirectly by first incorporating biotin nucleotides into the DNA by nick translation. Biotin has a very high affinity by ($K_a = 10^{15}$ l.mol^{-1}) for its binding protein avidin, or the bacterial equivalent, streptavidin. Thus a DNA or RNA probe labelled with avidin can be detected and quantified by adding avidin or streptavidin labelled with peroxidase or acridinium ester. Nick translation, for incorporating biotin into DNA, depends on two enzyme activities exhibited by bacterial DNA polymerase I, a 5'3'exonuclease and a 3'5'polymerase activity DNAase1 first induces nicks at random in the DNA. Biotin-11-dUTP + dATP,dGTP and dCTP + DNA polymerase I are added. Synthesis of DNA, and the incorporation of biotin-dUMP being at the 3'OH end, the exonuclease activity degrading the displaced DNA strand as it goes. Because of the random nature of the DNAase I, the DNA is essentially uniformly labelled with the biotin. The nick has been 'translated' towards the 5' end, and can be sealed using DNA ligase. An alternative biotin labelling method uses photobiotin to label DNA at random on exposure to light.

Kricka has established the feasibility of using chemiluminescence in recombinant DNA technology, to detect peroxidase by the enhanced luminol reaction, which can even be recorded on fast photographic film. As little as 1 pg of DNA was detectable in the well-used plasmid pBR322. I have found the isothiocyanate derivatives, e.g. of ABEI, can label DNA directly. Malcolm and colleagues have compared the detection of DNA probes labelled with peroxidase either using glutaraldehyde or biotin-11-dUTP and nick translation. The sensitivity was 5.4amol and 270 amol of DNA respectively. Alternatively N hydroxysuccinimide derivatives of acridinium esters can label the avidin, or an antibody to a hapten labelling the DNA probe. Gen-Probe have developed a very rapid, sensitive assay for the microbe *Chlamydia* using an acridinium ester tailed DNA probe to detect ribosomal rRNA (*ca* 10^4 copies per cell), and separation using magnetic beads. By coupling a chemiluminescent label to one nucleic acid strand and a fluor to the other, a homogeneous method for detecting hybridisation becomes feasible by detecting the disturbance of energy transfer when unlabelled nucleic acid strands compete for the labelled ones. The best candidates for the energy transfer pair in this strategy are a Ca^{2+}-activated photoprotein, e.g. aequorin or obelin, detectable in the tipomol (10^{-21}) range, i.e. better than ^{32}P, and their respective green fluorescent protein (see Chapter 9).

Chemiluminescence thus offers the possibility of removing the need for ^{32}P, for replacing nitrocellulose sheets or solid phase particles making quantification easier and more precise. By using energy transfer, there is the possibility of quantifying DNA recombinants in a homogeneous system. Ultimately, the labels need to be even better than ^{32}P, i.e. 10^{-18}–10^{-21} mol detection is required. Furthermore, because each phosphate in a nucleic acid molecule can be a ^{32}P the chemiluminescent label, be it organic, peroxidase, or a luciferase, must be coupled at each nucleotide to attain the same detection of the nucleic acid fragment made up of many kilobases.

8.4.3 New labels

To exploit the full potential of chemiluminescence in hetero- and homo-geneous assays new labels are required enabling, for example, enzyme-substrate and metal ion-chelates and protein–protein interactions in cells to be studied. Greater sensitivity could be achieved with present labels by increasing the number of labelled mols per mol of ligand. For example, imidoesters retain the charge on protein-NH_2 groups, thereby preventing precipitation when there are more than 10 mols of label per mol antibody. Polyamines such as polylysine attached to the antibody or the antigen provide multiple linking groups for labels. Acridinium esters linked at the N of the acridinium ring might be useful in energy transfer. But isoluminol derivatives still have poor quantum yields of 1% or so. They and acridinium esters, and peroxidase, work best at non-physiological pH, require H_2O_2 which is toxic to cells and are susceptible to interference from other agents in the assay. Bioluminescent luciferins have quantum yields in the range of 10–100% (see Chapter 3), can have fast kinetics at pH 7.3 (e.g obelin $k_{sat} = 4s^{-1}$, $t_{\frac{1}{2}}$ 0.18 s), can participate in energy transfer, and are protected from environmental factors by the luciferase. New luciferins are being discovered year by year, e.g. the tetrapyrroles in dinoflagellates and deep sea fish such as *Malacosteus* with its blue and red light emitting organs. Others await discovery. With modern cloning techniques no longer should the availiability of large amounts of luciferase be a problem. Many luminous animals contain organs where > 5% protein is involved in the luminous system. Bacterial and firefly luciferases, and aequorin, have already been cloned. The luciferins can be synthesised. Nature often knows best. She has had millions of years to evolve very efficient chemiluminescence, giving us access to systems with a wide range of physico-chemical properties. It is perhaps organisms from the deep oceans which offer greater promise for the future. Organisms in the sea also provide some of Nature's best examples of energy transfer, the topic of the last main chapter in this book.

8.5 RECOMMENDED READING

Books

Butt, W. R. (ed) (1984) *Practical immunoassay: the state of the art*. Marcel Dekker, New York and Basel.
Collins, W. P. (ed.) (1985) *Alternative immunoassays*. John Wiley, Chichester.
Kricka, L. J. & Carter, T. J. N. (1982) *Clinical and biochemical luminescence*. Marcel Dekker, New York and Basel.

Reviews

Campbell, A. K., Hallett, M. B., & Weeks, I. (1985) *Methods Biochem. Anal.* **31** 317–416. Chemiluminescence in cell biology and medicine.
Campbell, A. K., Roberts, P. A., & Patel, A. (1985) *Alternative immunoassays* pp 153–183, ed. Collins, W. P. John Wiley, Chichester. Chemiluminescence energy transfer: a technique for homogeneous immunoassay.
Schroeder, H. R., Boguslaski, R. C., Carrico, R. J., & Buckler, R. T. (1978) *Methods Enzymol* **57** 424–445. Monitoring specific protein binding chemiluminescence.
Weeks, I., Sturgess, M., Brown, R. C., & Woodhead, J. S. (1987) *Method Enzymol.* **133** 366–387 Immunoassays using acridinium esters.

Kricka, L. J. & Thorpe, G. H. G. (1987) *Method Enzymol.* **133** 404–420. Photographic detection of chemiluminescent and bioluminescent reactions.

Hummelen, J. C., Luider, T. M., & Wynberg, H. (1987) *Method Enzymol.* **133** 531–557. Stable 1,2-dioxetanes as labels for thermochemiluminescent immunoassay.

Papers

Carrié, M., Térouanne, B., Brochu, M., Nicolas, J. C., & Crastes de Paulet A. (1986) *Anal. Biochem* **154** 126–131. Bioluminescent immunoassays of progesterone. A comparative study of the different procedures.

Coyle, P. M., Thorpe, G. H. G., Kricka, L. J., & Whitehead, T. P. (1986) *Ann. Clin. Biochem.* **23** 42–46 Enhanced luminescent quantitation of horseradish peroxidase conjugates: application in an enzyme immunoassay for digoxin.

John, R. & Henley, R. (1987) *Lancet i* **446** Interference in commercial assays for thyrotropin.

John, R., Henley, R., Chang, D., & McGregor, A. M. (1986) *Clin. Chem.* **32** 2178–2183 Enhanced luminescence immunoassay: evaluation of a new, more sensitive thyrotropin assay.

Kricka, L. S. & Thorpe, G. H. G. (1987) *Method Enzymol.* **133** 404–420. Photographic detection of chemiluminescent and bioluminescent reactions.

Longeon, A., Henry, J. C., & Henry, R. (1986) *J. Immunol. Methods* **89** 103–109. A solid-phase lumininescent immunoassay for human chorionic gonadatropin using *Pholas dactylus* bioluminescence.

Weeks, I. & Woodhead, J. S. (1986) *Clin. Chim. Acta* **159** 139–145. Measurement of human growth hormone (hGH) using a rapid immunochemiluminometric assay.

Pazzagli, M., Kim, J. B., Messeri, G., Kohen, F., Bolleli, G. F., Tomasi, A., Salerno, R., Monetti, G., & Serio, M. (1981) *J. Steroid Biochem.* **14** 1005–1012. Luminescent immunoassays of cortisol. Development and validation of the immunoassay monitored by chemiluminescence.

Schroeder, H. R., Vogelhut, P. O., Carrico, R. J., Boguslaski, R. C., & Buckler, R. T. (1976) *Anal. Chem.* **48** 1933–1937. Competitive protein binding assay for biotin monitored by chemiluminescence.

Simpson, J. S. A., Campbell, A. K., Ryall, M. E. T., & Woodhead, J. S. (1979) *Nature* **279** 646–647. A stable chemiluminescent labelled antibody for immunological assays.

Térouanne, B., Nicolas, J. C., & Crastes de Paulet, A. (1986) *Anal. Biochem.* **154** 118–125. Bioluminescent immunosorbent for rapid immunoassays.

Velam, B. & Halmann, M. (1978) *Immunocytochemistry* **15** 331–333. Chemiluminescence immunoassay: a new sensitive method for determination of antigens.

Weeks, I., Beheshti, I., McCapra, F., Campbell, A. K., & Woodhead, J. S. (1983) *Clin. Chem.* **29** 1474–1479. Acridinium esters as high specific activity labels in immunoassay.

Weeks, I., Campbell, A. K., & Woodhead, J. S. (1983) *Clin. Chem.* **29** 1480–1483. Two-site immunochemiluminometric assay for human α-fetoprotein.

Weeks, I., Sturgess, M., Siddle, K., Jones, M. R., & Woodhead, J. S. (1984) *Clin. Endocrinol.* **20** 489–495. A high sensitivity immunochemiluminometric assay for human thyrotropin.

Whitehead, T. P., Thorpe, G. H., Carter, T. J. N., Groucutt, H. C., & Kricka, L. J. (1983) *Nature* **305** 158–159. Enhanced luminescence procedure for sensitive determination of peroxidase-labelled conjugates in immunoassay.

Energy transfer

Campbell, A. K. & Patel, A. (1983) *Biochem. J.* **216** 185–194. A homogeneous immunoassay for cyclic nucleotides based on chemiluminescence energy transfer.

Patel, A., Davies, C. J., Campbell, A. K., & McCapra, F. (1983) *Anal. Biochem.* **129** 162–169. Chemiluminescence energy transfer: a new technique applicable to the study of ligand–ligand interactions in living systems.

Patel, A. & Campbell, A. K. (1983) *Clin. Chem.* **29** 1604–1608. Homogeneous immunoassay based on chemiluminescence energy transfer.

William, E. J. & Campbell, A. K. (1986) *Anal. Biochem.* **155,** 249–255. A homogeneous assay for biotin based on chemiluminescence energy transfer.

DNA labels

Figueredo, H. & Malcolm, A. D. B. (1986) *Biochem. Soc. Trans.* **12** 640. A luminescent detection system for non-radioactively labelled DNA probes.

Matthews, J. A., Batki, A., Hynds, C. C., Kricka, L. J. (1985) *Anal. Biochem.* **151,** 205–209. Enhanced chemiluminescent method for the detection of DNA dot-hybridisation assays.

9

Chemiluminescence energy transfer

9.1 WHAT IS MEANT BY ENERGY TRANSFER?

How many people, I wonder, needing to release pent-up tensions will walk over to the piano, place their foot on the loud pedal, and play a resounding chord of C major, or some other favourite. The cacophony of sound continues even after the hands have been removed from the keyboard, so long as the loud pedal remains depressed. The keen ear will discern particular notes resonating in sympathy with harmonics in the original chord. This is an example of the direct transfer of energy, via longitudinal sound waves.

Direct energy transfer is also easy to demonstrate with transverse waves, exemplified by light. A familar example occurs in photoluminescence, be it fluorescence or phosphorescence. Here, light emitted by the donor is absorbed by the acceptor. As a result the acceptor is excited, and then remits light at a longer wavelentgh. Many children have been intrigued by such examples of direct energy transfer, which can be demonstrated with materials available from toy shops. Small discs coated with phosphorescent paint can convert the bedroom ceiling into an imaginary planetarium. Switch off the light and the various-sized glowing spots give the appearance of stars in the sky on a clear night. My daughter even has a pair of phosphorescent slippers!

But this chapter is not concerned with types of energy transfer such as these, which occur directly through the radiative medium itself, perhaps rather contemptuously often referred to as 'trivial' energy transfer. The more interesting, and sometimes more puzzling phenomenon, involves the transfer of energy from a donor to an acceptor via a *non*-radiative mechanism. An important consequence is that several of the key parameters which determine the efficiency of such *non-radiative* energy transfer are different from those determining the overall efficiency of *radiative* energy transfer. Thus, the efficiency of the radiative transfer of light generated in chemiluminescence depends on four parameters: the optical geometry and efficiency of light transfer through the medium, the absorbance coefficient of the acceptor within the spectral emission of the donor (D), and the fluorescence quantum yield of the acceptor (A). When the initial source of light is chemilumines-cent the complete radiative energy transfer process can be described therefore as:

$$\text{substrates} + \text{catalyst} \rightarrow D^* + \text{other products} \tag{9.1}$$

$$D^* \rightarrow hv_D \tag{9.2}$$

$$A + hv_D \rightarrow A^* \tag{9.3}$$

$$A^* \rightarrow hv_A \tag{9.4}$$

hv_A will be $< hv_D$, and the overall quantum yield (ϕ_0) of the complete reaction;

$$\text{substrates} + \text{catalysts} + A \rightarrow \text{products} + D + A + n_1 hv_D + n_2 hv_A \tag{9.5}$$

together with the ratio of n_1 to n_2, will depend on the overall chemiluminescence quantum yield of reactions $9.1 + 9.2$ (ϕ_D), the fluorescence quantum yield of the acceptor (ϕ_A), and the components of Beer Lambert's law, namely spectral overlap, extinction (absorbance) coefficient, and path length. Thus,

$$\phi_o < \phi_A \cdot \phi_D \tag{9.6}$$

In contrast, the efficiency of luminescence resulting from non-radiative, or radiationless, energy transfer depends on the nature of the transfer process. This can occur in one of three ways:

(1) via resonance between separated atoms or molecules;
(2) via direct electron interactions requiring electron transfer and/or atomic or molecular orbital overlap, such as occurs during a collision between two atoms or molecules;
(3) via exciton transfer, found usually only in solids.

Radiationless energy transfer, for our purposes is defined as:

the transfer of energy from one atom, or group of atoms, in an electronically excited state to another without the direct transfer of a photon.

The acceptors themselves become electronically excited, and emit a photon when decaying back to ground state. The species initially excited is the *donor*, or *sensitiser* as it is sometimes known. The final emitting species is the *acceptor*, or the *sensitised* species.

In principle, luminescence energy transfer does not depend on the mechanism by which the excited donor is produced, whether by the physicochemical processes involved in photo-, sono-, tribo-, electro-, radio-luminescence, or by a chemical reaction as in chemiluminescence. However, this may not always be the case in chemiluminescence energy transfer for, as we have seen in Chapter 1, the nature of the excited state product generated by a chemical reaction may be different from that produced by photo-excitation of the chemiluminescence product. Another interest-ing feature of some, but not all, types of chemiluminescence energy transfer is that the acceptor can actually play a part in the reaction generating the excited state donor.

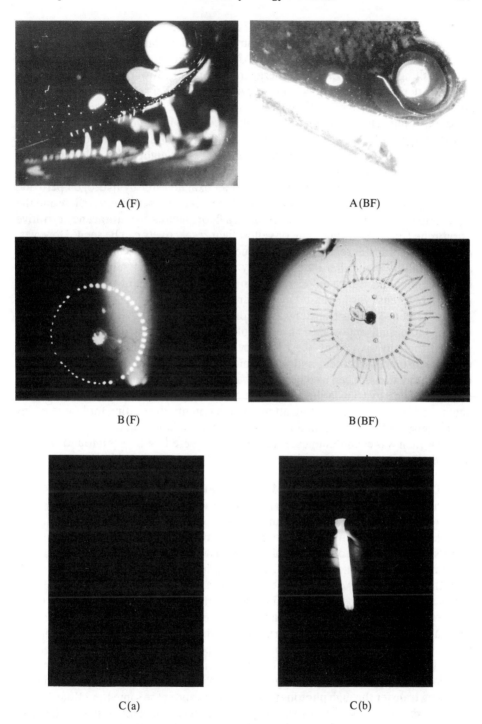

Fig. 9.1 — Some examples of energy transfer: A Malacosteus; B Obelia; C Light sticks, (a) no fluor, (b) + fluor; F = fluorescence; B/F = bright field.

Particularly spectacular end results of chemiluminescence energy transfer are a shift in colour of the emission or a conversion of a 'dark' to a 'light' reaction compared with that of the donor alone. Unlike 'trivial' energy transfer, the colour shift resulting from non-radiative energy transfer may be towards the red or towards the blue, depending on the mechanism involved. A further spectacular difference in some types of non-radiative energy transfer is a dramatic increase in overall quantum yield, resulting in much more light being emitted. This is particularly well illustrated by the 'light sticks' (Fig. 9.1) available commercially from sailing and toy shops, and often seen adorning waitresses and striptease dancers in night clubs. A luminous golf ball is now available on the market. The primary chemiluminescent reaction in these light sticks and golf balls is the oxidation of an oxalate ester by hydrogen peroxide (Fig. 9.2). On its own, the quantum yield of the oxalate reaction is very low and the light emission is invisible, but on addition of a fluor, such as an anthracene derivative or rubrene (Fig. 9.3), bright blue or yellow light respectively can be seen. However, not all energy transfer reactions result in such a dramatic increase in quantum yield, if at all. Similarly in another type of energy transfer, $KMnO_4$ + sulphide generates a weak chemiluminescence unless riboflavin is present as an energy transfer acceptor, when light intensity is increased more than 300-fold.

Fluorescence energy transfer was first discovered in the 1920s, and has provided a tool of increasing importance in cell biology since Stryer and Haugland in 1967 showed that, by estimating the efficiency of energy transfer, it could be used to measure distances between, and within, molecules. The existence of chemiluminescence energy transfer can also be traced back to the 1920s and 1930s. However, this example of energy transfer has only been extensively investigated within the last 20 years, and is only now attracting attention as a potentially exciting tool for studying the chemistry of living cells and for use in the clinical laboratory.

How then was chemiluminescence energy discovered? Where is it found? Is there more than one type? How can it be used to study biological problems?

9.2 THE DISCOVERY OF CHEMILUMINESCENCE ENERGY TRANSFER

The phenomenon of energy transfer involving transverse electromagnetic waves was first predicted by Franck in 1922, and reported in the following year by Cario and Franck, who observed it in the vapour phase between excited mercury atoms as the donors and thallium atoms as the acceptors. The mixture of mercury and thallium vapour produced an emission spectrum containing the emission lines characteristic of both atoms when excited by light at the mercury absorption line. Since thallium atoms are unable to absorb light at this wavelength they could only be excited, Cario and Franck argued, as the result of a transfer of energy from excited mercury atoms to some of the thallium atoms. Furthermore, they were able to rule out radiative ('trivial') energy transfer since significant absorption of light emitted from mercury atoms by thallium was impossible. They were therefore dealing with a new phenomenon, involving the transfer of energy between electronically excited states, and not involving transfer through photons themselves. At about the same time Kautsky and co-workers first envoked 'sensitised' chemiluminescence to explain the mechanism, and the spectral properties, of the oxidation of calcium silicide by HCl. In 1938 Helberger observed a red emission when the porphyrin magnesium phthalocyanine

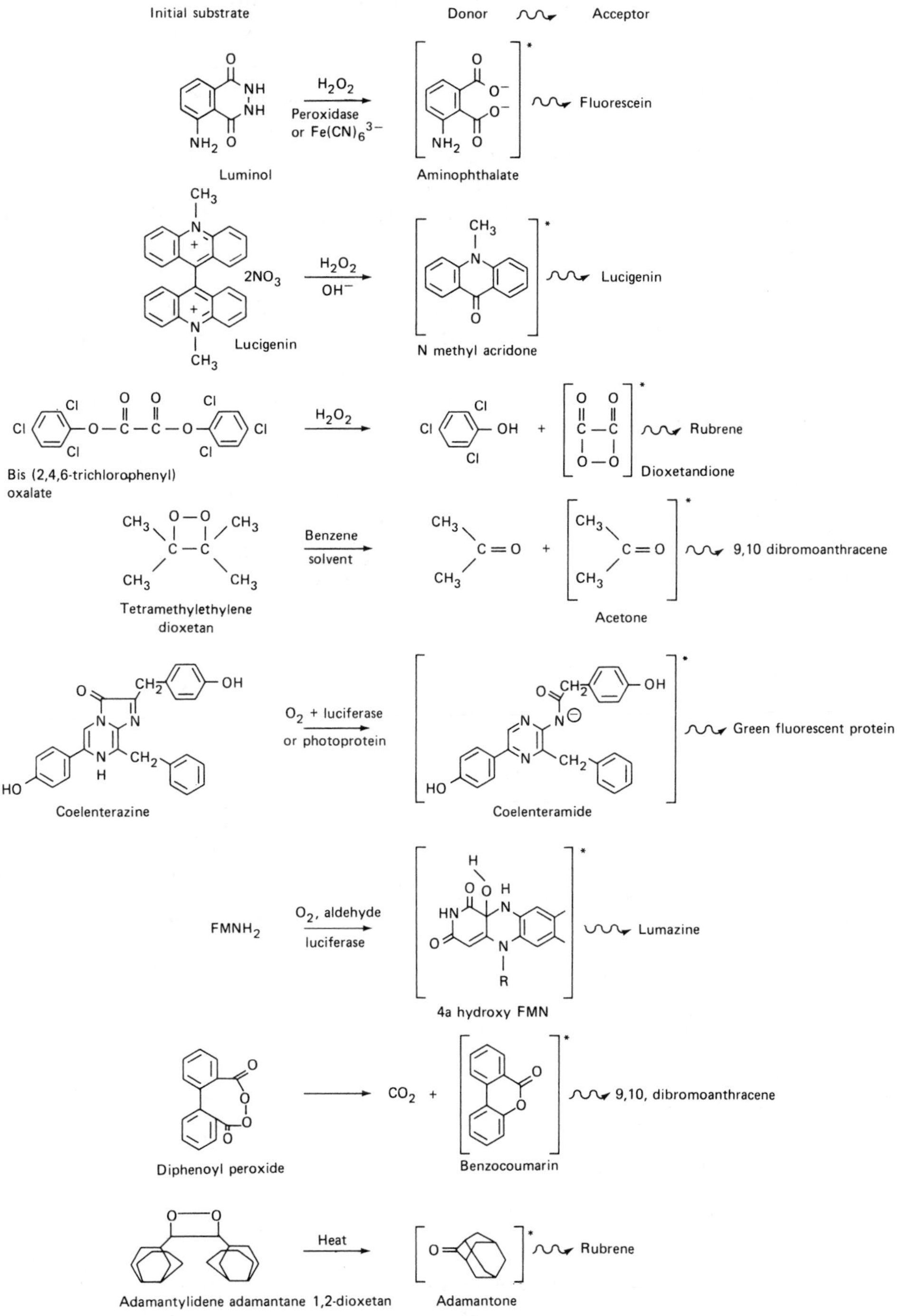

Fig. 9.2 — Energy transfer chemiluminescent donors.

was added to hot tetralin (Fig. 9.2), a reaction later shown to involve tetralin hydroperoxide and radical reactions in the energy transfer process.

The identification and characterisation of chemiluminescence energy transfer was thus heralded by five important discoveries. Firstly, the observation of fluorescence energy transfer by Cario and Franck in 1923 led to a search for a similar phenomenon with chemiluminescence, involving orbital resonance. Secondly, the interpretation by Kautsky of the calcium silicide reaction as chemiluminescence energy transfer. Thirdly, the investigation of radical generating reactions, leading to the discovery of electron transfer, or electron exchange, chemiluminescence energy transfer. Fourthly, the observation of red chemiluminescence produced by mixing H_2O_2 and NaOCl leading to the idea of energy pooling between two singlet oxygen molecules. Finally, the observation by Davenport and Nicol in 1955 that several luminous hydrozoa contained a brightly green fluorescent substance close to the same site producing the light. This lead to the discovery of chemiluminescence energy transfer in living organisms.

Chemiluminescence energy transfer has since been found in solids, liquids, and gases. The donor and acceptor can be in separate molecules or within the same molecule. It occurs in man-made compounds and in many luminous organisms (Table 9.1). What then are the donors and acceptors in these natural and synthetic examples of non-radiative chemiluminescence energy transfer?

Table 9.1 — Some examples of chemiluminescence energy transfer

Chemiluminescent reaction	Fluorescent acceptor (emitter)	Colour change
Solid		
Siloxene oxidation	eosin	red→green
	rhodamine	red→yellow
Liquid		
luminol oxidation	fluorescein	blue→green-yellow
Renilla bioluminescence	green fluorescent protein	blue→blue-green
Photobacterium phosphoreum	lumazine protein	blue (490)→blue (470)
Vibrio fischeri (Y1 strain)	yellow fluorescent protein	blue (490)→yellow (540)
oxalate ester oxidation	anthracene	no light→blue
	rubrene	no light→red
Gas		
O. generation	metal vapour, e.g. Na	nothing (IR)→yellow
singlet oxygen generation	$^1O_2$–$^1O_2$ dimol	IR→red
$^1O_2$ conjugation with aromatic hydrocarbon	rubrene	weak red→strong red
$^1O_2$+>C=C< (ethyl vinyl ether)	rubrene	weak blue→red

9.3 SOURCES
9.3.1 Man-made
Six classes of chemiluminescent reaction have been used to generate excited state donors for man-made chemiluminescence energy transfer (Table 9.2):

(1) Oxidation reactions, using either H_2O_2 or other oxygen containing oxidants.
(2) Stable peroxide decomposition.

Table 9.2 — Examples of man-made chemiluminescence energy transfer

Chemiluminescent reaction	Excited state donor	Acceptor or emitter
1. Oxidation reactions using an oxygen based oxidant		
luminol + H_2O_2	3-aminophthalate	fluorescein
ABEI + H_2O_2	4-aminophthalate	fluorescein
lucigenin + H_2O_2	N-methyl acridone	lucigenin
siloxene $(Si_6H_6O_3)_n$+ H_2O_2	Si oxide	siloxene (hydroxy) rubrene
methyl ethyl ketone	biacetyl	anthracene derivatives
2. Peroxide decomposition		
tetralin autoxidation	tetralone	Zn tetraphenyl porphyrin
tetralin hydroperoxide	tetralone	rubrene
3. Peroxyester reactions		
oxalyl Cl + H_2O_2	C_2O_4	diphenylanthracene
oxalate esters + H_2O_2	C_2O_4	diphenylanthracene
1-phenyl peroxyacetate	acetophenone-acetic acid	rubrene
dicyclohexylperoxycarbonate		dibromoanthracene
diphenyl peroxide	benzocoumarin	rubrene
4. Dioxetan or dioxetanone decomposition		
tetramethyl dioxetan	acetone	diphenylanthracene
dimethyl-1,2-dioxetanone	acetone	rubrene
5. Reactions with singlet oxygen	$^1O_2$	rubrene
	$^1O_2$	eosin
	$^1O_2$	Zn tetraphenylporphrin
6. Oxidoreductions not involving oxygen (radical annihilation)		
trisbypyridine ruthenium chelates + hydrazine	$Ru(bpy)^{3+}$	$Ru(bpy)_3^{2+}$
electrochemiluminescence	rubrene$^+$	rubrene$^-$
	dimethylanthracene	tri-p-tolyl amine

(3) Reactions of peroxyesters, stable or formed during the reaction.
(4) Decomposition of dioxetans or dioxetanones.
(5) Reactions producing singlet oxygen.
(6) Oxidoreduction reactions not involving oxygen.

The first five of these require oxygen; the last does not since it involves radical annihilation, the radicals being generated electrically (electrogenerated or electro-chemiluminescence) or chemically (chemically induced electron exchange lumines-cence). The first four produce excited state donors which are usually carbonyls; the other two need not.

The energy transfer acceptor can be a *non*-fluorescent chromophore, in which case the acceptor simply quenches the chemiluminescence. However, in this chapter we shall be dealing mainly with acceptors which are fluors, and thus re-emit after energy transfer has occurred. These fluorescent acceptors include hydrophilic compounds such as fluorescein, as well as substances only soluble in hydrophobic solvents such as the polycyclic hydrocarbons rubrene and diphenyl anthracene (Fig. 9.3). The efficiency of the acceptor depends not only on the nature of the fluor but on the type of energy transfer. Thus, fluorescein is a good transfer acceptor from excited state phthalic acid derivatives formed from phthalazinediones, but a very poor acceptor in the dicyclohexylperoxycarbonate system.

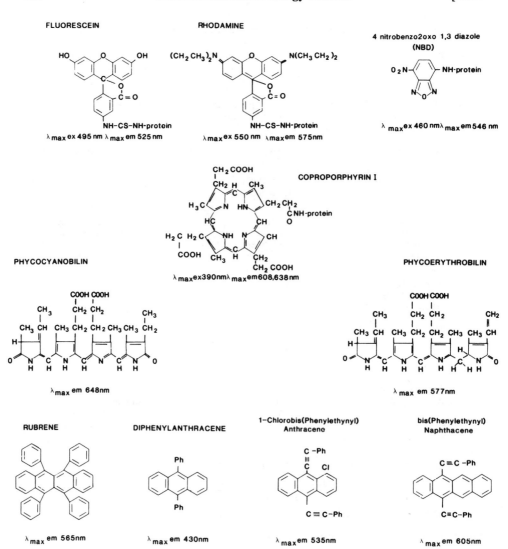

Fig. 9.3 — Some synthetic energy transfer acceptors.

The acceptor is usually a different substance from the initial substrate generating the excited state donor, but there are at least three examples where the initial substrate can also act as the energy transfer acceptor, so long as it remains at a sufficient concentration: lucigenin, unoxidised siloxene (or hydroxy derivative), and rubrene in radical annihilation. Furthermore, the acceptor need not necessarily be in the same phase as the donor. Energy transfer can occur at solid–liquid, solid–gas, or gas–liquid interfaces. Thus, liquid eosin and rubrene act as acceptors from solid siloxene, whereas gaseous singlet oxygen can react with solid fluors.

Energy transfer requires close association between donor and acceptor. This can be achieved in one of four ways:

(1) Non-covalent binding of donor to acceptor (intermolecular transfer).
(2) Covalent link between donor and acceptor (intramolecular transfer).
(3) Molecular collisions.
(4) High concentrations of donor and/or acceptor.

The first of these has been generated in the test tube between *Renilla* luciferin–luciferase and its green fluorescent protein (see section 9.6.2), and has been demonstrated between the excited product of the isoluminol derivative aminobutyl isoluminol (ABEI) on an antigen and fluorescein on an antibody, as well as between ABEI biotin and fluorescein avidin, the naturally occurring high affinity binding protein of biotin. Non-covalent molecular complexes also play a key role in the energy transfer systems involving peroxide and peroxyester reactions, singlet oxygen and in any chemiluminescence involving radical annihilation.

White and colleagues have synthesised a number of phthalhydrazide derivatives (Fig. 9.4), where the excited product, phthalic acid, normally very poorly chemiluminescent, transfers its energy intramolecularly to (a) unreacted hydrazide, or (b) an anthracene derivative, or (c) an *N*-methyl acridone, thereby markedly enhancing the overall quantum yield. Intramolecular energy transfer also plays an important part in the primary chemiluminescent reaction in several types of bioluminescence, including that of the firefly.

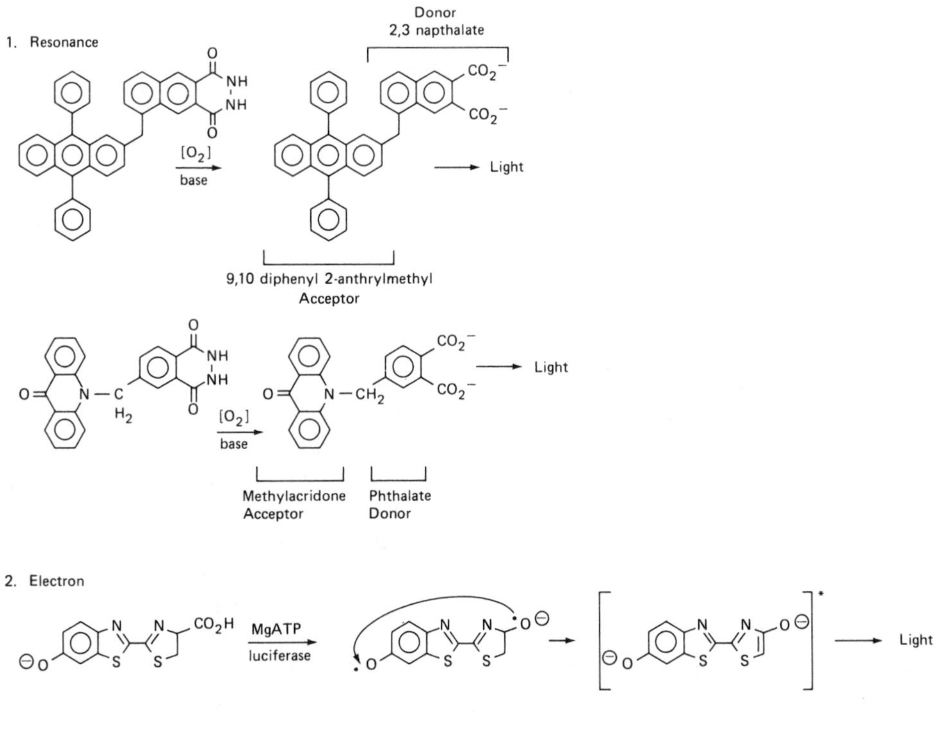

Fig. 9.4 — Intramolecular chemiluminescent energy transfer.

When donor and acceptor concentrations are greater than 1–10 mM, they may be close enough for energy transfer to occur without the formation of a molecular complex. This appears to explain the long known puzzle that lucigenin chemiluminescence is bluish-green (λ_{max} 510 nm), whereas the excited state product of lucigenin, N methyl acridone, ought to produce blue light (λ_{max} 460 nm). Lucigenin is itself fluorescent. Its oxidation in alkaline hydrogen peroxide or in dimethyl formamide plus H_2O_2 is relatively slow, and requires mM concentrations of lucigenin to produce visible light. The concentration of lucigenin is therefore sufficiently high for it to act as an acceptor from its product N-methyl acridone, formed on initiating the chemiluminescent reaction. Aryl acridinium esters, which are blue fluorescent initially anyway, react much faster than lucigenin and energy transfer from product to substrate is rarely seen.

9.3.2 Bioluminescence

Go out on a research ship with the necessary gear for collecting deep sea organisms and you will find a plethora of species emitting blue or green light. You may also be lucky enough to find one or two of the very few species of fish which emit red light (see Chapter 3). In several of these luminous species, non-radiative energy transfer plays an important role in determining the colour of the light emitted by the animal; in others absorbing filters affect their spectral characteristics (Fig. 9.5A,B).

Around 1000 metres below the surface live the hatchet fish. As we saw in Chapter 3, these luminous fish are able to alter the colour of the light they emit from that inherent in the chemiluminescence reaction, but do so by a 'trivial' mechanism. In the animal, the light passes through a filter pigment which absorbs part of the red end of spectral emission the resulting bluer light has a very similar spectrum to that of the dim ambient light at these depths (sea water absorbs red preferentially over blue light; see Chapter 3). This mechanism enables the fish to camouflage its silhouette against predators. Another fish, *Valencienellus* has a fluorescent pigment, an unusual porphyrin, not in the same cells as the chemiluminescent reaction responsible for light emission, which does a similar job. A similar phenomenon may occur in some squid photophores. Important as these spectral changes appear to be in the organism, they do not fall within our definition of energy transfer.

Walk out on the beach at night and examine carefully the brown seaweeds in the rock pools, or those washed up on the edge of the shore, and you have a chance to find one of Nature's genuine examples of non-radiative chemiluminescence energy transfer. The hydroid *Obelia geniculata* grows commonly on *Laminaria* and other seaweeds around the British coasts, and throughout the world. The isolated photoprotein obelin, containing the chemiluminescent reaction responsible for light emission in these organisms, emits blue light (λ_{max} 475 nm) when it binds Ca^{2+}. But in the hydroid, and in its jelly fish stage (Fig. 9.1B), there is a group of green fluorescent proteins which contain a chromophore (Fig. 9.6) acting as an energy transfer acceptor. The excited chromophore on the green fluorescent protein becomes the actual light emitter, the organism producing blue–green light (λ_{max} *ca.*

508 nm). This spectral shift can be demonstrated by using a chemiluminometer containing two photomultiplier tubes with 475 nm and 510 nm interference filters in front of them (Fig. 9.7). The ratio of chemiluminescence of *Obelia* at 510/475 nm is about 7, whereas with isolated obelin it is approximately 1. Ward and Cromier, who

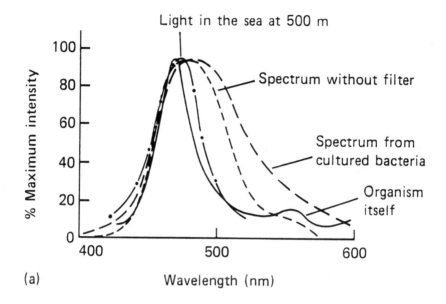

(a)

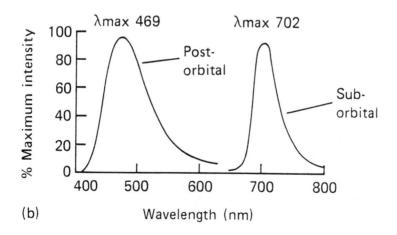

(b)

Fig. 9.5 — Colour shifts in bioluminescence. A, Bacterial emission from the deep sea fish *Opisthoproctus,* passed through absorbing filter. Adapted from Denton (1985) with permission of the Royal Society. B, Possible energy transfer in red emitting organ of the deep sea fish *Malacosteus* with permission Widder *et al.* (1984) *Science* **225,** 512–514.

(a) Coelenterates (analogue of this structure is green fluorescent)

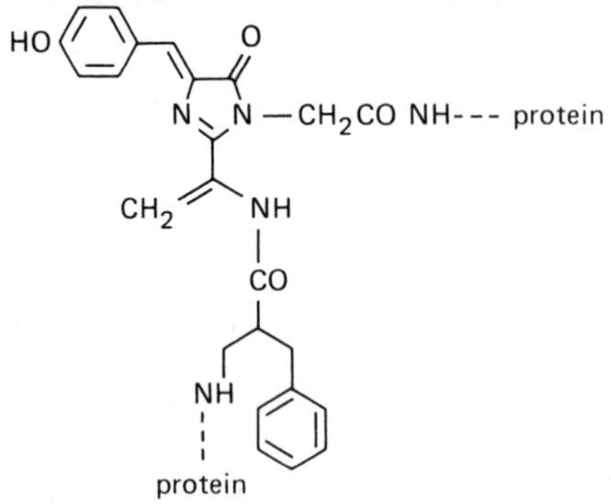

(b) Lumazine in bacteria (blue fluorescent)

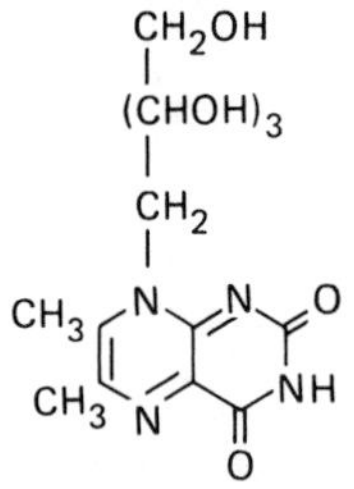

6,7-dimethyl-8-(1'-D-ribityl) lumazine
(an intermediate of riboflavin biosynthesis)

(c) Flavins in YI strain of *Vibrio fischeri* and others

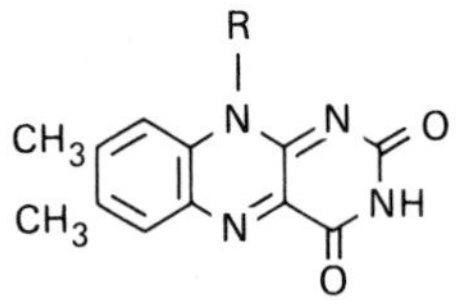

R = CH$_2$ (CHOH)$_3$ CH$_2$O PO$_3^{2-}$ = flavin mononucleotide (yellow-green fluorescent)
R = CH$_2$ (CHOH)$_3$ CH$_2$O PO$_3-$ AMP = flavin adenine dinucleotide (yellow-green fluorescent)
NB FMNH$_2$ and FADH$_2$ are non-fluorescent
R = CH$_2$ (CHOH)$_3$ CH$_2$OH = riboflavin (yellow-green fluorescent)
R = nothing i.e. 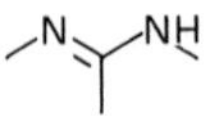 = lumichrome (blue fluorescent)

Fig. 9.6 — Structures of proposed energy transfer acceptors in bioluminescence.

A

B

Fig. 9.7 — Dual wavelength chemiluminometer for energy transfer. A, PM tubes at 180°; B. PM tubes at 90°, minimising reflection from one interference grating to another.

have made a considerable contribution to our understanding of energy transfer in coelenterates, have worried that in *Obelia* and *Aequorea* at least some of such spectral shifts might be due to 'trivial' radiative energy transfer. Given the concentration of intracellular green fluorescent protein, and the speed of the non-radiative energy transfer process, this is perhaps not very likely *in vivo*. Furthermore, in the related anthozoan, the sea pansy *Renilla*, radiationless energy transfer between the excited oxyluciferin on the luciferase and the fluor on the green fluorescent protein has been definitively demonstrated. Not only is there a spectral shift, but also an increase of some three fold in overall quantum yield.

Interestingly, energy transfer only occurs in some luminous coelenterates; the others do not contain any green fluorescent protein (Table 9.3). Particularly puzzling is the medusa *Phialidium hemisphaericum* (Hydroid: *Clytia johnstoni*). At Plymouth, U.K., it has no green fluor (chemiluminescence at 510/475 nm = 0.8), but in other parts of the world appears to have a green fluor, the jelly fish emitting light with a peak at 508 nm.

Table 9.3 — Luminous coelenterates with and without energy transfer

With (bioluminescence emission maximum >500 nm)

Hydrozoa:	*Obelia geniculata, dichotoma, longissima, commissuralis, bicuspidata, Phialidium gregarium, Clytia edwardsi, Clytia bakeri, Aequorea forskalea, Diphyes, Campanularia calceolifera, Lovenella gracilis*
Scyphozoa:	none known
Anthozoa:	*Renilla reniformis, kollikeri, mülleri, Pennatula phosphorea, Cavernularia obesa, Stylatula elongata, Acanthoptilum gracile, Umbellula huxeyi, Parazoanthus lucificum, Ptilosarus guernyi, Veretillum cynomorium, Virgularia mirabilis*

Without (bioluminescence emission maximum <500 nm)

Hydrozoa:	*Hippopodius hippopus, Vogtia glabra, Phialidium hemisphearicum, Halistaura cellularia*
Scyphozoa:	*Periphylla periphylla, Atolla wyvillei, Pelagia noctiluca*
Anthozoa:	*Acanella arbuscula, Iridogorgia, Chelodonisis aurantiaca, Acanthogorgia granulata, Candidella imbricata*
Ctenophora:	*Mnemiopsis, Beroë, Pleurobrachia* spp

Evidence based on one or more of:
1. Organism emission >500 nm = energy transfer.
2. Green fluorescent cells.
3. Isolation of green fluorescent protein.
4. Demonstration *in vitro* with isolated proteins.

Intermolecular energy transfer has also been clearly demonstrated by Lee and co-workers in certain species of luminous bacteria which contain a blue fluorescent protein containing lumazine (6,7-dimethyl-8-ribityl lumazine) as acceptor (Fig. 9.6), and by Nealson and co-workers in a yellow strain (Y1) of *Vibrio fischeri* which contains a yellow fluorescent protein with a flavin chromophore.

Fluors, apparently in the same compartment as the chemiluminescent reaction, have been observed in several other luminous organisms (Table 9.4). These fluors may explain not only differences between the chemiluminescence spectra of the isolated chemical reaction *in vitro* compared with that in the intact animal, but also peculiar biomodal and temporal changes in spectra. The latter occur in several luminous squid, arthropods, and fish, including the shrimp-like amphipod *Parapronoë crustulum* and the deep sea fish *Searsia*. This fish has been studied by Widder and Herring, and has a bimodal spectrum with peaks at 408 and 478 nm respectively. Initially, the emission is virtually all at 478 nm, with just a shoulder at 408 nm. Chemiluminescence at the shorter wavelength peak gradually increases with time until it predominates. Furthermore, the 'cells' released from the searsiid, which are responsible for the chemiluminescence, are also strikingly fluorescent. This fluorescence also changes with time, transforming from blue–green to blue as each cell chemiluminesces.

Table 9.4 — Chemiluminescence energy transfer in some luminous organisms

Phylum	Organism	Resulting colour shift
A. Established		
Intramolecular		
Arthropoda	*Photinus pyralis* (firefly)	Blue→red
Intramolecular		
Bacteria	*Photobacterium phosphoreum*	blue→bluer
	Vibrio fischeri (Y1)	blue→yellow
Cnidaria	*Obelia geniculata*	blue→blue-green
	Aequorea forskalea	blue→blue-green
	Renilla reniformis	blue→blue-green
	Yellow anthozoan	blue→yellow
B. Possible		
Fungi	*Lampteromyces japonicus*	blue→yellow-green
Mollusca	*Latia neritoides*	blue→green
Arthropoda	*Phrixothrix*	yellow-green→red
	Meganyctiphanes norvegica	blue→blue
Chordata	*Searsia*	blue→bluer
	Malacosteus niger	blue→red

Particularly striking examples of a colour shift are those luminous animals with more than one light organ, emitting different colours. The deep sea stomiatoid fish *Malacosteus*, *Aristostomias*, and *Pachystomias* have many small photophores scattered over the body and two pairs of relatively large light organs. The smaller pair, behind the eyes (postorbital), emit blue light (λ_{max} 470 nm), the other large pair, just underneath each eye (suborbital), emit red light (λ_{max} 705 nm). Although this difference between the two pairs of organs has not been shown definitely to be caused by energy transfer, the red organ does contain a red fluor very similar in spectral and chemical properties to a phycobiliprotein. These latter proteins contain linear tetrapyrrole chromophores which act as a highly efficient energy transfer system. The phycobiliproteins have been found previously only in blue–green, red, and brown algae. The chlorophyll of green plants absorbs light poorly in the region 500–600 nm, with the result that these plants are unable to utilise nearly one-third of the visible spectrum of photosynthesis. Certain algae, on the other hand, contain protein micelles called phycobilisomes which are highly organised complexes of phycobiliproteins with their phycobilin chromophores arranged so as to provide an energy transfer pathway to chlorophyll with more than 90% efficiency. We have found that a red-fluorescent protein, but blue in ambient light very similar, apparently, to phycocyanobilin, can be extracted from the red organ of *Malacosteus niger*. An important difference between this protein and the phycobiliproteins is that it is 'phosphorescent', which may relate to the chemiluminescent reaction itself, as yet unidentified.

Tempting as it is to speculate on the possible inter- or intramolecular energy transfer systems which might exist in *Malacosteus* and its relatives, between the chemiluminescent reaction of the blue organ and the red fluor, such changes in colour need not necessarily be due to energy transfer. Excited carbonyls, generated from the spontaneous breakdown of dioxetan or dioxetanone intermediates, formed as part of a chemiluminescent reaction normally emit blue light. When conveniently linked to an aromatic group, so that the orbitals are conjugated, the colour of the light emitted will be dictated by the chromophore, and can be green, yellow, or red. This is not usually regarded within the definition of energy transfer as conceived here unless the excited group is separated from an acceptor without conjugation. For example the aliphatic chains in the compounds synthesised by White and colleagues (Fig. 9.4) separate donor from acceptor and demonstrate intramolecular energy transfer.

A further confusion can occur in bioluminescence, because of the ability of the environment at the active centre of the luciferase to alter the spectrum of the chemiluminescence, or because of different ionisation states of the excited state product emitting different coloured light (see Chapter 3; section 3.4). For example, the chemiluminescent spectrum in luminous radiolarians, and in coelenterates not containing any green fluorescent protein, is due to oxidation of an imidazolopyrazine (Fig. 9.2), and the emission maximum can be anything from 440 nm to 490 nm. The emission from fireflies can vary from green to yellow, depending on the species. These colour shifts are caused by the luciferase. When studying these latter systems *in vitro* low pH or the binding of heavy metals can shift the emission of the oxyluciferin, a benzothiazole, from yellow–green to red. Several other luminous beetles from the same order, *Coleoptera*, also have two types of light organ, one emitting redder light than the other. This occurs in two members of the super-family *Elateroidea* (click beetles), *Pyrophorus* and *Photophorus*. A particularly spectacular example is found in the South American railroad worm, *Phrixothrix tiemanni*, belonging to the superfamily *Cantharoidea*. It was discovered by traveller-naturalists in 1809. The females have red 'headlights' on the head and 11 pairs of yellow–green organs on its body. As with the fireflies and click beetles, no separate fluor has been found to explain these spectral shifts. At present, it must be assumed that they are caused by the environment either within each luciferase, or within the cell. However they may also involve an interaction with the intramolecular energy transfer within the oxyluciferin responsible for light emission. Thus, spectral differences between species with the same primary chemiluminescent reaction, and other species anomalies, can be the result of energy transfer but need not necessarily be so. Radiationless chemiluminescence transfer can therefore have any one of three roles to play in bioluminescence (Table 9.5):

(1) *Intra*molecular energy transfer may be crucial in the generation of the excited state responsible for light emission in the primary chemiluminescence reaction, e.g. the firefly and coelenterate oxy-luciferins.
(2) *Inter*molecular energy transfer may occur between oxyluciferin and unoxidised luciferin, without specific binding of donor to acceptor when intracellular concentrations are high enough.
(3) *Inter*molecular energy transfer may occur between the primary excited product of the luciferin and a fluor, e.g. certain luminous coelenterates, some luminous bacteria, euphausiid shrimps, and the fresh water limpet *Latia*. In some cases, such as the coelenterates, the fluor plays no direct role in the chemiluminescent reaction. However, in others, for example the euphausiids, the fluor plays a direct role, in addition to acting as an energy transfer acceptor.

The function of these three types of energy transfer in bioluminescence is either to ensure an efficient chemiluminescence reaction producing visible light, or to produce light of the colour necessary for it to fulfil its function in the organism (see Chapter 3).

Table 9.5 — Evidence for energy transfer in bioluminescence

Evidence	Example
1. Mechanism from dioxetan breakdown	firefly oxyluciferin
	coelenterate oxyluciferin
	bacterial FMN hydroperoxide
2. Presence of fluors	
(a) necessary for chemiluminescence *in vitro*	euphausiid shrimp fluor
	freshwater limpet purple protein
(b) necessary for spectral shift	blue fluorescent protein in *P. phosphoreum*
	yellow fluorescent protein in *V. fischeri*
	green fluorescent protein in *Renilla*
(c) present in same cells as chemiluminescent system	blue and yellow fluors in particular luminous bacteria
	green fluor in some coelenterates
	? luciferin in searsiid fish
	? phycobiliprotein in Malacosteidae (stomiatoid) fish
	ergostatetraeone and riboflavin in luminous fungi
3. Chemiluminescence spectra	
(a) variations between species	bacteria with and without blue or yellow fluor
	coelenterates with and without green fluor
(b) variations within same species	*Phialidium* with or without green fluor
	two light organs in stomiatoid fish
(c) bimodal and temporal changes	some echinoderms (ophiaroids)
	some gorgonians
	some crustaceans, e.g. Pleuromamma, Paraponoë, Sergestes
	Cranchiid squid
	Searsiid fish

However, spectral shifts on their own are insufficient to *prove* the existence of energy transfer. How then do we establish its existence *in vitro*, and in living organism itself (Table 9.5)? In order to define the criteria to do this it is first necessary to differentiate the different types of chemiluminescence energy transfer (see section 9.4).

9.3.3 Ultraweak and indicator-dependent chemiluminescence

In Chapter 6 we examined several oxidative processes in cells which can generate an ultraweak chemiluminescence, from endogenous substrates. These included peroxidase, oxidase, and oxygenase reactions, and processes such as lipid peroxidation, drug detoxification, and cell division. Addition of a substrate, capable of generating chemiluminescence when it reacts with the oxidants involved in these reactions, results in a larger light signal, and is referred to as indicator-dependent chemiluminescence. In principle, the excited states generated by oxidation of either endogenous or added compounds should be capable of energy transfer to endogenous fluors such as flavins, porphyrins and, in plants, the chlorins of the chlorophylls. The first of these has been detected in the fungus *Blastocladiella emersonii*. Aerobic oxidation of added phenyl acetaldehyde by endogenous peroxidase generates excited state benzaldehyde, which then transfers its energy to green–yellow emitting flavins in the cells. In contrast, the red emission of chlorophyll has been detected following lipid peroxidation in chloroplasts.

Interestingly, Cilento and colleagues have demonstrated energy transfer from several weak chemiluminescent reactions *in vitro* using chlorophyll solubilised in micelles of cationic, anionic, or neutral detergents. In this environment the fluorescence of chlorophyll is more efficient and the energy transfer process is favoured. Chlorophyll a emits with a peak at around 680 nm, chlorophyll b at around 660 nm. Good donors for energy transfer to chlorophyll are (a) peroxidase catalysed reactions generating excited carbonyls via dioxetan or dioxetanone intermediates, (b) enediol oxidation to an excited carbonyl via electron transfer, and (c) excited quinones from catechol oxidation catalysed by enzymes such as catechol oxidase or tyrosinase (Fig. 9.2). An example of the first group is the oxidation of

$$\text{CHO}$$
$$|$$

phenyl—CH —CO_2Et occurring in the roots of the plant *Datura inoxia*. The second is exemplified by the oxidation of the hydrazide of isonicotinic acid, or dihydroxyfumaric acid, both catalysed by horseradish peroxidase. The hydrazide is of interest as it is a drug which can both be tuberculostatic and carcinogenic. Catechol oxidation generating chemiluminescence is well illustrated by the oxidation of adrenalin to its pink oxidation product, adrenochrome — a reaction first observed by Vulpian, a mentor of Louis Pasteur, who in 1856 commented on seeing extracts of adrenals turn pink that it was 'tout à fait remarkable'. No doubt he would have been even more surprised to learn that the reaction he was observing could also produce light!

9.4 A CLASSIFICATION

Energy transfer phenomena are therefore divided into two main groups (Table 9.6A and B). One is radiative or 'trivial', requiring the transfer of a photon from donor to acceptor; the other is non-radiative, occurring without the direct transfer of a photon

Table 9.6 — Classification of energy transfer

A. Radiative (trivial): direct transfer of photon from donor to acceptor
B. Non-radiative: no direct transfer of a photon
1. Exciton transfer: only occurs in rigid materials, i.e. solids and glasses
2. Resonance or coulombic energy transfer (Förster): no atomic or molecular orbital overlaps necessary, occurs up to *ca.* 10 nm, can be intra- or inter-molecular
3. Electron exchange (collisional): requires orbital overlap between donor and acceptor, without net electron transfer
4. Electron transfer: requires orbital overlap between donor and acceptor, with net electron transfer
5. Chemical reaction: requires fusion or fission of chemical bond resulting in transfer of atom(s) from donor to acceptor
6. Energy pooling: requires formation of an excimer or exciplex, the actual light emitter

from donor to acceptor. Non-radiative energy transfer can then be subdivided into six categories, based on the nature of the non-radiative mechanism. Of these, all but exciton transfer have been observed in chemiluminescence (Table 9.7) where the excited donor is formed via a chemical reaction.

The concept of excitons was first proposed by Frenkel in 1936, and is usually found only in rigid materials such as ionic crystals. Nevertheless, chemiluminescence energy transfer could, in principle, occur by this mechanism at solid–liquid or solid–gas interfaces, where many chemiluminescent reactions have been observed. Of the others only one, resonance energy transfer, does not require electronic orbital overlap. This mechanism can occur when the excited donor and acceptor are up to about 100 Å (10 nm) apart. This condition can be achieved within one molecule

Table 9.7 — Examples and classification of non-radiative chemiluminescence energy transfer

Type of energy transfer (Table 9.6)	Excited donor	Acceptor
Resonance (B2)		
intramolecular	phthalate	*N*-methyl acridone
	phthalate	anthracene
intermolecular	N methyl acridone	lucigenin
with substrate	siloxene product	siloxene
	phthalate	phthalhydrazide
intermolecular	*Renilla* oxyluciferin	green fluorescent protein
with a separate fluor	siloxene product	rhodamine
Electron exchange (B3)	? FMN peroxy intermediate	lumazine
Electron transfer (B4)		
intramolecular	phenolic O in firefly luciferin	CO_2 on firefly luciferin
intermolecular	C_2O_4 from oxalate ester oxidation	polyaromatic hydrocarbons
	$C_2O_4^-$	rubrene$^+$ (from electrode)
Chemical reaction (B5)	$^1O_2$	rubrene
Energy pooling (B6)	$^1O_2$	$^1O_2$

containing separate donor and acceptor (Fig. 9.4), or by specific binding, for example when ABEI-antigen binds to fluorescein-antibody. But resonance energy transfer can also occur in free solution when the concentrations of donor and/or acceptor are high enough, and may be aided by diffusion. The concentration of donor or acceptor has usually to be at least mM for this type of energy transfer to be detected in free solution. All of the other mechanisms require donor and acceptor to be within at least 2–5 Å (0.2–0.5 nm) of each other. This will only occur in free solution when the concentrations of donor and acceptor are greater than molar! Most chemiluminescent reactions are usually studied at concentrations several orders of magnitude less than this, particularly since several donors and acceptors are not soluble at this concentration. Thus, electron transfer, electron exchange, and energy pooling (Table 9.6) require the formation of molecular or atomic complexes following collisions. In some cases, the complex itself becomes the emitter, in which case it is known as an *excimer* or an *exciplex*. It is an *excimer* when the two molecules are chemically identical, e.g. O_2 dimol emission or rubrene$^+$·rubrene$^-$ in electrogenerated chemiluminescence, but an *exciplex* if the molecular species in the excited complex are different, e.g. rubrene$^+$. anthracene$^-$.

Some chemiluminescent reactions involving energy transfer use a combination of two or more of the different mechanisms to complete the non-radiative process. For example, several examples of electrogenerated chemiluminescence and chemically initiated electron exchange chemiluminescence require both electron transfer and energy pooling (see section 9.7). Singlet oxygen, on the other hand, can react chemically with an alkene double bond to form a dioxetan whose excited carbonyl product may then undergo resonance or electron energy transfer. Alternatively $^1O_2$ may react with an aromatic hydrocarbon such as rubrene to form an adduct whose breakdown results in excitation of the rubrene (see section 9.8).

The results of chemiluminescence energy transfer will be a change in overall quantum yield, and a change in spectrum of the emission. Occasionally, a change in some other physical parameter such as reaction kinetics or polarisation may also occur. The latter is exemplified when siloxene, a solid, is oxidised by H_2O_2. The endogenous red chemiluminescence of siloxene produces polarised light. Polarisation is lost, however, when energy transfer occurs to acceptors in free solution such as eosin or rubrene.

How then can we distinguish these various types of energy transfer, and perhaps harness them as analytical tools?

9.5 WHAT TYPE OF ENERGY TRANSFER IS IT?

9.5.1 The criteria

Add a small spatula end of fluorescein to a μM–mM solution of luminol in 0.1 M NaOH with a drop or two of H_2O_2. Then add an equal volume of 1–10 mM potassium ferricyanide, and the flash will glow immediately yellow–green instead of the usual blue chemiluminescence of luminol alone. Theoretical arguments may tell us that, at the concentrations of fluorescein and luminol used, the excited donor, aminophthalate, and the acceptor, fluorescein, will be close enough for resonance energy transfer to take place. Model systems may show that energy transfer can occur in free solution from donor to acceptor when the two are close enough, and that the speed of energy transfer is expected to be in the ps–fs range compared to the 1–10 ns for 'trivial' transfer via direct photon emission from a singlet excited state. Yet how can we be sure that the observed colour change is truly an example of non-radiative, as opposed to radiative, or 'trivial', energy transfer? Why don't the excited aminophthalate molecules, produced by the chemiluminescent reaction of luminol, simply emit photons which are absorbed by the fluorescein molecules which then re-emit just as they do in a fluorimeter?

This problem of distinguishing radiative from non-radiative energy transfer confronted the early workers studying this aspect of fluorescence, who went to some lengths to confound the sceptics. Not only were they able to show that energy transfer was non-radiative, but also whether or not it required atomic or molecular collisions, or complexes, to occur. In the original experiments of Cario and Franck this proved a little difficult with Hg as donor and Th as acceptor. However, when Na was used as acceptor the result was clear-cut. Excitation of Na, using a lamp with a line for exciting Hg, was impossible, and no yellow light was observed. However, addition of Na to Hg vapour, followed by excitation of Hg, produced the familiar Na line emission. This could only have occurred by a non-radiative process, over distances greater than those for normal atomic separations.

One of the first observations of non-radiative fluorescence energy transfer in solution was made by Perrin and Choucroun in 1929, using phenosafamine as donor and tetrabromoresorcifim as acceptor. Förster in 1949 was able to observe the phenomenon more quantitatively by detecting the reduction in trypoflavine (donor) fluorescence which occurs in the presence of rhodamine (acceptor). Only the quenching of trypoflavine fluorescence was observed. No rhodamine emission was reported. These experiments, therefore, identify one important criterion which can be applied to non-radiative chemiluminescence energy transfer, namely a decrease in the donor emission. This should be matched by an increase at the emission maximum if the acceptor is a fluor. On the other hand, a quenching of the chemiluminescence, together with a spectral distortion, will occur if the acceptor is a non-fluorescent chromophore, its excitation energy being lost via non-radiative processes. In itself, though, this is insufficient to prove non-radiative, as opposed to radiative, transfer. This requires a careful concentration dependence of the transfer to be established. Bowen and co-workers in the early 1950s established this for fluorescence energy transfer using 1–chloroanthracene as donor and perylene as acceptor. At a constant ratio of both components the absorption of exciting light should be a constant fraction, if radiative transfer was occurring. Yet as the *net* concentration of the two components was increased, keeping the ratio of donor to acceptor constant, there was an increase in perylene fluorescence together with a decrease in that from 1-chloroanthracene. Furthermore, the total light yield increased as the energy transfer became more efficient because 1-chloroanthracene has, by itself, a much lower fluorescence quantum yield than the energy transfer acceptor, perylene. A similar enhancement in quantum yield was used by Ward and Cormier to establish energy transfer in the chemiluminescence system of the sea pansy *Renilla* (see section 9.6.2). Similarly the energy transfer system established by combining oxalate ester chemiluminescence with fluorescent acceptors, such as anthracene derivatives or rubrene, cannot be 'trivial' on this basis. Without a fluor oxalate ester oxidation produces no visible light (Fig. 9.1).

Unfortunately, not all chemiluminescence energy transfers systems produce such a dramatic increase in overall quantum yield as that found with oxalate ester chemiluminescence. Sometimes a substantial decrease may occur, even when the acceptor is a fluor. Therefore, six experimental criteria should be examined to distinguish non-radiative from radiative chemiluminescence energy transfer. But not all will be satisfied in every system. The mechanism of energy transfer affects which particular criteria are satisfied. One advantage we have in studying chemiluminescence over those studying fluorescence is that the fluorescent acceptor rarely has any significant chemiluminescence. Thus, any increase in light emission at wavelengths near the acceptor emission maximum, on addition of the acceptor to the chemiluminescent reaction, must involve energy transfer of some sort, be it radiative or non-radiative. With fluorescent organic compounds in solution it is very difficult not to excite some of the acceptor molecules at the donor excitation wavelength.

The six parameters to examine as criteria for distinguishing non-radiative from radiative chemiluminescence energy transfer, where the acceptor is a fluor, therefore are:

(1) *Chemiluminescence spectrum*
Addition of the acceptor should result in an increase in light emission at the fluorescence emission of the acceptor, together with a decrease at the emission maximum of the donor. The spectral shapes of donor and acceptor should still be identifiable when non-radiative transfer has occurred, whereas radiative transfer leads to much distortion of the donor emission spectrum. Sometimes non-radiative transfer leads to a shift in emission towards the blue, as opposed to the usual red shift. This occurs, for example, with the blue fluorescent protein in some luminous bacteria shifting the peak emission from 490 nm to 472 nm. Such 'blue shifts' cannot occur by radiative transfer.

(2) *Chemiluminescence quantum yield*
An increase in overall quantum yield (photons emitted/substrate molecules reacting) is proof of non-radiative energy transfer. This occurs either when the acceptor has an increased fluorescence quantum yield compared with that of the excited state donor alone, or when the acceptor increases the yield of excited state donors by influencing the chemiluminescence reaction itself. This latter process can be particularly relevant in electron transfer processes. Unfortunately not all non-radiative energy transfer reactions lead to an increase in overall quantum yield. Radiative transfer always produces some reduction in overall quantum yield.

(3) *Effect of volume*
Increasing the reaction volume increases the pathlength of light through a solution containing the acceptor, thereby increasing the chance of radiative transfer. However, at constant concentration, volume changes should not affect the proportion of molecules undergoing non-radiative energy transfer.

(4) *Effect of concentration of donor and acceptor*
Depending on the nature of the non-radiative energy transfer process increasing or decreasing donor and acceptor concentrations, with or without a change in their ratio, will have predictable effect on the efficiency of transfer (see sections 9.6 and

9.7). In particular, the effect of changes in the mean distance between molecules can be predicted, together with law of mass action effects if the acceptor plays a 'catalytic' role. Effects of changes in concentration of the acceptor on radiative energy transfer can be predicted via Beer-Lambert's law. For example, non-radiative intramolecular energy transfer, or intermolecular transfer where binding of donor and acceptor occur, e.g. within an antibody-antigen complex, can be demonstrated at sub μM concentrations where absorption of light by the acceptor through radiation from the donor is negligible.

(5) Effect of physical properties of the acceptor
Predictable effects of changes in the extinction coefficient, fluorescence quantum yield, excitation and emission characteristics and fluorescence life times, can provide supportive evidence for non-radiative energy transfer (see sections 9.6 and 9.7).

(6) Effect of medium viscosity
This has little or no effect on the efficiency of radiative transfer, but can markedly affect the polarisation of the light emission and the efficiency of certain non-radiative processes. This is particularly relevant to energy transfer dependent on the formation of charge transfer complexes whose stability is enhanced in viscous solvents. Hence the use of viscous phthalate ester solvents for oxalate ester chemiluminescence energy transfer.

A combination of these experimental criteria, together with an identification of the mechanism (Table 9.6B) enables *non-radiative* chemiluminescence energy transfer to be distinguished from a *radiative* transfer.

9.5.2   The problem of electron spin
What happens to electron spin during the transfer of energy between excited states?

The Pauli exclusion principle, proposed in 1925, led not only to the development of the concept of electron spin, but also to the realisation that the crossing between different multiplicities, that is processes involving a change in spin, from singlet (S) to triplet (T) or vice versa, are much slower than those not involving such intersystem crossing, i.e. S$\rightarrow$ S or T$\rightarrow$ T. In the limiting case inter-system crossing is 'forbidden'.

Molecular vibrations, and the time spent by molecules in transition states during a chemical reaction, are typically in the range of 10 fs–1 ps (10^{-14}–10^{-12} s). Intersystem crossing from the singlet to triplet is at least 100 times slower than this and, for atomic molecules at least, may be a million times slower. The decay of an excited triplet to ground state singlet (phosphorescence) may be even slower, at least 10^{-5} s, and is sometimes many seconds. This is ample time for quenchers in the reaction mixture, such as oxygen in the triplet state itself, to act, and thus drastically reduce the quantum yield from such triplet states. Thus, except under special circumstances (see Chapter 1; section 1.5), true phosphorescence from triplet excited states would not be expected to play a major role in efficient chemiluminescent reactions.

Yet not only do singlet$\rightarrow$ singlet and triplet$\rightarrow$ triplet transfers occur in chemiluminescence energy transfer but also the so-called 'forbidden' processes of singlet$\rightarrow$ triplet or triplet$\rightarrow$ singlet (Table 9.8). How does this come about? Can this apparent paradox be resolved?

Table 9.8 — Electron spin conversion and change during chemiluminescence energy transfer

Excited spin		Example	
Donor before transfer	Acceptor after transfer	Donor	Acceptor
singlet	singlet	rubrene aryloxalates→'C_2O_4	9,10 diphenylanthracene rubrene
triplet	triplet	†rubrene methylethylketonebiacetyl	tetramethylphenylenediamine anthracene derivatives
singlet	triplet	$Ru(bpy)_3^{3+}$	$Ru(bpy)_3^+$
triplet	singlet	†acetone from (TMD)	Eu chelates or dibromoanthracene
		phthalate	N-methyl acridone or anthracene

† May involve a two-step process involving first triplet–triplet transfer then intersystem crossing within the excited acceptor.

There are in fact at least seven reasons why there may be a change in spin states from donor to acceptor as a result of energy transfer:

(1) The multiplicity of the complete reaction, which is what really matters in quantum theory, may still be conserved, even though the spin state of the individual components change, e.g.

$$D^*(T) + A(S) \rightarrow D(S) + A(T)^* \tag{9.7}$$

D = donor; A = acceptor; T = triplet; S = singlet; * = excited state.

This need not be very slow.

(2) The chemiluminescent reaction may favour the formation of the triplet excited state.

   Normally the first triplet state is of lower energy than the first singlet excited state. Thus less activation energy is required to form the triplet (see Chapter 1; section 1.5). The breakdown in aliphatic dioxetans particularly favours formation of the triplet state. For example, the ratio of quantum yields, ϕ_T/ϕ_S, for tetramethyl dioxetan → acetone* is more than 100. Aromatic substitution, however, as occurs in luciferins, may increase the proportion of excited products in the singlet spin state.

(3) The triplet state in the donor or acceptor may be stabilised by 'spin orbit coupling' between the excited state and another molecule or atom very close to it. This can be intra- or intermolecular. Heavy atoms found in dibromoanthracene (Fig. 9.3) and europium chelates can do this. 9,10 dibromoanthracene-2-sulphonate ions (DBAS), and flavins are particularly good acceptors from triplet state acetone, and xanthene dyes are efficient acceptors from triplet aliphatic or aromatic aldehydes. Electrochemiluminescence with, for example, rubrene may involve initially excitation to the *second* triplet excited state, followed by triplet–triplet annihilation to produce light.

(4) The local environment may stabilise donor or acceptor molecules in the triplet state. This could occur, for example, at the active centre of a luciferase, or in particular solvents, and by reducing 'quenchers' such as oxygen.

(5) The speed of transfer via intersystem crossing may still be acceptable, particularly if donor and acceptor molecules are brought together by diffusion or specific binding.

(6) The stable ground state of some molecules is triplet. An example of this is oxygen.

(7) Spin restrictions may not apply to certain energy transfer mechanisms. Restrictions apply in dipole–dipole resonance transfer, but not when radical reactions and electron transfer occur. In fact, triplet states are common in these latter processes, particularly those involving a biradical intermediate.

Three key parameters therefore characterise the different types of chemiluminescence energy transfer (Tables 9.4–9.9):

(1) Radiative or non-radiative.
(2) Electron transfer or not.
(3) Change in electron spin or not.

Let us now therefore examine in more detail the two main classes of non-radiative mechanism: resonance and electron transfer chemiluminescence energy transfer, to see how they occur.

9.6 RESONANCE ENERGY TRANSFER

This type of energy transfer occurs via dipole–dipole resonance. It is found in living organisms as well as with synthetic compounds. It can occur both within or between atoms of molecules in gases, liquids, and solids, as well as within or between molecules in the same phase (Table 9.9).

Some biologists are put off by equations! The one describing resonance energy transfer was worked out for fluorescence by Förster in the late 1940s, after some earlier attempts by Perrin in the 1920s and 1930s. It is a rather daunting-looking equation (Equations 9.9. and 9.10). Yet an examination of its individual components

Table 9.9 — Example of resonance (Förster) chemiluminescence energy transfer

Initial substrate	Excited donor	Fluorescent acceptor
phthalazine dione (luminol, ABEI)	aminophthalate derivative	fluorescein
lucigenin	N-methyl acridone	lucigenin
siloxene	oxidation product	eosin, rhodamine
coelenterate photoprotein	oxyluciferin on photoprotein	green fluorescent protein
Renilla luciferin–luciferase	oxyluciferin	green fluorescent protein

not only help us to understand the conditions required for optimal efficiency of resonance energy transfer, but also enable experiments to be designed to establish whether this particular mechanism is operating in a given phenomenon. Förster's equation also provides the basis for applying chemiluminescence resonance energy transfer as an analytical tool, and for using this phenomenon to monitor distances between atoms and molecules in living systems, following the pioneering work of Stryer and his colleagues using fluorescence resonance energy transfer.

What then is the Förster equation? How was it derived and what are its key components which allow, or restrict, resonance energy transfer in chemiluminescence?

9.6.1 The Förster equation

J. Perrin in 1925 was the first to attempt a mathematical treatment of resonance energy transfer. He predicted that the critical interaction distance (d) between donor and acceptor was $\lambda/2\pi$ for fluorescence, λ being the wavelength excitation maximum of the acceptor. In 1932 F. Perrin developed a more quantum mechanical treatment estimating d as

$$d = \lambda/\pi \sqrt{t/\tau} \tag{9.8}$$

where t = mean interval between successive molecular collisions and τ = mean excitation time.

Assuming t to be 10^{-14}–10^{-13}s for fluorescein excitation at $\lambda \doteqdot 500$ nm, he estimated the critical distance to be 150–250 Å (15–25 nm). But this is equivalent to a concentration of some 20–100 μM, whereas the fluorescence energy transfer process detected by loss of fluorescence polarisation, which can be observed in a viscous solvent like glycerine, occurred significantly only at concentrations around 1–10 mM. Förster decided in 1948 to modify Perrin's approach by considering two additional factors, both imperfections in the resonance process. One was the imperfect overlap between the emission spectrum of the donor and the excitation spectrum of the acceptor; the other, the fact that donor and acceptor would not always be in perfect orientation. Förster argued further that the interaction energy in dipole–dipole interactions depended on the inverse of the third power of the molecular separation distance (d), unlike straightforward electrostatic interactions which obey the inverse square law. Since the probability of energy transfer occurring was also proportional to the square of the dipole–dipole interaction energy, the efficiency of the complete process should therefore decrease with the *sixth* power of the distance.

The full Förster equation, converted for chemiluminescence energy transfer, therefore, is:

$$\text{energy transfer efficiency } (E) = d^{-6}/(d^{-6} + R_o^{-6}) \text{ or } R_o^6/(R_o^6 + d^6) \tag{9.9}$$

$$\text{and the rate of energy transfer } (k_T) = A \times J \times d^{-6} \text{ s}^{-1} \tag{9.10}$$

where d represents in Ångstroms (1Ångstrom = 0.1 nm) the distance between the centres of the chemiluminescent donor and the fluorescent acceptor.

$$R_0 = (J K^2 \phi n^{-4})^{1/6} \times 9.7 \times 10^3 \text{ Å} \tag{9.11}$$

incorporating the following spectroscopic variables:

J = spectral overlap integral (Fig. 9.8)

$$= \int C(\lambda)\, \varepsilon(\lambda)(\lambda)^4 \mathrm{d}\lambda \bigg/ \int C(\lambda)\, \mathrm{d}\lambda \tag{9.12}$$

$C(\lambda)$ = chemiluminescence emission of donor at wavelength λ.
$\varepsilon(\lambda)$ = absorption coefficient of acceptor at λ.
K^2 = the orientation factor between donor and acceptor, varying between 0. and 4, 2/3 for random orientation.
ϕ_0 = fluorescence quantum yield of chemiluminescence energy donor in the absence of acceptor.
n = refractive index of the medium between donor and acceptor.
A = constant = $K^2 n^{-4} k_F \times 8.71 \times 10^{23}$. \hfill (9.13)

where k_F = rate constant for photon emission by the excited donor formed through chemiluminescence reaction.

Equations 9.9. and 9.10 thus take into account the two processes which may compete with energy transfer, namely quenching of the donor and light emission from the donor without transfer. These are taken into account through ϕ_0 and k_F respectively. Since singlet excited state lifetimes are of the order of 1–10 ns, the transfer process must be in the ps–fs range to beat the simple 'fluorescence' decay of the donor. This will always be true for singlet–singlet transfer. Even when a change in spin state occurs the transfer may still be fast enough to be observed (see section 9.5.2).

In general, there are too many unknowns for the Förster equation itself to be used in calculating from first principles chemiluminescence energy transfer. However, an empirical estimate of the efficiency of transfer can be made by considering the relationship between acceptor concentration and 'sensitised' emission:

$$\frac{1}{I} = \frac{a}{(A)} + b \tag{9.14}$$

$$b/a = k_{ET}\, M^{-1} \tag{9.15}$$

where I = light intensity at the acceptor emission wavelength
(A) = concentration of acceptor
a, b = constants at constant donor (D) concentration.

Thus, by varying the concentration of acceptor (A), a plot of $1/I$ against (A) enables the rate constant for energy transfer (k_{ET}) to be estimated as the ratio of the intercept at the y-axis (b) and the slope of the plot. This has been used by Cilento and colleagues to demonstrate the efficiency of energy transfer to chlorophyll in organic micelles compared to that in aqueous media (section 9.3.3). k_{ET} in the micelle was as high as $10^6\ M^{-1}$, but was some 40 times less in water.

A parameter often ignored is the refractive index, related to the dielectric constant of the medium. The nature of the solvent, its polar or apolar nature, can thus influence the efficiency and rate of energy transfer processes occurring by dipole–dipole resonance. The dielectric constant within proteins varies greatly from that in water. Thus the protein environment between donor and acceptor may increase the efficiency of this type of energy transfer in bioluminescence.

In spite of the practical difficulties involved in using the Förster equation to describe chemiluminescence energy transfer quantitatively, it does help us to identify how to identify conditions for optimal energy transfer. This requires selection of the optimal conditions as follows:

(1) *Distance between donor and acceptor*
The closer the better, and not greater than 100–120 Å (10–12 nm) apart.

(2) *Spectral overlap between donor and acceptor (J integral)*
The greater the overlap between the normal donor emission, and the excitation spectrum of the acceptor the more the transfer.

(3) *Orientation between donor and acceptor*
Random orientation ($K^2 = 2/3$) is adequate, but the best is where the orbital dipoles are aligned and parallel ($K^2 = 4$). At right angles K^2 is zero and thus there is no transfer (Eqn. 9.10 $k_T = 0$).

(4) *The refractive index (n) of the medium between donor and acceptor*
Pure water has a refractive index of 1.33, and a dielectric constant of 80. As the refractive index decreases, equivalent to a decrease in dielectric constant (e.g. less polarisable solvents), energy transfer increases. Conversely a more polarisable solvent reduces transfer between dipoles.

(5) *Spin restrictions*
These are not dealt with by the Förster equation, but as we have seen (section 9.5.2) some fluors are particularly good acceptors from triplet excited state donors, others more efficient at accepting energy from singlets.

Two other factors affect the intensity of light, measured at the acceptor's emission, relative to that of the donor. These are the relative fluorescence quantum yields, that of the excited donor (ϕ_D) in the absence of the acceptor and that of the acceptor (ϕ_A). The greater ϕ_A/ϕ_D the more light at the acceptor emission maximum.

The distance between the two centres (d) is dependent on concentration if the transfer is intermolecular and takes place in free solution. It is influenced by molecular size if the transfer is intramolecular or within a molecular complex. Considering an oversimplified static model for molecules in a cubic lattice, where the side of the cube is 'a' cm:

$$\text{volume of cube} = a^3/1000 \; l \tag{9.16}$$

average volume of molecules per cube = 4/3, therefore

$$\text{number of molecules per litre} = 4000/3a^3 \tag{9.17}$$

$$\text{but number of molecules per litre also} = M \times 6.023 \times 10^{23}$$

where M = molarity in moles l^{-1}
    Therefore

$$a = \sqrt[3]{\frac{4000}{(3 \times M \times 6.023 \times 10^{23})}} = 1.2^3\sqrt{M} \text{ cm} \tag{9.18}$$

If $M = 1$ mM, then $a = 12$ nm (120 Å).

Förster estimated, using a more correct model than this, that separation of fluorescein molecules by 50 Å (5 nm) corresponded to a concentration of 3.2 mM. Equation 9.9 on the other hand shows that for fluorescent acceptors with extinction (absorbance) coefficients (ε) in the range 10^4–10^5 $M^{-1}.cm^{-1}$ (e.g. for fluorescein *ca* $\varepsilon = 5 \times 10^4$) energy transfer is only significant when $d < 80$ Å (8 nm). However, if better fluors were to be used, e.g. phycobiliproteins or aromatic hydrocarbons, with ε's in the range 10^5–5×10^5 $M^{-1} cm^{-1}$, then the separation distance could extend to 100–150 Å (10–15 nm). The theoretical maximum extinction coefficient is about 10^6 $M^{-1} cm^{-1}$.

These simple calculations explain why energy transfer can occur at mM concentrations in free solution between, for example the excited product of luminol chemiluminescence, 3-aminophthalate, and fluorescein, or between N methyl acridone and lucigenin.

A similar example of intermolecular energy transfer between an excited state product and residual substrate molecules can also be observed with the parent compound of luminol, phthalic hydrazide, i.e. luminol without the $- NH_2$ group. Oxidation of phthalic hydrazide produces phthalate, which is not fluorescent, either in aqueous or protic solvents. However, in aprotic solvents a weak yellow luminescence is detectable, the result of resonance energy transfer from the excited phthalate to the anion of the initial substrate, phthalic hydrazide.

Two ways of solving the problem of the distance separating the donor and acceptor molecules are to link them together either by a covalent bridge or non-covalently through a high affinity ligand–ligand binding reaction. Resonance energy transfer can now be observed at nM–μM concentrations, several orders of magnitude below those required to bring molecules in free solution close enough.

White and co-workers succeeded in generating intramolecular chemiluminescence energy transfer from phthalhydrazide donors covalently linked by a methylene bridge to either N methylacridone or diphenylanthracene as fluorescent acceptors (Fig. 9.4). Though the chemiluminescence quantum yields of some of these compounds were still less than that of luminol, the evidence for intramolecular energy transfer was clear:

(1) The chemiluminescence quantum yield for the coupled compound was much greater than that without the fluor. For example, phthalhydrazide has no significant chemiluminescent, yet coupled to N methylacridone chemiluminescence was observed, ϕ_{CL} *ca.* 8% that of luminol. Similarly, the chemiluminescence quantum yield of 2,3 naphthalic hydrazide was 0.06, relative to luminol, but 0.26 when covalently linked to diphenylanthracene.

(2) The chemiluminescence spectrum matched the fluorescence spectrum of the acceptor, either N methyl acridone or diphenylanthracene.

(3) The light yield of the coupled derivatives was linear over the range 1–100 μM, yet no increase in quantum yield was detectable when the donor and acceptor molecules were mixed together in free solution, uncoupled, at a concentration of 100 μM.

Antibodies of the IgG class are globular molecules of molecular weight 150 000, equivalent to a molecular radius of about 4 nm. They have two antigen binding sites with affinity constants in the range 10^8–10^{11} M^{-1}. Thus, significant antibody–antigen complex formation occurs even when the concentrations of the two components are as low as nM–pM (10^{-9}–10^{-12} M). Resonance energy transfer can occur between a chemiluminescent donor on the antigen and fluorescent acceptor on the antibody at concentrations some six orders of magnitude below those necessary to produce energy transfer between free, uncomplexed, molecules. This means that the antibody–antigen complex can be quantified by a change in colour (9.19), even in the absence of any significant effect on quantum yield (Fig. 9.9). Using the phthalhydrazide aminobutylethylisoluminol (ABEI) as the chemiluminescent donor and fluorescein as the acceptor, energy transfer has been established between antigens molecular weight 315–150000 and their respective antibodies (Fig. 9.10). The antigens include nucleotides, cyclic nucleotides, steroids and proteins.

$$
\begin{array}{cccc}
\text{Ag} & + & \text{Ab} \rightleftarrows \text{AgAb} & \\
| & & | \quad | \quad | & \\
\text{L} & & \text{F} \quad \text{L} \quad \text{F} &
\end{array}
\qquad 9.19
$$

$\downarrow$ oxidant $\downarrow$ oxidant

Ag   AgAb
| | \
L* $\Rightarrow$ blue light L* $\rightarrow$ F $\Rightarrow$ green light
  (460 nm)   (525 nm)
  no transfer   transfer

Addition of unlabelled antigen will displace AgL from the antibody–antigen complex and thus increase the ratio of blue to green light. This can be conveniently measured by a dual wavelength chemiluminometer with two photomultiplier tubes (Fig. 9.7).

The phenomenon also occurs between ABEI-biotin and its binding protein avidin, labelled with fluorescein (Fig. 9.11). However, there seems to be also a direct interaction between the donor and acceptor leading to quenching (a decrease in quantum yield), as well as a shift in ratio of blue to green light. Resonance energy transfer offers exciting possibilities for quantifying other ligand–ligand interactions in biology, such as hormones, neurotransmitters, and drugs with their receptors, RNA or DNA hybridisation, and even enzyme–substrate complexes.

Two other factors, in addition to distance, influence the extent of resonance energy transfer between a given donor and an acceptor, the spectral overlap and the orientation between them.

The greater the spectral overlap, or *J* integral (Fig. 9.8), between the chemiluminescence spectrum without the fluor and the excitation spectrum of the fluor, the greater the transfer. A rule of thumb is to examine how close are the peaks of chemiluminescence emission and the fluorescence excitation. For example, ABEI with a peak emission at about 460 nm is close enough to the 495 nm excitation peak of fluorescein, but too far from the 540 nm excitation peak of rhodamine, to produce significant energy transfer. However, this simple guide does not take into account spectral shape.

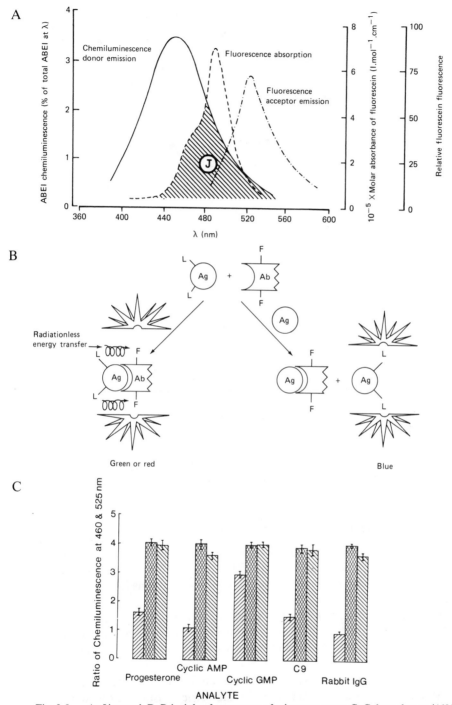

Fig. 9.8 — A: *J* integral. B: Principle of energy transfer immunoassay. C: Colour change (460/
525 emission ratio) on ABEI-antigen binding to fluorescein-antibody.
Hatched areas: first=fluorescein antibody, second=unlabelled antibody, third=fluorescein
non-immune IgG. Reprinted by permission from Campbell and Patel (1983). *Biochem J.* **216,**
185–194. © 1983 The Biochemical Society, London.

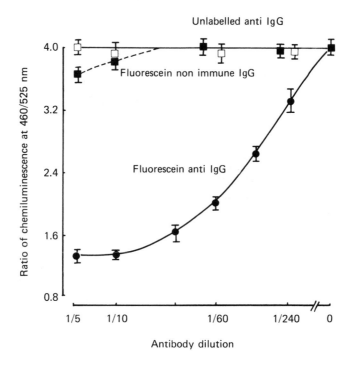

Fig. 9.9 — ABEI antigen binding to fluorescein antibody. Antigen = rabbit IgG; antibody = goat antirabbit IgG. Reprinted by permission from Campbell & Patel (1983) *Biochem J.* **216,** 185–194. ©1983 The Biochemical Society, London.

The orientation factor K^2, is defined by the equation:

$$K^2 = (\cos\alpha - 3 \cos\beta \cos\gamma)^2 \tag{9.20}$$

where α = angle between the donor and acceptor transition moments
        β = angle between the donor moment and the line joining the centres of the donor and acceptor molecules
        γ = angle between the acceptor moment and the line joining the centres of the donor and acceptor molecules.

Ideal orientation is when donor and acceptor orbitals are aligned and parallel, i.e. α, β and $\gamma = 0$ or 180°, since $\cos 0^0$ or $\cos 180° = \pm 1$ $K^2 = 4$. However, if donor and acceptor are perpendicular to each other, $\cos 90° = 0$, and thus $K^2 = 0$ and no energy transfer occurs.

For random orientation $K^2 = 2/3$. Generally we have no control on it. However, evolution may have deliberately selected molecular interactions which optimise this factor, e.g. in phycobilisomes for fluorescence energy transfer in blue–green and red algae.

The final factor in the Förster equation, often forgotten, is the refractive index. The refractive index is really related to the dielectric constant of the medium. This can be dramatically affected by the environment at the active centre of proteins and the interacting groups between macromolecules such as proteins, glycoproteins, nucleic acids, and lipids.

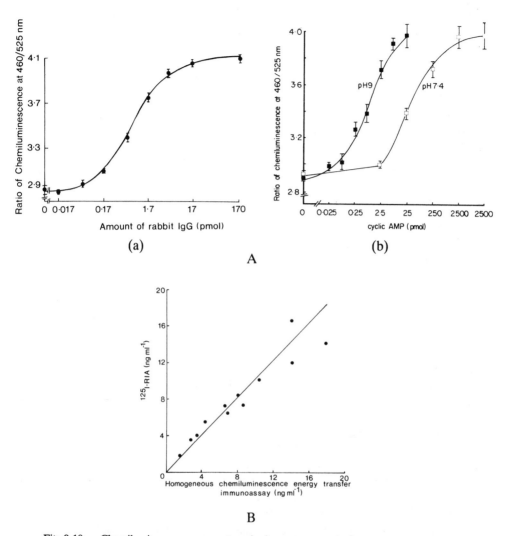

Fig. 9.10 — Chemiluminescence energy transfer immunoassay. Antigen = rabbit IgG–ABEI or cyclic AMP-ABEI. A: Standard curve, (a) IgG, (b) cyclic AMP. B: Correlation of energy transfer with RIA from Campbell *et al.* (1985). Chemiluminescence energy transfer immunoassay. In: Alternative Immunoassays pp. 153–183. Ed. Collins, W. P. © 1985 Reprinted by permission of John Wiley & Sons Ltd.

Thus the four main parameters to optimise for the best resonance energy transfer are: distance, spectral overlap, orientation, and dielectric constant between donor and acceptor molecules.

9.6.2   Evidence in bioluminescence

Of the many organisms where resonance energy transfer might explain abnomalies in quantum yield and spectra (Table 9.5), the most convincing evidence exists in luminous coelenterates containing green fluorescent proteins (Tables 9.3 and 9.10; Fig. 9.5). This is a family of proteins with molecular weights 30–80 000, with a fluor

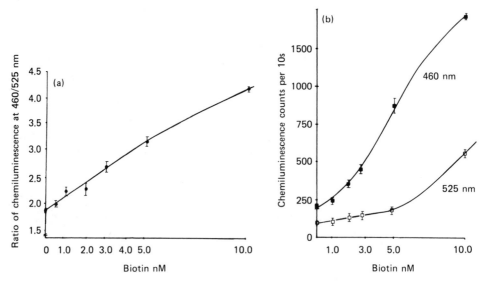

Fig. 9.11 — Quenching in ligand–ligand energy transfer. ABEI-biotin binding to avidin-fluorescein, from Williams & Campbell (1986) *Anal Biochem*. **155**, 249–255 with permission of Academic Press.

tightly bound to the protein. A structure for this fluor has been published by Shimomura (Fig. 9.6). However it is not sufficiently fluorescent to be the acceptor. An alternative has been proposed by McCapra, which has the acceptor excitation peak at around 470 nm and fluorescence emission at about 510 nm. All the green fluorescent proteins have a peak excitation at about 460 nm, very close to the emission peak of the photoproteins (460–480 nm), and a fluorescence emission at about 506 nm very close to the bioluminescence emission peak of 508–510 nm of the whole organism. Some of the green fluorescent proteins, e.g. from *Aequorea* but not from *Obelia*, also have an excitation peak at about 395 nm. This may be the result of a different ionisation state of the fluor within the protein. Using image intensification to match the locality of green fluorescence in *Obelia* with the bioluminescence,

Table 9.10 — Properties of some coelenterate fluorescent proteins

Protein source	mol. wt. $(\times 10^{-4})$	$\varepsilon\,M^{-1}\,cm^{-1}$	λ_{max} ex nm	λ_{max} em nm	ϕ_F
Aequorea	34, 53 (dimer), 69	4.3×10^4	395–400, 469	506–509	0.7–0.9
Obelia	? at least 2	?	469	506–509	?
Phialidium	?	?	487	497	?
Renilla	52 (dimer)	4.8×10^4	498	508–509	0.8

N.B. GFP's are acidic; some are monomers others dimers.
       Stokes radius of emitters: 20 $K \doteqdot 1.9$ nm 35 $K \doteqdot 3$ nm.

Morin and Reynolds were able to show that the fluor and the photoprotein occur within the same cell. The question therefore arises, is the wavelength shift from blue, in the purified chemiluminescent system, to blue–green in the organism the result of 'trivial' radiative, transfer, or non-radiative resonance transfer?

There are some other anomalies in the coelenterates. The hydrozoan jelly fish *Phialidium* from the eastern coast of the United States emits blue–green light and has a green fluorescent protein. However, specimens common during the summer at Plymouth in the United Kingdom have no detachable green fluorescence, nor any significant spectral shift between the chemiluminescence of the intact organism and that of the Ca^{2+}-activated photoprotein extracted from it. Another, even more remarkable, example of heterogeneity is found in some luminous anthozoans which can emit green or even yellow light (Table 9.4), in contrast to blue–green in *Renilla*. One part of the colony emits blue to blue–green light whereas other parts emit yellow light. This can be explained chemically by differences in fluor content in the luminous cells. However, the biological significance of this peculiar spectral perturbation is unclear.

Using the luciferase from *Renilla*, which, unlike its jelly-fish relatives, does not contain a photoprotein, Ward and Cormier demonstrated a three-fold increase in quantum yield, together with appropriate wavelength shift, at μM concentrations in the presence of its own green fluorescent protein. This could only be occurring if the two proteins associated to form a complex. A similar wavelength shift and spectral sharpening, but without a large change in quantum yield, can be observed between the photoproteins aequorin and obelin and their green fluorescent proteins, or flavins, bound to a solid matrix such as an ion exchange column. However, no such energy transfer has been demonstrated at μM concentrations in free solution. Whether binding between the photoprotein and fluor does not occur because the isolated protein has an N terminal fragment clipped off it compared to the native sequence, obtained from the cDNA (see Chapter 3), is not known. However, even if the concentrations within the cell are in the mM range, then the wavelength shift is still likely to result from a non-radiative mechanism, because the resonance process is very fast (ps) relative to photon emission from excited state (ns) (see Table 1.2).

The lumazine protein in the bacterium *Photobacterium phosphoreum* has a fluorescence excitation peak at 440 nm, and an emission at 470 nm. It shifts the peak in the bioluminescence spectrum from 490 nm to 470 nm. This might be thought to be the wrong way for resonance transfer. However, the possibility exists that there is an unstable oxy or hydroxyflavin intermediate formed at the active centre of the luciferase which would normally emit at 440 nm, but which undergoes intramolecular energy transfer in the absence of the fluor to emit at 490 nm, or can transfer the fluor. In one strain of *Vibrio fischeri* there is a flavin fluor which causes the organism to emit yellow light (λ_{max} 545 nm).

Deep sea searsiid fish excrete a particulate luminous cloud when stimulated. Herring, Widder and colleagues have observed that the bioluminescence spectrum changes with time, from a peak at 478 nm, to one at 408 nm. This coincides with a change in fluorescence, in the cellular clusters, from blue-green to blue. An obvious explanation for this is, like the cases of lucigenin and phthalhydrazide, that there is a sufficient concentration of the initial substrate to act as an energy transfer acceptor from the excited state product of the chemiluminescent reaction. If this is so then the initial chemiluminescence spectrum should be similar to the fluorescence spectrum of the luciferin. Unfortunately, however, the blue-green fluorescent emission spectrum of the substrate does not match the initial emission nor does the blue fluorescence emission of the product match the final 408 nm chemiluminescence emission.

9.6.3   Limitations of the Förster equation

Förster derived his equation to explain fluorescence energy transfer, basing it on the probability of going from $D^* + A$ to $D + A^*$. He assumed firstly that energy transfer was slow relative to nuclear motions, and secondly that transfer occurred between singlet states from the lowest vibrational and rotational energy state of the donors, in line with the Franck–Condon principle. However, several of these parameters may not hold for chemiluminescence. In particular, triplet excited states can exist, and since chemiluminescent reactions can generate excited state molecules in high vibronic energy levels, this may affect the energetics and efficiency of energy transfer. A further problem is that several parameters in the Förster equation are difficult to control when studying chemiluminescence energy transfer, in particular the refractive index, or dielectric constant of the environment through which transfer takes place, and the orientation factor. Diffusion of donor and acceptor may also enhance or reduce transfer, particularly if these are membrane bound as they appear to be in some luminous worms, squid, and fish. Nevertheless resonance energy transfer chemiluminescence offers some exciting possibilities as an analytical tool and a spectroscopic ruler, as well as providing a fascinating problem for the evolutionary biochemist to unravel.

9.7   ELECTRON TRANSFER REACTIONS

9.7.1   Chemically initiated electron exchange luminescence (CIEEL)

Quite a mouthful! Whilst the concept of resonance transfer in chemiluminescence can be traced back to the 1920s, the concept of electron transfer is much more recent. During the 1960s a number of chemiluminescent reactions were discovered where the addition of a fluor not only increased the overall light yield and shifted the spectral emission to that of the fluor, but also had a catalytic effect on the chemiluminescent reaction itself. This could not be explained by resonance energy transfer, where the fluorescent acceptor has no effect on the chemiluminescent reaction generating the excited state donor (NB: an exception to this rule may occur in bioluminescence if the fluor interacts directly with the luciferase, or the chemiluminescence reaction, in addition to acting as the energy transfer acceptor). Peroxides and dioxetanones, or reactions generating them, featured prominently as the chemiluminescence substrates in this new type of energy transfer. The mechanism, first suggested by Linschitz in 1961 to explain the chemiluminescence of tetralin peroxide catalysed by porphyrin, is that a charge transfer complex forms allowing electron exchange between the product of the chemical reaction, the donor, and the fluorescent acceptor, leading to excitation of the fluor which then emits.

Schuster distinguished three types of chemiluminescence from peroxides (AB), exemplified by compounds generating dioxetanones (Fig. 9.13):

(1) Direct chemiluminescence

$$AB \;\rightarrow\; B + A^* \;\rightarrow\; h\nu \tag{9.21}$$

(2) Indirect chemiluminescence

$$AB \rightarrow B + A^* \xrightarrow[+F]{} A + F^* \rightarrow h\nu \tag{9.22}$$

(3) Activated chemiluminescence

$$AB + F \rightarrow (AB.^- F.^+) \rightarrow B + (A.^- F.^+) \rightarrow A + F^* \rightarrow h\nu \tag{9.23}$$

Note . = radical; F = fluor.

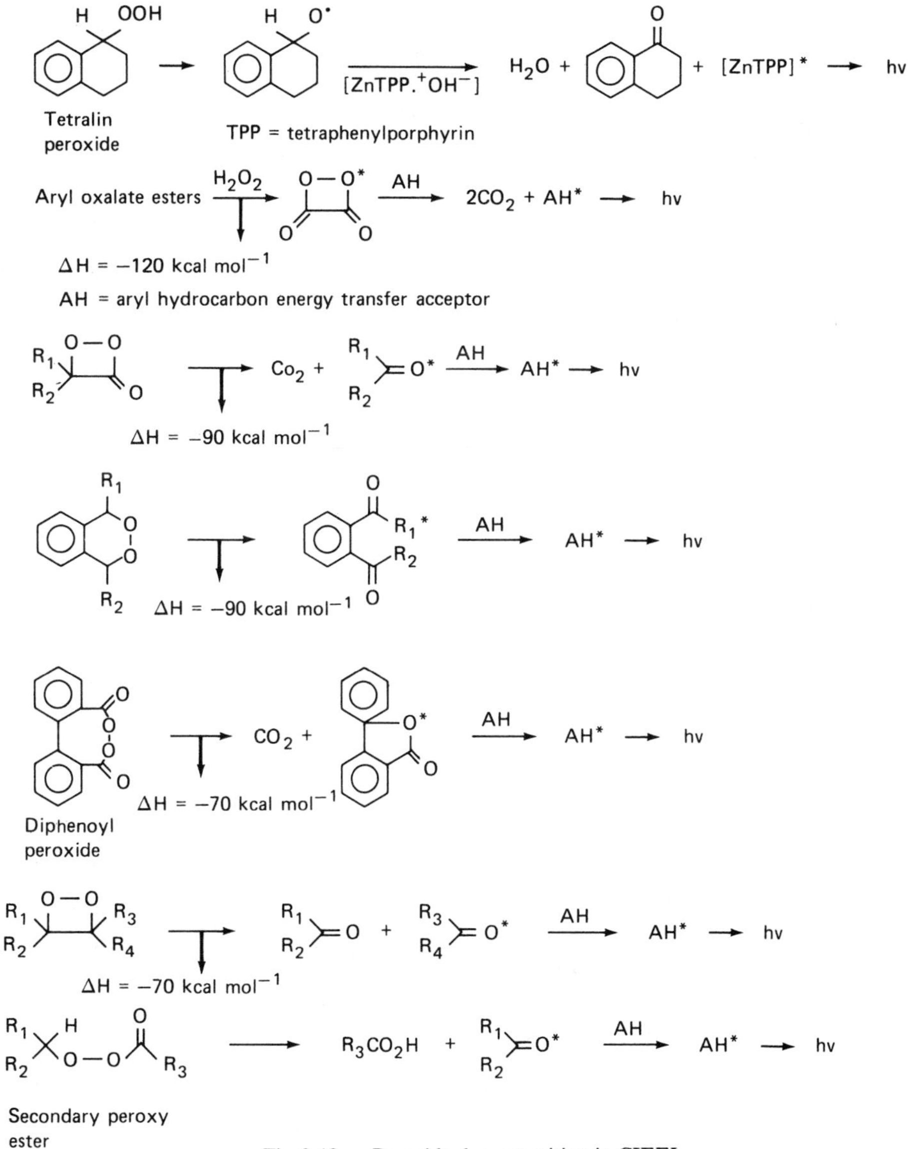

Fig. 9.12 — Peroxide decomposition in CIEEL.

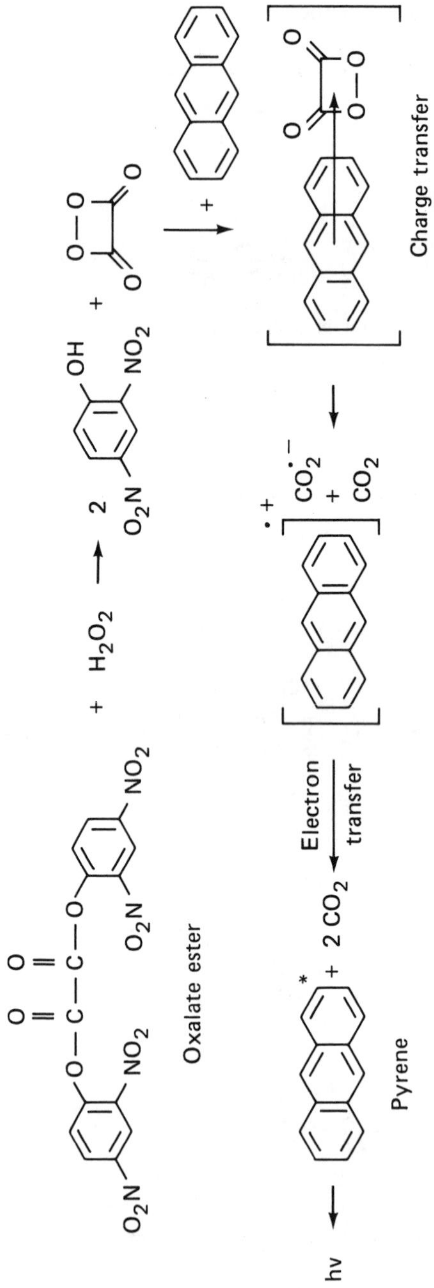

Fig. 9.13 — Oxalate ester CIEEL.

The latter two both involve energy transfer, 'indirect' via resonance or collisional transfer, 'activated' requiring electron transfer through a charge transfer complex.

The first clear evidence for 'activated' chemiluminescence occurring via electron transfer was obtained by studying the decomposition of a diacyl peroxide, diphenoyl peroxide, to benzocoumarin and CO_2 (Fig. 9.12). The problem was that benzocoumarin was not fluorescent and thus was incapable of excitation by the reaction, so no 'direct' chemiluminescence was observed. Only when aromatic hydrocarbon fluors such as rubrene, perylene, or 9,10 diphenylanthracene were added was a bright chemiluminescence seen. This could not be explained by indirect chemiluminescence, occurring via resonance or collisional energy transfer, since the phenomenon appeared to disobey the simple rules of electron spin (see section 9.5.2), for example when using standard triplet state acceptors such as biacetyl or 9,10-di*bromo*anthracene. This goes to the second excited triplet, and then to the first excited singlet.

Two key observations confirmed that a resonance energy transfer process was not operating, and that electron exchange could provide the explanation (Eqns. (9.24)–(9.26)). Firstly, the observed rate constant (k_{obs}), and intensity of light emission for the breakdown of diphenoyl peroxide (DPP), were linearly related to the concentration of the fluorescent activator added. Thus the observed rate constant was the sum of components from two distinct, albeit related, reactions, (a) + (b):

(a) no interaction with fluor

$$\text{diphenoyl peroxide (DPP)} \xrightarrow{k_1} \text{benzocoumarin (BC)} + CO_2 \qquad (9.24)$$

BC = benzocoumarin

(b) interaction with fluor

$$\text{DPP} + \text{F} \xrightarrow{k_2} \text{BC} + CO_2 + \text{F}^* \rightarrow h\nu \qquad (9.25)$$

therefore

$$k_{obs} = k_1 + k_2 \text{ (F)} \qquad (9.26)$$

(F) = concentration of fluor.

This would not be the case with resonance transfer. Secondly, not only was k_2 dependent on the nature of the fluor, but also the lower the oxidation potential (E_{ox}) of the fluor, a measure of its ability to loose an electron, the greater was k_2. In fact, a plot of $\log_e k_2$ versus E_{ox} for six aromatic activating fluors, with E_{ox} from 0.82–1.36 V, was linear. This then provided the clue to the molecular mechanism, the reaction had to be occurring through a bimolecular complex where the fluor gave up one electron to the chemiluminescent substrate. Schuster and Koo proposed the complete reaction to be:

$$\text{DPP} + \text{F} \rightarrow (\text{DPP}^{\cdot -}\text{F}^{\cdot +}) \rightarrow (\text{BC}^{-}\cdot, CO_2, \text{F}^{+}\cdot) \rightarrow \text{BC} + CO_2 + \text{F}^* \rightarrow h\nu$$
$$(9.27)$$

The compounds in brackets represent molecular complexes within a solvent cage. Thus the efficiency of these electron transfer reactions is increased in solvents which stabilise these 'charge transfer' complexes. In other words, particular organic solvents can increase the quantum yield of these reactions. For example with diphenoyl peroxide the lowest chemiluminescence quantum yields are found in polar, nonviscous solvents such as acetonitrile and methylene chloride. On the other hand, using dimethyl phthalate as solvent, which has the same dielectric constant as methylene chloride but is more viscous, the quantum yield is increased three-fold. Similarly, phthalate esters are usually the solvent of choice for oxalate esters in light sticks. However, such solvent effects are not always as predicted. Koo and Schuster found that the highest quantum yield with diphenoyl peroxide as donor and perylene as fluorescent activator was obtained with diethyl ether as solvent, which has both a low dielectric constant and low viscosity; whereas in benzene, which is less polar and more viscous, where the yield should have been better than in diethyl ether, it was, in fact, five times lower. This is because dielectric constant and viscosity are not the only solvent parameters which can influence the efficiency of electron transfer chemiluminescence.

At least eight criteria combine to distinguish electron transfer chemiluminescence, or *chemically initiated electron-exchange chemiluminescence* (CIEECL) as it has come to be called, as an energy transfer process, from direct chemiluminescence not involving energy transfer or from resonance energy transfer.

(1) Chemiluminescence quantum yield and light intensity are usually very low in the absence of the fluorescent activator and, like resonance transfer, the fluor determines the spectrum of the light emitter.
(2) The rate of the chemiluminescent reaction is enhanced by activators which are good electron donors, but which must be fluorescent if they are to increase the overall quantum yield. Amines and transition metals are good electron donors and catalyse dioxetanone decomposition, but they do not enhance the light yield as they are not fluorescent.

Thus, the complete reaction can be sub-divided into two competing reactions:

$$A \xrightarrow{k_1} \text{products (i.e. no fluorescent acceptor)} \tag{9.28}$$

$$A + F \xrightarrow{k_2} \text{products} + F^* \text{ (i.e. fluorescent)} \tag{9.29}$$

$$\text{rate of product formation} = k_{obs} A \tag{9.30}$$

$$k_{obs} = k_1 + k_2 (F) \tag{9.31}$$

k_2 is made up itself of two components because of two rate limiting steps:

$$A + F \underset{K_{CTC}}{\rightleftharpoons} (AF) \xrightarrow{k_{CTC}} (A.^- \ F.^+) \xrightarrow[\text{fast}]{\text{very}} F^* \tag{9.32}$$

formation of   formation of
complex   charge transfer complex, CTC
(reversible)   (irreversible)

K is an equilibrium constant; k a rate constant, therefore

$$k_2 = k_{CTC} \cdot K_{CTC} \tag{9.33}$$

$$\text{and } k_2 = \text{constant} \times \exp(-\alpha(E_{ox} - E_{red} - e^2/\varepsilon R_0)/RT) \tag{9.34}$$

where α = constant, related to the transfer coefficient of electrode reactions
                  and is between 0 and 1, e.g. for diphenoyl peroxide $\alpha = 0.3$
        E_{ox} = oxidation potential of activating fluor $(F - e \rightleftharpoons F^{\cdot+})$
        E_{red} = reduction potential of the peroxide $(A + e \rightleftharpoons A^{\cdot-})$
        e = the charge of the electron
        ε = dielectric constant of the solvent
        R_0 = distance between ions in the charge transfer complex at the
                  transition point
        R = the universal gas constant
        T = temperature in degrees K.

(3) The efficiency of the energy transfer reaction does not conform to the Förster equations (Eqns. 9.9–9.13), rather the best fluors for CIEECL are those with the lowest oxidation potential (i.e. the greatest ability to donate electrons). $\log_e k_2$ (Eqn. 9.34) is linearly related to E_{ox}, and the fluorescent efficiency is not influenced by the J integral (Fig. 9.8).

(4) Formation of charge transfer complexes are required, necessitating close association (1–5 Å) of electron donor and acceptor, whereas resonance transfer can still be significant when the donor and acceptor molecules are up to 80–120 Å apart.

(5) Effects of different solvents depend on their ability to stabilise, or conversely to destabilise, charge transfer complexes. Viscous, non-polar solvents are usually, but not inevitably, the best.

(6) Charge transfer complexes are less susceptible to quenching by substances like oxygen, which rapidly quench triplet excited states in resonance transfer, since the complex is protected within a solvent cage.

(7) The energetics are predicted by the number of electron volts (eV) required to generate a photon (1 eV per molecule are equivalent to about 23 kcal.mol^{-1}, and 71 kcal mol^{-1} are necessary for a 400 nm photon; see Chapter 1; section 1.5).

(8) Radical reactions are not susceptible to the same spin restrictions found in resonance energy transfer. In fact here formation of triplet states may actually be favoured, even though they may not be the actual light emitter.

Since the mid-1970s a range of peroxy compounds have been found which are capable of undergoing chemically initiated electron-exchange luminescence (Table 9.11, Fig. 9.13). The first such reactions to be discovered required formation of charge transfer complexes between dissimilar molecules, e.g. C_2O_4 from oxalate ester oxidation and anthracene or rubrene. However, a number of intramolecular examples of the phenomenon have since been found. Furthermore, intramolecular electron transfer has been proposed to play an important role in the chemiluminescent reactions of fireflies, coelenterates, and several other examples of bioluminescence. Inorganic examples also exist, such as the reaction of ruthenium chelates when reduced by hydrazine or aqueous base. This latter reaction has been extensively investigated in another form where the radicals making up the charge transfer complex are formed not chemically but electrically. This, then, is known as electrogenerated chemiluminescence (EGC) or electrochemiluminescence (ECL).

Table 9.11 — Chemically initiated electron-exchange luminescence (CIEEL)

Substrate/donor	Acceptor
Intramolecular	
>C=O from dioxetanone	benzothiazole nucleus of firefly luciferin
>C=O from dioxetanone	open imidazolopyrazine nucleus of coelenterazine
Intermolecular	
tetralin peroxide	Zn tetraphenylporphyrin
aryl oxalate esters	polyaromatic hydrocarbons (e.g. rubrene, anthracene)
dimethyl dioxetan	polyaromatic hydrocarbons
secondary peroxy esters	polyaromatic hydrocarbons
hydrazine	Ru(bipyridyl)$_3$

9.7.2   Electrochemiluminescence

Although the ability of electrodes to generated chemiluminescence has been known for over 100 years, extensive investigation of this phenomenon did not occur until the mid 1960s and early 1970s, particularly as a result of the experiments of Chandross, Faulkner, and co-workers.

In electrochemiluminescence, radical ions are formed either sequentially at one electrode, or simultaneously at two separate electrodes. Cyclic voltammetry is a particularly convenient method for doing this. The potentials for generating maximum $+$ve and $-$ve radical ions are thus identified, corresponding to their respective oxidation potentials (9.35). The potential is then switched for a period of 10 μs–10 ms to that necessary to form $R_a{\cdot}^-$, and then to that necessary to form $R_b{\cdot}^+$. The two radicals of opposing charge then diffuse together and react, thereby annihilating the charges and generating an excited state which emits a photon.

Thus the primary process in electrochemiluminescence is:

$$\text{two electrical potentials} \rightarrow R_a{\cdot}^- + R_b{\cdot}^+ \rightarrow (\text{CTC}) \rightarrow R_a^* + R_b \qquad (9.35)$$

where CTC = charge transfer complex.

R_a and R_b may be the same substance, e.g. rubrene, or different substances, e.g. rubrene and tetramethylphenylene diamine (TMPD), see Fig. 9.14 for structures. In the latter case the yellow light of rubrene is emitted.

$$\text{TMPD}{\cdot}^+ + \text{rubrene}{\cdot}^- \rightarrow \text{TMPD} + {}^*\text{rubrene} \rightarrow \text{yellow light} \qquad (9.36)$$

In fact, either ion precursor may be the emitter. Thus,

$$\text{rubrene}{\cdot}^- + \text{p-benzoquinone}^+ \rightarrow \text{p-benzoquinone}^* \rightarrow h\nu \qquad (9.37)$$

Sometimes, both emission spectra are seen:

$$\text{thianthrene (TA)} + 2{,}5\text{diphenyl-1,3,4-oxadiole (DPD)} \qquad (9.38)$$

$$TA.^{+} + DPD.^{-} \left\langle \begin{matrix} \rightarrow {}^{1}TA^{*} + DPD \\ or \\ \searrow TA + {}^{1}DPD^{*} \end{matrix} \right.$$

1 = singlet state
* = excited state emitter

Emission from both TA and DPD is observed. The population spread depends on the energy required to produce the excited singlet of either reactant.

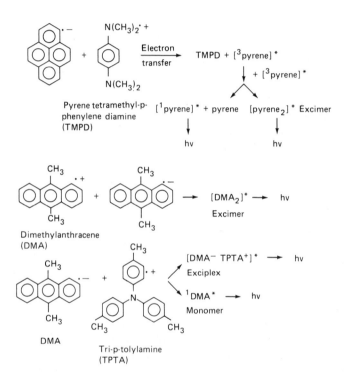

Fig. 9.14 — Electron transfer CL.

When the energy of radical annihilation is *sufficient* to produce R_a (Eqn. (9.35)) in a primary emitting state then this is known as the 'S' route. Since efficient triplet emission is rare in chemiluminescence, without energy transfer, then R_a is usually expected to be in the singlet spin state. This is true for a number of examples (Table 9.12), including rubrene and diphenylanthracene plus thianthrene. However, one well-studied example of the S route where triplet state primary emission does occur is that of the ruthenium chelates. Here chelates such as bipyridyl stabilise the triplet state:

$$Ru\,(bipyr)_3^{+} + Ru(bipyr)_3^{3+} \rightarrow Ru(bipyr)_3^{2+} + [{}^{3}(Ru(bipyr)_3^{2+}] \rightarrow h\nu \ (610 \ nm)$$

(9.39)

Table 9.12 — Electrochemiluminescence

S route e transfer
    rubrene$\bar{\cdot}$ + rubrene$\overset{+}{\cdot}$
    9,10 diphenylanthracene$^-$ + thianthrene$^+$
    Ru(bipyr)$_3^+$ + Ru(bipyr)$_3^{3+}$
    $C_2O_4^-\cdot$ + rubrene$^+$

T route e transfer
    rubrene$^-$ NNN'N' tetramethyl-*p*-phenylenediamine$^+$

Excimer formation
    (pyrene)$_2$

Exciplex formation
    benzophenone$^-$ + tri-*p*-tolyl amine$^+$

In contrast to the 'S' route, when radical annihilation is *energy deficient*, and therefore unable to generate an excited singlet state directly, then an excited triplet formed from the charge transfer complex may be able to generate a singlet for efficient light emission by triplet–triplet annihilation. This is the so-called 'T' route involving both radical ion and triplet–triplet annihilation processes:

$$\text{electrochemical potential} \rightarrow R_a\cdot^- + R_b\cdot^+ \tag{9.40}$$

$$R_a\cdot^- + R_b\cdot^+ \rightarrow CTC \rightarrow {}^3R_a{}^* + R_b \tag{9.41}$$

$$ {}^3R_a{}^* + {}^3R_a{}^* \rightarrow {}^1R_a{}^* \rightarrow {}^1R_a + h\nu \tag{9.42}$$

CTC = charge transfer complex

This 'T' route is well illustrated when rubrene = R_a and tetramethylphenylene-diamine (TMPD) = R_b.

So far we have considered two types of electrochemiluminescence, where the charge transfer complex generates one of the components of the reaction in an excited state, be it singlet or triplet. There is a third type where a molecular complex is itself the emitter. Evidence for this can be obtained by using photoluminescence to identify the bimolecular emitter. If the excited complex emitter is formed from two like molecules it is known as an *excimer*, but if formed from two unlike molecules it is an *exciplex*.

For example, the electrochemiluminescence of pyrene (Py) + tetramethylphenylenediamine (TMPD) produces a bimodal spectral emission with peaks at *ca.* 400 nm from ^{1}Py*, and *ca* 480 nm from the Py$_2$* *excimer*:

$$Py\bar{\cdot} + TMPD\overset{+}{\cdot} \rightarrow {}^3Py^* + TMPD \tag{9.43}$$

$$ {}^3Py^* + {}^3Py^* \Big\langle \begin{array}{l} {}^1Py^* + Py \\ Py_2{}^* \end{array} \tag{9.44}$$

1 and 3 represent singlet and triplet states respectively. In contrast dimethylanthra-cene (DMA) + tri-p-tolylamine (TPTA) can lead to (DMA.TPTA) *exciplex* emission:

$$DMA.^- + TPTA^+ \rightarrow (DMA.^-TPTA^+)^* \begin{cases} TPTA + {}^1DMA^* \rightarrow h\nu \\ DMA + TPTA + h\nu \end{cases} \quad (9.45)$$

Support for this latter mechanism comes from the fact that the singlet/exciplex emission ratio increases exponentially with temperature. Even intramolecular exciplexes are possible, e.g. dimethylaniline (D) coupled via a methylene bridge $(CH_2)_n$ to anthracene (A):

$$A.^-(CH_2)_nD + A(CH_2)_nD.^+ \rightarrow A(CH_2)_nD + {}^1[A^-(CH_2)_nD^+]^* \rightarrow h\nu \quad (9.46)$$

This occurs via the T route, since it is 'energy deficient'.

What about the energetics and kinetics of electrochemiluminescence? Where does the energy come from, and what controls the reaction rates?

The energetics of the phenomenon are described through the standard electrochemical potentials (E):

$$R_a + e^- \rightleftharpoons R_a.^- \qquad E_a^0 \qquad\qquad (9.47)$$

$$R_b.^+ + e^- \rightleftharpoons R_b \qquad E_b^0 . \qquad\qquad (9.48)$$

One area of argument is whether one should use ΔG or ΔH in the energetic equations. In other types of chemiluminescence it is ΔH which actually provides the *energy* for the excited state:

$$\text{for } R._a^- + R_b^+. \rightarrow R_a^* + R_b \qquad\qquad (9.49)$$

$$\Delta H^0 = E_a^0 - E_b^0 .$$

For the reactions we have been considering ΔH^0 is apparently quite large, some 1–4 eV per molecule, equivalent to 25–100 kcal.mol^{-1}. But for a photon at 400 nm (violet) 71 kcal.mol^{-1} are required, compared with 41 kcal.mol^{-1} at 700 nm (red). Thus, several of the radical annihilations will be 'energy deficient'. Hence the frequency of the T route. Although the triplet state has a lower energy than the singlet, pooling the energy of two triplets will, however, provide sufficient energy for singlet excitation, and thus a visible photon.

The energy potential of radical ion and triplet annihilation, together with singlet and triplet state levels, determine whether chemiluminescence can occur at all. Without the necessary energy it cannot. But it is the rate of reaction which determines the light intensity ($dh\nu/dt$). The more molecules reacting per second the brighter the emission.

$$\text{Light intensity} = I = dh\nu/dt = N.\phi_s.\phi_F . \qquad\qquad (9.50)$$

where N = number of electrons transferred per second during the potential step (Eqn. 9.35), i.e. R_a^-. already formed when using cyclic voltammetry.

ϕ_s = efficiency of producing the emitting state (equivalent to ϕ_{ex} in a simple chemiluminescent reaction)

ϕ_F = efficiency of light emission from excited state (equivalent to ϕ_F in a simple chemiluminescent reaction)

ϕ_S and ϕ_F in the S route are independent of time.

for the S route $R_a.^- + R_b.^+ \rightarrow (R_a.^- R_b.^+) \rightarrow {}^1R_a^* + R_b$ (9.51)

$$N = AD^{\frac{1}{2}} \frac{C_P}{t_f^{\frac{1}{2}}} [a(t_r/t_f)^{-\frac{1}{2}} - b]$$ (9.52)

where A = electrode area
       D = diffusion coefficient of the precursor (R_b)
       C_P = concentration of the precursor (R_b)
       t_f = time of second potential step
       t_r = time after start of the reaction
       when signal strength is greatest $t_r/t_f \leqslant 0.2$.

a and b are constants requiring evaluation. The reaction zone may be only 10 nm wide.

Single chemiluminescent transients formed by the S route will have

$$I \propto (t_r/t_f)^{-\frac{1}{2}}$$ (9.53)

The T route, on the other hand, involves second order triplet annihilation so one would expect:

$$I \propto N_2$$ (9.54)

The electron transfer rate constant (k_{et}) is predicted by the Marcus theory as:

$$k_{et} = z K\rho \exp(-\Delta H_a/RT)$$ (9.55)

where z = solution-phase collision frequency (*ca.* 10^{11} 1 mol^{-1} s^{-1}).
       K = adiabaticity factor, $= 1$ if the reaction takes place on a single potential energy surface
       ρ = coordinate-dependent factor, sensitive to fluctuations in the internal distances within the activated complex
       ΔH_a = activation energy necessary to bring the separated reactants to the distorted transition state

$$\Delta H_a = (w^r + w^P)/2 + \wedge/4 + \Delta H^\circ/2 + (\Delta H^\circ + we^P - w^r)^2/4\wedge$$ (9.56)

where w^r and $w^P =$ work necessary to bring the reactant and product pairs to the reaction distance

$\wedge$ = a reorganisation parameter relating to bonding and solvation configurations $= 0.4$–0.8 eV in aprotic solvents

Electrochemiluminescence (ECL) is susceptible not only to electrical potential but also to the magnetic field, because of interactions between electron and nuclear spin. It has provided an important experimental approach to understanding many electron transfer reactions in chemiluminescence, including chemically-induced electron transfer luminescence (CIEEL). A nice example of this comes from experiments by Chang and colleagues who were trying to generate the elusive dioxetandione, the proposed intermediate in oxalate ester chemiluminescence (Fig. 9.13), by oxidation of the oxalate ion using electrolysis. Attempts to intercept excited states produced by spontaneous decomposition of the hypothetical dioxetandione with substances such as rubrene, only produced chemiluminescence emission when both the interceptor (R) and the oxalate (C_2O_4) were oxidised, and not when the oxalate was oxidised. The proposed reaction therefore was:

$$R \quad + C_2O_4 \quad \rightarrow C_2O_4.^- + R.^+ \tag{9.57}$$
$$\text{rubrene} \quad \text{oxalate}$$

$$C_2O_4.^- \rightarrow CO_2 + CO_2.^- \text{ (a very powerful reductant)} \tag{9.58}$$

$$CO_2.^- + R \rightarrow CO_2 + R^{\cdot-}. \tag{9.59}$$

$$R.^+ + R.^- \rightarrow R + R^* \rightarrow h\nu \tag{9.60}$$

NB: oxidation = losing an electron, reduction $\equiv$ gaining an electron.
The wide range of compounds (Fig. 9.14) capable of undergoing electrochemiluminescence is a reflection of three features of the organic fluors involved:

(1) The very large range of fluors available with redox potentials suitable for generating enough energy to produce excited states.
(2) Redox reactions are simple and fast.
(3) Many fluors have high fluorescence quantum yields.

9.8 ENERGY POOLING

In a sense, the coming together of two or more atoms of molecules under any circumstances to provide energy for chemiluminescence is an 'energy pooling' reaction. However, the term 'energy pooling' is usually restricted, in chemiluminescence, to reactions where more than one atom or molecule, all in an electronically excited state, come together, pool their energy, and emit it as a photon. Photon emission emanates either from the molecular complex or from one of the components of a complex released as an excited monomer. In the former case the emission therefore arises from excimers or exciplexes. Thus we can have:

$$\text{eximer} \qquad A^* + A^* \rightarrow (AA)^* \rightarrow h\nu \tag{9.61}$$

$$\text{monomer} \qquad A^* + A^* \rightarrow (AA)^* \rightarrow A + A^* \rightarrow h\nu \tag{9.62}$$

$$\text{exciplex} \qquad A^* + B^* \rightarrow (AB)^* \rightarrow h\nu \tag{9.63}$$

$$\text{monomer} \qquad A^* + B^* \rightarrow (AB)^* \rightarrow A + B^* \rightarrow h\nu \tag{9.64}$$

where * represents an excited state
The energy of A^* and B^* alone is insufficient for visible photon emission, though it may produce infrared emission. For example:

$$\text{monomer } {}^1\Delta g \, O_2 \rightarrow h\nu \text{ (1238 nm)} \tag{9.65}$$

$$\text{dimol } ({}^1\Delta g \, O_2 . \, {}^1\Delta g \, O_2) \rightarrow 2 \, \Sigma_g^- \, {}^3O_2 + h\nu \text{ 705, 634 nm} \tag{9.66}$$

In the previous section the 'T' route in electrochemiluminescence involved pooling from triplet states:

$${}^3R + {}^3R \rightarrow R + {}^1R^* \rightarrow h\nu \; . \tag{9.67}$$

It is helpful to exclude from the definition of 'energy pooling' radical annihilation itself, as well as energy transfer within a charge transfer complex such as occurs in chemically included electron exchange chemiluminescence, since both involve electron transfer rather than collective pooling.

9.9 SOME POSSIBILITIES IN BIOLUMINESCENCE

9.9.1 Luminous bacteria

Ever since the publication of the range of emission spectra of luminous bacteria by Eymers and Schouwenburg in 1937 the chemical explanation for the difference in the colour of light emitted by the various species has been a source of much argument. Their emissions range from 470–505 nm, except one, a Y1 strain of *Vibrio fischeri* discovered by Ruby and Nealson in 1977, which has a bimodal blue (490 nm) and yellow (545 nm) spectrum. The proposed function of the wavelength shift from 495 to 472 nm is that the bluer transmission is particularly suitable for deeper waters (optimum wavelength for transmission of sunlight through seawater = 476 nm). In contrast 495 nm emission would be suitable for shallow waters. Two particular problems have emerged:

(1) λ_{max} for intact *Photobacterium phosphorum* and *leiognathi* is 472 nm, whereas both the purified luciferase and other intact bacteria, such as the common strains of *Vibrio fischeri* and *harveyi*, emit with a peak at about 495 nm (range 487–500 nm).

(2) The precise nature of the primary excited state responsible for bacterial chemiluminescence has still not been completely agreed, even though the 4a hydroxyflavin intermediate isolated by the Hasting's group is a convincing candidate (see Chapter 3, section 3.4).

In 1962 Terpeska observed some unusual fluorescence in the extracts of some bacteria. Fifteen years later Gast and Lee isolated a 21 000 molecular weight blue fluorescent protein from *Photobacterium phosphoreum* with a fluorescence emission maximum at 476 nm. This protein contains a chromophore, 6,7 dimethylribityl lumazine (Fig. 9.6), which has an affinity constant for the protein of about 50 nM. The protein shifts the peak of fluorescence emission of the lumazine from 491 nm to 476 nm on binding, with a ϕ_F of *ca.* 0.6 and a fluorescence lifetime of 14 ns. Could this then be an energy transfer protein?

Addition of the blue fluorescent protein to its own luciferase plus the other reactants (O_2, long chain aldehyde, NADH, FMN, and oxidoreductase) *in vitro* not only shifts the chemiluminescence emission from 495 to 472 nm, but also increases the overall yield of light emission. The blue fluorescent protein also increases the maximal intensity, by more than four times. The normal quantum yield for bacterial luciferase *in vitro*, in the absence of the blue fluorescent protein, is about 0.1 So for a 10-fold increase in quantum yield it would be necessary for the primary chemilumi-nescent reaction to be 100% efficient. The effect of the blue fluorescent protein can be detected at concentrations down to 0.6 μM, far too low for radiative ('trivial') transfer to be the explanation for the spectral shift and increased quantum yield. The effect can also be observed with luciferases from bacteria which do not contain the blue fluorescent protein, though higher concentrations of the fluor are required.

Photobacteria leiognathi also contains a similar 21 000 molecular weight blue fluorescent protein which shifts the chemiluminescence spectrum *in vitro*. Similarly, Leismann and Nealson have shown that their Y1 strain of *Vibrio fischeri* contains a yellow fluorescent protein, also around 20 000 molecular weight, but with flavin as the fluor which is more loosely attached than is lumazine to its protein. The yellow fluorescent protein produces a large enhancement of the *in vitro* chemiluminescence and generates a bimodal spectrum (λ_{max} 485 and 540 nm) of the light emission. Individual strains of photobacterium vary greatly in their content of the lumazine protein. Those with the least amounts have *in vivo* and *in vitro* chemiluminescence emission maxima closest together. Under conditions where Y1 does not produce sufficient yellow fluor, the bacterium emits only blue light (λ_{max} 485 nm).

This all fits in nicely with an energy transfer system. But there are two puzzling aspects as yet unexplained:

(1) The wavelength shift from 495 to 472 nm in *Photobacterium phosphoreum*, and some other species, induced by the blue fluorescent protein is the wrong way for a conventional resonance transfer from a 495 nm primary emitter by the Förster mechanism.

(2) The fluorescence excitation maxima for the fluorescent proteins are wrong for them to be excited by a 495 nm primary emitter, the *in vitro* emission from the luciferase. For *P. phosphoreum* the fluorescent excitation maxima of the lumazine protein are about 270 nm and 418 nm ($\varepsilon = 4000M^{-1}.cm^{-1}$), and for the Y1 yellow fluorescent protein they are 270 nm and 380 nm, nowhere near the 495 nm necessary for resonance transfer.

There are two possible ways to resolve this problem. Either the energy transfer involves an unusual electron exchange, or transfer process, to excite the fluor. Alternatively, the primary excited state in the absence of the fluor is really a flavin intermediate which would like to emit at around 400–420 nm, but which via intramolecular energy transfer when bound to the luciferase produces an excited state which actually emits at 495 nm. This would require the luciferase reaction to generate a normally weakly emitting species with an energy of some 300 kJ mol^{-1} or 71 kcal mol^{-1}. Interaction with the fluor could then enable the '400–420 nm' intermediate to transfer its energy under conditions compatible with the Förster equation, so that the fluor then emits. This has lead McCapra to look for such an intermediate, different from that of Hastings. In an ingenious experiment, using a fluor which he synthesised attached to a long chain aldehyde, small wavelength shifts were observed, consistent with a '400–420 nm' excited state being the primary excited state generated by the bacterial luciferase when the efficiency of different fluors was compared. Thus, as in luminous coelenterates, both intra- and inter-molecular chemiluminescence energy transfer may play a role in the family of luminous bacteria, though only some species contain the fluorescent acceptor proteins necessary for intermolecular transfer.

9.9.2 Some puzzles
As we have seen there are many peculiarities in bioluminescence spectra (Tables 3.12 and 9.5), including spectral differences between different species in the same phylum or class, as well as bimodal and temporal changes in an individual's

spectrum. Some can be explained by energy transfer, others simply by the effect of the environment surrounding the oxyluciferin at the active centre of the luciferase. Three particular examples of living light stand out as possibilities for new energy transfer systems requiring investigation: luminous fungi, the euphausiid shrimp, of which nearly all species are luminous, and the remarkable freshwater limpet *Latia neritoides* discovered by Gray in 1849 and Suter in 1890 in the North Island of New Zealand. Although chemically quite distinct they have a common property that the *in vitro* chemiluminescence cannot mimic that *in vivo*. Either chemiluminescence cannot be generated at all *in vitro*, or the *in vivo* spectrum cannot be produced *in vitro* without the addition of a fluor from the organism concerned.

The fungal system, studied by Wassinck in 1970s, like luminous bacteria, uses NAD(P)H as an electron source. However, as shown in the luminous mushroom *Lampteromyces japonicus*, a fluorescent steroid found in the fungus, ergostatetrae-none, may be necessary to convert the *in vitro* spectrum (λ_{max} 490 nm) to that *in vivo* (λ_{max} 530 nm). This fluor when protonated has a fluorescence emission at 530 nm, and can quench the excited singlet of acetone formed by the decomposition of some dioxetans. But it can act as an energy transfer acceptor in tetramethyldioxetan chemiluminescence. More recently Goto and colleagues have identified riboflavin as the probable emitter in this species.

In the late 1960s Shimomura and Johnson isolated a protein from the euphausiid *Meganyctiphanes norvegica*, existing in two high molecular weight forms (360 000 and 900 000) and a fluor, molecular weight *ca.* 1000. The protein was defined as a photoprotein since it appeared to contain the primary chemiluminescent chromo-phore (the luciferin) tightly bound to it. This was consumed during the reaction, rather than acting catalytically as a conventional luciferase. Oxygen was also required. The fluor was necessary for efficient emission *in vitro*, but was not apparently consumed. Its fluorescence emission maximum was 476 nm, the same as that of the *in vivo* chemiluminescence.

Unlike these two previous groups of organisms the luciferin from *Latia* has been identified. It is an aliphatic aldehyde which oxidises to the acid (Fig. 3.14), with a chemiluminescence quantum yield < 1%. However, addition of a 'purple protein', observed by Suter in violet light when he found the animal in 1890, increases the light yield from the reaction, but only when preincubated with the luciferin and not with the luciferase. The purple fluor is recycled; stoichiometric amounts are not required. The problem is that the absorbance maxima of the purple protein are 568 and 620 nm, and its fluorescence emission peaks are 570 and 640 nm, yet the bioluminescence emission is at 536 nm. Could the fluor be a tetrapyrrole? Time will tell.

In conclusion, intramolecular energy transfer plays an important part in the primary chemiluminescent reaction of several groups of luminous organisms. Inter-molecular energy transfer, on the other hand, has evolved as a mechanism for enhancing light emission by increasing the net chemiluminescence quantum yield, and for producing the colour of light emission necessary for it to fulfil its biological function (see Chapter 3, section 3.6). In many others there are sufficient hints for the enthusiast to pursue.

9.10 BIOMEDICAL APPLICATIONS

9.10.1 Principles

Chemiluminescence energy transfer has provided the chemist with an important tool for quantifying the formation of singlet and triplet states, for example in dioxetan decomposition using diphenylanthracene for singlets (S–S transfer) and dibromoanthracene for triplets (T–S transfer). There are five main areas where chemiluminescence energy transfer could have exciting applications in biology (Tables 9.13 and 9.14):

(1) Detection and quantification of ligand–ligand interactions without the need to separate bound and free ligand.
(2) Spectroscopic ruler to measure distances within and between biological molecules.
(3) Detection of excited states in living systems which otherwise would have low or negligible chemiluminescence quantum yields.
(4) Measurement of enzymes and metabolites capable of generating or reacting with donors or acceptors in the energy transfer system.
(5) Detection and quantification of fluors in chromatographic systems. Here, energy transfer would be detected by a decrease in light emission at the donor maximum and an increase at the acceptor emission maximum, or by a change in ratio of the emission at the two wavelengths.

Whilst the original experiments of Stryer and Haugland in 1967 have led to many applications of fluorescence energy transfer as a spectroscopic ruler over distances up to 100 Å (10 nm), uncertainties in parameters of the Förster equation, such as the orientation factor, mean that we have yet to see any biological applications of chemiluminescence energy transfer to measure distances within or between molecules. Electron transfer reactions could not be used for this purpose as they act only

Table 9.13 — Some applications of chemiluminescence energy transfer

Type of energy transfer (Table 9.6)	Donor	Acceptor	Application
Resonance (B2)	ABEI	fluorescein	homogeneous immunoassay
	ABEI	fluorescein	homogeneous biotin assay
Electron transfer (B4)	C_2O_4	rubrene, anthracene	light sticks, golf balls
	peroxalate	anilonaphthalene sulphonic acid	H_2O_2 detection
	C_2O_4	dansyl	amino acid detection on tlc or hplc
	C_2O_4	fluorescamine	catecholamine detection on tlc or hplc
	C_2O_4	polyaromatic hydrocarbon	polyaromatic hydrocarbon detection
Electrogenerated chemiluminescence (B4)	oxalate	Ru bipyridine	oxalate assay
Chemical reaction (B5)	$^1O_2$	rubrene	$^1O_2$ detection
Energy pooling (B6)	$^1O_2$	$^1O_2$	$^1O_2$ detection

Table 9.14 — 12 — Biomedical analytes detected by chemiluminescence energy transfer

Energy transfer system	Analytical application	Analyte
A. Resonance		
ABEI-fluorescein	homogeneous immunoassay	cyclic, AMP, steroids, C9, IgG
ABEI-fluorescein	homogeneous ligand–ligand	biotin–avidin
cells + fluor	excited state detection	neutrophil activation
B. Electron transfer		
oxalate ester (peroxyoxalate)	fluor in liquid chromatography	dansyl, fluorescamine
oxalate ester	H_2O_2	glucose oxidase
oxalate ester	enzyme label in immunoassay	T_4, steroids
$Ru(pyr)_3^{3+}$ electro-chemiluminescence	urine metabolites	oxalate

over much shorter distances (<5 Å) than those in resonance transfer. Since the radius of a typical globular protein is 10–150 Å (1–15 nm) for a molecular weight range of 10 000–300 000, then the only type of energy transfer applicable to the study of ligand–ligand interactions such as hormone–receptor or antibody–antigen interactions would be that occurring via the resonance, or Förster mechanism. A further problem with many electron transfer reactions such as the oxalate esters, or peroxyoxalates as they are often called, is that they are poorly soluble in aqueous solvents. They usually work best in non-polar solvents which stabilise charge transfer complexes. However, one feature of electron transfer reactions which may work for, rather than against, electron transfer as an analytical tool is the time scale of the reactions relative to resonance transfer.

Resonance (Förster) transfer can occur only once when the primary excited state has been formed. Formation of this excited state will occur within the time scale of chemical reactions, but once formed will lose its energy very quickly within ns for singlet to ground state or ps–ns (10^{-12}–10^{-9} s) for intersystem crossing. Furthermore, any triplet states which are formed will be quenched very rapidly. The energy transfer process itself takes place over a fs–ps (10^{-15}–10^{-12}) time scale. Thus, whilst molecular diffusion may sometimes aid resonance transfer, for example in biological membranes, resonance transfer can only detect *low* ($<\mu$M) concentrations of an analyte if specific binding of the analyte to a ligand occurs. This ligand must be covalently linked to one of the components of the energy transfer process, namely donor or acceptor. In contrast, the formation of charge transfer complexes, or radical and triplet state annihilation, in electron transfer reactions, can occur only when the reaction moieties have diffused together. Since, without energy transfer, the quantum yield is usually very low, e.g. oxalate esters without a fluorescent acceptor (Fig. 9.1), electron transfer reactions can be used to detect quite low concentrations of either donor or acceptor.

Let us therefore see how the two main types of energy transfer, resonance and electron, can be used as an analytical tool.

9.10.2 Measurement of ligand–ligand interaction by resonance transfer

Quantification of ligand–ligand interactions plays a fundamental part in the full understanding of biological phenomena. Just think, for example, of the importance of interactions between metal ions and proteins, between hormones or drugs and their receptors, between enzymes and substrates, between antibodies and antigens, and between RNA, DNA and protein. Some of the complexes formed through these interactions, for example antibody–antigen reactions, have slow dissociation rates. They can be studied by separately bound and free ligand without significantly disturbing the equilibrium. However, with fast equilibrating systems ($K_a < 10^7 \, M^{-1}$) this is impossible. Even with systems where it is possible, the separation step introduces imprecision, is laborious, and cannot be applied when wishing to measure chemical events inside live cells. Attempts to use fluorescence, through enhancement, polarisation, or energy transfer, or the use of other labels such as enzymes or esr, have not provided a universally applicable method which is both homogeneous and has the equivalent sensitivity of heterogeneous assays using radioactive substances such as [125]I.

A homogeneous immunoassay system has been developed by ourselves, using a chemiluminescent labelled antigen (CLAg) and a fluorescent labelled antibody (FAb). Formation of the antibody–antigen complex brings the donor and acceptor close enough for resonance transfer to take place (see Chapter 8 section 8.2.6). Addition of the analyte displaces the equilibrium, shifting the chemiluminescence spectrum back towards that of the initial label:

$$\text{blue light} \ll\!\!\wedge\!\!\wedge \text{CLAg} + \text{FAb} \rightleftharpoons (\text{CLAg} - \text{FAb}) \wedge\!\!\wedge\!\!\gg \text{green or red light}$$
$$(9.68)$$

Since the decrease in the blue emission and increase in the green emission turned out to be small, it was necessary to measure the ratio of chemiluminescence at blue and green wavelengths simultaneously, using a dual wavelength chemiluminometer (Fig. 9.7). This also increased precision, reducing mixing artefacts on addition of the oxidant and catalyst of the chemiluminescent reaction. In principle, several blue emitting donors and several green, yellow, or red emitting fluorescent acceptors could be used (Table 9.15). Selection of the appropriate pair is made on the basis of three chemical parameters, together with four characteristics attempting to optimise parameters defined by the Förster equation.

Chemical parameters to be satisfied
(1) Antigen and antibody must be labelled with a chemiluminescent and fluorescent substance respectively, to high specific activity and without damaging the chemiluminescent or fluorescent properties of the energy transfer donor and acceptor, nor the affinity of antibody–antigen binding.
(2) The conditions for chemiluminescence (e.g. high pH and H_2O_2) must not disturb antibody–antigen binding, nor affect the fluor. The pH must be <9 to be sure not to break the Ab–Ag complex.
(3) The chemiluminescent donor and fluorescent acceptor should not be affected by substances, e.g. quenchers, in the sample.

Energy transfer parameters to be optimised
(1) High quantum yield of chemiluminescence and fluorescence (ideally $\phi > 10\%$).
(2) High extinction coefficient for fluorescent acceptor ($\epsilon > 10^4 \, M^{-1} \, cm^{-1}$).

Table 9.15 — Possible energy transfer pairs in homogeneous immunoassay

Chemiluminescent donor	λ_{max} (nm)	
aminobutylethyl isoluminol (ABEI)	ca. 460	
aryl acridinium esters	ca. 460	
pholasin (*Pholas* luciferin)	490	
aequorin	465	
obelin	475	
Fluorescent acceptor	λ_{max} ex (nm)	λ_{max} em (nm)
dansyl	340	490
4 chloro-7-nitro-benzo-2-oxa-1,3-diazole (NBD)	465	530
fluorescein	495	525
rhodamine	550	575
coproporphyrin	410	620
green fluorescent protein	465	508
phycoerythrin R	485	575

(3) Good spectral overlap (J integral, Fig. 9.8) between chemiluminescent donor and fluorescent acceptor.

(4) Good Stokes shift, though if the shift is into the yellow or red (λ em fluor $> ca.$ 540 nm) it may be necessary to use a red-sensitive photomultiplier in the chemiluminometer (see Chapter 2).

We have no control on the refractive index, nor on the distance between donor and acceptor (d) or the orientation factor (K^2) in the Förster equation. The latter two will depend on where the donor and acceptor are on the antigen or antibody. We ensure that there is at least 3–4 mols fluor per mol of antibody, coupled by using the isothiocyanate derivative of the fluor. Much greater labelling can be achieved, without antibody precipitation, by using imidate ester coupling.

Three chemiluminescent donors (ABEI, pholasin, and aryl acridinium esters) and four fluorescent acceptors (NBD, fluorescein, rhodamine, and coproporphyrin) have been investigated (Fig. 9.3). The only pair where significant energy transfer has been detected are ABEI labelled antigens bound to fluorescein labelled antibodies. Dansyl groups, and many other fluors, have not been investigated, because they are excited in the UV, a long way from the emission maxima of the chemiluminescent donors. The lack of any detectable energy transfer using NBD as acceptor was surprising but may be explained by its low extinction coefficient. Pholasin, the protein bound luciferin from the bivalve *Pholas dactylus* (Fig. 3.1), should be a good donor to fluorescein, since its emission maxima is at about 490 nm. However, attaching antigens to pholasin without large losses of chemiluminescence has proved difficult. Lack of detectable energy transfer with aryl acridinium esters was somewhat surprising. There are, however, two problems which may explain this. Firstly, they work best at pH 13 where antibody–antigen complexes dissociate within ms.

Triggering of the chemiluminescence must therefore be carried out at pH<9, well below the optimal pH. Secondly, the excited N methylacridone product of the acridinium ester reaction dissociates from the antigen, though diffusion away from the potential acceptor should be slow relative to the rate of energy transfer.

The evidence that the ABEI antigen–fluorescein–antibody couple is truly resonance energy transfer is still somewhat circumstantial. The reduction in light emission at 460 nm together with the increase at 525 nm were small (Fig. 9.10). Furthermore, in another ligand–ligand system, ABEI-biotin bound to fluorescein-avidin, the donor and acceptor can be close enough for direct interactions to occur resulting in a very large quenching of the ABEI chemiluminescence (Fig. 9.11). This occurs in spite of a detectable increase in the ratio of light emission at 525/460 nm. Nevertheless, both avidin and antibody dilution curves, together with standard curves, can be established (Figs 9.10 and 9.11) to measure analytes, molecular weight range 315–150 000, including cyclic nucleotides, nucleosides, steroids, protein antigens, and IgG. The assays are at least as sensitive as their radioactive counterparts, and analysis of analytes from biological extracts correlated well between the chemiluminescence energy transfer assay and that using radioactivity.

At high fluorescein antibody concentrations some wavelength shift is observed with non-immune antibody. This helps to explain why the assay works at all. As we have already seen, μM–mM concentrations of fluorescent acceptor are necessary for Förster resonance energy transfer to occur in free solution. Since the antibody and antigen reagents in an immunoassay are usually less than 10 nM, and sometimes in the pM range, only within the antibody–antigen complex will the excited state donor be close enough to the fluorescent acceptor for energy transfer to occur.

Chemiluminescence energy transfer within antibody–antigen complexes has been detected so far over the temperature range 0–37°C. In principle, this phenomenon could also be used to detect the 'on' and 'off' rates of antibody–antigen binding (Fig. 9.15).

Possible applications of resonance energy transfer in the future include not only high affinity single site and two site immuno- and protein binding assays for important clinical indicators such as creative kinase MB in the assessment of myocardial infarction, but also low affinity ligand–ligand interactions such as hormone–receptor and enzyme–substrate binding, measurement of free analytes without disturbing the equilibrium, and DNA and RNA recombinants. In the last of these the chemiluminescent donor and the fluorescent acceptor could be either on separate nucleic acid strands or on the same strand. In the former the loss of energy transfer when individual labelled strands bind with unlabelled ones should be quantitatively related to the number of base pairs involved. In the latter, binding in a new environment should alter the separation and orientation of the donor and acceptor molecules, and thus increase or decrease energy transfer. The possibility of replacing Southern blotting by a homogeneous gene detection system is an exciting prospect.

Chemiluminescence resonance energy transfer offers exciting opportunities also to the molecular cell biologist wishing to quantify chemical events in living cells. We have developed methods, in addition to micro-injection, for incorporating the energy transfer reagents into intact cells. These are based on reversible osmotic swelling, cell fusion, and release from micropinocytotic vesicles (Fig. 4.3). These

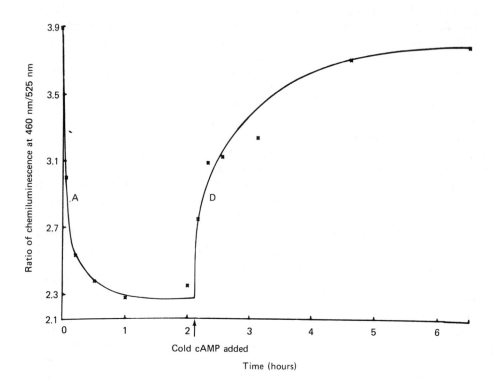

Fig. 9.15 — Association (A) and disassociation (D) of antibody. Antigen = ABE1-cyclic AMP. Antibody = fluorescein rabbit anticyclic AMP IgG. By permission of Academic Press.

experiments have, however, highlighted three problems associated with the use of ABEI–fluorescein energy transfer couple inside cells.

(1) ABEI chemiluminescence is best in alkali, no good for cells. It has a low ϕ_{CL} at pH 7.4 (<1% of that at pH 11).
(2) In order to trigger ABEI chemiluminescence efficiently mM concentrations of H_2O_2 are required. Not only are these potentially harmful to cells, but also simply adding H_2O_2 externally to a cell does not cause an equivalent rise in *intracellular H_2O_2*.
(3) The energy transfer assay system needs to be improved with a better J integral and a larger Stokes shift.

This last problem might be solved by using phycobiliproteins as the acceptors. These highly efficient fluors not only have a large J integral and Stokes shift with the donor emission, but also have large extinction coefficients. Bioluminescent energy transfer pairs such as aequorin- or obelin-green fluorescent protein, with their high quantum yields and lack of requirement for H_2O_2 offer exciting possibilities for the future.

9.10.3 Applications of electron transfer
There are three assay systems based on electron transfer reactions, with potential biomedical applications:

(1) Detection of fluors in liquid chromatography.
(2) Measurement of H_2O_2.
(3) Measurement of oxalate in urine using ruthenium chelate electrochemiluminescence.

The first two have used the oxalate ester bis-(2,4,6,-trichlorophenyl) oxalate (TCPO) (Fig. 9.2). The reaction requires H_2O_2 and needs at least 50% organic solvent, e.g. ethyl acetate, mainly to stop TCPO precipitation. Fluorescent detectors in present use for hplc are often unstable, have high backgrounds, and have limited sensitivity. Several compounds can act as the fluorescent acceptors in the electron transfer reaction. Other metabolites can be converted to a fluorescent acceptor either pre- or post-column, e.g., using dansylation of amino acids, fluorescein to label the NH_2 of catecholamines, ethenylation of nucleotides, or o-phthalaldehyde. Pre-column derivatisation is best, the chemiluminescent components being added in a small volume to the column effluent so that the electron transfer reaction takes place quickly (<5 s) relative to column flow. Dansyl derivatives can be detected down to 10 fmol with linearity over three orders of magnitude by this method. f–amol (10^{-15}–10^{-18} mol) detection may be possible in the future, providing a much simpler, less expensive, and more sensitive detector than the conventional fluorescent one. The fluorophors must, of course, have a low oxidation potential, if efficient electron transfer is to occur within the charge transfer complex (see section 9.7).

We have seen in Chapter 5 that the TCPO energy transfer system can be used to detect pmol (10^{-12} mol) quantities of H_2O_2, as well as enzymes or metabolites coupled to the production or breakdown of H_2O_2. A variant of this is to use glucose oxidase as an analyte label (e.g. to thyroid hormone or steroid) and then to use TCPO, plus 8-anilonaphthalene-1-sulphonic acid (ANS) as acceptor, to detect the enzyme label in a separation type immunoassay.

9.10.4 Detection of excited states in cells
We saw in Chapter 6, and in this chapter (section 9.3.3), that an intriguing product of some cellular processes is the generation of a molecular species, in an electronically excited state. In addition to photochemical reactions, these include peroxidase, oxidase, and oxygenase catalysed reactions associated with lipid peroxidation, detoxification, and eicosanoid synthesis. Other phenomena generating excited states include cell division and egg fertilisation. Sometimes an endogenous ultraweak chemiluminescence is detectable, sometimes not. Energy transfer offers the exciting opportunity of learning more about these excited states, and the discovery of new ones, their origin, their nature, and their function. Addition of a fluor, e.g. flavin, fluorescein, dibromoanthracene, or chlorophyll, can increase light emission through energy transfer. Acceptors such as dibromoanthracene provide evidence for triplet states versus singlets. Imaging of the light may provide further clues as to the intracellular site of generation. Alternatively, fluors can be added externally to cells to distinguish intra- from extra-cellular generation of excited states. Atoms and molecules in electronically excited states have a chemistry quite distinct from that of

the ground states. Their identification and characterisation in cell activation and cell injury provide an opportunity to open up a new field for exploration, crucial to understanding the relationships between molecular and holistic events in living systems.

So chemiluminescence energy transfer not only poses many stimulating questions for the basic scientist, but also provides a potentially unique tool for the cell and molecular biologist to exploit recent advances in biotechnology. So far, few have yet to find wide application in either the research or clinical laboratory. Is this because the right biological or clinical questions are not being asked? Are we being imaginative enough in our exploitation of chemiluminescence? The final chapter in this book attempts to give the reader a wider perspective on chemiluminescence as a whole, to see how the marriage of its scientific and technological features can provide a phenomenon which is both fun to work with and rewarding to use.

9.11 RECOMMENDED READING
Books
Adam, W. & Cilento, G. (eds) (1982). *Chemical and biological generation of excited states*, Academic Press, New York.
Phillips, G. O. (ed.) (1965). *Energy transfer in radiation processes*. Elsevier, Amsterdam.

Reviews
Campbell, A. K., Roberts, P. A., & Patel, A. (1985) In: *Alternative immunoassays*, pp 153–183. Ed. Collins, W. P. John Wiley, Chichester.
Faulkner, L. R. & Glass, R. S. (1982) In: *Chemical and biological generation of excited states*, pp 197–277. Eds Adam, W. & Cilento, G., Academic Press, New York. *Electrochemiluminescence*.
Glazer, A. N. (1984) *Biochim. Biophys. Acta* **768** 29–51. Phycobilisome: a macromolecular complex optimized for light energy transfer.
Gorman, A. A. & Rodgers, M. A. S. (1981) *Q. Rev. Chem. Soc.* **10** 205–231. Singlet molecular oxygen.
Kahn, A. U. (1985) *Singlet oxygen*, pp 39–79. Ed. Frimer, A. A. CRC Press Inc., Florida.
Schuster, G. B. & Thorn, K. A. (1982) In: *Chemical and biological generation of excited states*, pp 229–247. Eds Adam, W. & Cilento, G., Academic Press, New York. Chemically initiated electron-exchange luminescence.
Stryer, L. (1978) *Ann. Rev. Biochem.* **47** 817–846. Fluorescence energy transfer as a spectroscopic ruler.
Ward, W. W. (1979) *Photochem. Photobiol. Rev.* **4** 1–58. Energy transfer processes in bioluminescence.

Historical
Cario, G. & Franck, J. (1923) *Z. Phys.* **17** 202–212. Sensitized fluorescence of gases (in German).
Förster, T. (1948) *Ann. Phys.* **2** 55–75. Zwischenmoleculare Energiewanderung und Fluoreszenz (translation available on request).
Förster, T. (1959) *Disc. Faraday Soc.* **27** 7–17. Transfer mechanisms of electronic excitation.

Papers

Campbell, A. K. & Herring, P. J. (1987) *Comp. Biochem. Physiol.* **86B** 411–417. A novel red fluorescent protein from the deep sea luminous fish *Malacosteus niger*.

Campbell, A. K. & Patel, A. (1983) *Biochem. J.* **216** 185–194. A homogeneous immunoassay for cyclic nucleotides based on chemiluminescence energy transfer.

Gast, R. & Lee, J. (1978) *Proc. Natl. Acad. Sci. US* **75** 833–837. Isolation of the *in vivo* emitter in bacterial bioluminescence.

Herring, P. J. (1983) *Proc. Roy. Soc.* **B220** 183–217. The spectral characteristics of luminous marine organisms.

Lee, J., Okane, D. J., & Visser, A. J. W. G. (1985) *Biochem.* **24** 1476–1483. Spectral properties and function of two lumazine proteins from *Photobacterium*.

Leisman, G. & Nealson, K. H. (1982) *Flavins and flavoproteins*, pp 83–386. Eds Massey, V., & Williams, C. H., Elsevier North-Holland, Amsterdam. Characterisation of a yellow fluorescent protein from *Vibrio (Photobacterium) fischeri*.

Levine, L. D. & Ward, W. W. (1982) *Comp. Biochem. Physiol.* **72B** 77–85. Isolation and characterization of a photoprotein, "phialidin", and a spectrally unique green-fluorescent protein from the bioluminescent jelly fish *Phialidium gregarium*.

Lucia, A. do Nascimento, T. O., & Cilento, G. (1985) *Biochim. Biophys. Acta* **843** 254–260. Induction of chemiluminescent processes in the fungus *Blastocladiella emersonii* by exposure to enzyme-generated triplet benthaldehyde.

Morin, J. G. & Hastings, J. W. (1971) *J. Cell Physiol.* **77** 313–318. Energy transfer in a bioluminescent system.

O'Kane, D. J. & Lee, J. (1987) *Methods Enzymol.* **133** 435–449. Purification and properties of lumazine proteins from *Photobacterium* strains.

Patel, A., Davies, C. J., Campbell, A. K., & McCapra, F. (1983) *Analyt. Biochem.* **129** 162–169. Chemiluminescence energy transfer: a new technique applicable to the study of ligand–ligand interaction in living systems.

Roberts, D. R. & White, E. H. (1978) *J. Am. Chem. Soc.* **92** 4861–4867. Energy transfer in chemiluminescence III Intramolecular triplet–singlet transfer in derivatives of 2,3,-dihydrophthalazine-1,4-dione.

Ward, W. W. & Cormier, M. J. (1979) *J. Biol. Chem.*, **254** 781–788. An energy transfer protein in coelenterate bioluminescence. Characterization of the *Renilla* green-fluorescent protein.

Applications

Bohne, C., Campa, A., Cilento, G., Nassi, L., & Villablanca, M. (1986) *Anal. Biochem.* **155** 1–9. Chlorophyll: an efficient detector of electronically excited species in biochemical systems.

Williams, E. J. & Campbell, A. K. (1986) *Anal. Biochem.* **156** 155 249–255. Homogeneous assay for biotin based on chemiluminescence energy transfer.

Campbell, A. K. & Patel, A. (1983) *Bichem. J.* **216** 185–194. A homogeneous immunoassay for cyclic nucleotides based on chemiluminescence energy transfer.

Imai, K. (1987) *Methods Enzymol.* **133** 435–449. Chemiluminescence detection system for high-performance chromatography.

Rubinstein, I., Martin, C. R., & Bard, A. J. (1983) *Anal. Chem.* **55** 1580–1582. Electrogenerated chemiluminescent determination of oxalate.

Sigvardson, K. W. & Birks, J. W. (1983) *Anal. Chem.* **55** 432–435. Peroxalate chemiluminescence of polycyclic aromatic hydrocarbons in liquid chromatography.

10

Hills and horizons

In art nothing worth doing can be done without genius, in science even a very moderate capacity can contribute something to a supreme achievement.

Bertrand Russell, 1917

10.1 THE CLIMB

10.1.1 The philosopher's stone?

The ingredients of a chemiluminescent reaction may not be able to turn metals into gold, but they do result in a remarkable and striking transformation. They convert a completely dark gas, liquid, or solid, invisible in a darkened room, into light.

There is little doubt that scientists learn from each other very quickly. As a result, small advances and discoveries can 'contribute to a supreme achievement'. Yet the quotation from Russell highlights a lack of understanding about the creative forces, the imagination, ingenuity, and critique required even to achieve a small step forward in science. We live in an age where natural science, the study and perception of natural phenomena by experiment and hypothesis, is often confused with what is usually referred to as technology. The dictionary will inform you that something 'technical' pertains to skill in the *arts*. Yet most of us think of technology as the industrial application of science to the benefit of the community.

Scientific enquiry into chemiluminescence gives us an opportunity not only of learning something solid about a fascinating and aesthetically pleasing natural phenomenon, but also of viewing the interfaces between the arts and science on the one hand, and science and technology on the other.

Our perception, and appreciation, of the universe is dominated by light. Evolutionary forces have given us a keen eye, and the photon is a fundamental particle throughout the universe, transferring energy and communicating events between atoms just a few nanometres apart as well as those separated by millions of miles of space. But what questions should we ask to find out more about a phenomenon which generates light? Our scientific training provides us with a logic, our inventiveness enables us to make the tools necessary to record and measure the phenomenon, to mimic it and even to exploit it for human good, but without a question we cannot begin. It is in our inquisitiveness, and in our desire to communicate the understanding of ourselves and the natural world, that we find ourselves close to the artist.

Chemiluminescence may not be the philosopher's stone but it can certainly transform a rather dull looking object or problem into a shining one.

10.1.2   What questions to ask?

Light is energy, so the first question is, where does the energy come from? In Chapter 1 we saw that the term 'luminescence' was first coined by a German physicist, Eilhardt Wiedemann, in 1888, to distinguish such phenomena producing 'cold light', from incandescence, where high temperature makes a body first red hot and eventually white hot. Incandescence has been used by man for centuries as a source of illumination: the sun, a burning torch, a candle, an oil lamp, a light bulb. Yet several of these also contain a chemiluminescent component. Strip lights based on electroluminescence have replaced street and office filament lights. An explanation of the energetics of both luminescence and incandescence requires a quantum approach. Are luminescence and incandescence, therefore, really distinct phenomena? The requirement for an electronically excited state in luminescence would appear to answer yes. What then is the difference beteeen the various types of luiminescence, photo- (fluorescence and phosphorescence), chemi-, tribo-, electro-, radio-, pyro, and thermo?

My first objective therefore was to identify sources of light which were caused by a chemical reaction. We now assume that all bioluminescence is chemiluminescence, but could there be organisms using triboluminescence? Conversely, is a chemical reaction the real source of light in some other types of luminescence, e.g. tribo-, sono-, and electro-? By investigating the chemistry of both artificial reactions and living light I have tried to show how the true chemical nature of the luminescence can be established.

Having separated out the chemical sources of luminescence, two questions then arise:

(1) What is so special about a chemiluminescent reaction compared with other, 'dark', chemical reactions?
(2) What physical and chemical parameters determine the two most obvious characteristics of the phenomenon, brightness and colour?

A number of specific and detailed questions then follow concerning the chemistry and biology of the chemiluminescent reaction:

(1) Does it occur in Nature? If so, where, and can it be reconstituted, and mimicked, under controlled conditions in the laboratory?
(2) What components are required to generate the light? What is their function in the reaction, i.e. chemiluminescent substrate, oxidant, cofactor, catalyst, trigger, and how do they act?
(3) What are the steps in the pathway generating the final intermediate which generates the excited state emitter, and what is this final step?
(4) What is so special about 'oxygen' that it is the oxidant for most organic chemiluminescence, and all bioluminescence?
(5) Are there chemical families in living light, or is each group of luminous organisms unique?

(6) What are the mechanisms controlling, and the functions of, living light?
(7) What has been the relevance of chemiluminescence to the chemical and biological evolution of the earth?

Finally, does chemiluminescence have any distinctive properties which can be exploited in chemical analysis, biotechnology, and biomedicine? Are these features unique, and how does chemiluminescence compare with other methods of analysis such as fluorescence and radioactivity?

10.1.3 How to find some answers

To answer these questions we needed a combination of perceptive observations, the right tools, and definitive experimental design. Two features have been crucial, firstly the construction of a sensitive and accurate device for detecting and quantifying chemiluminescence, and secondly, methods for establishing the chemical basis of light production. We saw in Chapter 2 that, prior to this century, many important observations were made about chemiluminescence using the naked eye, photography, or thermopiles. The invention of the photomultiplier tube during the 1920s gave much greater scope. Now it was possible to analyse the time course of a very wide range of light intensities, even those invisible to the naked eye, and to record accurately their spectra. The development of imaging devices and intensifiers enabled the site of origin of the light emission, particularly within living cells, to be identified. Chemical and biochemical separation techniques, such as ultracentrifugation and chromatography have enabled components of chemiluminescent reactions and the products to be separated, and then identified by chemical analysis and spectroscopy. The ingenuity of the synthetic chemist has been of great value in identifying bioluminescent reactions and the key structural determinants required for chemiluminescence to occur. New chemiluminescent compounds have, as a result, been invented. Genetic engineering has provided us with the means of cloning bioluminescent proteins, enabling the amino acid sequences to be easily obtained, and manipulated, thereby providing the means of probing the puzzles of the evolutionary origins of bioluminescence.

A glance at the index of any standard textbook of chemistry, biochemistry, or biology will reveal the study of chemiluminescence to be a Cinderella science. Man has no light organ! There seems an enormous gulf between the luminescence of deep sea animals, and artificial low-pressure gas phase chemiluminescent reactions. Should chemiluminescence be regarded as a single phenomenon at all? Both the sea and the sky are blue, but this does not mean that they should be shielded and examined under the same scientific umbrella. Can a reaction producing such a dim light that we cannot see it be of any real significance?

The exciting thing about studying chemiluminescence is that to answer the most interesting and penetrating questions we have to cross interdisciplinary boundaries. We need physics to understand the energetics of the excited state, electronics to measure it, chemistry to discover its cause and reproduce it, biochemistry to examine living light, cell biology, ethology, and ecology to discover its function, and molecular biology and geology to follow its evolutionary history. Chemiluminescence is a true Natural Science.

We have followed the paths of explorers to lands and seas revealing hitherto unknown luminous creatures. We have penetrated environments requiring sophisticated collecting gear, ships for deep sea animals, rockets and balloons for atmospheric chemiluminescence. Have we discovered anything new on our way up to the summit?

10.2 THE SUMMIT

10.2.1 Discoveries and surprises

Animals producing their own light have fascinated the layman and scientist alike since the ancient Chinese and Greek writers first described them. However, serious scientific investigation into the phenomenon did not really begin until the late 17th century. In the early studies there was much confusion about the origin of phenomena producing light. Was it physical or chemical? Was it living or inorganic? It is now clear that far from being an unusual phenomenon in Nature, chemiluminescence is quite common. It is found in the atmosphere, in many flames, in a great range of living organisms, and can easily be produced in the laboratory with both inorganic and organic compounds (Table 10.1). So this is perhaps our first surprise; chemiluminescence, far from being a laboratory curiosity is not rare at all, but is associated with several familiar phenomena: living light, *after* an electric spark, cold flames and the blue colour in hot ones, certain atmospheric glows.

By the mid–late 19th century the extraordinary 'shining flesh', recognised in cadavers on some battlefields and in butchers' stores for centuries, was recognised as being of bacterial origin. Exploration of hitherto unknown regions of the earth had begun in earnest. Darwin, in his account of the *Beagle* expedition, from 1831 to 1836, describes many luminous creatures. Hardy and others on the first *RRS Discovery* expeditions early in this century began to reveal the plethora of luminous animals in the deep oceans. Research over the past 30 years has established that all living light so far investigated is chemical light. Our next surprise — because of our spasmodic encounters with terrestrial living light we have been misled into believing it is rare. In the sea, it is common, and chemiluminescence in living organisms plays a major role in the ecology of the deep oceans, which cover more than half our planet.

Table 10.1 — The natural occurrence of chemiluminescence

Type		Location	Example
1.	Living light (bioluminescence)	air land sea	male fireflies 'shining flesh' jelly fish
2.	Atmospheric	air after-glow upper atmosphere	following electric discharge atomic recombination
3.	Flames	mines marshes 'cold' flames coal gas	fire damp will o' the wisp ether blue portion (not the yellow)
4.	Ultraweak	living organisms air smoke	phagocytic cells animal breath cigarettes
5.	Synthetic	inorganic laboratory organic laboratory	several radical reactions several oxidations

A further feature of some living light, and artificial chemiluminescence, is the ability to change the colour, and intensity, of the light by adding a fluor to the reaction. This, under the appropriate conditions, acts as an energy transfer acceptor and thus enables us to define two types of chemiluminescence: direct and sensitised.

The final surprise with respect to occurrence was the discovery that many, if not all, eukaryotic cells are capable of producing an ultraweak chemiluminescence, invisible to the naked eye yet detectable by a sensitive photomultiplier tube. But the light appears to have no function, or does it?

The identification of phenomena, both natural and artificial, which are truly chemiluminescent, has enabled several familiar light emissions to be excluded from our group (Table 10.2). Thus, red hot coal, the yellow in a gas cooker flame, fish in a home aquarium, and many man-made objects such as a television screen and phosphorescent paints are not cause by chemiluminescence. Phosphorescence is not chemiluminescence!

Three key features of chemiluminescence distinguish it from phosphorescence and fluorescence, more correctly types of photoluminescence. Firstly, the energy for light emission in chemiluminescence is chemical, whereas in photoluminescence it is photon absorption. Secondly, when an atom or molecule fluoresces it remains chemically unchanged and can be re-excited immediately, once light emission has occurred. In contrast, in chemiluminescence each molecule reacts only once to form an excited state. Thirdly, the excited emitter formed by the chemiluminescent reaction is structurally different from the initial substrate. This is still true in energy transfer processes, where the primary excited state acts as the energy donor.

Elucidation of the chemical pathways in natural and artificial chemiluminescence has established three special features which combine to make chemiluminescent

Table 10.2 — Light emission not chemiluminescence

Example	True classification
1. *Natural*	
red hot embers	incandescence
yellow portion of coal gas flame	incandescence
electric spark	electroluminescence
salt in a flame	pyroluminescence
crushing raw sugar crystal	triboluminescence
fluorescent pigments in tropical fish	photoluminescence
radium	radioluminescence
2. *Artificial*	
light bulb	incandescence
street lamp	electroluminescence
TV screen	electroluminescence
Sellotape	triboluminescence
self-sealing envelope	triboluminescence
phosphorescent paint	photoluminescence

reactions distinct from those which do not produce light. Firstly, they are highly exothermic; at least 70 kcal per mole of reactant is required for blue photons. Secondly, the actual light emitter must itself be a fluor, or strictly capable of electronic excitation. Thirdly, in order to channel the chemical energy rapidly into an electronically excited state, either electron transfer followed by radical annihilation or thermal decomposition of a peroxy-intermediate occurs. The discovery of the four-membered

$$\begin{array}{c} O-O \\ | \quad | \\ -C-C- \\ | \quad | \end{array}$$

dioxetan ring intermediate decomposing to two

carbonyls, one excited, was a major advance in understanding many types of organic chemiluminescence. There have been two particular chemical surprises in bioluminescent reactions. Firstly, all require oxygen in some form as an oxidant, oxidation to a carbonyl being the energy source. Even when ATP hydrolysis occurs, as in luminous beetles, some squid, and a millipede, this is not the energy for chemiluminescence. Secondly, whilst each group of luminous organisms has its own luciferin some species, quite unrelated, use the identical substance. For example, jelly fish, certain squid, shrimp, copepods and fish all seem to use the same imidazolopyrazine as luciferin. Whether these form part of a food chain, or whether the luciferin has appeared separately several times during evolution is not known.

The brightness and colour of a chemiluminescent reaction are not always dictated entirely by the chemistry inherent in the reaction itself. For example, the active centre of a luciferase has profound effects on both reaction rate and colour. Furthermore, these two parameters can also be altered by energy transfer.

At first sight, there appear to be a multiplicity of functions for bioluminescence in different species. However, these can be rationalised into three main groups: obtaining food, defence, and reproduction (see Chapter 3). Symbiotic relationships play a role in a few. The common occurrence of chemiluminescence in marine organisms implies that the phenomenon has played an important part in biological evolution. The importance of non-living chemiluminescence to the development of the earth is less clear. Its occurrence in the upper atmosphere suggests that it not only has a function in protecting the earth's surface from ultraviolet and other radiations but that it has played a part in the chemical evolution of the atmosphere as a whole.

10.2.2　A new light on medicine

Man has certainly exploited chemiluminescence in some ingenious ways: as a torch or lamp, for fishing, for decoration and courting, for scientific or medical purposes, by military personnel and for fun (see Table 4.1). Light sticks are available commercially. Fishermen in certain parts of the world have learnt to mimic the blue chemiluminescence of the angler fish. Attempts have been made to use it to detect nuclear submarines as they spark off deep sea luminous organisms. Yet chemiluminescence also has some important scientific applications. It provides the chemist and biologist with a fascinating model for studying the chemistry of the excited state. It can be used as a model for substances such as drugs and anaesthetics, which either interact with the excited state itself or with molecules responsible for its formation. But perhaps the application which has had the greatest impact on research and biotechnology is its use as an analytical tool (Table 10.3). Three key features stand

Table 10.3 — Biomedical applications

Application	chemiluminescent system
1. ATP to assess biomass and cell viability	firefly
2. Enzymes and metabolites in tiny tissue and fluid samples	firefly bacteria
3. Toxic oxygen metabolites	luminol pholasin
4. Concentration of free calcium in live cells	jelly fish photoproteins
5. Non-isotopic labels in immunoassay and recombinant DNA technology	luminol-peroxidase acridinium esters
6. Gene expression in genetic engineering	bacterial lux genes firefly luciferase gene

out: firstly, the unique sensitivity with which efficient chemiluminescent compounds can be detected in relatively large volumes, down to one cell and a tipomol (10^{-21} mol); secondly, the ability to illuminate chemical events in living cells; and thirdly, the fact that they are non-radioactive, yet detectable as sensitively as ^{125}I, and ^{32}P.

To measure chemical changes in living cells methods have had to be developed to get the chemiluminescent indicator into the cell itself. Five ways have been established to achieve this: injection through a micropipette, transient permeation of the cell membrane by swelling or by making a hole (e.g. electroporation), fusion of the cell with a membrane sack, internal release from micropinocytotic vesicles, incorporation of genetic material using a plasmid or virus (i.e. transfection) (see Chapter 4 for details).

The ability to image and intensify the light emission enables the site of the chemiluminescence to be located within a cell, a gel or a well in a microtitre plate. Three bioluminescent systems, from the firefly, bacteria, and jelly-fish, together with two synthetic chemiluminescent compounds, luminol and acridinium esters, have been widely used to exploit these features (Table 10.3). But have we actually discovered anything from them? Are they of any use at the patient's bedside? Is this simply technology transfer, or are they really able to measure things which cannot be measured in any other way?

The firefly luciferin–luciferase reaction is now the method chosen for measuring ATP (see Chapter 5). It has even been used to search for life on Mars! It can detect ATP down to 10–100 amol (10^{-17}–10^{-16} mol). The smallest eukaryotic cell (*ca.* 6 μm diameter) contains about 0.1 fmol ATP, the intracellular concentration in most healthy cells being 5–10 mM. Thus, single cells can be analysed. This method has use not only in the research laboratory but also in the assessment of biomass in sewage plants, and in microbiological contamination of foods and textiles. The key feature is that live cells contain ATP, dead ones contain virtually none, yet both may contain DNA and protein antigens.

Yet the results obtained from use of the bacterial oxidoreductase–luciferase to measure NAD(P)H and related metabolites or enzymes have been disappointing. Yes, plenty of assays have been established, but few have yet to be used to provide real insights into a cell biological or clinical problem. In contrast, much has been learnt from using chemiluminescence to monitor toxic oxygen metabolites (see Chapter 6) or intracellular Ca^{2+} (see Chapter 7).

Cellular and acellular sources of oxygen metabolites have been identified. Stimuli, both physiological and pathological, have been characterised. Environmental factors, drugs, and scavengers have been investigated as potentiators or inhibitors of their action on cells and tissues. Using pholasin, one cell at a time can be studied to see if, and when, the cell switches on and off. Neutrophils, a potent source of reactive oxygen metabolites, infiltrate the joints of patients with rheumatoid arthritis and the heart following a heart attack. Chemical changes found in these tissues are consistent with release of oxygen metabolites causing destruction of the host's own tissue. Oxidases, and oxidants, have been found in patient serum. These may provide indirect indicators of neutrophil activation, and be useful at the bedside in assessing the risk of complications following a heart attack. The key question now is, how close to the primary cause of the disease is the production of these metabolites, or are they simply an epiphenomenon?

Ca^{2+}-activated photoproteins from luminous jelly fish have played a major role in identifying types of cell activation provoked by a rise in intracellular free Ca^{2+}. These include muscle contraction and chemotaxis, secretion from intracellular granules, cell aggregation, switching off cell–cell communication through gap junctions, certain types of cell division, transformation, and intermediary metabolism. In spite of the ingenious invention of fluorescent Ca^{2+} indicators by Tsien, much new information is still being learned about intracellular Ca^{2+} using photoproteins. Because of their larger size the latter are particularly valuable in the investigation of Ca^{2+} in cell injury. Is a rise in intracellular Ca^{2+} a cause or a consequence of injury? Is the rise a friend or foe? By using photoproteins in cells we have begun to obtain the answers to these questions. Four ranges of cytoplasmic free Ca^{2+} have been defined: resting (approx. $0.1 \, \mu M$); stimulated ($1–5 \, \mu M$); reversibly injured ($5–30 \, \mu M$); $>50 \, \mu M$ irreversibly injured. Too much Ca^{2+} in the cell will cause irreversible injury and even death, yet under some circumstances the rise in free Ca^{2+} following injury can help the cell protect itself from attack. Membrane pore formers such as the terminal complex of complement, bacterial, and other toxins, T cell perforins, and viral proteins, let Ca^{2+} in to the cell and release it internally within seconds. This Ca^{2+} activates a protection mechanism aimed at removing the potentially lethal complex by vesiculation. If removal occurs before the lethal threshold then the cell will recover. A rise in intracellular free Ca^{2+}, provoked for example by a component of the immune system, can thus cause oxygen metabolite and myeloperoxide release from neutrophils in arthritic joints or in the heart, provoke insulin secretion in diabetes where β cells still exist, stimulate thyroid hormone secretion in thyroid disease, and in multiple sclerosis initiate breakdown of the myelin sheath which covers healthy nerve axons in vertebrates. The challenge now is to use the *in vitro* observations to identify parameters which can be used to test these hypotheses in the patient.

The isolation of mRNA and the cloning of genes responsible for bioluminescent proteins has opened up new areas for exploiting chemiluminescence in cell biology and biotechnology.

We have seen that de Wet and colleagues have transfected the firefly luciferase gene into protoplasts of the plant *Daucus carota* using electroporation and have produced transgenic tobacco plants of *Nicotiana tabaccum* using a tumour-inducing plasmid from *Agrobacterium tumeficans*. The plants lit up when 'watered' with luciferin providing a novel approach to quantifying organ-specific gene expression, and the development of plants resistant to herbicides and pathogens. Firefly luciferase cDNA has also been transfected into monkey CV1 cells in culture. We have succeeded in expressing the mRNA for the Ca^{2+} activated photoprotein obelin in human neutrophils following incorporation by liposome fusion, whilst Ulitzer and colleagues have genetically engineered bacteriophages selective for individual bacterial species, such as *Salmonella*, containing the bacterial lux genes. Bacteria transduced by the phage light up, enabling 1 cell to be detected in a 'soup'.

But chemiluminescence is not only of use in the investigation of cellular physiology and the mechanisms underlying the pathogenesis of disease. It provides the clinician with rapid, non-isotopic diagnostic methods for immediate use in patient care. Immunoassay and recombinant DNA technology are probably two of the most important analytical techniques to have evolved in clinical medicine this century. The former is well established, the latter poised to have its impact. Radioisotopes are hazardous and legislation is increasingly restricting their use. Many non-isotopic labels have been developed to replace ^3H and ^{125}I, particularly enzymes and fluors. Chemiluminescent labels, under optimal conditions, can be more sensitive, more rapidly detected, and measured with a simpler apparatus. Commercially based chemiluminescence kits for the assay of hormones, foetal proteins, and drugs are already in use in many laboratories. Whether all of these are better or worse than their absorbance of fluorescence competitors is often the source of much debate and polemics. Often the key to assay success lies in the antibody and not the label. The use of luminol-enhanced peroxidase or acridinium esters for detecting DNA–DNA or DNA–RNA hybrids on Southern or Northern blots respectively, or on solid phase, looks promising. However, there is some way to go if they are to compete with the amol (10^{-18} mol) detection of ^{32}P, present in each nucleotide of a nucleic acid molecule.

Chemiluminescence has thus taken us from the ocean, through the research and development laboratory, to the patient's bedside where clinical decisions are taken. What does the future hold? What can we see from the summit?

10.3 THE VIEW
10.3.1 New horizons
What more is there to learn about chemiluminescence? What more can chemiluminescence tell us about the chemistry and biology of exited states, and about living processes?

Much has still to be learnt about why radical annihilation and dioxetan decomposition generate an excited state, when other exothermic reactions do not. Quantum theory has helped us rationalise and predict many phenomena at the subatomic level. But do we now need something new? The controversy over the particle versus wave theory of light is not fully resolved. How big is a quantum anyway? Perhaps we should examine the real validity of the probabilistic view of nature and the universe. Is everything really left to chance, or is throwing the dice just an excuse for man's ignorance and narrowness of perspective? One poorly studied area which should reap rewards is polarisation of light in chemiluminescence, a property of light difficult to comprehend at the level of photons.

We saw in Chapter 3 that luminous species representing nearly 700 genera from 16 phyla have been found. Yet why, of the hundreds of land snail genera, are only two luminous? Why are there no luminous vertebrates, apart from fish? Why are there no luminous sea urchins or higher plants, or are there? Why no freshwater luminous animals and so few red emitters? Why is bioluminescence so common in the sea, yet relatively rare on land? The deep oceans, the tropical rain forests, and other poorly explored habitats no doubt have some surprises for us. But even of those luminous organisms that we know, in only a few do we really understand the chemistry or biochemistry (Table 10.4). We discovered that there are two oxidative reactions producing bioluminescence, one involves oxidation of an aldehyde to an acid, the other an oxidative cleavage to form CO_2. The chemiluminescent reactions in lampyrid beetles, bacteria, and jelly fish are known, but we do not yet know the structure of the luciferin in the bivalve piddock *Pholas*, nor that in a host of arthropods, echinoderms, tunicates, and fish. The chemistry of luminous bacteria, dinoflagellates, and jelly fish appears uniform throughout each phylum. Yet in the great phylum Arthropoda we already know that there are at least three different chemistries, and of those in only a few of the beetles (the fireflies and glow-worms) has it been fully characterised. Of our five chemical families of luciferin, aldehyde, imidazolopyrazine, benzothiazole, linear tetrapyrrole and flavin (Fig. 3.11), no doubt several of the unknowns will join one of them. However, there are likely to be others. I argue this because of the differences in properties that we already know of. For example, tunicate luminescence is lost on freeze-thawing or homogenisation. In some squid, polynoid worms, and searsiid fish the luminous system appears to be membrane bound.

Of the chemistries that we know, emphasis is now swinging away from the luciferin back to the luciferase. Genetic engineering techniques enable us to clone the gene(s), find the amino acid sequence quickly, discover the luciferin and oxygen sites using site-directed mutagenesis, and compare the active centre with other luciferases, and non-bioluminescent proteins. However, we still know virtually nothing about the biosynthesis of the known luciferins. These are all unusual compounds. Apart from the linear tetrapyrroles and the relationship to haem, chlorophyll, or phycobilin products, the others bear little resemblance to known compounds in non-luminous organisms. How can the same chemistry be present in organisms as diverse as jelly fish, shrimp, and fish?

The cell biologist has much to learn about the control mechanism responsible for eliciting a flash or glow, and the biologist still has little direct evidence for function, particularly in deep sea organisms which are difficult to watch in their natural habitat. Submersibles are now available for ecologists to go many hundreds of metres below the ocean surface. More ingenuity is needed in the design of ethological experiments which can be conducted in the laboratory, particularly under high pressure and in the cold.

There is still much to learn about the role of chemiluminescence in shaping the

Table 10.4 — Chemistries known and unknown in living light

Group containing luminous species	Phylum
1. *Known*	
bacteria	Bacteria
dinoflagellates	Dinophyta
radiolarians	Protozoa
hydroids, jelly fish, sea pansies	Cnidaria
sea combs, sea jellies	Ctenophora
some gastropods	Mollusca
a few squid	Mollusca
oligochaete worms	Annelida
ostracods	Arthropoda
copepods	Arthropoda
decapod shrimp	Arthropoda
beetles	Arthropoda
a few fish	Chordata
2. *Unknown, not proven or incomplete**	
*fungi	Eumycota
sponge	Porifera
nudibranch gastropods	Mollusca
bivalves	Mollusca
many squid	Mollusca
nemertine worms	Nemertea
polychaete worms	Annelida
pycnogonids (sea spiders)	Arthropoda
flies	Arthropoda
amphipods	Arthropoda
*euphausiid shrimp	Arthropoda
penaeid shrimp	Arthropoda
*diplopods (millipedes)	Arthropoda
chilopods (centipedes)	Arthropoda
collembola (springtails)	Arthropoda
ectoproctids	Ectoprocta
sea mat	Bryozoa
hemicordates	Hemichordata
crinoids, holothurians, starfish	Echinodermata
tunicates	Chordata
many fish	Chordata

chemistry and biology of the earth's surface and its atmosphere. How stable an ecosystem is the deep ocean anyway? Perhaps living light can give us some clues.

The discovery of new biological chemiluminescent reactions, together with the invention of new synthetic ones, will, no doubt, provide scope for developing assays for substrates not yet couplable to chemiluminescence. However, there is still much to be done with existing systems. Many of the major advances in biochemistry over the past century have been based on studying molecules and structures isolated from living cells. From this conventional 'grind and find' approach, we have learnt about enzyme characteristics, protein and nucleic acid structure, the sequence of reactions in pathways, and mechanisms by which they can be regulated, the location of substances and reactions within organelles. The founder of British biochemistry, Frederick Gowland Hopkins, argued that it was acceptable to study reactions in broken cell preparations provided that the *event* which occurred in the cell also occurred in the test tube. If we are to push forward the frontiers of modern biology it is clear that we need to re-assess this argument. There are three limitations to studies in cell and tissue extracts:

(1) Many events, particularly those associated with cell activation and injury, do not occur readily or at all once the cell membrane is broken; for example, fluid and vesicle secretion, endocytosis, cell division and transformation, a rise in intracellular Ca^{2+}, a change in membrane potential.
(2) Tissue homogenisation destroys the individuality and heterogeneity of each cell, or sub-populations within a multicellular tissue.
(3) Once the cell is broken the intracellular location of events, the polarity of organelles, and intracellular co-ordination are destroyed.

The challenge of modern cell biology is to develop ways of quantifying and visualising chemical events within, and emanating from, live cells, not only in the cytoplasmic compartment but within organelles such as the nucleus, mitochondria, and the endoplasmic reticulum. These chemical events must take us from the intracellular signalling system right through to the end response. Thus we can measure intracellular Ca^{2+} with photoproteins, ATP secretion from platelets, nerve terminals, or injured cells, using firefly luciferin–luciferase, and neurotransmitters such as acetylcholine with acetylcholine esterase and choline oxidase+luminol or pholasin. There are, however, three groups of intracellular signals we need to measure:

(1) Ions: Ca^{2+}, H^+, Na^+, and possibly K^+ and Zn^{2+}
(2) Nucleotides: cyclic AMP, cyclic GMP, AMP, $GTPMg^{2-}$ and $ATPMg^{2-}$
(3) Phospholipid derivatives: inositol tris- and tetraphosphates
inositol sugars, diacylglycerol

The problem is highlighted by inositol trisphosphate (IP_3) which releases Ca^{2+} from the endoplasmic reticulum in the activation of many cells and inositol tetrakisphosphate (IP_4) which opens Ca^{2+} channels in the cell membrane. To measure these, at present, the cell must be labelled with ^{32}P or ^3H inositol, and IP_3 and IP_4 then extracted. This does not yield the absolute concentration of IP_3 or IP_4 in the cell. Furthermore, the inositol lipids in cells such as neutrophils are very difficult to label.

Other chemical events requiring measurement in the live cell include intermediary metabolism, reactive oxygen metabolite production, DNA, RNA and protein synthesis and breakdown, receptor–agonist coupling, intracellular signal-protein activator (e.g. protein kinase and calmodulin) coupling.

My attempt to solve this problem is to link a bioluminescent indicator, prepared in sufficient quantities by cloning, to the substance to be measured. Using a fluorescent antibody or binding protein the ligand–ligand interaction can be quantified by a colour shift resulting from energy transfer (see Chapter 9). Of course, the indicator must be incorporated into the cell. Transfection of cDNA, or 'microinjection' of mRNA into cells (Fig. 10.1), and the production of transgenic cells or animals offers exciting possibilities for getting bioluminescent indicators into cells. This may also help in cloning, since eukaryotic cells such as yeast or invertebrates may allow the glycosylation which does not occur following expression in bacteria.

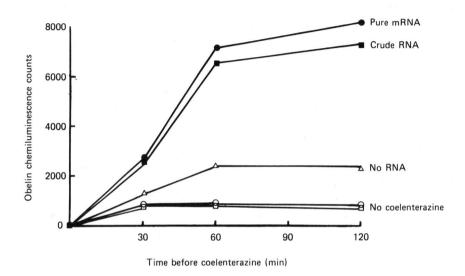

Fig. 10.1 — Formation of Ca^{2+}-activated photoprotein obelin from mRNA inside human neutrophils. Reprinted by permission from Campbell & Patel (1988) *Biochem. J.* **252,** 143–149 ®1988. The Biochemical Society, London in press, with permission. Crude RNA or mRNA incorporated into cells using liposome fusion.

Despite many discoveries in basic science, in physics, in chemistry, in the molecular and physical basis of biology, the cause of many major diseases remains a mystery. The biggest killer in the West is ischaemic heart disease, one of the commonest causes of illness arthritis. The causes are essentially unknown. The precise molecular basis of cancer is still not established, nor do we fully understand the primary cause and development of inflammatory and immune-based disease, where the body's defence system attacks itself. These include rheumatoid arthritis, diabetes, and multiple sclerosis. A wide spectrum of inherited disorders, including muscular dystrophy and cystic fibrosis, still pose serious problems for many families, and some unsuspecting parents. If we are to understand fully the pathogenesis and natural history of such diseases at the cellular and molecular level we need to identify the initiating agent, e.g. an abnormal gene, a virus, bacterium, or poison, and how, in molecular terms, this provokes the clinical manifestations of disease. Chemiluminescence, as well as fluorescence and nmr, provide the research scientist with the tools for defining the cell and molecular biology of cell injury in the test tube. From this, sites of potential 'intervention' can be identified, and parameters defined which can be measured in patient material to test whether the mechanism discovered in vitro also occurs *in situ*. It may be possible to move forward to explain the course of the disease, and backwards to find its route cause. Chemiluminescence lies at the key interface between basic science and clinical medicine.

As an example, is neutrophil infiltration in inflammatory disease and following a heart attack a friend or foe? What is the nature and action of oxidants and oxidases released by those cells? How does this relate to the dichotomy; oxygen is necessary for life yet it is also toxic. Bioluminescent proteins may not only provide indicators for measuring oxygen and its derivatives, and their gradients in cells and tissues, but also may provide a model for their interaction with proteins.

But do we have the right framework for interpreting experiments with intact cells? I think not.

10.3.2   Chemisymbiosis — a new light on life and death?

What is the difference between a bag of chips and a sack of potatoes? Chemical analysis may tell you that they are the same, yet put the chips into the soil — nothing. The potatoes will grow. Show a biochemist what is obviously a live cell and ask him to define life, chemically. Thermodynamically everything is in flux, no reactions are at true equilibrium. The cell is respiring, metabolising, capable of division, contains enzymes, carbohydrates, nucleic acids, ATP, ion gradients across the cell membrane, and so on. Yet none of these really explains the difference between a live cell and a bag of chemicals. Similarly it is obvious when a cell is dead, yet impossible to define it biochemically by using a simple analytical approach.

Activation of a cell by a hormone or electrical impulse acting at the cell membrane provokes an event within the cell: movement, vesicle fusion, metabolism, division, or transformation. An intracellular messenger is required to transmit the information that the cell has received a signal, to the reaction or structure within the cell which is to be activated. Intracellular Ca^{2+} is certainly one such messenger. Yet cyclic AMP and cyclic GMP often change during the process as well. The cell also contains an activating mechanism independent of a rise in Ca^{2+}, using protein kinase C activated by agents such as diacyl glycerol. Why do we have all these different intracellular signals in the same cellular process? What are the reasons for internal release of Ca^{2+} and movement of Ca^{2+} through the cell membrane during the activating reaction?

What is the unit of life? Is it a protein molecule, a piece of DNA, a chromosome maybe? No, it is a cell.

The *chemisymbiotic* hypothesis (Fig. 10.2) is a quantal approach to cell activation and cell injury. It is an attempt to develop a new conceptual framework, based on chemical thresholds, in order to understand how reactions and events within a cell are coordinated, and in particular to rationalise the molecular basis of cell activation and cell injury at the level of individual cells. The ultimate goal is to explain the difference between life and death at the molecular level. The corner stone of the hypothesis is that there are a discrete number of states in which the live cell can exist. Further, it is proposed that there are chemical thresholds determining the rapid transition from one state to another. The transition state is inherently unstable and the cell cannot remain there. The chemisymbiosis between the intracellular signalling system and energy balance determines whether a cell will transform from one

CHEMISYMBIOSIS

A QUANTAL HYPOTHESIS FOR CELL ACTIVATION AND CELL INJURY

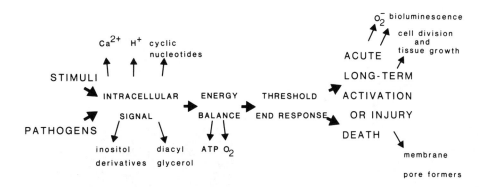

Fig. 10.2 — The chemisymbiotic hypothesis.

state to another. If this chemisymbiosis breaks down to chaos, the cell is biochemically dead since this chaos will be irreversible.

Consider a population of cells responding to a stimulus, or a dose of pathogen, eliciting only half the maximum response possible. Have all of the cells been activated to produce 50% of their maximum response, or have only half of the cells been switched on? Is the time course of response of a large population of cells explained by individual cells switching on at different times? In other words, is there a chemical threshold between the resting and activated state (Fig. 10.3)? How many states of activation are there? Similarly, is there a chemical threshold at which a cell transforms from being alive to being dead?

Biochemists used to measuring pathway rates, K_m's and K_d's, tend to think of biological phenomena occurring in a graded manner. The trouble is that biochemical analysis is usually the mean of thousands, or even millions, of cells. When we examine cells one at a time we find a different story. As we saw in Chapter 4, many phenomena clearly occur via thresholds (Table 4.4) for example: secretion via vesicle–membrane fusion, cell division, and transformation; or cell aggregation. Also a luminous cell flashes or remains invisible. But what about intermediary metabolism in hepatocytes or reactive oxygen metabolite production in phagocytes? We have developed two approaches to defining chemical thresholds for reactive oxygen metabolite production, and thus ultraweak chemiluminescence, in neutrophils:

(1) 2'7'dichlorofluorescin as an intracellular indicator in the granules, quantified by the fluorescence activated cell sorter and photon counting microscopy. The former counts the number of activated cells, the latter gives us the time course in a single cell.

(2) Pholasin as an indicator for single neutrophils, isolated from a cell sorting apparatus. The results show (Fig. 10.4) not only a heterogeneity in response

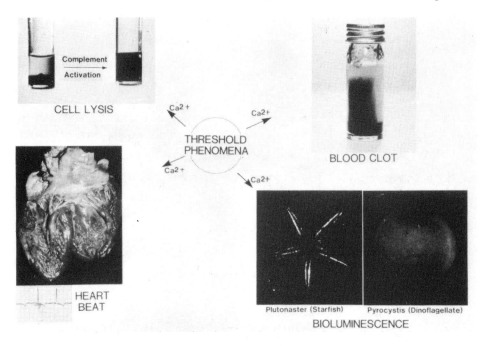

Fig. 10.3 — Chemical thresholds in cell activation and injury.

between individual cells, but that there is indeed a threshold for activation in each cell. The lag time before activation of the end response is detected varies considerably from cell to cell, between different stimuli, and some cells do not switch on at all, even at the maximum dose of stimulus. At a dose required to produce a 50% response in a population of some 100 000 cells, only half the cells become switched on. Adenosine, a secondary regulator in these cells, appears to inhibit by reducing the number of cells which switch on, by altering the lag before the threshold is reached, and by reducing the magnitude of the eventual response once the threshold has been achieved.

This re-examination of the cell biology of cell activation means we have to rethink our approach to the production and action of the intracellular signals linking stimulation to the end response. Furthermore, cell injury induced by 'pore formers' such as complement also involves at least three chemical thresholds in individual cells, in this case permeability to Ca^{2+} and other ions, permeability to small molecules, and finally permeability to macromolecules, lysis (Fig. 10.4C). The key question now is what is the molecular basis of these chemical thresholds in cell activation and cell injury? There are at least five possibilities:

(1) Gradients of the primary substrate.
(2) Intracellular signal gradients or 'clouds'.
(3) Organelle thresholds.
(4) Critical receptor occupancy.
(5) Concerted action of regulatory proteins or other macromolecules with those responsible for the event.

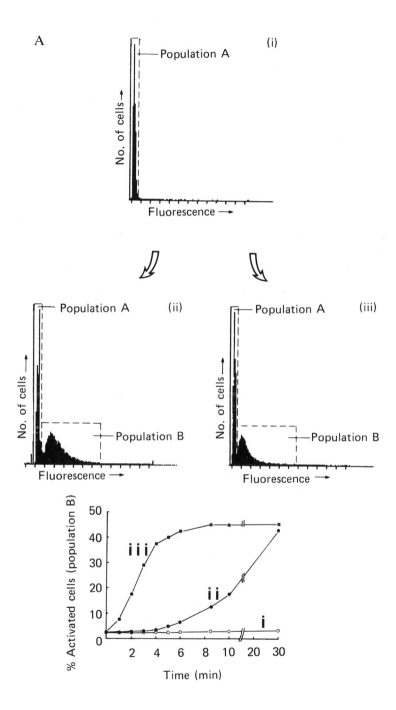

Fig. 10.4A.

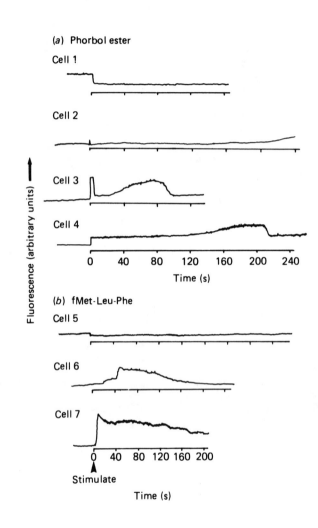

Fig. 10.4 B

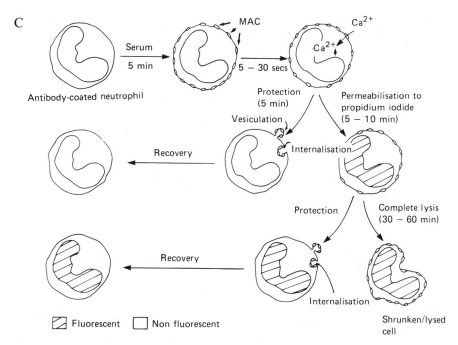

Fig. 10.4 — Threshold activation of neutrophils. Reprinted by permission from Patel *et al.* (1987) *Biochem. J.* **248** 173–180, ®1987 The Biochemical Society, London. A: Number of cells activated counted in fluorescence activated cell sorter using 2'7 chlorofluorescin as indicator of oxygen metabolite production; (i) unstimulated; (ii) chemotactic peptide; (iii) phorbol ester. B: Time course of individual cells observed by micro fluorimetry. C: Permeability thresholds in cells attacked by membrane pore formers, such as the membrane attack complex of complement, T cell perforins, bacterial and intal toxins, from Morgan, Luzio & Campbell (1986) *Cell Calcium* **7** 399–411, with permission of Churchill Livingstone. Permeability determined using propidium iodide as a fluorescent nuclear stain. With permission of Blackwell Scientific Publications Ltd. from Patel *et al.* (1987) *Immunol* **60,** 135–140.

The oxygen concentration in air-saturated buffer is about 220 μM at 37°C, in arterial blood around 100 μM, but around cells *in situ* may be as low as 1–10 μM or less. Oxygen gradients, generated by the balance between diffusion and utilisation, occur extracellularly and within cells. Oxygen has three main cellular functions: energy production, detoxification and killing, and production of regulators, e.g. the eicosanoids. The K_m's for the oxidases and oxygenases responsible for these range from sub-μM to around 50 μM. Could an O_2 gradient generate a molecular threshold in the cell?

We already know that when Ca^{2+} is released inside the cell it diffuses only a limited distance, because of buffering by binding proteins and the mitochondria. If the Ca^{2+} cloud does not reach the site where the cellular event for end-response occurs, then the cell will not activate.

A live cell, such as a hepatocyte, contains about 1000 mitochondria. Each one is so small that the number of free H^+ or Ca^{2+} ions inside is only about 5–10. The chemiosmotic hypothesis of Mitchell assumes, incorrectly, a giant mitochondrian as a macrothermodynamic system. But each mitochondrion is a unit. The obvious way to resolve this problem is to propose that the proton electrochemical potential responsible for providing the energy link between NADH oxidation and ATP synthesis is a threshold phenomenon in each 'unit'. Different respiration rates reflect different numbers of mitochondria switched on. Similarly different secretory rates from cells reflect variations in the number of vesicles reaching, and fusing with, the cell membrane.

The final possibility for generating a threshold in a cell is the concerted action of a critical number of molecules either at the cell surface, e.g. receptors, or within the cell, e.g. calmodulin. We know that there are only a few thousand receptors on the surface of each cell. At the muscle end plate the rate of receptor occupancy must be high enough to generate an action potential. Within the cell, metabolite pathways are now thought to exist as enzyme complexes rather than isolated reactions, and a critical number of Ca^{2+} binding proteins may be required to provoke a gel–sol transition in a particular region of the cell, or throughout the cytosol.

How can we study these threshold events? Indeed, what do we need to measure, and how sensitive a method do we need? We need to measure and locate both the end response and the intracellular signals in single living cells, and link them to the energy balance, namely ATP and O_2 (Fig. 10.2).

Most eukaryotic cells, apart from a few giant ones such as nerve and muscle, are 5–100 μm in diameter. The $ATPMg^{2-}$ concentration is about 5–10 mM, the content some 0.1–10 fmol. A bacterium is smaller, 1 μm, equivalent to a mitochondrion, with an ATP content of 1–10 amol. The concentration and amount of intracellular signals are, in general, some 10^3–10^5 less than this. In the resting cell cyclic AMP, cyclic GMP, Ca^{2+}, and IP_3 are around 0.1 μM, and some 1–10 μM in the stimulated cell. So, for single cell analysis we need to detect attomol quantities (10^{-18} mol). For end response, O_2 utilisation or ATP we need to measure in the fmol (10^{-15}) region. Thus, chemiluminescent techniques for single cell analysis offer exciting prospects for demonstrating, and unravelling, the mechanism underlying, chemical thresholds in single cells.

The symbiotic relationship between chemical events provides a communication mechanism which is the key to understanding the transition between rest and activity, or life and death.

10.3.3 A perspective on biochemical evolution

The solar system is about 5000 million years old, the earth some 4500 million years old. Life began in a reducing atmosphere, about 3500 million years ago. Oxygen producing microbes appeared 2300 million years ago. When did bioluminescence first appear? The earliest prokaryotes have been found in rocks some 3500 million years old, and coelenterates some 600 million years ago, but when did luminous bacteria and jelly fish first develop?

The fossil record is not of much help. Lampyrid beetles, including a very well preserved firefly, have been found in amber from the Oligocene (38–26 million years ago) and the Miocene (26–7 million years ago) epochs. Luminous fish, with recognisable photophores, have been found in rock from the Oligocene epoch. These include stomiatoids and luminous fish similar to modern *Chauliodus, Cyclothone, Sternoptyx, Argyropelecus* (hatchet fish), *Gonostoma, Photichthys,* and many myctophids (lantern fish). Yet this is equivalent to viewing but the last six days of evolution where life began at the beginning of the year, man appearing some four hours before midnight on the 31 December. During the Pre-cambrian era (3000–700 million years ago) the biochemical pattern of life was set; chirality, the genetic code, 20 amino acids, and four bases in triads as the letters of life. During this period, the main cell types of life, bacteria, the archaebacteria, and the eukaryotes, developed, as did meiosis, mitosis, photosynthesis, the first plants, an oxygen atmosphere, and the first metozoans. The next 400–500 million years, the Paleozoic era, saw an explosion in metozoans, plants, and invertebrates, and the origin of the vertebrates. The Mesozoic era, 245–66 million years ago, saw the origin of mammals. So by the Cenozoic era, 66 million years ago to the present, where most of the fossil evidence for luminous organisms can be found, biological evolution was well advanced. Clearly, bioluminescence must have appeared before this, but how can we trace its origins? How many sites of origin were there? Examination of the individual structures and phylogenetic distribution of the five known chemical families of luciferin, aldehydes, imidazolopyrazines, benzothiazoles, linear tetrapyrroles and flavins, suggests that bioluminescence must have appeared independently, at least nine times during evolution (bacteria, gastropod snails, earthworms, beetles, jelly fish, ostracods, dinoflagellates, euphausiids, stomiatoid fish), more if organisms as diverse as jelly fish, squid, and fish are not involved in a food chain as the luciferin source. Imidazolopyrazines are particularly prevalent. Are there others formed from different tripeptides? The luciferins and luciferases of Phengodid and Elaterid beetles cross react with those of the Lampyrids, but not apparently with those of the fungus gnat *Arachnocampa*, even though the latter uses ATP. Thus there may another ten origins of 'unknown' reactions (Table 10.4).

The puzzle for the biochemist is the spasmodic occurrence of bioluminescence, even within one genus, let alone a phylum. There are some 2000–4000 luminous beetles, but one discovered by Lloyd seems be be gradually losing its ability to produce light. There are several species of fungi and fish which are luminous in one part of the world and not in another. However, it is difficult to believe in this negative mechanism as the reason why certain species are *non*-luminous; rather one would expect a positive mechanism for developing it in those that are.

There are five particular evolutionary puzzles about living light:

(1) How did the chemical reactions appear?
(2) What is the origin of the catalysts, the luciferases?
(3) What is the origin of any energy transfer chromophores and proteins involved in final light output?
(4) How did the cell biology evolve; the photocytes, the organelles containing components of the chemiluminescent reaction, the control mechanism triggering the cells to light up?
(5) What was the origin and evolutionary development of the ancillary structures such as filters, lenses, and reflectors, often associated with photophores and light organs?

It is not too difficult to imagine the evolution of the cells containing the chemiluminescent components, and the ancillary structures, from precursor cells by conventional Darwinian–Mendelian selection. Structural similarities between several photocytes and other cells in the organism, together with their embryological origin, often but by no means always ectodermal, have already given us some clues. Similarly the modern biochemical approach to protein evolution, based on working out an evolutionary clock from amino-acid changes arising from base mutations, giving rise to protein families, and the concept of selfish DNA, neutral mutations and the production of new proteins from recombination of exons from different sources, will help us to understand the evolution of luciferases and photoproteins. Gene transfer between the symbiotic luminous bacteria in the gut of the pony fish and the host may have occured with enzymes like superoxide-dismutase. Cloning and sequencing has already given us some information. The coelenterate photoproteins belong to the family of Ca^{2+} binding proteins with 'EF hands', like calmodulin. Other biochemical clues include oxygen carrying and detoxification, the peroxidase activity of several luciferases, e.g. earthworms and the piddock *Pholas*, the 'luciferase' activity of albumin with coelenterazine may help us to find the origin of the luciferin active centre of coelenterate photoproteins. The fact that glow-worms narcotise their prey and that the luciferase is inhibited by general anaesthetics may point the way to its origin in beetles. Other functions for the luciferin or its precursors requiring investigation include hormonal, toxin or waste products.

However, the origin of the chemical reactions in living light is much more of a puzzle. Here we have an evolutionary threshold. For conventional Darwinian–Mendelian principles to act the luminescence must have an advantage, however small, to the propagation of the species to be selected. Because of the wide range of ultraweak chemiluminescent reactions it is not difficult to imagine a step-by-step improvement of the quantum yield and kinetics resulting in light visible to a potential responder. But how can these steps prior to the visibility threshold be selected for?

Similarly, it is easy to see how 6% of a modern eye is better than one with only 5% capacity, and would thus be selected for. But how did our primitive visual sensor take the quantal leap from 0 to 1%? Gene duplication, balanced by neutral mutation? Perhaps. Bioluminescence is a biochemical example of a problem which Darwin attempted to tackle in the sixth edition of 'The Origin', published in 1872. Critics asked 'How can a structure such as an insect wing evolve by natural selection when the incipient structure serves no useful function?' Darwin suggested that the precursor, non-functioning, wing initially fulfilled another function e.g. balance or reproduction, resulting in selective advantage. Thus the precursors of visible bioluminescence could have been oxidative reactions used in digesting food, oxygen processing and protection or oxidative killing producing a weak chemiluminescence as a by-product.

Bioluminescence provides the molecular evolutionist with a phenomenon which exhibits three features common to many other biological proceses, but whose evolutionary origin is unknown. Firstly, it is a threshold phenomenon. Secondly, two components, the protein and chromophore, had to evolve together in order to generate light. Other such pairs, whose evolutionary marriage is a mystery, include hormone and neurotransmitters with their receptors, and oxygen binding and redox pigments with their respective proteins. Thirdly, the universal requirement for oxygen, or one of its metabolites, in bioluminescence gives us proteins as models for the direct interaction of oxygen with polypeptide chains, providing an insight into the primitive mechanisms underlying the evolution of oxygen carriers and the detoxification of oxygen. Both were essential for the successful evolution of the modern eukaryotic and prokaryotic cell.

There are essentially three philosophical approaches to understanding evolution: the mechanistic–materialistic, the holistic, and the dialectic. The enigma facing modern biologists is how to marry the micro-evolutionary mechanisms based on molecular genetics and biochemical clocks, with macro-evolution at the level of the appearance of phenomena, species, and phyla. It is my belief that if we ask the right questions about living light we may have the chance of learning much about chemical and biological thresholds in evolution, and of tackling one of the central problems in contemporary biology, the incompatability of the molecular approach to evolution with Darwinian–Mendelian selection.

10.3.4 Biotechnology — new lamps for old?

It might appear at first sight that all we have done in developing chemiluminescence labels in immunoassay and recombinant DNA technology is to replace old lamps by new ones. But in this case, the principle on which the lamp works is different. Furthermore, the glasses through which we view it have different characteristics. Chemiluminescence can detect easily substances in the atto–femto (10^{-18}–10^{-15}) mol range. The challenge now is to achieve tipomol (10^{-21}) sensitivity, which will enable us to measure things invisible to other techniques. By coupling 10–100 chemiluminescent molecules, detectable in the sub-tipo region, it should be possible to detect a single antibody, RNA, or DNA molecule.

To reach the horizons of biotechnology by the use of chemiluminescence, a combination of innovative chemistry and electronics is required. More efficient chemiluminescent labels, with colours covering the whole visible range (400–750 nm), are required. Nature has given us some already in living light. We need to discover more about their chemistry, and clone them. Better methods for coupling chemiluminescent probes to proteins, small molecules, and nucleic acids are required. Again, genetic engineering offers some fascinating possibilities. By coupling good fluors like phycobiliproteins or bioluminescent energy transfer proteins, to one ligand and the chemiluminescent label to the other homogeneous assays for antibody–antigen, RNA–DNA, DNA–DNA reactions should be achievable in free solution without the need for gel or particle separation and blotting. A further intriguing possibility is to generate chemiluminescence polarisation as a homogeneous method for ligand–ligand interactions.

Three improvements in light detection are needed to realise the potential of sub-tipomol and homogeneous chemiluminescence analysis:

(1) A better photon detector than the modern photomultiplier tube, with a flat response over the entire visible range with a quantum efficiency approaching 100%. Photomultipliers are at best 10–20% efficient, and then only with blue light.

(2) Colour and polarised image intensifiers to visualise energy transfer and polarised light from chemiluminescent reactions. The pseudo-colour images from present intensifiers are misleading. They represent intensity, and not colour, thresholds.

(3) A robust small battery operated detector equivalent in sensitivity to a photomultiplier tube, or a new, ultrasensitive photographic film with single photon efficiency.

Cooled photodiodes are red sensitive, and are improving year by year, but the obvious alternative is to build biological imagers and detectors. The potential markets in clinical laboratories are already thousands of millions of dollars a year. Patient tests are increasingly being carried out on the ward, in specialised clinics, and in the doctor's surgery. Faster results are needed if immediate clinical decisions are to be taken. Computers and televisions are highly complex and sophisticated electronic devices, yet they are usable by anyone in the family at the press of a button. High tech for low tech people, using chemiluminescence, offers exciting opportunities for the biosensor industry in the years ahead.

10.4 CODA

I grew up just a few miles from the village where Charles Darwin lived after his return from *The Beagle*, Downe. Now I find myself in another, but quite different, environment of inspiration, Ynys Môn, just a few miles from where I was born. Wales is a small principality yet full of history, culture and music, of industrial success and savagery, and of great natural beauty. The Welsh are fiercely proud of their language and I hope, therefore, some of my fellow countrymen will appreciate the insertion of a Welsh word into the English language, the tipomol.

Two places in Britain have been a particular source of inspiration for me, Plymouth and North Wales. I look out of the window now, and see Snowdon living up to its English name. How soon will it be before the glow-worms reappear there with their 'amorous fire'? Many famous naturalists, from the 16th century onwards, have also found inspiration here, and have tried to express this in writing as 'A tour of Wales'. Being close to Nature gives one a breadth of perspective on life rarely achievable with mere 'artefacts', however clever and perceptive these may appear. I have tried to use chemiluminescence as an agent to explore, and extend such a view of Nature, one which allows us to use molecular dissection to understand and appreciate the whole. Because one is never entirely cut off from the community, the opportunity is there, I hope, to make this a reality for others.

For me, science is a means of developing an intimacy with natural things, without losing sight of their inherent beauty and elegance. Rather the more one learns about the molecular basis of life the more awesome natural phenomena become. As if this weren't enough, how rewarding also it is to see an idea or a discovery evolve to

Fig. 10.5 — By Alder from The Sunday Times 12th April 1987 © Times Newspapers.

something practical from which large numbers of people can benefit or find enjoyment. But it takes a long time to solve big problems, or even to be awakened to what the problem actually is. I still have to think carefully before writing the word 'chemiluminescence' (Fig. 10.5)!

The study of chemical reactions which produce light gives us an opportunity of converting an apparently insoluble enigma about a major natural phenomenon into a fascinating, but soluble, problem with years of pleasure, stimulation, and practical rewards to be enjoyed. At school, we all, I am sure, learned Boyle's Law, but how many are aware that he pioneered some of the earliest controlled experiments into living light and other luminous phenomena. His wisdom from the past surely sums our faith in the future.

Light is so noble a thing . . . I cannot but hope that something would be produced, to the discovery of nature, not only of light but divers other bodies, and perhaps also of good use to human life.

Robert Boyle, 1681

Appendix I

A glossary of some terms associated with luminescence

These are not necessarily intended as 'dictionary' definitions, rather a guide as to how I have used these terms in this book.

Anodoluminescence: Radioluminescence caused by positive 'rays' (α particles).

Apoprotein: protein without its prosthetic group e.g. apo-photoprotein.

Bioluminescence: Visible light emission from luminous organisms. All known examples are biological chemiluminescence. Also used to describe reactions extracted from luminous organisms e.g. in bioluminescent assays.

Candoluminescence: Luminescence in incandescent solids.

Cathodoluminescence: Radioluminescence caused by cathode 'rays' (β particles).

Charge-coupled device: a solid state device (pixels per square cm) which produces a charge when hit by a photon. The array is then reset by a computer which reconstructs the image.

Charge-transfer complex (CTC): Complex between two molecules of opposite charge (A^+B^- or A^+A^-). The resulting energy can give rise to luminescence.

Chemically initiated electron-exchange luminescence (CIEEL): Luminescence resulting from a process involving electron exchange between atoms or molecules, initiated by a chemical reaction.

Chemienergisation: Process where energisation of an atom or molecule arises from a chemical reaction.

Chemiexcitation: Process where excitation of an atom or molecule arises from a chemical reaction, i.e. the key step in a chemiluminescent reaction.

Chemiluminescence: The emission of light (strictly electromagnetic radiation in the UV, visible and IR) as a result of a chemical reaction. The enthalpy of reaction gives rise to an atom or molecule in a vibronically excited state, which then emits a photon on decaying to ground state.

Chemiluminescent reaction: The complete reaction of an atom or molecule to form products, producing light. One of the steps will involve *chemiexcitation*.

Chemiluminometer: Apparatus for detecting and recording the light from a chemiluminescent reaction.

Chromatophore: A coloured cluster of cells.

Chromophore: A coloured substance. The term is most familiar when the substance absorbs light in a particular part of the visible spectrum. The colour is made up of light not absorbed, but see *luminescent chromophore*.

Conjugation: 'mating' of two bacterial cells leading to transfer of some of the DNA from one cell to the other.

Counter-illumination: In bioluminescence, light emission emitted to counter the ambient light, and thus act as a camouflage.

Crystalloid: Crystal-like inclusion found in some luminous cells.

Crystalloluminescence: An old term for triboluminescence when solutions crystallise.

Dichroic: Substance or solution appearing to have two colours depending on the angle at which it is observed.

Dim chemiluminescence: Chemiluminescence from a high quantum yield reaction, but which is invisible to the naked eye because of slow kinetics or low reactant concentrations. It is, however, detectable by a chemiluminometer.

Dioxetandione:

$$
\begin{array}{c}
\text{O}\diagdown \qquad \diagup \text{O} \\
\quad \text{C} - \text{C} \\
\quad | \qquad | \\
\quad \text{O} - \text{O}
\end{array}
$$

The proposed key intermediate in oxalate ester chemiluminescence.

Dioxetan:

$$
\begin{array}{c}
| \qquad | \\
- \text{C} - \text{C} - \\
| \qquad | \\
\text{O} - \text{O}
\end{array}
$$

The key intermediate in many organic chemiluminescent reactions. Spontaneous (thermal) decomposition produces two carbonyl products, one in an electronically excited state. Dioxetan*e* is used by several authors but dioxetan is strictly correct, analogous to dioxan.

Dioxetanone:

$$
\begin{array}{c}
\text{O}\diagdown \qquad | \\
\quad \text{C} - \text{C} - \\
\quad | \qquad | \\
\quad \text{O} - \text{O}
\end{array}
$$

A key intermediate in the chemiluminescence of several bioluminescent reactions, (e.g. beetle, jelly fish). Spontaneous decomposition yields CO_2 and an electronically excited carbonyl.

Dipole–dipole resonance: resonance energy transfer.

Direct chemiluminescence: Chemiluminescence where light emission arises directly from a product of the reaction.

Dynode: The amplifying component of a photomultiplier tube. Electrons hitting it release larger numbers of secondary electrons.

Einstein: The number of photons equivalent to 1 mole (i.e. 6.023×10^{23} photons).

Electrochemical luminescence (EL): Chemiluminescence resulting from electrolysis.

Electrochemiluminescence (ECL): Chemiluminescence, in solution, resulting from high energy electron transfer reactions.

Electrogenerated chemiluminescence (EGCL): Electrochemiluminescence.

Electroluminescence: Luminescence resulting from an electric current or discharge.

Electron exchange: Exchange of electrons between atoms or molecules, usually close enough for molecular orbital overlap. If one of the pair is an electronically excited state, it may thereby transfer its energy to an electron in the other atom or molecule.

Electronically excited state: An atom or molecule where an electron is raised into an orbital of higher energy. It is thus unstable. One mechanism for loss back to ground state is photon emission (luminescence).

Electron transfer: The transfer of an electron from one molecule to another. Its occurrence during radical annihilation or charge transfer complexes provides a mechanism for both generating an electronically excited state or for transferring energy from one atom or molecule to another.

Energy transfer: The transfer of energy from one atom or molecule to another. In luminescence, it involves transfer from an excited state donor to an acceptor which itself becomes excited, and then emits. It is usually used only for non-radiative processes.

Excimer: A charge-transfer complex between two molecules of the same chemical structure (A^+A^-). Neutral complex (AA) is also possible.

Exciplex: A charge-transfer complex between two molecules of different chemical structure (A^+B^-). Neutral complex (AB) is also possible.

Excited state: Electronically excited state as used in this book.

Extracellular: Outside a cell.

Filter: In a luminous organism this refers to a layer of pigmented cells or an acellular pigment, through which the light passes. The colour of the light emanating from the organism is thus different from that from the photocytes, because of absorption by the pigment.

Fluorescence: A form of photoluminescence where the electronically excited state is generated by absorption of light (UV or visible). The emission is from the same spin state as the ground state (usually singlet to singlet). The life time of a fluorescent atom or molecule is very fast, usually 1–30 ns, though some longer-lived species are known.

Fluorophore: A fluorescent substance.

Förster mechanism (or energy transfer): Resonance energy transfer.

Franck–Condon principle: The principle by which absorption of energy by a chemical bond, e.g. from a photon, is so fast that it occurs without a change in separation between the nuclei, i.e. in a potential energy well diagram the path to the excited state is vertical.

Galvanoluminescence: An old term, a form of electroluminescence, when solutions are electrolysed.

Hydroid: The fixed state of a hydrozoan jelly fish (phylum Cnidaria). N.B. Some hydroids have no free-floating stage.

Image intensifier: An instrument for amplifying, or intensifying an image received by its detector. N.B. A simple TV camera on its own is not an intensifier since it does not involve amplification of the initial signal received, it just records it.

Incandescence: Emission of light as a result of raising a body to high temperatures. The energy for light emission is essentially kinetic, involving interactions between rather than within, an atom or molecule (as in luminescence). The emission obeys the laws of black body radiation.

Indicator-dependent chemiluminescence: Chemiluminescence resulting from the reaction of cellular metabolites with an added chemiluminescent compound, the indicator.

Indirect chemiluminescence: Sensitised or energy transfer chemiluminescence.

Intermolecular: Between two or more molecules.

Intersystem crossing: Transition of an electron in an atom or molecule from one spin state to another.

Intracellular: Inside a cell.

Intramolecular: Within one molecule.

Iridocyte: A reflecting cell, found within many photophores or luminous organs.

Iridophore: Reflecting structure within a photophore or luminous organ.

Lens: In the eye, a transparent structure focusing light on the retina. In a photophore or luminous organ, a transparent structure directing the light from the organism.

Light intensity (I): The amount of light received, from a source, per unit time per unit solid angle, i.e. $I=dhv/dt$ received. Qualitatively intensity is used to describe the brightness of a luminous source, without defining the precise solid angle observed.

Light organ: Multicellular large structure emitting light in a luminous organism.

Living light: Bioluminescence.

Luciferase: The protein catalyst, i.e. the enzyme, in a bioluminescent reaction. The term is normally used only when the luciferin is isolated separately from the luciferase, to distinguish it from a *photoprotein*.

Luciferin: The chemiluminescent substrate in a bioluminescent reaction. In this book, I have used the term exclusively as the small organic moiety responsible for the chemiluminescence. Sometimes it may be protein bound as in a photoprotein.

Luminescence: The emission of light resulting from atoms or molecules produced in electronically excited states. Originally defined by Wiedemann to distinguish phenomena producing light which were not caused by incandescence, i.e. high temperatures.

Luminescent chromophore: A coloured substance capable of luminescence.

Luminescent reaction: A chemical reaction producing light, i.e. chemiluminescence.

Luminometer: Apparatus for detecting and quantifying light from a luminescent phenomenon. However some commercial instruments constructed for chemiluminescence are given this abbreviated description.

Luminous organism: An organism emitting visible light from an endogenous source.

Lumiphore: A substrate capable of luminescence.

Lux gene: A gene within an operon responsible for producing bioluminescence, e.g. the marine bacterium *Vibrio fischeri* has seven lux genes:

lux A and B code for the α and β subunits of the luciferase respectively

lux C, D, and E code for polypeptides responsible for producing the long chain aldehyde

lux I and R control the transcription of the whole operon (lux genes C, D, A, B, E).

Lyoluminescence: An old term for triboluminescence when crystals dissolve.

Measurement of light: *Luminous* refers to light received by a detector

> *radiant* refers to light emitted by the source
>
> *flux* Φ (luminous or radiant) refers to the amount of light (as energy or photons) per unit time
>
> *intensity* I (luminous or radiant) refers to the amount of light (energy or photons) per unit solid angle (Ω) from the source unit solid angle=steradian.

Thus total flux $\Phi = 4\pi I$

For other terms and units such as exitance, yield, density, exposure, see Chapter 2.

Medusa: Jelly fish

Molar units: 1 mole or mol=6.023×10^{23} molecules

> 1 millimol=1 mmol=10^{-3} mol
>
> 1 micromol=1 μmol=10^{-6} mol
>
> 1 nanomol=1 nmol=10^{-9} mol
>
> 1 picomol=1 pmol=10^{-12} mol
>
> 1 femtomol=1 fmol=10^{-15} mol
>
> 1 attomol=1 amol=10^{-18} mol
>
> †1 tipomol=1t mol=10^{-21} mol
>
> †1 impossomol!!=1 imol=10^{-24} mol

†proposed; see text

Organochemiluminescence: Luminescence, bio, or ultraweak, from living material: cells, tissues or organisms.

Oxyluminescence: An old term for chemiluminescence resulting in oxidation of substances, usually organic.

Phosphorescence: A form of photoluminescence where the light emission is prolonged after removal of the exciting light source, because of intersystem crossing. Light emission is typically from triplet excited to singlet ground state and can last from a few milliseconds to many seconds. In the older scientific literature, and to the modern layman, phosphorescence is often used to describe any form of light emission in the absence of a light source, including bioluminescence.

Photoblast: The precursor cell of a photocyte.

Photocathode: The light sensitive detector in a photomultiplier tube, which releases electrons when hit by photons. It is maintained at a negative voltage relative to the other end of the photomultiplier tube (i.e. cathode negative and anode earthed or cathode earthed and anode positive).

Photochemical reaction: A chemical reaction initiated by light.

Photochemiluminescence: Chemiluminescence initated by light.

Photocyte: A luminous cell, in a photophore or luminous organ.

Photodiode: A solid state, light sensitive semiconductor. It can replace the photo-multiplier tube in a chemiluminometer or spectrometer, but is usually less sensitive.

Photogen: A term coined by Shimomura and Johnson to embrace the chemilumines-cent substrate in all bioluminescent reactions. It therefore includes the conventional luciferins and the active moiety in photoproteins. However, in this book I have defined the term luciferin as this generic term.

Photogogikon: A term coined by Shimomura and Johnson to describe the actual light emitter in a bioluminescent reaction. It is not necessarily the final product of the reaction, but rather can be an intermediate molecule or complex.

Photoluminescence: Luminescence resulting from absorption of ultraviolet, visible or near infra-red light, conventionally called fluorescence or phosphorescence.

Photomultiplier tube: A device for converting photons into electrons, and then amplifying these to produce a pulse of charge. The light detector, the photocathode, generates the primary electrons. Secondary electrons are then generated by the amplifying dynode chain. The overall amplification gain can be up to 10^7. Photomultiplier tubes are the most common light detectors found in a chemiluminometer.

Photon: The fundamental particle, or quantum, or light. Its energy (E) is described by Einstein's equation:

$E=h\nu$ where $h=$the Planck constant, and $\nu=$frequency.

Photo-organelle: An intracellular organelle capable of generating luminescence.

Photophore: A minute light organ made up of a discrete cluster of luminous cells (photocytes), and often ancillary structures such as reflectors, a lens and filters. The term is commonly used when the luminous organism, e.g. a squid, a shrimp or a fish, contains multiple, small light organs.

Photoprotein: A protein-luciferin (or photogen in Shimomura and Johnson nomenclature) complex, where the luciferin is tightly bound, often cova-lently. Thus the photoprotein is isolated as the complex, and light emission is directly proportional to the amount of protein present.

Photoreceptor: A structure in an organism capable of responding to light. It is usually a sensory nerve ending.

Piezoluminescence: Triboluminescence.

Post-orbital: Posterior to the eye.

Pyroluminescence: Luminescence resulting from atoms produced at high tempera-tures, e.g. in a flame.

Quantum yield (ϕ): In luminescence, the fraction of atoms or molecules which produce a photon. In chemiluminescence the overall quantum yield (ϕ_{CL}) can be subdivided into three components:

$\phi_{CL}=\phi_C \cdot \phi_{EX} \cdot \phi_F$

chemical quantum yield (ϕ_C)= the fraction of molecules reacting through the chemiluminescent pathway

excitation quantum yield (ϕ_{EX})=the fraction of reacting molecules which generate an excited state

fluorescence quantum yield (ϕ_F)=fraction of excited molecules which actually emit a photon

Radical: An atom or molecule with an one or more unpaired electrons.

Radical annihilation: Reaction of two or more radicals resulting in loss of the unpaired electrons.

Radical ion: A radical with a net charge.

Radioluminescence: Luminescence arising from radioactive decay or atoms or molecules subjected to bombardment by subatomic particles (α=anodoluminescence, β=cathodoluminescence, γ rays, X-rays).

Reflector: A structure, for example in a luminous organ or photophore, which reflects light.

Resonance energy transfer: The transfer of energy from one molecule to another by dipole-dipole resonance. Also known as 'Förster' transfer.

Sensitised chemiluminescence: Chemiluminescence where the actual light emitter is not the primary excited product of the chemiluminescent reaction. The light emitter in sensitised chemiluminescence is excited as a result of energy transfer.

Singlet oxygen ($^1O_2$): Molecular oxygen where the outer pair of electrons are in the singlet spin state (i.e. of opposite spin), designated Δ when they are in the same 2p orbital and Σ when in different 2p orbitals. Ground state oxygen is triplet ($^3O_2$).

Singlet spin state (S): Molecule where the outer electrons are of opposite spin.

Sonoluminescence: Luminescence resulting from exposure of materials to sound waves, usually ultrasonic.

Sub-orbital: Below the eye.

Thermoluminescence: Luminescence, due to release of energy from excitons, when materials are subject to mild heat, i.e. below the temperature for incandescence. The excitons are usually generated by prior irradiation of the solid.

Transduction: transfer of a bacterial gene, e.g. lux gene, from one bacteria to another using a phage.

Transfection: acquisition of new genetic material into eukaryotic cells from added DNA, introduced as cDNA by electroporation, plasmids or viruses.

Transformation: acquisition of new DNA by bacteria.

Transgenic cell: incorporation of new DNA e.g. luciferase genes, in to the host genome of a eukaryotic cell, using for example a retrovirus.

Triboluminescence: Light emitted when a material is stressed to the point of fracture.

Triplet spin state (T): Molecule where the outer electron pair are of the same spin. They must therefore be in separate orbitals.

Ultraweak chemiluminescence (UWCL): Chemiluminescence, invisible to the naked eye, resulting from a very low quantum yield reaction. It may be cellular or acellular, indigenous or indicator-dependent. Many examples of ultraweak chemiluminescence have not been proven to be very low quantum yield reactions, and may turn out to be dim chemiluminescence.

Units of light measurement: Energy: joules (10^7 ergs) or watts (joule $\cdot$ s^{-1})
Quanta: photons or photons $\cdot$ s^{-1}
See Chapter 2 for candelas, lux, phots, lumens, stilbs, lamberts, foot candles, etc.

Vibronically excited state: Molecules in an electronically excited state, including high vibrational and rotational energy levels.

Appendix II

Living light

A. LUMINOUS GENERA

This is a list of genera containing known luminous species. All species within a genus are not necessarily *luminous*. In other genera either no luminous species have been recorded or the evidence is weak. Organisms with luminous bacterial light organs are indicated by (B). Other specific names in brackets refer to former names. *Some classifications leave Fungi in the plant kingdom.

	Family	Genus
Kingdom MONERA		
Phylum: BACTERIA (=SCHIZOMYCOTA)		
Order Eubacteriales	Vibrionaceae (Bacteriaceae)	Alteromonas, Lucibacterium
		Photobacterium, Vibrio(Beneckea),
	Enterobacteriaceae	Xenorhabdus (Achromobacter)
*Kingdom FUNGI		
Phylum: EUMYCOTA (fungi) Class		
Hymenomycetes (Basidiomycetes)		
Order Agaricales	Polyporaceae	Polyporus,
	Agaricaceae	Mycena (Filoboletus and Poromycena), Pleurotus
		(incl. Lampteromyces Omphalotus Clitocybe)
		Armillaria, Omphalia, Locellina, Marasmius,
		Dictyopanus, Panus
Kingdom PLANTAE		
Phylum: CHROMOPHYCOTA (DINOPHYTA)		
Class Dinophyceae (dinoflagellates)		
Order Gymnodiniales	Gymnodiniaceae	Gymnodinium
	Polykrikaceae	Polydrikos
Order Noctilucales	Noctilucaceae	Noctiluca
Order Pyrocystales	Pyrocystaceae	Pyrocystis (=Dissodinium)
Order Peridiniales	Peridiniaceae	Proroperidinium (=Peridinium)
	Gonyaulacaceae	Gonyaulax, Pyrodinium, Alexandrium
	Ceratiaceae	Ceratium
	Heteraulacaceae	Heteraulacus (=Gonoidoma)
	Pyrophacareae	Fragilidium, Pyrophacus
Kingdom ANIMALIA		
Subkingdom: PROTOZOA		
Phylum: Sarcomastigophora		
Class Polycystina (Actinopoda)		
Order Spumellarida (Radiolaria)	Sphaerozoidae	Collozoum, Rhaphidozoum, Sphaerozoum
	Collosphaeridae	Collosphaera, Acrophaera, Siphonosphaera
	Thalassicollidae	Thalassicolla
	Thalassothamnidae	Cytocladus
Class Phaeodaria		
Order Phaeosphaerida	Aulosphaeridae	Aulosphera
Phylum: PORIFERA (sponges)		
Class Calcarea		
Order Sycettida	Grantiidae	Grantia
Phylum: CNIDARIA (hydroids)		
Class Hydrozoa (hydroids and jellyfish)		

	Family	*Genus*
Order Athecata (=Anthomedusae)	Pandeidae	*Stomotoca, Leuckartiara*
	Tubulariidae	*Euphysa*
Order Thecata (=Leptomedusae)	Campanulariidae	*Campanularia, Obelia, Clytia (polyp)=Phialidium (medusa), Tima*
	Lovenellidae	*Lovenella*
	Mitocomidae	*Halistaura*
	Aequoridae	*Aequorea*
	Phialuciidae	*Octophialucium (=Octocanna)*
Order Limnomedusae	Rhopalonematidae	*Colobonema, Crossota*
Order Narcomedusae	Cuninidae	*Solmissus, Cumina*
	Aeginidae	*Aeginura, Aegina*
Order Siphonophora	Hippopodiidae	*Vogtia, Hippopodius*
	Apolemidae	*Apolemia*
	Diphyidae	*Diphyes, Chelophyes, Sulculeolaria*
	Abylidae	*Abyla, Abylopsis, Bassia, Ceratocymba*
	Prayidae	*Praya, Rosacea, Amphicaryon, Maresearsia, Nectopyramis*
	Agalmidae	*Agalma, Halistemma, Nanomia*
	Forskaliida	*Forskalia*
	Rhizophsida	*Periphylla*
	Periphyllidae	*Rhizophysa*
Class Scyphozoa (jellyfish)	Atollidae	*Atolla*
Order Semaeostomeae	Pelagiidae	
Order Semaeostomeae	Pelagia	
	Ulmaridae	*Poralia*
Class Anthozoa (sea anemones, sea fans, sea cactus)		
Sub-class Octocorallia (=Alcyonaria)		
Order Gorgonacea (sea fans)	Isididae	*Isidella (=Acanella), Keratoisis, Lepidisis*
	Primnoidae	*Primnoisis, Thourella*
	Chrysogorgiidae	*Iridogorgia*
	Acanthogorgiidae	*Acanthogorgia*
Order Alcyonacea	Alcyoniidae	*Anthomastus*
Order Pennatulaceae	Veretillidae	*Veretillum, Cavernularia*
	Renillidae	*Renilla*
	Funiculinidae	*Funiculina*
	Umbellulidae	*Umbellula*
	Virgulariidae	*Virgularia, Acanthoptilum, Stylatula*
	Pteroeididae	*Pteroeides*
	Pennatulidae	*Pennatula, Ptilosarcus*
	Protoptilidae	*Distichoptilum*
	Stachyptilidae	*Stachyptilum*
Sub-class Zoanthinaria		
Order Antipatharia	unassigned	*unidentified*
Order Zoanthiniaria	Parazoanthidae	*Parazoanthus*
	Eipzoanthidae	*Epizoanthus*
	Savaliidae	*Gerardia*
Phylum: CTENOPHORA (comb jellies)		
Class Tentaculata		
Order Cydippida	Haeikeliidae	*Haeckelia*
	Lampeidae	*Lampea*
	Mertensiidae	*Callianira, Mertensia*
Order Cydippida	Pleurobrachiidae	*Pleurobrachia, Hormiphora*
Order Lobata	Bolinopsidae	*Bolinopsis, Deiopea, Mnemiopsis*
	Leucotheidae	*Leucothera (=Eucharis)*
	Ocytopsidae	*Ocyropsis*
	Eurhampheidae	*Eurhamphea*
Order Cestida	Cestidae	*Cestus, Velamen*
Class Nuda		
Order Beroida	Beroidae	*Beroë*
Phylum: NEMERTEA (RHYNCHOCOELA=NEMERTINI) (Ribbon worms)		
Class Enopla		
Order Hoplonemertea	Emplectonematidae	*Emplectonema*
Phylum: NEMATA (NEMATODA) (Roundworms)		*Heterorhabditis (B)*
Phylum: MOLLUSCA		
Class Gastropoda (snails)		
Order Mesogastropoda	Planaxidae	*Planaxis*
Order Nudibranchia	Polyceratidae	*Plocamopherus, Kaloplocamus*
	Phyllirrhoidae	*Phyllirrhoë*
Sub-class Pulmonata		
Order Basommatophora	Latiidae	*Latia*
	Planaxidae	*Planaxis*
Order Stylommatophora	Helicarionidae	*Quantula (=Dyakia* and *Hemiplecta)*
Class Bivalvia (bivalves)		
Sub-class Lamellibranchia		
Order Myoida (Adepodonta)	Gastrochaenidae	*Gastrochaema (Rocellaria)*
	Pholadidae	*Pholas*

	Family	Genus
Class Cephalopoda (octopus, cuttlefish and squid)		
Sub-class Coleoidea		
Order Octopoda	Bolitaenidae	*Bolitaena, Eledonella, Japetella, Callistoctopus*
Order Sepioidea	Spirulidae	*Spirula*
	Sepiolidae	*Semirossia (B), Sepiolina, Sepiola (B), Euprymma (B), Inioteuthis, Rondeletiola (B), Nectoteuthis, Iridoteuthis, Stoloteuthis, Heterotenthis (B)*
Order Teuthoidea		
Sub-order Myopsida	Loliginidae (B)	*Logilo, Uroteuthis, Doryteuthis, Uroteuthis*
Sub-order Oegopsida	Ommastrephidae	*Ommastrephes, Symplectoteuthis, Ornithoteuthis, Hyaloteuthis, Dosiducus,*
	Lycoteuthidae	*Lycoteuthis, Selenoteuthis, Oregoniateuthis, Nematolampas, Lampadioteuthis,*
	Enoploteuthidae	*Enoploteuthis, Abraliopsis, Abralia, Watasenia, Ancistrocheirus, Thelidioteuthis, Pyroteuthis Pterygioteuthis*
	Octopoteuthidae	*Octopoteuthis Taningia*
	Onychoteuthidae	*Onychoteuthis Chaunoteuthis*
	Cycloteuthidae	*Cycloteuthis Discoteuthis*
	Gonatidae	*Gonatus*
	Histioteuthidae	*Histioteuthis (=Calliteuthis)*
	Bathyteuthidae	*Bathyteuthis*
	Ctenopterygidae	*Ctenopteryx*
	Brachioteuthidae	*Brachioteuthis*
	Batoteuthidae	*Batoteuthis*
	Chiroteuthidae	*Chiroteuthis Chiropsis*
	Mastigoteuthidae	*Mastigoteuthis Echinoteuthis*
	Grimalditeuthidae	*Grimalditeuthis*
	Cranchiidae	*Cranchia, Leachia, Drechselia, Liocranchia, Liguriella, Pyrgopsis, Taonius, Egea, Phasmatopsis, Belonella, Megalocranchia (=Vossoteuthis and Verriliteuthis), Helicocranchia, Ascocranchia, Galiteuthis, Corynomma, Bathothauma, Mesonychoteuthis, Sandalops, (=Anomalocranchia), Teuthowenia, Vampyroteuthis*
Order Vampyromorpha	Vampyroteuthidae	
Phylum ANNELIDA (worms)		
Class Polychaeta (bristle worms)		
Order Chaetopterida	Chaetopteridae	*Chaetopterus Mesochaetopterus*
Order Cirratulida	Cirratulidae	*Tharyx (=Heteroccirrus), Dodecaceria*
Order Phyllodocida	Polynoidae	*Polynoë, Lagisca, Gattyana, Harmothoë, Polyennoa, Lepidonotus, Malmgrenia, Acholoë, Eunoë, Lepidasthenia, Hesperonoë*
	Syllidae	*Odontosyllis, Pionosyllis, Syllis*
	Nephtyidae	*Aglaophamus*
	Tomopteridae	*Tomopteris*
Order Eunicida	Onuphidae	*Onuphis*
Order Flabelligerida	Flabelligeridae	*Flabelligera, Pherusa (=Stylarioides)*
	Fauveliopsidae	*Flota*
Order Terebellida	Terebellidae	*Polycirrus, Thelpus*
Class Oligochaeta (earthworms)		
Order Haplotaxida (Opisthopora)	Lumbricidae	*Eisenia*
	Megascolecidae	*Microscolex (=Eodrilus), Parachilota, Lampito, Pontodrilus Fletcherodrilus, Eutyphoeus, Ramiella, Octochaetus, Digaster, Diplocardia, Diplotrema, Spenceriella, Megascolex*
Phylum: ARTHROPODA		
Sub phylum CHELICERATA		
Class Pycnogonida (sea spiders)		*Colossendeis*
Sub phylum CRUSTACEA		
Class Ostracoda		
Order Myodocopida	Cypridinidae	*Cypridina, Pyrocypris, Vargula*
Order Halocyprida	Halocypridae	*Conchoecia*
Class Maxillopoda		
Sub-class Copepoda		
Order Cyclopoida	Oncaeidae	*Oncaea*
Order Calanoida	Metridinidae	*Metridia, Pleuromamma, Gaussia*
	Lucicutiidae	*Lucicutia*
	Heterorhabdidae	*Heterorhabdus, Hemihabdus, Disseta, Heterostylites*
	Augaptilidae	*Euaugaptilus, Centraugaptilus, Augaptilus, Haloptilus, Heteroptilus, Pachyptilus*
	Megacalanidae	*Megacalanus*
Order Harpacticoida	Aegisthidae	*Aegisthus*
Class Malacostraca		
Order Mysidacea (shrimp)	Lophogastridae	*Gnathophausia*
Order Amphipoda	Scinidae	*Scina, Acanthoscina*
	Pronoidae	*Parapronoë*

	Family	*Genus*
	Lysianassidae	*Cyphocaris, Danaella Thoriella, Chevreuxiella*
	Lanceolidae	*Megalanceola*
Order Euphausiacea (krill)	Euphausiidae	*Euphausia, Nyctiphanes Meganyctiphanes, Stylocheiron, Nematoscelis, Thysanopoda, Thysanoessa, Nematobrachion, Tessarabrachion, Pseudeuphausia*
Order Decapoda (shrimp)		
Sub-order Natania		
Section Caridea	Oplophoridae	*Oplophorus, Acanthephyra, Systellaspis, Notostomus, Meningodora, Hymenodora, Ephyrina*
	Pandalidae	*Stylopandalus (Parapandalus), Heterocarpus,*
	Pasiphaeidae	*Glyphus*
	Thalassocarididae	*Thalassocaris, Chlorotocoides*
Section Penaedia	Solenoceridae	*Hymenopenaeus, Hadropenaeus, Mesopenaeus, Solenocera*
	Sergestidae	*Sergestes, Sergia*
Class Diplopoda (millipedes)		
Sub-class Chilognatha		
Order Polydesmida	Xystodesmidae	*Motyxia (=Luminodesmus)*
Order Spirobolida	Spirobolidae	*Spirobolellus, Dinematocrius*
Class Chilopoda (centipedes)		
Order Geophila	Himantariidae	*Stigmatogaster*
	Oryidae	*Orya Orphaneus*
	Geophilidae	*Geophilus Scolioplanes*
Class Insecta		
Sub-class Apterygota		
Order Collembola (spring-tails)	Onychiuridae	*Onychiurus (=Lipura, Aphorura, Anurophorus)*
Order Coleoptera (beetles)		
Super-family Elateroidea		
	Elateridae	
	Sub-family Pyrophorinae	
	Tribe Pyrophorini	*Nyctophyxis, Noxlumenes Cryptolampros, Sooporanga, Pyroptesis Hapsodrilus, Ptesimopsia, Phanophorus Hypsiophthalmus, Pyrearinus, Fulgeochlizus, Opselater Ignelater, Deilelater, Vesperelater, Pyrophorus, Lygelater*
	Unknown tribe	*Photophorus*
	Sub-family Campyloxeninae	*Campyloxenes*
	Sub-family Melanactinae	*Melanactes*
	Super-family Cantharoidea	
	Homalisidae	*Homalisus*
	Telegeusidae	*Telegeusis?*
	Phengodidae	*Phengodes, Zarhipis, Cenophengus, Mastinocerus, Phrixothrix, Diplocladon, Rhagophthalmus, Falsophrixothrix, Cydistus, Dioptoma, Pseudophengodes, Stenophrixothrix*
	Lampyridae	
	Tribe Lampyrini	*Alectron, Diaphenes, Lampronetes, Lampyris, Lychnuris, Microlampyris, Micophyus, Nyctophila, Ovalampis, Paraphausis, Petalacmis*
	Tribe Pleotomini	*Calyptocephalus, Ophoelis, Phaenolis, Pleotomus, Roleta*
	Tribe Lamprocerini	*Lamprocera, Lucernuta, Lucio, Lychriacris, Tenaspis*
	Tribe Cretomorphini	*Aspisoma, Cassidomorphus, Cratomorphus, Micronaspis, Pyractomena*
	Tribe Photinini	*Callopisma, Calotrachelum, Dadophora, Dilychnia, Ellychnia, Erythrolychnia, Jamphotus, Lamprigera, Lamprohiza, Lucidina, Lucidota, Lucidotopsis, Macrolampis, Microdiphot, Mimophotinus, Olivierus, Phausis, Phosphaenus, Phosphaenopterus, Photinoides, Photinus, Platylampis, Pristolycus, Pseudolychnus, Pyractomena, Pyropyga, Robopus*
	Sub-family Amydetinae	*Amydetes, Cladodes, Dodacles, Drypletytra, Ethra, Ledocas, Magnoculus, Photoctus, Pollaclasis, Psilocladus, Scissicanda, Vesta*
	Sub-family Photurinsae	*Bicellorycha, Photuris, Presbyolampis, Pyrogaster*
	Sub-family Luciolinae	
	Tribe Luciolini	*Bourgeoisia, Colophotia, Lampyroidea, Luciola, Pteroptyx, Pyrophanes*
	Tribe Curtosini	*Curtos*
	Tribe Ototretini	*Brachylampis, Ototreta*
	Sub-family Matheteinae	*Ginglymocladus, Matheteus*
	Sub-fmily Pterotinae	*Harmatelia, Pterotus*
	Sub-family Rhagophthalminae	*Dioptoma, Ochotyra, Mimotyra, Rhagophthalmus*
Order Diptera (flies)	Mycetophilidae	*Keroplatus, Orfelia (=Platyrua), Arachnocampa (=Bolitophila)*

	Family	Genus
Phylum: BRYOZOA (ECTOPROCTA) (sea mats)		
Class Gymnolaemata		
Order Cheilostomata	Membraniporidae	*Membranipora (Acanthodesia)*
Phylum: HEMICHORDATA		
Class Enteropneusta (acorn worms)	Ptychoderidae	*Balanoglossus, Ptychodera, Glossobalanus*
Phylum: ECHINODERMATA		
Class Crinoidea (sea lilies)		
Order Isocrinida	Isocrinidae	*Annacrinus*
Order Comatulida	Pentametrocrinidae	*Thaumatocrinus*
	Thalassometridae	*Thallassometra*
Class Holothuroidea (sea cucumbers)		
Order Aspidochirotida	Synallactidae	*Scotothuria (Galatheathuria), Paelopatides, Paroriza, Pseudostichopus, Mesothuria*
Order Elasipodida	Laetmogonidae	*Laetmogone, Benthogone, Pannychia*
	Elpidiidae	*Kolga Peniagone, Ellipinion, Scotoanassa, Scotoplanes*
	Psychropotidae	*Psychropotes, Benthodytes*
	Pelagothuriidae	*Enypniastes (Pelagothuria* and *Planktothuria)*
Class Asteroidea (sea stars, starfish)		
Order Paxillosida	Astropectinidae	*Plutonaster, Dytaster*
	Benthopectinidae	*Benthopecten, Pectinaster, Pontaster*
Order Spinulosida	Pterasteridae	*Hymenaster*
Order Forcipulatida	Asteriidae	*Cryptasterias, Hydrasterias*
	Zoroasteridae	*Zoroaster*
	Brisingidae	*Brisinga, Brisingella Freyella*
Class Ophiuroidea (brittle stars)		
Order Ophiurida	Ophiomyxidae	*Ophioscolex*
	Ophiacanthidae	*Ophiacantha, Ophiomitra, Ophiomitrella Ophioplinthaca*
	Amphiuridae	*Amphiura, Acrocnida, Amphipholis*
	Ophiotrichidae	*Ophiothrix*
	Ophiocomidae	*Ophiopsila*
	Ophiodermatidae	*Ophiarachnella*
	Ophionereididae	*Ophionereis*
	Ophiuridae	*Ophiura, Homalophiura, Ophiomusium*
	Ophiactidae	*Ophiopholis*
CHORDATA		
Sub-phylum UROCHORDATA (TUNICATA)		
Class Appendicularia (Larvacea)		
Order Copelata	Oikopleuridae	*Oikopleura, Stegosoma*
Class Thaliacea		
Order Pyrosomatida	Pyrosomatidae (?B)	*Pyrosoma, Pyrosomella, Pyrostremma*
Order Salpida	Salpidae	*Cyclosalpa*
Sub-phylum CRANIATA (=VERTEBRATA)		
Pisces (fish)		
Class Chondrichthyes (=Elasmobranchii) (cartilaginous fish)		
Order Squaliformes (Selachii)	Squalidae	*Isistius, Euprotomicrus, Squaliolus, Etmopterus (Spinax), Centroscyllium*
Class Osteichthyes (bony fish)		
Infra-class Teleostei		
Super-order Elopomorpha		
Order Anguilliformes (Elopiformes)		none
	Saccopharyngidae	*Saccopharynx*
	Congridae (B)	*Lumiconger*
Super-order Clupeomorpha		
Order Clupeiformes	Engraulidae	*Coilia*
Super-order Protacanthopterygii		
Salmoniformes	Opisthoproctidae (B)	*Opisthoproctus, Winteria, Rhynchohyalus*
	Alepocephalidae (smooth heads)	*Xenodermichthys, Photostylus, Rouleina, Microphotolepis,*
	Platytroctidae (Searsiidae)	*Searsia, Searsioides, Platytroctes, Holtbyrnia, Paraholtbyrnia Platytroctegen, Normichthys, Persparsia Barbantus, Pellisolus, Sagamichthys, Maulisia, Mentodus, Microrictus*
	Gonostomatidae	*Bonapartia, Cyclothone, Gonostoma, Triplophos, Margrethia*
Order Stomiiformes	Sternoptychidae (hatchet fish)	*Argyropelecus, Polyipnus, Sternoptyx, Thorophos, Danaphos, Diplophos, Araiophus, Maurolicus, Sonoda, Valenciennellus, Argyripnus,*
	Photichthyidae	*Yarrella, Phoththys, Polymetme, Pollichthys, Woodsia, Ichthyococcusic, Vinciguerria*
	Astronethidae (snaggle tooths)	*Astronesthes, Cryptostomias, Borostomias, Heterophotus, Rhadinesthes, Neonesthes*
	Melanostomiatidae (scale less dragon fish)	*Chirostomias, Eustomias, Flagellostomias, Opostomias, Melanostomias, Odontostomias*

	Family	Genus
		Leptostomias, Grammatostomias, Photonectoides, Tactostoma, Echiostoma, Thysanactis, Trigonolampa, (Bathysphaera) Bathophilus, Photonectes, Pachystomias
	Stomiatoidea (Malacosteidae) (loose jaws)	*Malacosteus, Aristostomias, Photostomias, Ultimostomias,*
	Chauliodontidae (viper fish)	*Chauliodus*
	Stomiidae (scaley dragon fish)	*Stomias, Macrostomias*
	Idiacanthidae (black dragon fish)	*Idiacanthus*
Order Aulopiformes	Chlorophthalmidae (B)	*Chlorophthalmus*
	Paralepididae	*Lestidium, Lestrolepis*
	Evermannellidae	*Coccorella*
	Scopelarchidae	*Scopelarchoides, Benthalbella*
Order Myctophiformes	Neoscopelidae	*Neoscopelus*
	Myctophidae (lantern fish)	*Protomyctophum (incl. Hierops), Electrona, Metelectrona, Hygophum Benthosema, Diogenichthys, Myctophum, Symbolophorus, Loweina Tarletonbeania, Gonichthys, Centrobranchus, Lobianchia, Diaphus, Notolychnus, Taaningichthys, Lampadena, Lampanyctodes, Stenobrachius, Triphoturus, Lampanyctus (incl. Parvilux), Lepidophanes, Ceratoscopelus, Gymnoscopelus (incl. Nasolychnus) Lampichthys, Notoscopelus (incl. Pareiophos), Hintonia, Scopelopsis, Dorsadena, Bolinichthys, Gymnoscopelus Idiolychnus, Krefftichthys.*
Super-order Paracanthopterygii		
Batrachioidoformes	Batrachoididae	*Porichthys*
Lophiiformes (angler fish)	Ogcocephalidae (batfish)	*Dibranchus*
	Melanocetidae (B)	*Melanocetus*
	Oneirodidae (B)	*Oneirodes, Chaenophryne, Puck, Danaphryne, Pentherichthys, Lophodolos, Tyrannophryne, Phyllorhinichthys, Microlopichthys, Ctenochirichthys, Chirophryne, Leptacanthichthys, Bertella, Dolopichthys,*
	Diceratiidae (B)	*Diceratias, Paroneirodes*
	Centrophrynidae (B)	*Centrophryne, Spiniphryne*
	Gigantactinidae (B)	*Gigantactis, Rhynchactis*
	Himantolophidae (B)	*Himanotolophus*
	Ceratiidae (B)	*Ceratias, Cryptosaras*
	Thaumatichthyidae (B)	*Thaumatichthys, Lasiognathus*
	Linophrynidae (B)	*Linophryne, Photocorynus, Haplophyryne (Edriolychnus), Acentrophryne, Borophryne*
Order Gadiformes	Moridae (B)	*Physiculus, Brosmiculus, Gadella, Tripterophycis*
	Steindacheriidae (Merlucciidae) (B)	*Steindachneria*
	Macrouridae (B) (rat tails, grenadiers)	*Coelorhynchus Nezumia, Malacocephalus Hymenocephalus, Odontomacrurus, Ventrifossa Trachonurus, Cetonurus Sphagemacrurus, Mesobius Lepidorynchus (incl. Pramacrurus), Haplomacrourus, Idiolophorhynchus*
Super-order Acanthopterygii		
Order Beryciformes	Monocentridae (B)	*Monocentris, Cleidopus*
	Anomalopidae (B) (flashlight fish)	*Anomalops, Photoblepharon, Kryptophanaron*
	Trachichthyidae	*Paratrachichthys*
Perciformes	Apogonidae	*Apogon, Archamia, Siphamia (B), Epigonus, Howella, Rhabdamia, Rosenblattia, Florenciella,*
	Acropomatidae (B)	*Acropoma*
	Leiognathidae (B) (pony fish)	*Leiognathus, Gazza, Secutor*
	Pempheridae	*Pempheris, Peraprianthus*
	Chaismodontidae	*Pseudoscopelus*
	Sciaenidae	*Collichthys*

Adapted from Herring, P.J. (ed.) (1978) *Bioluminescence in action*. Academic Press, London and New York and Herring, P. J. (1987) *J. Biochem. Chemilum.* **1** 147–163. Systematic distribution of bioluminescence in living organisms. See also Harvey (1952) *Bioluminescence*. Academic Press, New York.

B. MAJOR NON-LUMINOUS GROUPS

This is a list of phyla and classes where either no luminous species have been recorded or evidence is weak.

1. Phyla with no known luminous genera

Kingdom VIRUS
Kingdom MONERA
- ARCHAEBACTERIA
- CYANOPHYCOTA (CYANOPHYTA) (blue-green algae)
- PROCHLOROPHYCOTA

Kingdom PLANTAE
- MYXOMYCOTA (slime moulds)
- RHODOPHYCOTA (red algae)
- CHLOROPHYCOTA (green algae)
- EUGLENOPHYCOTA, PHODOPHYTA, CRYPTOPHYTA, PRYMNESIOPHYTA, CHRYSOPHYTA, EUSTIGMATOPHYTA, RHAPHIDOPHYTA — algae
- PHAEOPHYTA
- BRYOPHYTA, PSILOPHYTA, LYCOPODIOPHYTA (LYCOPHYTA), EQUISETOPHYTA, FILICOPHYTA, PINOPHYTA, MAGNOLIOPHYTA, SPHENOPHYTA, PTEROPHYTA, PROGYMNOSPERMOPHYTA, PTERIDOSPERMOPHYTA, CYCADOPHYTA, GINKGOPHYTA, CONIFEROPHYTA, GNETOPHYTA, ANTHOPHYTA — 'higher' plants

Kingdom ANIMALIA
- LABYRINTHULATA, APICOMPLEXA, MICROSPORA, MYXOZOA, ASCETOSPORA, CILIOPHORA — protozoa

PLACOZOA
PLATYHELMINTHA (flat worms)
GNATHOSTOMULIDA
MESOZOA
ROTIFERA
GASTROTRICHA
KINORHYNCHA
NEMATOMORPHA
ACANTHOCEPHALA
ENDOPROCTA
PRIAPULIDA
LORICIFERA
SIPUNCULA
ECHIURA
TARDIGRADA
PENTASTOMIDA
ONYCHOPHORA
PHORONIDA
BRACHIOPODA (lamp shells)
CHAETOGNATHA (arrow worms)
POGONOPHORA (beard worms)
ENTOPROCTA

2. Some classes with no known luminous species but where other classes in same phylum contain luminous species

Kingdom FUNGI
Phylum: EUMYCOTA
      Pyrenomycetes (Ascomycetes)? Xylaria
Kingdom ANIMALIA
Phylum: SARCOMASTIGOPHOZA
      Rhizopoda ⎤
      Sporozoa   ⎥ protozoa
      Cnidosporidia ⎥
      Ciliata   ⎦
Phylum: PORIFERA (sponges)
      Hexactinellida
Phylum: NEMERTEA (RHYNCHOCOELA=NEMERTINI) (ribbon worms)
      Amopla
Phylum: MOLLUSCA
      Monoplacophora
      Amphineura
      Scaphopoda
Phylum: ANNELIDA (worms)
      Myzostomaria
Phylum: ARTHROPODA
      Merostemata

Arachnida (spiders)
Cephalocarida
Branchiopoda
Remipedia
Symphyla
Pauropoda
Phylum: BRYOZOA (ECTOPROCTA) (Sea mats)
Phylactolaemata
Phylum: HEMICHORDATA
Pterobranchia
Planktosphaeroidea
Phylum: ECHINODERMATA
Echinoidea (sea urchins)
Phylum: CHORDATA
Amphibia
Reptilia
Aves
Mammalia

Note: This list is based on Herring, P. J. (1978). His more recent list does not refer to all these classes.

C. SOME COMMON NAMES

By their very nature common names have an inherent imprecision. In the following list I have included both names specifically used to describe a particular luminous species as well as group names which, although they do not specifically refer to luminous species, are often used to describe a group which does contain such organisms. I have indicated the phylum, class, order, or family in brackets. In the column headed 'systematic name' I have given either a specific name, or the genus only, in which case some or all of the species are luminous and are referred to under the common name. *'et al..'* indicates that other species and/or genera are also included.

*indicates that the common name usually refers to a species or group which are all luminous. See also Appendix IIA.

Common name	*Systematic name*
Phylum: *Eumycota* (fungi) Class:	
Basidiomycetes	
mushroom	*Panus stypticus* et al.
*moonnight mushroom or Tsukiyo-taka (Japan)	*Pleurotus japonicus*
mould, e.g. on a leaf	*Omphalia flavida* et al.
honey, honey-tuft or boot-lace fungus	*Armillaria mellea*
*foxfire (on decaying wood)	*Armillaria mellea*
*Jack-my-lantern (USA)	*Clitocybe illudens*
candle-snuff fungus or stag's horn	*Xylaria hypoxylon*

Common name	Systematic name
Phylum: Cnidaria	
sea fir or hydroid	*Obelia geniculata* et al.
jelly fish or medusa	*Pelagia noctiluca* et al.
sea fan (order Gorgonacea)	*Isidella* et al.
sea pen (order Pennatulacea)	*Pennatula* spp et al.
sea pansy (family Renillidae)	*Renilla reniformis* et al.
sea cactus (family Veretillidae)	*Cavernularia* spp.
sea anenome (order Zoanthiniaria)	*Parazoanthus* spp et al.
Phylum: Ctenophora (comb-jellies)	
sea comb or sea walnut	*Mnemiopsis leidyi* et al.
sea jelly	*Beroë albens* et al.
sea gooseberry	*Pleurobrachia*
Phylum: Mollusca	
sea slug (order Nudibranchia)	*Plocamopherus* spp
land snail	*Quantula hemiplecta*
freshwater limpet	*Latia neritoides*
*common piddock (Class: Bivalvia)	*Pholas dactylus*
octopus (Order Octopoda)	*Japetella* spp et al.
cuttle fish (Order Sepioidea)	*Sepiolina* spp et al.
squid (Order Teuthoidea)	*Histeoteuthis* spp et al.
*firefly squid (hotaru-ika Japan)	*Watasenia scintillans*
lantern squid (Chochin-ika in Japan)	*Inioteuthis inioteuthis*
Phylum: Annelida	
polychaete worm	*Chaetopterus variopedatus* et al.
scale worm	*Acholoë astericala* et al.
*fireworm	*Odontosyllis enopla*
Bermudean or Palolo worm	*Odontosyllis enopla*
earthworm (Class Oligochaeta)	*Diplocardia longa* et al.
Phylum: Arthropoda	
sea spider (Class Pycnogonida)	*Collosendeis* spp
*sea firefly (Class Ostracoda)	*Vargula* (formerly *Cypridina*) *hilgendorfi*
opossum shrimp (order Mysidacea)	*Gnathophausia ingens*
euphausiid shrimp (North Atlantic krill)	*Meganyctiphanes norvegica* et al.
krill (Antarctic krill)	*Euphausia superba* et al.
decapod shrimp or prawn	*Acanthophyra* spp et al.
millipede (Class Diplopoda)	*Motyxia (Luminodesmus) sequoia*
centipede (Class Chilopoda)	*Scoliophanes* spp et al.
springtail (Class Insecta order Collembola)	*Onychiurus spp et al.*
lantern fly (order Fulgonidae)	*Fulgora laternaria* (?self-luminous)
spark fly, star of Eve, Eye of Confucius, candlefly (order Fulgoridae)	*Pyrops candel aria* (?bacteria)
beetle (order Coleoptera)	*Photinus pyralis* et al.

Common name	*Systematic name*
click beetle (family Elateridae)	elaterid beetle
*fire beetle (family Elateridae)	*Pyrophorus* spp et al.
*cucujus (West Indies and South America)	*Pyrophorus* spp et al.
*fireworm (family Elateridae)	larva of *Melanactes*
lightning bug (family Elateridae)	*Photophorus* spp
glow-worm (family Lampyridae)	*Lampyris noctiluca* et al.
firefly (family Lampyridae)	*Photinus pyralis* et al.
snow-worm	larva of Lampyrid on snow
starworm or diamond worm (family Phengodidae)	*Diplocladon hasseltii* (Malaysia)
railroad worm (family Phenogodidae)	*Phrixothrix tiemani* larva (South America)
fungus gnat (order Diptera)	*Keroplatus sesioides* et al.
*glow-worm fly (family Mycetophilidae)	*Keroplatus sesioides* et al.
New Zealand glow-worm (family Mycetophilidae)	*Arachnocampa luminosa*

Phylum: Hemichordata

acorn worm (Class Enteropneusta)	*Balanoglossus* spp

Phylum: Echinodermata

sea lily (Class Crinoidea)	*Thaumatocrinus* spp et al.
sea cucumber (Class Holothuroidea)	*Laetmogone* spp et al.
starfish (Class Asteroidea)	*Plutonaster* spp et al.
brittle or snake stars (Class Ophiuroidea)	*Ophiopsila californica* et al.

Phylum: Chordata

*fire cylinder or fire-body (Class Thaliacea)	*Pyrosoma* spp
salp	*Cyclosalpa* spp
shark (family Squalidae)	*Squaliolus luticandus* et al.
anchovy (family Engraulidae)	*Coilia dussemieri*
*hatchet fish (family Sternoptychidae)	*Argyropelecus* spp et al.
*lantern fish (family Myctophidae)	*Myctophum spinosum* et al.
toad fish (family Batrachoididae)	*Porichthys notatus* et al.
Atlantic toad fish	*Opsanus tau*
*midshipman fish	*Porichthys notatus* and *margaritatus*
Californian singing fish	*Porichthys notatus*
*angler fish (order Lophiiformes)	*Melanocetus pelagicus*
batfish (family Ogcocephalidae)	*Dibranchus atlanticus* et al.
morid cod (family Moridae)	*Physiculus japonicus* et al.
merluccid hake (family Merlucciidae)	*Steindachneria* spp

Common name	*Systematic name*
rat tails or grenadiers (family: Macrouridae)	*Malococephalus laevis* et al.
knight, pineapple or pinecone fish (family: Monocentridae)	*Monocentris japonicus* et al (popular food in Japan)
port and starboard light fish (New South Wales)	*Monocentres (Cleidopns) gloria-maris*
*flashlight fish (family Anomalopidae)	*Photoblepharon palpebratus* et al.
slimehead (family Trachichtyidae)	*Paratrachichthys prosthemius*
cardinal fish (family Apogonidae)	*Apogon ellioti* et al.
pony fish or slipmouths (family Leignothidae)	*Leiognathus equulus* et al. (food for natives in Java)
drums or croakers (family: Sciaenidae)	*Collichthys lucidus* et al.

Appendix III

How to seek out living light

Most people have seen by night, at some time or other, firefly or glow-worm displays, or dinoflagellate luminescence in the sea. For those more interested in learning about the natural history, or in the scientific investigation of living light, five questions need to be considered:

(1) Where to look?
(2) When to look?
(3) How to look and find?
(4) How to collect?
(5) How to bring back specimens alive, or biochemically intact, and preserve them for long periods?

Where to look?
The answer to this lies in two subsidiary questions.

(1) Which geographical location?
(2) Where at a particular site to look?

As we saw in Chapter 3 (section 3.2.2, Tables 3.1–3.3) examples of living light can be found all over the world. There are several famous locations such as the Waitomo caves in New Zealand with the spectacular fungus gnat larvae *Arachnocampa luminosa*, with their luminous 'fishing' lines. Several 'phosphorescent' bays light up regularly each year, and several regions in the tropics are known for firefly displays, such as synchronous flashing in trees found in New Guinea (Table 3.2). Yet look on any beach in the world and you are likely to find some luminous species. Bacteria glowing blue on a piece of decaying flesh or a dying invertebrate, dinoflagellates on the surface of rock pools and sea itself, hydroids on brown seaweeds or under rocks, various polychaete worms in the mud or on the bottom of pools. Dive at night in the coral reefs of the Arabian Sea or Indian Ocean and you may be lucky enough to see flashlight fish. Virtually any fine-net sample from the deep oceans will contain luminous species, but the surface and midwater also harbour many luminous organisms. Some species are of worldwide distribution, for example bacteria and the hydroid *Obelia*. Others are only found in particular regions. For example the rock boring bivalve *Pholas*, the common piddock, is essentially a European species, at its most northerly point on the southern coasts of Britain. Tapping the rocks at night produces a spectacular starlight display as each burrow emits a luminous secretion.

Living light on land is more unevenly distributed and less easy to find than in the sea. Luminous bacteria in the soil are more difficult to find than their marine relatives. Fireflies and glow-worms have a worldwide distribution apart from the Arctic or Antarctic, and they are localised. The numbers also vary greatly. A firefly meadow may contain some 10 000 specimens, yet a group of glow-worms localised to one or two quiet valleys in Snowdonia Wales, may number just a dozen or so. Luminous fungi are also found all other the world, but their often dim emission makes them less obvious than luminous animals, and individual strains are not always luminous. Honey-tuft fungus (*Armillaria*) is one of the commonest causes of tree rot in Britain.

If you are looking for a particular species then the books of Harvey and Herring (see Chapter 3, section 3.8) give many examples. Books on local fauna and from local natural history societies can be helpful. Many marine laboratories and other biological stations publish lists of fauna and flora which can be checked against the luminous genera listed in Appendix II section A. For example, the *Plymouth marine fauna* published by the Marine Biological Association contains many luminous dinoflagellates, coelenterates, polychaete worms, and starfish, together with sites.

Where to look in a location depends both on local knowledge and a knowledge of the biology of the organisms. *Pholas* lives in rock, but can't bore into granite or other igneous rocks. A geological map may pinpoint soft sedimentary rocky areas. Firefly and glow-worm larvae eat snails. Earthworms obviously live most of the time in the soil. Many other luminous organisms can remain hidden in the mud or under rocks, unless you search them out.

When to look?
Four factors need to be taken into account before deciding whether to go out and look:

(1) The season.
(2) By day or by night. If the latter, how bright is the moon?
(3) The tide.
(4) The weather.

Luminous organisms can be found at any time of year, by day or night and in any weather. However, terrestrial and marine surface species are most prevalent during the summer, i.e. June–September in the Northern hemisphere, particularly around the mating season. If you know what to look for, then many species can be found readily during the day. However, the luminescence is only really visible by night, ideally with little or no moon. Furthermore, many luminous species move upwards in the sea at night, or appear from day-time hiding, and only switch on their lanterns at night. The latter is controlled not only by acute regulators such as internal chemical or external mechanical stimuli, but also by a circadian rhythm and dark adaptation which only allows them to glow and flash by night.

The weather seems to affect more the human observer rather than the luminosity of the luminous organism, as I have found glow-worms glowing quite happily in a clump of bracken in the pouring rain. However, it is easier to observe in fine weather, and warmth does seem to bring certain species out. Furthermore, using nets, surface, or deep sea trawls, is very difficult in rough weather.

Some examples. The worm *Odontosyllis* produces spectacular displays during the summer mating season. The hydroid *Obelia* can be found all the year round, but only during June–September is the brown seaweed *Laminaria* or *Sacchoriza* thick, and carpetted with colonies. *Obelia* can be found in rock pools but grows best on seaweed never exposed at low water. Furthermore, only during the summer months are British coastal water samples full of *Obelia* medusae. The European glow-worm, *Lampyris noctiluca* appears to live through the winter as the larva, hidden underground. In July to August the new adult females cling to a blade of grass, thereby shining their lantern upwards to attract the male. By August–September new larvae can be found, also glowing.

Finally, in this section a word about tides. Littoral animals and rock borers are best found at low water. A tide table (British Admiralty Tide Tables vols I–III, = world tides) will tell you when the best spring tides are. The weekly cycle means that collecting, e.g. for *Pholas*, is difficult when there is a neap tide. The tidal falls vary enormously. The Severn estuary, which separates England from Wales, and harbours the rock borer *Pholas*, has one of the largest tidal falls in the world. But you may need to scuba dive.

How to look?
You will need some equipment.

(1) A torch to see (but don't loose your dark adaptation which can take up to 15 minutes to optimise), and a lantern to attract insects.
(2) Collecting gear includes: shovels to dig up mud; bolster hammer and chisel to break open rock; nets; insect nets on land; neuston (top 10 cm) net; plankton tow net; midwater trawls with closing cod ends to seal off a tube of water at a defined depth; and a benthic sledge (Fig. 3.2). The latter two require lifting gear only usually available on research ships. But you can make plankton nets from a bit of fine nylon mesh (see Hardy, A. (1956) *The open sea: its natural history* pp 42–43, published by Collins in The Fontana new Naturalist Series) and towed behind a small dinghy. See Baker *et al.* (1971) *J. Mar. Biol. Assoc.* **53** 167–184, Wild *et al.* (1985) *Deep Sea Res.* **32** 1583–1589, and Campbell & Herring (1987) *Comp. Biochem. Physiol.* **86B** 411–417 for details and examples of the use of deep sea trawls used on *RRS Discovery*, the largest British research ship. These nets can be opened and closed at defined depths by a sonar signal from the ship. Some nets are multiple, containing as many as three pairs of nets. The trawling time for each net is usually about two hours at two knots. Lights or bait on the nets may help. Some pelagic organisms, such as a few luminous squid, can be caught by rod and line. Also take a sieve, forceps, scissors, collecting jars and a bucket.

Visualisation of the luminescence
Most luminous organisms will light up if you kill them, but this would be a pity, and it destroys the luciferin. Many will light up on mechanical treatment, for example

dinoflagellates and hydroids; others glow spontaneously, e.g. bacteria and glow-worms. The latter can be seen on a clear night many tens of metres away. However, a handy device is a chemical gun containing KCl (0.5 M) to activate neuronal pathways on organisms under rocks. The results can sometimes be spectacular. Scaleworms, for example, light up this way, but are less easy to get to glow by mechanical treatment. Other chemical stimuli include H_2O_2, and neurotransmitters (octopamine, adrenaline, and 5 hydroxytryptamine).

How to collect?

We have already seen that to find the luminous animals a range of collecting gear may be necessary. Some will only be available on research ships. Many of the animals are fragile. Squashing them into a bottle at the end of a net may damage or destroy some of them. Furthermore, the deep sea is at several hundreds of atmospheric pressure and at about $+4°C$. Animals brought up from the depths suffer therefore major environmental shock unless retained within specialised devices. Handle the organisms with care if you want to bring them back alive. It is easy to squash a glow-worm if you try to get it to start glowing again when it has stubbornly gone out after capture. *Pholas* can be embedded in rock some 10–20 cm down. It is easy to smash the shells in over-eagerness to get the animal out!

Return to the lab or house

Provided that the organisms are alive and healthy it is relatively straightforward to bring them back intact. Marine organisms should be kept in cooled sea water and can be maintained, sometimes for weeks, in a cold room, if the water is changed regularly. Seaweed rots quickly, so *Obelia* should be removed, otherwise it is damaged. Similarly the polychaete *Chaetopterus* in its chitinous tube decays very quickly if simply humped in large numbers into a black bag for transportation.

Some organisms can be cultured. Hydroids grow out on glass slides in circulating fresh sea water maintained at 10–15°C and fed on brine shrimp *Artemia* or other crustaceans. Bacteria need to be cultured in large numbers to glow. Dinoflagellates are more difficult, but the Hastings group culture them routinely (see Appendix IV and *Meth Enz* vols 57 and 133 for bacterial and dinoflagellate culture conditions). The media and conditions, e.g. light and dark, temperature, and O_2 may be critical.

There have been very few systematic studies on how best to preserve the biochemistry of living light. Liquid N_2 is useful, but most go off on long term storage, even at $-70°C$. All exhibit a resting glow. The proteins are more stable than the chromophores. Many can be stored frozen at $-70°C$ for months, or as an acetone powder. This is not successful, however, for coelenterates (acetone powder no good), copepods and *Pyrosoma*. Even when acetone powders can regenerate light, this may not necessarily be the optimal extraction procedure e.g. *Pholas* and luminous beetles. When you don't know, investigate the following for optimal storage of light generating capacity.

(1) Freeze at $-70°C$, and freeze-dry (but keep dry with desiccant).
(2) Acetone powder.
(3) Homogenate frozen at $-70°C$, different pH's $\pm$ oxygen metabolite scavengers (e.g. ascorbate).

(4) $(NH_4)_2 SO_4$ precipitate.
(5) H^+/methanol to extract the luciferin.
(6) Argon or nitrogen to remove oxygen.

Always store at $-70°C$ if you can, as all exhibit a resting glow.

Fixation should be carried out in 10% formalin (in sea water if a marine species) buffered with borate. This will, of course, destroy the bioluminescence. Prior narcotisation in 0.36 $MgCl_2$ can be useful.

We saw in Chapter 3 that there is much for the professional scientist to learn about the biology and biochemistry of bioluminescence. Observing living light in its natural habitat is exhilarating, challenging, and great fun. Yet there is still plenty for the naturalist to contribute to the science of living light. There are five main areas where observations in the field can enhance our knowledge and understanding of biology.

(1) Distribution, both national and local, of particular species, and its relation to the environment, climate, and geology.
(2) The ecology, particularly the stability of particular colonies and the variation in appearance of the organisms, with weather and other factors.
(3) Factors determining the onset, duration, and switching off of the luminescence and their relationship to function, e.g. mating rituals.
(4) Food sources.
(5) The timing of the life cycle.
(6) Reproduction under artificial conditions. This requires careful documentation of the numbers of individuals of each species (sex and stage of life cycle), plus behavioural relationships with luminescence.

The value of these types of study, be they by professionals or amateurs, is well illustrated in what we now know about the biology of the luminous beetles (Order: Coleoptera).

The European glow-worm *Lampyris noctiluca* is found all over Europe from northern Scandinavia to the southernmost reaches of Europe. Some colonies, such as those in Denmark studied by Dreisig, contain several hundred individuals, whereas some British colonies may have just a handful. Their numbers appear to have diminished in recent years. But this, and the effect of modern farming methods, needs investigation. The adults appear usually in mid-June. The female glows continuously from the last three segments for some ½–2 hours and stops after copulation. All stages of life cycle glow. The eggs appear normally a day or two after mating, and need to be kept moist. They hatch within 3–4 days at 18–20°C. The adults die off by midsummer. The females rarely survive a few days after egg laying, but will glow until mating occurs. The newly hatched *instar*, as it is known, lives through the winter, and develops to the larva from April onwards through 2–3 outer coat changes. The larvae have only two tiny light organs on the last segment and glow when disturbed, but for only a few seconds-minutes. There is disagreement over the length of time from hatching to full adult. It is probably at least one year, and may require two winters and up to 34 months for the larva to form a pupa, which it enters early in June. The adult imago hatches some 10–13 days later. The larva appears to be the main feeder. It narcotises snails, particularly *Helix* spp, and slugs, with its pincers. Then it injects enzymes into the prey, and sucks out the digested juices over a day or so. In spite of this detail the variation of this pattern with geography, climate, and temperature is not well documented. The separation of colonies suggests inbreeding. And the occurrence of other luminous beetles and their biology in various parts of the world still needs much work, inspite of the outstanding contribution of the entomologist J. E. Lloyd from which we have learnt so much about firefly biology and sexual behaviour. For some interesting references see:

Dreisig, H. (1971) *J. Zool.* **165** 229–244. Control of the glowing of *Lampyris noctiluca* in the field (Coleoptera: Lampyridae).

Dreisig, H. (1974) *Ent. Scand.* **5** 103–109. Observations on the luminescence of the larval glow-worm, *Lampyris noctiluca*. L (Col. Lampyridae).

Dreisig, H. (1976) *Physiol. Ent.* **1** 123–129. Phase shifting the circadian rhythms of nocturnal insects by temperature changes.

Lloyd, J. E. (1971) *Ann Rev. Ent.* **16** 97–122. Bioluminescent communication in insects.

Newport, G. (1852) *J. Proc. Linn. Soc.* **1** 40–71. On the natural history of the glow-worm (*Lampyris noctiluca*).

Schwalb, H. H. (1961) *Zool. Jb. Syst.* Jena **88** 399–550. Beitrage zur Biologie der einheimischen Lampyriden *Lampyris noctiluca*. Geoffr. und *Phausis splendida* Lec. und experimentelle Analyse ihres Beutefang. und Sexualverhaltens.

One kind word from a naturalist, keen to preserve the natural beauty of living light. Remember, protect the environment, don't overcollect, and take care, particularly if you indulge in night diving. *Pholas* is diminishing on the French coast, and fireflies are being overcollected for commercial preparations of luciferase. Cloning may alleviate this problem. Good hunting!

Appendix IV

Some chemiluminescent demonstrations

A particular attraction of giving lectures about chemi- and bio-luminescence is that one can begin with a dramatic flash. It certainly helps the audience to remember the topic, but do they remember anything about the science? The newspapers and television in recent years contain more and more articles about science and technology. In spite of the beautiful presentations like the 'Ascent of Man' by Jacob Bronowski and the 'Living Planet' and many other programmes by Sir David Attenborough, the general public remains bemused. British television programmes such as 'Tomorrow's World' and 'Horizon', clever as they are, fail to get across the distinction between science and technology, and how the latter evolves from the former. The natural scientist is just as creative as a painter, a composer, or a writer. In my view there are two reasons why the popularists of science have really failed to get across the truly creative element in science. Firstly, the language of science is experiment, and not words. Michael Faraday realised this, but we lack modern exponents to demonstrate scientific principles by experiment. Secondly, there is too much faking on TV science programmes. I have come across this several times in programmes in which I have been directly involved. The director may think that it looks OK to record an event, and then record the sound to a fake demonstration, but to the audience it will lack reality, however good the acting!

Here are a few chemiluminescent demonstrations which can be both spectacular and aesthetically pleasing. I hope that you also will be able to use them to demonstrate a scientific idea or principle, or to pose a question to the audience. All of the reagents are either commercially available (see Appendix V) or easy to find in the field (see Appendix III). Take care, several demonstrations use potentially toxic or harmful reagents. You need to have a fully darkened room to see these demonstrations properly, but the audience need not be dark adapted.

(1) Luminol (5 amino-2,3-dihydro 1,4 phthalazine dione)

(a) *Glowing flask*
Add a large spatula end (about 50 mg) luminol to about 50–100 ml dimethyl sulphoxide in a 250 or 500 ml flask. It will dissolve quickly. Add about 50 g K or NaOH pellets, stopper the flask, and swirl or shake vigorously. Take care! Blue chemiluminescence will be seen immediately around the K or NaOH pellet. After 2–3 minutes shaking, the flask suddenly shines very brightly, blue. Addition of a few ml of *t*-butyl alcohol can shorten the time taken for luminescence to become very bright, but don't add too much.

(b) *The need for O_2*
Leave the flask in (a) during the lecture and you will see that bright chemilumine-sence is visible only at the surface. Shake again to reoxygenate, and the whole flask becomes brilliantly chemiluminescent again. Addition of 5–10 ml water quenches the reaction.

(c) *Catalysts*
These are best demonstrated in aqueous media. Add a small spatula end (about 10 mg) luminol to 5 ml 0.1 M NaOH in a test tube or small flask. Add 0.1 ml 1% H_2O_2. This will give a very faint glow to the dark adapted eye. Then inject, from a syringe, or just pour, 1–2 ml 10 mM potassium ferricyanide or 1 mg ml^{-1} microperoxidase. A bright blue flash lasting just a few seconds will be seen. Any haem containing compound, e.g. haemoglobin, blood, or horseradish peroxidase will 'catalyse' the reaction. Transition metals will also work. Dropping a British penny into the luminol/H_2O_2/NaOH solution will cause it to glow.

(d) *Enhancement*
Compare two tubes containing 10 mM luminol in 0.1 M tris pH 8.5 ± 10 μM p-iodophenol or firefly luciferin as enhancer. Start the reaction with 1 μg ml^{-1} horseradish peroxidase. Note the tube plus enhancer is much brighter, and the glow continues for several minutes.

(2) Chemiluminescent clock (be careful, this uses potassium cyanide)
Solution A   200 mg luminol dissolved in 10 ml dimethyl formamide
Solution B   100 ml concentrated ammonia (58%) in 1 litre water
Solution C   1.3 g $CuSO_4 \cdot 5H_2O$ in 446 ml solution B + H_2O to 500 ml
Solution D   *CARE!* 2 g KCN in 100 ml H_2O
Solution E   10 ml 30% H_2O_2 to 100 ml H_2O
Solution F   Add solution D to solution C with great care until the dark blue colour
              of C has disappeared. Then add about 5% v/v excess KCN (solution D).

The clock
Add 100 ml solution F (ammonium cyanide: Cu^{2+} complex) to a flask, and add 1 ml solution A (luminol). Mix well. Then add exactly 6 ml solution E (3% H_2O_2). Mixing starts the clock. The H_2O_2 consumes the CN^-. At the end point, approx $3\frac{1}{2}$ min, there is a blue flash as the Cu^{2+} is able to catalyse the luminol reaction, and the appearance of dark blue tetraamine Cu^{2+} complex.

(3)   Energy transfer
(a)   *Resonance*
(i)   Use luminol as in (a) and add a small spatula end (1–2 mg) of fluores-

cein   to produce a yellow-orange luminescence or rose

bengal   to produce a pink colour.

Alternatively add about 1 mg fluorescein to the tube in 1c prior to adding the catalyst. By altering the amounts of fluorescein the colour will go from blue to green-yellow.

(ii)   Acridinium salts
Add a small spatula end (*ca.* 5 mg) of lucigenin   or

acridinium ester

to 5 ml 0.1 M NaOH in a test tube. Add rapidly, or inject from a syringe, 1 ml 0.1% H_2O_2. The acridinium ester will give a bright blue, rapid flash of light, over within a few seconds. Lucigenin gives a longer, greenish-blue lumines-cence, because of the energy transfer between the product of the reaction and unreacted lucigenin molecules. Acridinium ester is best starting at acid pH to reduce pseudo-base formation.

(b)   *Electron transfer using oxalate esters*
(i)   A range of light sticks are available commercially.
(ii)   Synthesis of bis 2,4-dinitrophenyl oxalate.

Dissolve 370 g (approx 2 mole) of 2,4 dinitrophenol in 2 l reagent grade benzene, and dry by azeotropic distillation. Cool to 10°C under a N_2 atmosphere and add 100 g (approx. 1 mol) of freshly distilled triethylamine. Then add 70 g (approx 0.5 mol) of oxalyl chloride, with stirring, in an ice bath. Stir for 30 min, and maintain tempera-ture at 10–25°C. Stir yellow slurry for a further 3h and evaporate to dryness under reduced pressure. Remove triethylamine HCl by mixing the solid with 400 ml chloroform. Collect the solid on a scintered glass funnel, and wash further in the chlorform. Dry under vacuum. Recrystallise from nitrobenzene below 100°C. bis 2,4 dichlorophenyl oxalate can be prepared similarly from 2,4 dichlorophenol, but requires refluxing for 4–5 hours to complete the reaction. Also this oxalate ester is soluble in $CHCl_3$, so triethylamine HCl has to be removed by washing the crude solid with H_2O, and recrystallise from petroleum ether.

(iii) Oxalate ester chemiluminescence

This has to be carried out in an organic solvent which enhances charge transfer complex formation (see Chapter 9). Diethyl phthalate is a good one.

Dissolve a large spatula end (*ca.* 50 mg) of bis (2,4 dinitrophenyl) oxlate in 10 ml diethyl phthalate. Add a small spatula end of diphenyl anthracene (*ca.* 3 mg) for blue, bis (phenyl ethynyl) anthracene (*ca.* 10 mg) for green, or rubrene (*ca.* 3 mg) for yellow light. Then add 5 ml of diethyl phthalate containing 0.5 ml butyl alcohol + 0.3 ml 30% H_2O_2. Chemiluminescence is bright and decays rapidly.

For red chemiluminescence add a large spatula end (*ca.* 50 mg) of bis(2,4 dinitrophenyl) oxalate to a few ml of a saturated solution of rhodamine B (lactone form) in polyethylene glycol 400 containing 1 drop of 30% H_2O_2. Leave out the fluor in these reactions for comparison, to show no light is visible without an energy transfer acceptor.

McCapra invented a neat way of demonstrating this type of chemiluminescence which has been copied in the commercially available light sticks. An ampoule containing the H_2O_2 is inserted into a tube containing the oxalate ester + fluor, which is then sealed. On breaking the ampoule bright chemiluminescence is seen.

(4) Other fun demos

Blowing cigarette smoke through alkaline H_2O_2 and a chemiluminescent compound, and the oxidation of lophine (yellow light) or pyrogallol (pink light).

(5) Light pen

Painting a dry filter paper, impregnated with lucigenin, with alkaline peroxide produces visible writing. But a better one is to use a pen containing tetrakis dimethyl aminoethylene

$$(CH_3)_2N \diagdown \qquad \diagup N(CH_3)_2$$
$$C = C$$
$$(CH_3)_2N \diagup \qquad \diagdown N(CH_3)_2$$

which chemiluminesces green spontaneously in air and can be used to write sentences on a white board.

(6) Bioluminescence

Some of these are easily visible at only a short distance.

(a) *0.5 M KCl stimulation of Obelia and other bioluminesence.*

(b) *Luciferin — luciferase reaction.*

Grind a *Pholas*, firefly, or glow-worm light organ in 50 mM Na phosphate pH 7–8 and allow light to decay. Grind another *Pholas*, firefly, or glow-worm light organ in boiling water. Allow to cool and add back to the cooled first extract. Light will appear, but some ATP may be needed to increase light output from the firefly or glow-worm system.

(c)   *ATP requirement for beetle bioluminescence*
Homogenise 1–4 firefly or glow-worm tails in 3–4 mol 0.1 M tris, 10 mM Mg acetate
pH 8. Divide into four tubes added:

(i)   nothing, gives very dim yellow (firefly) or green (glow-worm) light.
(ii)   0.1 ml 10 μM ATP (neutralised) gives visible chemiluminescence.
(iii)   0.1 ml 1 mM ATP (neutralised) gives very bright chemiluminescence.
(iv)   0.1 ml mM GTP (neutralised) gives little or no chemiluminescence.

(d) Ca^{2+} *requirement for jelly fish and hydroid chemiluminescence.*
Rinse 100 *Obelia* hydroids from one frond of *Laminaria* in 50 ml 0.5 M NaCl, and
then homogenise in 5 ml 200 mM tris, 1 mM EDTA pH 7.5. Filter on a tea strainer.
Addition of 1 ml 50 mM $CaCl_2$ gives a bright blue flash.

(e)   *Culturing and O_2 requirement for luminous bacteria.*
Look for some decaying fish or invertebrate glowing on the beach at night, or a
luminous bacterial organ of a fish or squid (or obtain some bacteria from a culture
bank). Using an inoculation loop, infect Petri dishes containing sterile agar with the
following ingredients:

NaCl	2.5 g
$Na_3(PO_4)_2$	0.5
KH_2PO_4	0.2 g
$NH_4 H_2PO_4$	50 mg
bactopepton (Difco)	1g
glycerol	0.3 ml
bacto-agar	1 g

Make up to 100 ml and pH 7.3. After 3–4 days the plates will be glowing with
luminous bacteria. Blow Ar or N_2 over them, or put in a desiccator and evacuate, and
the light will dim, returning on readmission of air. For liquid cultures use:

NaCl	30 g
Na_2HPO_4	2.5 g
KH_2PO_4	2.1 g
$NH_4H_2PO_4$	0.5 g
$MgSO_4$	0.1 g
bactopepton (Difco)	10 g
glycerol	3 ml

in 1 litre pH 7.4, sterilise first. Stir continuously. After inoculation from a Petri dish
or agar slope luminescence will be very bright within 1–2 days or sooner. Stop
stirring, and the light will disappear, to reappear on agitation (i.e. $+ O_2$) Species =
Vibrio fischeri, V. harveyi, Photobacterium phosphoreum. The freshwater and soil
species obviously requires less salt.

   Chemi- and bioluminescent reactions offer the ingenious lecturer many opportunities for demonstrating principles not only about luminescence but also about
catalytic reactions in general.
   Good luck, take care and have fun!

Appendix V

Some commercial sources of materials and equipment

These notes are intended only as a guide. Many biotechnology companies are springing up which realise the potential of chemiluminescence analysis. Cloned material for the bacterial, firefly, and coelenterate photoprotein systems is likely to be available commercially soon. Many companies are developing the luminol enhanced peroxidase, acridinium ester, and coelenterate photoprotein systems as non-isotopic labels in both immunoassay and recombinant DNA technology, but at the time of writing only some immunoassays are commercially available.

A. Live luminous organisms

(1) *Bacteria*
Many culture banks for *Vibrio fischeri*, *Vibrio harveyi*, and *Photobacterium phosphoreum* e.g. American Type Culture Collection, 12301 Parklawn Drive, Rockville, Md 20852, USA.

(2) *Dinoflagellates*
Gonyaulax polyhedra and *Noctiluca miliaris* available from some culture banks but more difficult to grow than bacteria. Also found in the wild where red tides are found and dinoflagellate blooms. See papers of J. W. Hastings *et al.* for details, e.g. Meth Enzymol 133:307 (1987).

(3) *Coelenterates*
(a) *Obelia geniculata* (medusa *lucifera*)
Phialidium hemisphericum (hydroid *Clytia*)
from The Marine Biological Association Laboratory, Citadel Hill, Plymouth, Devon, UK.
Clytia spp (hydroid of *Phialidium*) from Panacea, Florida 32346, USA.
(b) Luminous anthrozoans
Renilla mulleri from Gulf Specimen Co. Inc, PO Box 237, Panacea, Florida 32346, USA.
Renilla kollikeri, *Ptilosarcus guerneyi*, *Stylatula elongata*, *Acanthoptilum gracile* from Pacific Bio-Marine Labs Inc, PO Box 536, Venice, California 90291, USA.
Renilla reniformis, contact Dr. M. J. Cormier, Department of Biochemistry, University of Georgia, Athens, Georgia 30602, USA.

(c) Luminous ctenophores
Mnemiopsis mercradii. Gulf Specimen Co.
Beroë ovata, Gulf Specimen Co.
(d) Wide range of frozen fish, crustaceans, molluscs, coelenterates, tunicates, starfish, annelids — contact the author or your nearest marine lab.

B.   Chemiluminescent reagents
(See (4) for addresses)

Reagent	Source
(1) *Bioluminescent materials*	
Bacterial system:	
luciferase (pure)	ALL, BCL, LKB, Lumac, Sigma
luciferase + oxidoreductase	ALL, BCL, LKB, Sigma
oxidoreductase	BCL
Tetradecanol and other aldehydes, NAD(P)H, FMN	Aldrich, Sigma
Firefly system	
firefly tails	Sigma
luciferase (crude)	Sigma (FLE50)
luciferase (pure)	ALL, BCL, CLEAR, Lumac, LKB, Sigma
luciferin	ALL, BCL, CLEAR, Lumac, LKB, Sigma
ATP	many biochemical companies including the above
Photoproteins	
aequorin	Sigma (poor quality)
	Professor J. R. Blinks (high quality)
coelenterazine	London Diagnostics
(2) *Synthetic*	
acridinium ester labels	*London Diagnostics*
ABEI, AHEI,	Sigma (ABEI), LKB
various fluors (fluorescein, anthracene derivatives, rubrene)	Aldrich, Sigma
iodophenol	Aldrich, Amersham
isoluminol	Aldrich, LKB, Sigma
light sticks	American Cyanamid Co. (? still in existence), Edmund Scientific Ventron Corp, Alfa Products, Orchard Crown Ltd
lophine	Aldrich
lucigenin	Aldrich, Sigma
luminol	Aldrich, ALL, LKB, Lumac, Sigma
oxalate esters	see light sticks
peroxidase	Sigma
pyrogallol	Aldrich
steroid-ABEI	LKB

Many of these are also sold by other chemical companies.

(3) *Assay kits*

bacterial contamination	Lumac, CLEAR
ATP	CLEAR, LKB, Lumac, Sigma
NAD(P)H	LKB, Lumac, Sigma
acetylcholine	LKB
creatine kinase (+ isoenzymes)	ALL, LKB, Ciba Corning

various immunoassays:
    Amersham (luminol enhanced per-
      oxidase) = Amerlite for thyroid
      hormones, tumour markers, drugs
    Ciba Corning (acridinium ester)
      = MagicLite, TSH and thyroid hor-
      mones, other hormones and drugs
    ALL (immobilized luciferase)
      = luminescence immobilized
      enzyme systems or LIES

DNA probe for bacteria (LEADERI)	Gen-Probe
Phagocytosis	ALL, Lumar

NB: A major problem with commercially available reagents has been lack of reproducibility in activity and purity. This is improving but should be checked.

For further information write to the following addresses.

(4) *Addresses*

Many of these companies have marketing divisions in many countries. Only US or UK addresses are given here.

Aldrich Chemical Co. The Old Brickyard, New Road, Gillingham, Dorset, SP8 4JL, UK. or 940 W St Paul Ave, PO Box 355, Milwaukee, W1 53201, USA

ALL (Analytical Luminescence Laboratories Inc), 31125 Via Colinas, Suite 905, Westlake Village, CA 91361, USA.

American Cynamid Co., Organic Chemicals Division, Bound Brook, NJ 08805, USA.

Amersham International PLC, Amersham, UK.

BCL Boehringer Mannheim House, Bell Lane, Lewes, East Sussex BN7 1LG, UK or Boehringer Mannheim, PO Box 50816, Indianapolis, 1N46250.

Professor J. R. Blinks, Department of Pharmacology, Mayo Medical School, Rochester, Minnesota 55901, USA.

Ciba-Corning Diagnostics Corp., 63 North Street, Medfield, MA 02052, USA or Halstead, Essex, CO9 2DX.

CLEAR (Cardiff Laboratories for Energy and Resources Ltd), Lewis Road, Unit M40 East Moors, Cardiff, CF1 5EG, UK.

Edmund Scientific, 402 Edscorp Building, Barrington, NJ 08007, USA.

Gen-Probe, 9880 Campus Point Drive, San Diego, CA 92121, USA

LKB-Wallac, Wallac Org. PO Box 10, SF–20101, Turku 10, Finland.

London Diagnostics, 10300 Valley View Road, Eden Prairie, MN 55344, USA.

Lumac BV, PO Box 31101, 6370 AC Schaesberg, The Netherlands or PO Box 28254, Titusville, FL 32780, USA.

Orchard Crown Ltd, 99 Lytham Road, Fulwood, Preston, Lancs PR2 2EN UK.

Packard Instrument Company, 2200 Warrenville Reach, Downers Grove, Illinois 60515, USA.

Sigma Chemical Co. Fancy Road, Poole, Dorset BH17 7NH or PO Box 14508, St Louis, MO 63178, USA.

Ventron Corp, Alfa Products, PO Box 244, 152 Andover Street, Danwers, MASS 01923, USA.

C. Equipment

(1) *Photomultipliers*
Centronics
Hamamatsu
Thorn EMI
RCA
Balzers (for interference filters only)

(2) *Imaging systems and intensifiers including CCD's and IPD's*
Astromed
Coralab
EG and G (Princeton Electronics)
Hamamatsu
ITL
RCA
Thorn EMI
Surface Science Laboratories

(3) *Chemiluminometers*
Alkemp (Automated luminescence analyser)
ALL
Amersham (Amerlite)
Applied Photophysics
Berthold (Biolumat)
Ciba-Corning
CLEAR (SpeedTech 2000, fully automated system and computer
Controlled, see Fig. 2.6).
Foss Electronic (BactoFoss)
EG and G (Princeton Electronics)
Hamilton (Unicon)
LKB Wallac (1250 and 1251)
Lumac (various)
Marwell International
Packard (Autolite and Picolite)
SAI, Skan
Thorn EMI (EHT, discriminator and counter)
Turner Designs
Vitatect

D. Addresses
Alkemp Corp, Clackamas, Oregon, USA.
ALL Inc, 31125 Via Colinas, Suite 905, Westlake Village, CA 91367, USA.
Amersham International, Amersham UK.
Applied Photophysics, Imperial College, University of London, London.
Astromed Innovation Centre, Cambridge Science Park, Milton Road, Cambridge, CB4 4GS, UK.
Balzers High Vacuum Ltd., Northbridge Road, Berkhamsted, Herts, HP4 1EN, UK.
Berthold, Berthold Laboratorum, Wildbat, West Germany.
Centronics, Ltd., Centronic House, King Henry's Drive, New Addington, Croydon, CR9 0BG, UK.
Ciba Corning Diagnostics, 63 North Street, Medfield, MA 02052, USA.
CLEAR, Unit M40 Lewis Road, East Moors, Cardiff, CF1 5EG, UK.
Coralab Research, The Innovation Centre, Cambridge Science Park, Cambridge, CB4 4GS, UK.
Foss Electric 69 Slangerupgade, DK3400, Hillerød, Denmark.
EG and G Brookdeal (incorporation Princeton Electronics), Princeton Applied Research Group, PO Box 2565, Princeton, NJ 08540, USA or Doncastle Road, Bracknell, Berks, RG12 4PG, UK.
Hamamatsu, Photonics Microscopy Inc., 2625 Butterfield Road, Suite 204–S, Oak Brook, Illinois 60521, USA. or Europa GMBH, Arzberger Str 10, 8036 Herrschinga. A., West Germany.
Hamilton Company, PO Box 10030, Reno, Nevada 89520, USA.
Instrument Technology Ltd. (ITL), 28 Castleham Road, St. Leonards-on-Sea, East Sussex, TH38 9NS, UK.
LKB Wallac, Wallac Oy., PO Box 10, 20101 Turku, Finland.
London Diagnostics, 10300 Valley View Road, Eden Prairie, MN 55344 USA.
Lumac BV, PO Box 31101, 6370 Ac Shaesberg, The Netherlands or PO Box 2805, Titusville, FL 32780, USA.
Marwell Instruments AB, Solva, Sweden.
Packard Instrument Co., Ltd., 2200 Warrenville Road, Downers Grove, Illinois 60515, USA.
RCA, 2000 Clements Bridge Road, Deptford, NJ 08096, USA or Sunbury-on-Thames, Middlesex, TW17 7HW, UK.
SAI Technology Inc., San Diego, California, USA.
Skan AG, Basel, Switzerland.
Surface Science Laboratories (SSL), 1206 Charleston Road, Mountain View, California 94043, USA.
Thorn EEM Electron Tubes Ltd., Bury Street, Ruislip, Middlesex, HA4 7TA, UK.
Turner Designs, Mountain View, California, USA.
Vitatect Corp, Alexandria, Virginia, USA.

Of this plethora of instrumentation which to choose? I would go for a digital system. The most sensitive and fully computerised is by CLEAR, though the Berthold, Ciba Corning, Hamilton, and Lumac single or multi-tube devices are good. For research purposes you may need to construct your own 'housing' in which case the electronics available from EG and G or Thorn EMI may be best. Several miniature photodiode chemiluminometers are also becoming available.

Index